Adaptive Dynamics of Infectious Diseases:
In Pursuit of Virulence Management

Emerging diseases pose a continual threat to public health. Fast multiplication and high rates of genetic change allow pathogens to evolve very rapidly. It is therefore imperative to incorporate evolutionary considerations into longer-term health management plans. The evolution of infectious diseases is also an ideal test bed for theories of evolutionary dynamics. This book combines both threads, taking stock of our current knowledge of the evolutionary ecology of infectious diseases, and setting out goals for the management of virulent pathogens. Throughout the book, the fundamental concepts and techniques that underlie the modeling approaches are carefully explained in a unique series of integrated boxes. The book ends with an overview of novel options for virulence management in humans, farm animals, plants, wildlife populations, and pests and their natural enemies. Written for graduate students and researchers, *Adaptive Dynamics of Infectious Diseases* provides an integrated treatment of mathematical evolutionary modeling and disease management.

ULF DIECKMANN is Project Coordinator of the Adaptive Dynamics Network at the International Institute for Applied Systems Analysis (IIASA) in Laxenburg, Austria. He is coeditor of *The Geometry of Ecological Interactions: Simplifying Spatial Complexity*.

JOHAN A.J. METZ is Professor of Mathematical Biology at the Institute of Evolutionary and Ecological Sciences at the University of Leiden, and Project Leader of the Adaptive Dynamics Network at IIASA. He is coeditor of *The Geometry of Ecological Interactions: Simplifying Spatial Complexity* and of *The Dynamics of Physiologically Structured Populations*.

MAURICE W. SABELIS is Professor of Population Biology at the Institute for Biodiversity and Ecosystem Dynamics at the University of Amsterdam. He is coeditor of *Spider Mites: Their Biology, Natural Enemies and Control*, of *Eriophyoid Mites: Their Biology, Natural Enemies and Control*, and of *Ecology and Evolution of the Acari*.

KARL SIGMUND is Professor at the Institute for Mathematics at the University of Vienna and also a Research Scholar with the Adaptive Dynamics Network at IIASA. He is the author of *Games of Life: Explorations in Ecology, Evolution, and Behaviour*, coauthor of *Evolutionary Games and Population Dynamics*, coeditor of *Dynamics of Macrosystems*, of *Dynamical Systems*, and of *Evolution and Control in Biological Systems*.

T0297717

Cambridge Studies in Adaptive Dynamics

Series Editors

ULF DIECKMANN
Adaptive Dynamics Network
International Institute for
Applied Systems Analysis
A-2361 Laxenburg
Austria

JOHAN A.J. METZ
Institute of Evolutionary
and Ecological Sciences
Leiden University
NL-2311 GP Leiden
The Netherlands

The modern synthesis of the first half of the twentieth century reconciled Darwinian selection with Mendelian genetics. However, it largely failed to incorporate ecology and hence did not develop into a predictive theory of long-term evolution. It was only in the 1970s that evolutionary game theory put the consequences of frequency-dependent ecological interactions into proper perspective. Adaptive Dynamics extends evolutionary game theory by describing the dynamics of adaptive trait substitutions and by analyzing the evolutionary implications of complex ecological settings.

The *Cambridge Studies in Adaptive Dynamics* highlight these novel concepts and techniques for ecological and evolutionary research. The series is designed to help graduate students and researchers to use the new methods for their own studies. Volumes in the series provide coverage of both empirical observations and theoretical insights, offering natural points of departure for various groups of readers. If you would like to contribute a book to the series, please contact Cambridge University Press or the series editors.

1. *The Geometry of Ecological Interactions: Simplifying Spatial Complexity*
   Edited by Ulf Dieckmann, Richard Law, and Johan A.J. Metz

2. *Adaptive Dynamics of Infectious Diseases: In Pursuit of Virulence Management*
   Edited by Ulf Dieckmann, Johan A.J. Metz, Maurice W. Sabelis, and Karl Sigmund

In preparation:

Elements of Adaptive Dynamics
Edited by Ulf Dieckmann and Johan A.J. Metz

Cambridge Studies in Adaptive Dynamics

Adaptive Dynamics of Infectious Diseases: In Pursuit of Virulence Management

Edited by

Ulf Dieckmann, Johan A.J. Metz, Maurice W. Sabelis, and Karl Sigmund

CAMBRIDGE UNIVERSITY PRESS
Cambridge, New York, Melbourne, Madrid, Cape Town, Singapore, São Paulo

Cambridge University Press
The Edinburgh Building, Cambridge CB2 2RU, UK

Published in the United States of America by Cambridge University Press, New York

www.cambridge.org
Information on this title: www.cambridge.org/9780521781657

First published 2002
This digitally printed first paperback version 2005

Typefaces Times; Zapf Humanist 601 (Bitstream Inc.)   *System* LATEX

A catalogue record for this publication is available from the British Library

ISBN-13 978-0-521-78165-7 hardback
ISBN-10 0-521-78165-5 hardback

ISBN-13 978-0-521-02213-2 paperback
ISBN-10 0-521-02213-4 paperback

Contents

Contributing Authors

Frederick R. Adler (adler@math.utah.edu) Department of Mathematics, University of Utah, Salt Lake City, UT 84112, USA

Viggo Andreasen (viggo@ruc.dk) Department of Mathematics, Roskilde University, DK-4000 Roskilde, Denmark

M. Ali Anwar (ali.anwar@zoology.oxford.ac.uk) Department of Zoology, University of Oxford, Oxford OX1 3PS, United Kingdom

Joost B. Beltman (Beltman@rulfsb.LeidenUniv.nl) Theoretical and Evolutionary Biology, Institute of Evolutionary and Ecological Sciences, Leiden University, NL-2311 GP Leiden, The Netherlands

Sebastian Bonhoeffer (bonhoeffer@eco.umnw.ethz.ch) Ecology and Evolution, Eidgenössische Technische Hochschule Zürich, CH-8092 Zürich, Switzerland

José A.M. Borghans (borghans@pasteur.fr) Biologie des Populations Lymphocytaires, Institut Pasteur, F-75105 Paris, France

John P. Burand (jburand@microbio.umass.edu) Department of Microbiology, University of Massachusetts, Amherst, MA 01003, USA

Alfons J.M. Debets (Fons.debets@genetics.dpw.wau.nl) Laboratory of Genetics, Wageningen University, NL-6703 HA Wageningen, The Netherlands

Rob J. de Boer (R.J.deBoer@bio.uu.nl) Theoretical Biology, Utrecht University, NL-3584 CH Utrecht, The Netherlands

Mart C.M. de Jong (m.c.m.dejong@id.wag-ur.nl) Wageningen University and Research Centre, NL-8200 AB Lelystad, The Netherlands

Giulio De Leo (deleo@dsa.unipr.it) Dipartimento di Scienze Ambientali, Università degli Studi di Parma, Parco Area delle Scienze, I-43100 Parma, Italy

Ulf Dieckmann (dieckman@iiasa.ac.at) Adaptive Dynamics Network, International Institute for Applied Systems Analysis, A-2361 Laxenburg, Austria

Andy Dobson (andy@eno.princeton.edu) Department of Ecology and Evolutionary Biology, Eno Hall, Princeton University, Princeton, NJ 08544, USA

Jonathan Dushoff (dushoff@eno.princeton.edu) Department of Ecology and Evolutionary Biology, Eno Hall, Princeton University, Princeton, NJ 08544, USA

Greg Dwyer (gdwyer@miway.uchicago.edu) Department of Ecology and Evolution, University of Chicago, Chicago, IL 60637, USA

Martijn Egas (egas@bio.uva.nl) Population Biology Section, Institute for Biodiversity and Ecosystem Dynamics, University of Amsterdam, NL-1098 SM Amsterdam, The Netherlands

Joseph S. Elkinton (elkinton@ent.umass.edu) Department of Entomology, Fernald Hall, University of Massachusetts, Amherst, MA 01003, USA

Sam L. Elliot (selliot@ic.ac.uk) Leverhulme Unit for Population Biology and Biological Control, NERC Centre for Population Biology and CABI Science, Imperial College at Silwood Park, Ascot, Berkshire, SL5 7PY, United Kingdom

Paul W. Ewald (pwewald@amherst.edu) Department of Biology, Amherst College, Amherst, MA 01003, USA

Sylvain Gandon (Sylvain.Gandon@ed.ac.uk) Laboratoire d'Ecologie CNRS-URA 258, Université Pierre et Marie Curie, F-75252 Paris Cedex 05, France (currently at the Institute for Cell, Animal and Population Biology, University of Edinburgh, Edinburgh EH9 3JT, United Kingdom)

Andy Goodman (andrew-goodman@student.hms.harvard.edu) Harvard Medical School, Boston, MA 02115, USA

Sunetra Gupta (sunetra.gupta@zoology.oxford.ac.uk) Department of Zoology, University of Oxford, Oxford OX1 3PS, United Kingdom

Michael E. Hochberg (hochberg@isem.univ-montp2.fr) Génétique et Environnement, ISEM – University of Montpellier II, F-34095 Montpellier Cedex, France

Rolf F. Hoekstra (rolf.hoekstra@genetics.dpw.wau.nl) Laboratory of Genetics, Wageningen University, NL-6703 HA Wageningen, The Netherlands

Robert D. Holt (predator@falcon.cc.ukans.edu) Natural History Museum, University of Kansas, Lawrence, KS 66045, USA

Luc L.G. Janss (l.l.g.janss@id.wag-ur.nl) Institute for Animal Science and Health (ID-DLO), NL-8200 AB Lelystad, The Netherlands

Arne Janssen (janssen@science.uva.nl) Population Biology, Institute for Biodiversity and Ecosystem Dynamics, University of Amsterdam, NL-1098 SM Amsterdam, The Netherlands

Andrew M. Jarosz (amjarosz@msu.edu) Plant Biology Laboratory, Michigan State University, E. Lansing, MI 48824, USA

David C. Krakauer (krakauer@lamarck.zoo.ox.ac.uk) Institute for Advanced Study, Princeton, NJ 08540, USA

Simon A. Levin (slevin@eno.princeton.edu) Department of Ecology and Evolutionary Biology, Eno Hall, Princeton University, Princeton, NJ 08542, USA

Marc Lipsitch (mlipsitch@hsph.harvard.edu) Department of Epidemiology, Harvard School of Public Health, Boston, MA 02115, USA

Margaret J. Mackinnon (m.mackinnon@ed.ac.uk) Institute of Cell, Animal and Population Biology, University of Edinburgh, Edinburgh EH9 3JT, United Kingdom

Angela R. McLean (angela.mclean@zoo.ox.ac.uk) Department of Zoology, Oxford University, Oxford OX1 3PS, United Kingdom

Johan A.J. Metz (metz@rulsfb.LeidenUniv.nl) Theoretical Biology Section, Institute of Evolutionary and Ecological Sciences, Leiden University, NL-2311 GP Leiden, The Netherlands & Adaptive Dynamics Network, International Institute for Applied Systems Analysis, A-2361 Laxenburg, Austria

Yannis Michalakis (Yannis.Michalakis@mpl.ird.fr) Centre d'Etudes sur le Polymorphisme des Microorganismes, UMR CNRS, F-34032 Montpellier Cedex 1, France

Julio Mosquera Losada (j.mosquera-losada@imag.wag-ur.nl) Institute of Agricultural and Environmental Engineering (IMAG-DLO), Mestbehandeling en Emissies, NL-6700 AA Wageningen, The Netherlands

Martin A. Nowak (martin.nowak@zoo.ox.ac.uk) Department of Zoology, University of Oxford, Oxford OX1 3PS, United Kingdom

Bas Pels (pels@bio.uva.nl) Population Biology Section, Institute for Biodiversity and Ecosystem Dynamics, University of Amsterdam, NL-1098 SM Amsterdam, The Netherlands

Bruce Rannala (brannala@ualberta.ca) Department of Medical Genetics, University of Alberta, Edmonton, Alberta T6G2H7, Canada

Andrew F. Read (a.read@ed.ac.uk) Institute of Cell, Animal and Population Biology, University of Edinburgh, Edinburgh EH9 3JT, United Kingdom

Maurice W. Sabelis (sabelis@science.uva.nl) Population Biology Section, Institute for Biodiversity and Ecosystem Dynamics, University of Amsterdam, NL-1098 SM Amsterdam, The Netherlands

Akira Sasaki (asasascb@mbox.nc.kyushu-u.ac.jp) Department of Biology, Faculty of Science, Kyushu University, Fukuoka 812-8581, Japan

Karl Sigmund (ksigmund@esi.ac.at) Institut für Mathematik, Universität Wien, A-1090 Wien, Austria & Adaptive Dynamics Network, International Institute for Applied Systems Analysis, A-2361 Laxenburg, Austria

Douglas R. Taylor (drt3@virginia.edu) Department of Biology, Gilmer Hall, University of Virginia, Charlottesville, VA 22904, USA

Louise H. Taylor (ltaylor@lab0.vet.ed.ac.uk) Centre for Tropical and Veterinary Medicine, University of Edinburgh, Easter Bush, Roslin, Midlothian EH25 9RG, United Kingdom

Minus van Baalen (mvbaalen@snv.jussieu.fr) Institut d'Ecologie, CNRS UMR 7625, Université Pierre et Marie Curie, F-75252 Paris Cedex 05, France

Claus Wedekind (cwedekind@ed.ac.uk) Institute of Cell, Animal and Population Biology, University of Edinburgh, Edinburgh EH9 3JT, United Kingdom

List of Boxes

This volume features an integrated series of boxes that systematically introduce the tools needed to analyze virulence evolution and to assess strategies of virulence management. Readers interested in the fundamental concepts and techniques used throughout the book are invited to turn to these boxes. Written in a didactic style, the material listed below also provides convenient points of departure for readers new to the field.

Notational Standards

Few things are as much a distraction as irregular changes of mathematical notation between the individual chapters of a book. While mathematicians have learned to cope with this, such changes pose serious problems for many other readers.

To allow for a better focus on the content of chapters and to highlight their interconnections, we have encouraged all the authors of this volume to adhere to the following notational standards:

S, I, R	Host population sizes or densities of susceptible, infected, and removed individuals
N	Total host population size or density ($N = S + I + R$ or $S + I$)
s, i	Proportion of susceptible and infected hosts ($s = S/N, i = I/N$)
b	Per capita birth rate of hosts
d	Per capita death rate of disease-free hosts
r	Intrinsic growth rate of disease-free hosts ($r = b - d$)
K	Carrying capacity of hosts
B	Population-level rate of host birth or immigration
R_0	Basic reproduction ratio
S_0	Value of S in the absence of infected hosts
N_0	Value of N in the absence of infected hosts
b_0	Value of b at low population density
d_0	Value of d at low population density
r_0	Value of r at low population density
α	Per capita disease-induced death rate of hosts
β	Infection rate constant
γ	Per capita recovery rate of hosts, from infected to removed hosts
θ	Per capita recovery rate of hosts, from infected to susceptible hosts
μ	Per capita removal rate of infected hosts ($\mu = \alpha + d + \gamma + \theta$ or $\alpha + d + \gamma$ or $\alpha + d$)
λ	Force of infection ($\lambda = \beta S$)
U	Population density of uninfected vectors
V	Population density of infected vectors
Z	Total population density of vectors ($Z = U + V$)
v	Proportion of infected vectors ($v = V/Z$)
χ	Per capita bite rate of vectors
F	Population-level rate of vector birth or immigration

f	Fitness in continuous time ($f = 0$ is neutral)
w	Fitness in discrete time ($w = 1$ is neutral)

t	Time
τ	Delay time
T	Duration of time period
a	Age

p	Probability (subscripted if necessary)
c	Cost-related constant
c_0, c_1, c_2	Arbitrary constants

$\cdot_{res}$	Trait value of resident individuals
$\cdot_{mut}$	Trait value of mutant individuals
$\cdot'$	Derivative
$\bar{\cdot}$	Average
$\cdot^*$	Equilibrium value

1

Introduction

Karl Sigmund, Maurice W. Sabelis, Ulf Dieckmann, and Johan A.J. Metz

Toward the end of the 1960s, by dint of science and collective efforts, humankind had managed to eradicate smallpox and to land on the moon. Accordingly, some of the best-informed experts felt that the time had come to close the book on infectious diseases, and that the colonization of interplanetary space was about to begin. Today, these predictions seem as quaint as the notion – also quite widespread at the time – that the Age of Aquarius was about to begin.

The subsequent decades have taught us to be less sanguine about the future. In 2001 we do not send out manned spacecraft to meet with extraterrestrials, but instead are shutting down obsolete space accommodation. And far from closing the book on infectious diseases, we find that books on infectious diseases still have to be written. Few experts believe, nowadays, that we are witnessing the beginning of the end of our age-old battle against germs. In 1999, for instance, the World Health Organization (WHO) launched an ambitious program, "Roll Back Malaria" – a battle cry that seems tellingly defensive. In the 1960s, optimists still entertained hopes that malaria could be wiped out altogether. And why not? It had worked for smallpox, after all.

Aside from the disappointments with malaria and other infectious diseases – alarming outbreaks of cholera or foot-and-mouth epidemics, for instance – we had to learn to come to terms with other baffling setbacks. New scourges such as acquired immunodeficiency syndrome (AIDS, which is killing humans by the millions), the prions pandemonium, or the humiliating effectiveness of bacteria in their arms races against pharmaceutical companies are but a few examples.

Not that scientific progress has come to a halt: far from it. But it has led us to a point at which we can see, much more clearly than before, a long and bumpy stretch of road extending before us, probably with many twists and turns hidden from view. Cartographers of yore would have inscribed the warning "there be monsters here". In this book we have tried to be a bit more specific, with the help of some of the most expert scouts in the field. However, infectious diseases are among the relatively uncharted realms in evolutionary biology, offering plenty of drama and scope for adventure – witness, for instance, the efforts to reconstruct the genome of the virus responsible for the 1918 Great Influenza Epidemic: monsters be here indeed!

A generation ago, medical doctors and biologists were brought up on what is nowadays called the "conventional wisdom". It holds that pathogens should evolve toward becoming ever more benign to their hosts, since it is selectively

Box 1.1 Notions of virulence

Virulence describes the detrimental effect of parasitic exploitation on the host (just as resistance characterizes the detrimental effect of host defense on the parasite). Virulence therefore arises from processes through which parasites exploit their host to further their own multiplication and transmission. This general definition is respected throughout the present book.

To unravel alternative, more specific notions of virulence, it is useful to distinguish diseases according to how the process of damage to the hosts unfolds:

■ *Killing the host.* For relatively harmful diseases, the exploitation of hosts often results in their death. In such cases, a large part of the parasite's tendency to inflict harm can usually be summarized in terms of the parasite-induced additional mortality rate of the hosts. Many chapters in this book focus on this case and therefore equate virulence with parasite-induced mortality.

■ *Impairing other life-history characters.* Other negative consequences of parasite exploitation gain in relative importance if infection only rarely leads to death. Such alternative detrimental impacts of the parasites – ranging from a decrease in host fecundity through a change in its competitive abilities to a mere plunge in its mobility or well-being – are important aspects of parasite virulence in their own right and can impact on its evolution. While changes in mortality and fecundity affect host fitness directly, to understand the contributions of other side-effects of host exploitation to both parasite and host fitness may require an in-depth consideration of relatively subtle mechanisms.

■ *Gaining entrance.* Especially in the plant world, the potential of a pathogen to inflict damage often strongly depends on whether or not there is a match between resistance genes in the host and genes in the parasite to overcome that resistance. Often little variation is found in the damage inflicted on hosts by different parasite strains once they have gained entrance to the host. The relative capacity of parasites to enter the host then becomes the key determinant of any detrimental effects. Plant pathologists thus tend to use the term virulence to refer to those capacities. In this book, the term "matching virulence" is used for this; in contrast to this, and when the need arises, the term "aggressive virulence" is used for the detrimental effects of the parasite's exploitation strategy.

■ *Local spreading.* When hosts are structured into local populations, the harm that pathogens can bring to these depends on their transmission within the local populations – which, in turn, depends both on the local transmission rate and on the damage inflicted on individual hosts. "Virulent" parasites may then be defined as those that quickly and relentlessly spread throughout a local population. Such a use of the word virulence correlates it with traits that affect the transmissibility of the pathogen.

In an agricultural setting, these last two aspects of virulence tend to be present together (with a farm's crop as the local population), which explains the different terminological tradition in the phytopathological literature compared to, for example, the medical literature. While the last three aspects of virulence listed above may all be attractive for defining virulence for particular systems, the goal of conceptual clarity compelled us, throughout the book, to use them only with further qualification.

advantageous for parasites to have efficient vehicles at hand for their transmission. Thus, the virulence of a pathogen (Box 1.1) was envisaged as an adaptive trait: all pathogens would eventually become avirulent if given enough time to evolve. This Panglossian view has not always been that conventional: indeed, it helped, in its day, to spread the idea that virulence is subject to evolution, very rapid evolution, in fact – and this was quite a revolutionary insight at one time. Of course, it was but a first step. Evolutionary biologists have since learned that constraints within the relationship between transmissibility and virulence can seriously upset the trend toward harmlessness (Box 1.2), and that competition between several strains of a pathogen within one host demand an altogether more complex analysis than the former optimization arguments offered. These insights have prompted the idea that it may be feasible to interfere with or even redirect the evolution of virulence to achieve some desired practical goals – such as low virulence in the parasites of crops, cattle, or humans, and high virulence in the parasites that control weeds and pests. This Darwinian approach gave rise to a new research program on virulence management (Box 1.3) and provides the basis for this book.

Many of the arguments on the adaptive dynamics of virulence have become so involved that they are easier to analyze mathematically rather than verbally. We have nevertheless tried in this book to keep the mathematical techniques down to earth, and to display the modeling techniques in "stand-alone" boxes which, in combination, offer a concise and coherent introduction to the theoretical approaches used in the book (see the overview on page xvi).

Our emphasis is on the connection of this theory with empirical data and experimental set-ups. It turns out, in fact, that the data prove quite hard to interpret without a clear understanding of the actual meaning of basic notions such as virulence and fitness. To a first approximation, fitness is reproductive success and virulence is the additional mortality caused by the pathogen (see Box 1.1). However, in many instances, such as for populations that are not well mixed but distributed in clumps, this first approximation is not adequate. Case studies from infectious diseases in humans, chestnut blight, senescence in fungi, rinderpest, and, of course, the celebrated myxoma virus in rabbits, all show how difficult it is to disentangle rival concepts and to assess different modeling approaches.

Like all good Darwinians, we look toward theory to guide us through the plethora of facts. So in this book the initial chapters set the stage by discussing the impact of alternative transmission modes and ecological feedbacks on the evolution of virulence (Part A). We then proceed systematically to analyze, first, the implications of host population structure for the evolution of virulence (Part B), second, the competition of pathogens within a host (Part C), and, finally, pathogen–host coevolution (Part D) and multilevel selection (Part E). We firmly believe that only when armed with these tools is there a reasonable chance of understanding the long-term effects of vaccines and drugs (Part F) and of successfully addressing the options and problems of virulence management (Part G).

Box 1.2 A simple example of virulence evolution and management

Here we illustrate how evolutionary theory can be used to suggest measures that will help manage the virulence of a pathogen. We start with some conventional assumptions about the disease under consideration.

Single-species assumptions

- Pathogens only survive in living hosts.
- Pathogens can enter disease-free hosts only through contact between these and infected hosts.
- Once in a host, pathogens multiply rapidly, so that the first infection determines the final impact.
- Within the hosts, pathogens compete only with their own offspring.
- The per-host disease-free death rate is constant.

Interaction assumptions

- The rate at which susceptible hosts become infected is proportional to the product of the density of infected and that of susceptible hosts (law of mass action). The proportionality constant, termed per-host disease transmission rate, increases with pathogen replication.
- Pathogen replication occurs at the expense of the host's resources, and this damage to the host, termed virulence, increases the per-host disease-induced death rate.
- The trade-off between the per-host transmission rate and the per-host disease-induced death rate conforms to a law of diminishing returns.

For pathogens to transmit they require living hosts, so pathogen fitness depends on the average survival time of the hosts. Thus too high a virulence is not expected to pay off. As a representative measure of pathogen fitness, we use the number of new infections produced per host over the period it survives and is infectious, known as the pathogen's basic reproduction ratio R_0 (see Box 2.2). As shown in Box 9.1, the pathogen strain with highest R_0 outcompetes all others.

The disease-induced death rate that maximizes R_0 can be found graphically, the rationale for which is given in Box 5.1. In the figure at the end of this box, the fixed disease-free death rate is plotted to the left of the origin, while the evolutionarily variable disease-induced death rate, or virulence, is plotted to the right. The thick trade-off curve describes the effect of virulence on the disease transmission rate. Figure (a) shows how, by drawing a tangent line from the point on the left to the trade-off curve on the right, the optimal level of virulence is found just below the tangent point. In this simple example, pathogens are therefore expected to evolve toward intermediate levels of virulence.

continued

Box 1.2 *continued*

This graphic construction immediately suggests two possible routes to managing virulence:

- Either we change the trade-off curve such that the tangent point shifts to the left, Figure (b);
- Or we decrease the disease-free host death rate and keep the trade-off curve in place, Figure (c).

Both options are expected to result in the evolution of reduced virulence levels. Moreover, the second option generates the interesting hypothesis that investment in host health – so as to promote the life span of the hosts *in the absence of the disease* – creates an environment in which pathogens evolve to become more benign.

Of course the model as discussed above is overly simplistic. The remainder of this book investigates the various intricacies that should be considered to capture a wider range of circumstances.

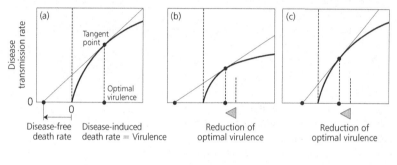

Whenever public health officials, veterinary epidemiologists, advisory plant pathologists, conservation biologists, or biocontrol workers want to devise strategies to manage the course of infectious diseases, they must bear in mind that they are merely adding one level of strategic action on top of other, age-old layers of strategic interactions. These have been devised through the programming by natural selection of both the pathogens and hosts – organisms that differ widely in scale, generation time, and life history, and that use individual variability and polymorphisms to fuel their arms races. If public health decisions are not based on a sound knowledge of these underlying tugs of war, they risk being counterproductive. Many human interferences, far from managing disease, have helped disease to manage us.

No doubt the next generations will know vastly more than we do now, but we hope that this book will offer no reason for them to deem us naively oversimplistic, as the 1960s appear to us now. To take Einstein's dictum to heart, we and all the contributors to this book have tried to present matters as simply as possible, but not simpler, and have endeavored to approach the complexity of our subject with the appropriate respect.

Box 1.3 A research program on virulence management

As a backbone for further research efforts, we outline a systematic sequence of steps to test hypotheses about virulence evolution and to probe options for virulence management:

1. Specify how the hosts are affected by the parasite's exploitation (effects of virulence).
2. Assess which of these effects influences parasite transmission (identification of trade-offs).
3. Spell out the ecological setting (e.g., which of the participants interact with each other, and how mixing takes place). Derive suitable representative measures for fitness given the ecological setting (e.g., R_0).
4. Analyze the adaptive dynamics of the ecological and evolutionary feedback processes.
5. Extract model predictions on how selection affects virulence and, in particular, how controllable epidemiologic parameters can be changed to select for reduced virulence.
6. Test these predictions theoretically (e.g., robustness of the model) and empirically.
7. Search for alternative explanations (e.g., multiple instead of single infection) and, if necessary, carry out tests to distinguish between the alternative mechanisms.

The chapters in the book follow this agenda and describe results for particular ecological settings. Given the diversity of relevant scenarios and the empirical uncertainty regarding some of their key components, it is evident that much research remains to be done in pursuit of this program.

Acknowledgments   Development of this book took place at the International Institute of Applied Systems Analysis (IIASA), Laxenburg, Austria, where IIASA's former director Gordon J. MacDonald and current director Arne Jernelöv have provided critical support. To achieve as much continuity across the subject areas as possible we organized two workshops in which the authors were brought together to discuss their contributions. The success of a book of this kind depends very much on the cooperation of the authors in dealing with the many points the editors are bound to raise, and we thank our authors for their patience. The book has benefited greatly from the support of the Publications Department at IIASA; we are especially grateful to Anka James, Martina Jöstl, Eryl Maedel, John Ormiston, and Lieselotte Roggenland for the work they have put into preparing the manuscript. Any mistakes that remain are our responsibility.

Part A
Setting the Stage

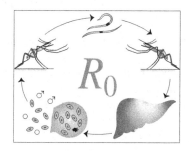

Introduction to Part A

Investigating options for virulence management is a multidisciplinary endeavor. To identify the most promising avenues, contributions from epidemiology, ecology, microbiology, genetics, and theoretical biology have to be integrated into a common perspective. That goal is an inspiration and challenge for this book as a whole.

Before diving into this complexity, some readers might appreciate a gentle start. Part A therefore introduces the essential ideas and concepts in this book and addresses the following questions:

- Is it realistic to expect measures of virulence management to succeed in practice?
- What are the epidemiological and ecological complexities that virulence management strategies ultimately may have to deal with?
- Which methods are suitable for assessing outcomes of virulence evolution and for predicting consequences of managerial interference?
- Which problems and dilemmas are bound to arise in the context of virulence management efforts?

Chapter 2 provides first suggestions of management options that can successfully influence the virulence of pathogens. Ewald and De Leo emphasize the critical importance of the mode of pathogen transmission for virulence evolution. They propose that, if pathogens can be transmitted from host to host along several routes, public health managers should be concerned primarily with those routes that are least dependent on the host's health. Taking waterborne transmission as an example, a model of diarrheal disease is presented. Maximization of the basic reproduction ratio shows that, when waterborne transmission prevails, evolutionarily stable levels of virulence tend to be high. Narrowing this transmission channel will therefore often select for less virulent pathogens.

Whereas Chapter 2 offers an optimistic view on the feasibility of virulence management for systems in which interventions are relatively easy and data are available, Chapter 3 concentrates on the opposite end of the scale. In their review of wildlife diseases, De Leo, Dobson, and Goodman flag some of the problems that arise from the distinction between micro- and macroparasites, from genetic diversity, and from coevolution. They make the important point that much of theory on the evolution of virulence has been developed for microparasites, even though macroparasites can have a major impact on host dynamics and community structure. The authors also stress that both micro- and macroparasites exert strong selection pressures on the host and that frequency-dependent selection plays an important role in the evolution of virulence. Moreover, they highlight that human populations expand and thereby come into contact with wildlife and their parasites: this creates the danger of parasites jumping over to humans, which in turn may lead to newly emerging diseases.

Chapter 4 explains why the traditional approach of predicting evolutionary outcomes by maximizing the basic reproduction ratio of a disease is not always appropriate. Since pathogens tend to affect their host environment in radical ways, selection pressures usually depend on the types of pathogens and hosts that are established in an infected population. In this chapter, Dieckmann outlines the theory of adaptive dynamics as a versatile toolbox for investigating the evolution and coevolution of pathogen–host interactions under conditions of frequency-dependent selection. Examples illustrate how classic methods and the new models presented here result in different predictions about the evolution of infectious diseases.

Decisions on virulence management strategies are fraught with dilemmas, as illustrated by the investigation of a model for the coevolution of virulence and recovery ability in Chapter 5. Van Baalen explains why there can be conflicts of interest between the individual host and the host population as a whole. Since selection tends to favor virulent parasites or those that can overcome host defenses, increased investment in the defense of individual hosts does not necessarily minimize the parasite load for the population as a whole. If more aggressive parasites are favored, hosts play "defense games" against each other, and thereby potentially trigger selection for a further increase of virulence. In the long run, hosts either pay heavily to defend themselves against a rare but extremely virulent parasite or they tolerate the parasite if it stays relatively benign. Human health care managers may thus be confronted with the ethical dilemma of creating either common-but-mild or rare-but-serious diseases.

The four chapters of Part A set the stage for this book by indicating the range of basic issues that have to be considered in the evaluation of strategies of virulence management: transmission routes (direct versus indirect; vertical versus horizontal), distinction between micro- and macroparasites, genetic diversity in host resistance and parasite virulence, frequency-dependent and reciprocal selection, multiplicity of evolutionarily stable virulence levels, and ethical dilemmas in medical epidemiology. Of course, many more aspects must be considered to assess and improve the match between models and epidemiological reality. That is what the remainder of this book is about.

2

Alternative Transmission Modes and the Evolution of Virulence

Paul W. Ewald and Giulio De Leo

2.1 Introduction: Historical Background

For most of the 20th century, medical scientists writing about the evolution of infectious diseases generally concluded that parasites are expected to evolve toward states of benign coexistence with their hosts (reviewed in Ewald 1994a). According to this line of reasoning, parasites that harm their hosts are harming their own long-term chances of survival, and are therefore at a disadvantage over evolutionary time. Theory developed since the 1980s emphasizes that this traditional viewpoint is based on faulty assumptions about the level at which natural selection acts. Specifically, natural selection is a process by which organismal variants that contribute more of their genetic instructions into future generations become increasingly represented in the gene pool of future generations. When applied to parasite virulence, the appropriate focus is therefore on the short-term competitive processes among parasite variants rather than on the characteristics that would allow a particular parasite species to persist most stably over the long term. According to this reasoning, by the time any variants reap such long-term benefits, they would already have been displaced by the variants that held the short-term advantage. Any increases in long-term survival of the parasite species associated with benignity are therefore of little if any relevance to the evolution of virulence if benign strains lose the short-term competition.

A large body of theory and empirical evidence now supports the idea that natural selection can favor evolution of parasitism toward virtually any position along a spectrum that ranges from commensalism to lethality (Fine 1975; Anderson and May 1981, 1982; Levin and Pimentel 1981; Levin *et al.* 1982; Ewald 1983, 1994a; Frank 1996c). Central to this theoretical framework is the trade-off concept, which proposes that the level of virulence to which a pathogen evolves is determined by a trade-off between the benefits and costs associated with increased host exploitation. In this case, benefits and costs are measured in units of evolutionary fitness, which quantify, at the genetic level, the passing on of particular genetic instructions relative to alternative instructions. At the organismal level, evolutionary fitness results from the differential survival and reproduction rates of organisms, in accordance with the definition of natural selection presented above. The fitness benefits associated with increased host exploitation are generated by the increased conversion of host resources into pathogen production and propagation. In models of virulence, fitness benefits are typically portrayed as a result of competition between genetically distinct variants within hosts (e.g., Van Baalen and Sabelis 1995a, see also Chapter 11). Competition between parasites in different hosts may also yield fitness benefits, if parasites in different hosts can be transmitted to susceptible hosts before competing strains reach these hosts, especially if the prior access to the host stimulates a defensive response that inhibits further transmission of competing variants (Ewald 1983, 1995). Fitness costs are typically accrued as a result of the negative effects that host illness and death exert on transmission. Fitness benefits thus influence the probability of a strain being transmitted per contact with susceptible hosts, whereas fitness costs influence the probability of an infected host contacting susceptible hosts.

Epidemiological modeling of these trade-offs is often based on so-called SIR models, in which the epidemiological process is divided into changes in susceptible S, infected I, and recovered-and-immune R subpopulations. Box 2.1 illustrates the general form of this kind of model, which in this case is built upon the first two of these classes, and is therefore referred to as an SI model. The epidemiological dynamics of these models are analyzed in the context of the basic reproduction ratio of infections, R_0 (see Box 2.2). This approach to epidemiological processes dates back to Ronald Ross's modeling of malaria at the beginning of the 20th century (Ross 1911; Kermack and McKendrick 1927; Macdonald 1952; see Heesterbeek and Dietz 1996 for a historical review).

During the last three decades of the 20th century, SI and SIR models were adapted to explain the major categories of transmission (for an overview of different transmission modes see Table 2.1). Levin and Pimentel (1981), for example, used this trade-off concept to illustrate how natural selection could favor intermediate levels of virulence in a vectorborne pathogen, the myxoma virus, employing a version of the mathematical model presented in Box 2.3. Similar approaches were applied to a broad array of transmission modes, often with expressions for evolutionary fitness explicitly incorporated into models (e.g., Dietz 1975; Anderson 1982; Anderson and May 1982; Levin *et al.* 1982; May and Anderson 1983a; Frank 1996c).

In this chapter we briefly review trade-offs for different modes of disease transmission (Section 2.2) and outline how their effect on virulence evolution was modeled in earlier studies (Section 2.3). To illustrate an alternative modeling approach, we then introduce a model that compares the lethality of disease when both waterborne and direct transmissions occur with that when only direct transmission occurs (Section 2.4). Section 2.5 discusses possible generalizations of our approach and concludes with ramifications for the goal of virulence management.

2.2 Virulence Depending on Transmission Modes

A major interest in the evolutionary trade-off approach stems from considering different modes of disease transmission. Virulence is predicted to be particularly high when the transmission mode allows pathogens to be readily transmitted, even when hosts are entirely immobilized by illness (Ewald 1983, 1994a). If pathogens

Box 2.1 SI models of directly transmitted diseases

In SI models, hosts are divided into two classes, susceptible and infective, occurring at densities S and I, respectively. A simple example is given by differential Equations (a) and (b), which represent changes in the densities of susceptible and infective hosts over time

$$\frac{dS}{dt} = B + \theta I - \beta SI - dS ,$$ (a)

$$\frac{dI}{dt} = \beta SI - (\alpha + d + \theta)I .$$ (b)

Here, B is the rate at which new susceptible hosts enter the population through birth or immigration, β is a transmission coefficient, d is the natural mortality rate for uninfected individuals, α is the disease-induced mortality rate (thus $\alpha + d$ is the total mortality rate for infected hosts), and θ is the recovery rate. Individuals move from the susceptible class into the infected class according to the rate at which infections are generated, βSI, and return to the susceptible class through recovery, θI; this implies that immunity is absent. New susceptible hosts arise at rate B; the population is diminished by natural and disease-induced mortality, dS and $(\alpha + d)I$. The equilibrium state of this system is determined by solving Equations (a) and (b) after setting the rates dS/dt and dI/dt equal to zero.

Assuming B to be constant is a little awkward biologically as it is expected to be a function of S and I, $B = b_S(S, I)S + b_I(S, I)I$. However, keeping B constant simplifies some of the calculations without affecting the evolutionary conclusions in any essential manner. [Sometimes the more extreme assumption is made that $B = dS + (\alpha + d)I$; in this way the model's total population density $N = S + I$ is kept constant, which simplifies the mathematical analysis even further.]

The term βSI is based on the principle of mass action, borrowed from chemistry (Dietz 1976). The validity of this assumption is somewhat controversial; for example, the transmission rate may not increase in direct proportion to the size of the susceptible population when transmission requires intimate direct contact, because the number of susceptible hosts that can be intimately contacted by an infected individual may be much smaller than the total number. Nevertheless, the mass action assumption is a useful starting point in the study of epidemiological dynamics.

Equations (a) and (b) can also be expressed in terms of proportions of the equilibrium population density in the absence of the infection, $N_0 = B/d$. With $s = S/N_0$ and $i = I/N_0$ this yields

$$\frac{ds}{dt} = B + \theta i - \beta si - ds ,$$ (c)

$$\frac{di}{dt} = \beta si - (\alpha + d + \theta)i .$$ (d)

It is important to realize that the transmission rate β in Equations (c) and (d) is defined differently from that in Equations (a) and (b), i.e., $\beta_{\text{proportions}} = \beta_{\text{densities}}/N_0$.

continued

Box 2.1 *continued*

The same applies to B, with $B_{\text{proportions}} = B_{\text{densities}}/N_0 = d$; all the other coefficients do not change in meaning.

To achieve a better match between models and reality it may sometimes be necessary to subdivide the classes of susceptible and infective hosts on the basis of inherent differences between host individuals (such as differences in age). It may also be useful to add further classes to the model, like recovered-and-immune individuals (SIR models) or free-living stages of parasites (see Box 2.3).

Table 2.1 Categories of transmission modes that have been studied using SIR or SI models.

Transmission mode	Description	Dependent on host mobility?	Example
Direct	Propagules are transmitted directly from one host to another through the air or by physical contact.	Yes	Common cold, measles
Vectorborne	Propagules are transported between hosts by a second species of host, the vector (e.g., a mosquito).	No	Malaria
Waterborne	Propagules are transmitted through water. In humans they typically cause diarrhea and can be transmitted by alternative modes: directly (person-to-person) or indirectly (person-to-food-to-person).	No	Cholera
Sit-and-wait	Propagules are shed into the environment where they remain until picked up by another host. Their greater durability in the external environment distinguishes them from directly transmitted pathogens.	No	Smallpox
Attendant-borne	Propagules are picked up by attendants, generally on hands, and transmitted to susceptible hosts without infecting the attendants.	No	Hospital-acquired *Staphylococcus aureus* and *Escherichia coli*
Respiratory	Propagules are transmitted by droplets expelled by sneezing; their transmission can be classified as direct or sit-and-wait, depending on their durability in the external environment.	Variable	Common cold, measles, smallpox
Venereal	Propagules are directly transmission by sexual contact; this is a subcategory of direct transmission.	No	Syphilis

Box 2.2 The basic reproduction ratio for infectious diseases

One of the most informative features of SI and SIR models (where R refers to the class of recovered-and-immune individuals) is the basic reproduction ratio of an infectious disease. Denoted by R_0, this is defined as the number of new infections generated from an existing infection when that infection is introduced into a population that comprises entirely susceptible hosts.

R_0 can be calculated by multiplying the rate at which new infections are generated by an infected host by the average duration of an infection. For the model specified in Box 2.1, the rate at which new infections are generated is βN_0 (where N_0 equals B/d, the equilibrium population density of susceptible hosts when $I = 0$). The average duration of an infection is $1/(\alpha + d + \theta)$ and we thus obtain

$$R_0 = \frac{\beta}{\alpha + d + \theta} N_0 . \tag{a}$$

If $R_0 > 1$, the infection spreads until susceptible hosts become so rare ($s = 1/R_0$) that an infected individual will, on average, infect only one susceptible host throughout its life. By contrast, if $R_0 < 1$, the infection cannot become established in the population.

It is clear that R_0 is a key variable to be considered for eradicating a disease: a classic epidemiological question is how can systems be influenced so as to bring R_0 below 1. This importance of R_0 has led to all sorts of exercises in which various properties of the disease dynamics, such as the equilibrium number of susceptible hosts S^*, are expressed in terms of R_0 instead of through the underlying and more mechanistic rate coefficients (Dietz 1975; Anderson and May 1981). For example, for the model in Box 2.1, we obtain $S^* = N_0/R_0$. A first benefit of such a relation is that we can use simple observables at the population level to estimate R_0 robustly, instead of having to estimate many separate rate coefficients at the individual level. Second, we can use such a relation to predict how intervention strategies influence R_0. For example, again for the model in Box 2.1, if we effectively vaccinate a fraction p of the instream of newborns B, we decrease R_0 by a fraction p. Therefore, to eradicate a disease through vaccination an effective vaccination coverage of $p > 1 - 1/R_0$ is required.

In models that exclude within-host interactions between disease strains (as results from multiple infections or when recovery does not result in full immunity), R_0 can serve as a convenient yardstick to assess the evolutionary dynamics of a disease: strains with a higher R_0 outcompete strains with a lower R_0. This implies that the introduction and fixation of mutational variation increases the R_0 value of a disease until the evolutionary options for further increase are exhausted. For more details on the evolutionary implications of R_0 calculations see Boxes 5.1 and 9.1.

When there exists a fixed trade-off, $\beta = \beta(\alpha)$, between the transmission coefficient β and the disease-induced mortality rate α, R_0 often attains a maximum for some intermediate value of α. The corresponding evolutionarily stable level of virulence, α^*, can be found by substituting $\beta = \beta(\alpha)$ in Equation (a) and setting $dR_0/d\alpha|_{\alpha=\alpha^*} = 0$. This procedure has the following biological interpretation. At first glance, pathogens always benefit from a high transmission coefficient. As

continued

Box 2.2 *continued*

the transmission coefficient is positively correlated with the production rate of propagules within hosts, a high transmission coefficient carries the expense of increased virulence. This virulence may express itself as pathogen-induced mortality or illness, either of which may reduce the transmission coefficient. When such a trade-off occurs, the virulence α affects the basic reproduction ratio R_0 both directly and indirectly by affecting the transmission coefficient β. As a result, R_0 first increases with virulence for low values of α and then, beyond $\alpha = \alpha^*$, decreases with virulence for high values of α (see Figures 2.4 and 2.5). Hence, there is an intermediate virulence $\alpha = \alpha^*$ that optimizes host exploitation (Anderson and May 1991). For $\alpha > \alpha^*$, disease-induced effects on the host lead to reduced transmission. For $\alpha < \alpha^*$, low propagule production leads to reduced transmission. Under the simplifying assumptions stated above this optimal level of virulence represents an evolutionarily stable strategy (ESS): because of reduced fitness, any mutant that deviates from this strategy cannot invade a host population infected by pathogens with virulence α^* (Maynard Smith 1982; Van Baalen and Sabelis 1995a).

---

are transmitted by biting arthropod vectors, for example, host mobility is not necessary for transmission. The costs incurred by exploitation of hosts therefore rise more slowly, as a function of that exploitation, for vectorborne pathogens than for directly transmitted pathogens. The level of host exploitation at which the fitness costs to the pathogen rise more rapidly than the fitness benefits (the "ESS" in Box 2.2) should therefore be higher for vectorborne pathogens than for directly transmitted pathogens. Insofar as the level of host exploitation is positively associated with virulence, natural selection is expected to favor a higher level of virulence among vectorborne pathogens than among directly transmitted pathogens; this expectation accords with the observed differences in lethality between vectorborne and directly transmitted pathogens of humans (Figure 2.1; Ewald 1983, 1994a).

Aspects of human behavior and culture can similarly generate "cultural vectors," which, like biological vectors, can transport pathogens from immobilized hosts. Waterborne transmission of diarrheal pathogens offers an example. If water supplies are not adequately protected, the washing of materials that carry pathogens from an immobilized infected individual can cause contamination and thereby infect large numbers of other people. This example emphasizes that fitness benefits accrued by pathogens through host exploitation probably also vary differentially for the different modes of transmission. When diarrheal pathogens are transmitted by water the benefits associated with increasing exploitation are expected to saturate at a far greater level of host exploitation than when transmission is solely by direct contact. For waterborne transmission the asymptote would be reached only when each person in the population that is exposed to potentially contaminated water has a probability of infection from the infected individual equal to one.

Box 2.3 Modeling for vectorborne transmission

The model specified in Boxes 2.1 and 2.2 can be adapted to vectorborne transmission of vertebrate hosts. For this we introduce new classes for the infected and uninfected vectors, occurring at densities V and U, respectively, and modify the mass action assumption to account for a vector taking a fixed number of blood meals per unit of time (Anderson 1982).

Keeping the notation of Box 2.1 and with $S + I = N$ this yields the following system

$$\frac{dS}{dt} = B + \theta I - \psi \phi V \frac{S}{N} - dS \,, \tag{a}$$

$$\frac{dI}{dt} = \psi \phi V \frac{S}{N} - (\theta + d + \alpha) I \,, \tag{b}$$

$$\frac{dU}{dt} = F - \psi \frac{I}{N} U - \omega U \,, \tag{c}$$

$$\frac{dV}{dt} = \psi \frac{I}{N} U - \omega V \,. \tag{d}$$

The rate of transmission per unit of area to a susceptible vertebrate host is the product of the rate at which an individual vector bites hosts ψ, the probability of transmission by a bite from an infected vector to a vertebrate ϕ, the density of infected vectors V, and the probability that the victim is still disease free, S/N. An uninfected vector is assumed to become infected by a single bite. Therefore the rate per unit of area for uninfected vectors to become infected is the product of the rate at which vectors bite ψ, the probability that a bitten vertebrate is infected, I/N, and the density of uninfected vectors U. The rate per unit of area at which new, uninfected, vectors enter the system is denoted by F. Infected and uninfected vectors die at a per capita rate of ω per unit of area.

As can be seen from Equations (c) and (d), the total density of vectors, $Z = U + V$, converges to $Z_0 = F/\omega$. It is therefore possible to replace Equations (c) and (d) by setting $V = v Z_0$ and $U = (1 - v) Z_0$ together with

$$\frac{dv}{dt} = \psi \frac{I}{N} (1 - v) - \omega v \,. \tag{e}$$

Determining the basic reproduction ratio for this model we obtain

$$R_0 = \frac{\psi^2 (Z_0/N_0) \phi}{\omega (\theta + d + \alpha)} \,, \tag{f}$$

where Z_0/N_0 denotes the number of vectors per host after the density of vectors has equilibrated and before the density of hosts changes from its disease-free state. As before, $N_0 = B/d$ is the equilibrium density of hosts in the absence of the disease. Equation (f) suggests various interventions that could, in theory, eradicate malaria by making $R_0 < 1$. In particular, the quadratic effect of ψ on R_0 indicates that reducing biting rate might be surprisingly effective.

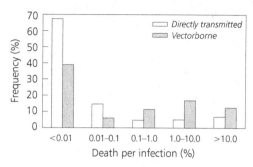

Figure 2.1 Mortality of vectorborne and directly transmitted pathogens of humans. Percentages correspond to the percentage of all species of pathogens in the transmission category that fall within the mortality category. Details of calculations are given in Ewald (1983).

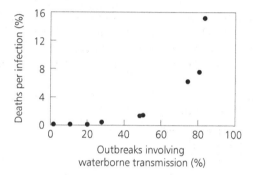

Figure 2.2 Mortality of diarrheal bacteria of humans as a function of their tendencies to be waterborne. Details are given in Ewald (1991a).

These low costs and high benefits of virulence lead to the hypothesis that waterborne transmission should be associated with particularly high virulence. This hypothesis has been tested by determining whether, for diarrheal pathogens of humans, the degree of waterborne transmission relative to direct transmission positively correlates with the lethality of untreated infections. It was shown that the two variables correlated significantly (Figure 2.2). Literature-based tests and more recent experimental assays of virulence indicate that associations between virulence and waterborne transmission also occur when the taxonomic focus is narrowed: temporal and geographical variations in waterborne transmission help explain geographic variations in virulence within genera of diarrheal pathogens as well as within a particular species, *Vibrio cholerae* (Ewald 1991a, 1994a, and Chapter 28).

Analogous arguments have been applied to hospital-acquired infections and pathogens of agricultural plants, and the initial testing of these ideas confirmed the central predictions in both cases. For hospital-acquired infections, the lethality of *Escherichia coli* was positively correlated with the duration of attendant-borne transmission (Figure 2.3); in the case of agricultural pathogens, virulence of plant

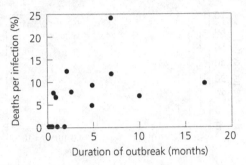

Figure 2.3 Mortality of *E. coli* infections in hospitals as a function of the duration of the outbreak. *Source*: Ewald (1988), Ewald (1991b).

viroids positively correlates with the degree of transmission on agricultural utensils (Ewald 1988).

This kind of trade-off argument also suggests that increased durability in the external environment should favor increased virulence, because increased durability allows greater reliance on the mobility of susceptible hosts for transmission, a mode termed "sit-and-wait" transmission (Ewald 1987a). As expected from this hypothesis, variation in durability in the external environment explains a significant amount of the variation in lethality among human respiratory tract pathogens (Walther and Ewald, unpublished).

Of course, the comparative nature of these tests leaves room for alternative explanations regarding causation. Although the predictions were generated from consideration of host mobility, other uncontrolled variables could be correlated with such transmission and might be responsible for the association. Alternatively, the identified transmission modes could be causing the associations indirectly; for example, by generating greater within-host genetic variation, and thereby causing greater within-host competition and hence favoring evolution of increased virulence. Such alternative explanations need to be developed and evaluated, and none has yet been tested empirically. It is possible, for example, that vectorborne or waterborne transmission is associated with a greater within-host genetic variation than directly transmitted pathogens, and this genetic variation could favor increased virulence. There is no empirical evidence for this association, but it is a feasible alternative hypothesis. In contrast, there is empirical evidence that vectorborne and waterborne transmission can occur more efficiently from immobilized hosts (e.g., Prescott and Horwood 1935; Levine *et al.* 1976; Waage and Nondo 1982; Day *et al.* 1983). The immobilization argument is therefore at a slightly more advanced stage of testing than alternative hypotheses because its central assumption has empirical support.

2.3 Effects of Transmission Mode on Virulence

The preceding considerations emphasize the importance of distinguishing between modes of transmission in theoretical analyses. With regard to the arguments based on host mobility, the important theoretical distinction is between one mode that depends on host mobility and one that does not. This distinction is particularly apparent in the comparison of vectorborne with direct transmission, but it also occurs in the other cases. For waterborne transmission, the distinction requires a separation of transmission through water from transmission through direct contact between people (or through modes of indirect contact that require mobility of infected individuals, such as contamination of food by food handlers).

The subtlety of these distinctions has led to ambiguity in the literature, both in verbal and mathematical treatments of the hypothesis. Resolution of these ambiguities is a necessary step toward identification of the appropriate avenues for further development of theory about the evolution of virulence. With regard to mathematical models of waterborne transmission, Van Baalen and Sabelis (1995b) concluded that purification of water supplies would not cause the evolution of reduced pathogen virulence, but their model did not separate the transmission modes that require host mobility (e.g., transmission by direct contact) from those that do not (i.e., waterborne transmission). Their model therefore suggests that improvements in hygiene will not cause an evolutionary reduction in virulence if transmission occurs through only one mode. Although this result is interesting, it does not represent evidence against the hypothesized link between waterborne transmission and the evolution of virulence. To evaluate this link appropriately, the two modes of transmission must be specified in the model because, according to the waterborne transmission hypothesis, the increase in levels of host exploitation (and hence virulence) is greater in the presence of waterborne transmission than in its absence.

Similarly, with regard to the effects of pathogen durability in the external environment, Bonhoeffer *et al.* (1996) concluded that increased durability does not result in an evolutionary increase of virulence, but their model did not distinguish the two categories of transmission central to the sit-and-wait hypothesis. Transmission that is attributable to the mobility of susceptible hosts ("sit-and-wait" transmission, dependent on durable stages in the external environment) must be separated from the direct transmission attributable to the mobility of infected individuals.

2.4 Model of Virulence Evolution and Waterborne Transmission

In light of these considerations, we adapted standard SI models to incorporate the trade-offs associated with two or more transmission modes. Our goal is to assess whether levels of virulence depend on variations in the relative weights of transmission via mobility-dependent and mobility-independent modes. The model contrasts waterborne transmission with direct transmission, but it is readily modifiable to conform to other alternatives, such as those pertaining to the contrast between sit-and-wait transmission and direct transmission.

Transmission and recovery rates as functions of virulence

We assume that the production of pathogens within infected hosts increases with the pathogen-induced mortality rate, or virulence α, and that this relationship is

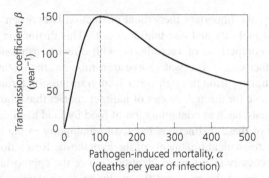

Figure 2.4 Transmission coefficient β as a function of pathogen-induced mortality α for direct transmission based on Equation (2.1).

independent of transmission mode. We assume that the transmission coefficient β first rises as a function of α because increased host exploitation increases both α and the probability of infection per contacted host (a component of β), thus making increased virulence beneficial when virulence is low. We incorporate a cost of host immobility on direct transmission by assuming that the transmission coefficient β eventually declines as a function of α. For large values of α the transmission coefficient β remains positive, reflecting that infectious contact is possible even when hosts become immobilized by the pathogen (e.g., through susceptible hosts caring for or visiting the immobilized patient). A derivation of the relation between β and α from first principles is very complicated and is therefore beyond the scope of this chapter. As a first step toward assessing the effect of such a functional relationship, we assume the dependence of β on α is given by

$$\beta(\alpha) = \frac{c_0 \alpha}{c_1 + \alpha^2}. \tag{2.1}$$

Parameter c_0 measures increased infectivity per unit increase in α (pathogen-induced mortality, our indicator of virulence). Parameter c_1 is introduced so that the functional association between β and α captures the fundamental trade-off on which the theory of transmission mode and virulence is based, namely that β first rises as a function of α and then declines gradually after passing through a maximum. The functional relationship specified by Equation (2.1) is illustrated in Figure 2.4.

In addition to the transmission rate, we also consider that recovery rate depends on virulence. In particular, we assume that the recovery rate θ is inversely proportional to the pathogen-induced mortality rate, $\theta(\alpha) = c_2/\alpha$.

Rate equations for direct and waterborne transmission

All waterborne diarrheal pathogens of humans can also be transmitted by other routes that depend on host mobility. To assess the effects of waterborne transmission, models therefore need to incorporate both waterborne transmission and direct transmission.

If only direct transmission can occur, the dynamics of susceptible and infected subpopulations are given by Equations (a) and (b) in Box 2.1, in which the constants β and θ are now functions of α. When both direct and waterborne transmissions can occur, changes in densities of susceptible and infected subpopulations S and I, as well the density of waterborne pathogens W, are described by the following equations

$$\frac{dS}{dt} = B + \theta(\alpha)I - \beta(\alpha)SI - \beta_W WS - dS , \tag{2.2a}$$

$$\frac{dI}{dt} = \beta(\alpha)SI - [\alpha + d + \theta(\alpha)]I + \beta_W WS , \tag{2.2b}$$

$$\frac{dW}{dt} = \rho(\alpha)I - mW . \tag{2.2c}$$

Susceptible hosts are infected by waterborne pathogens at a rate given by $\beta_W WS$, that is, by the product of the density W of pathogens in the water, the waterborne transmission coefficient β_W (which measures the probability of generating an infection in a susceptible host per pathogen in the water), and the density S of the susceptible subpopulation. Equation (2.2a) shows that the subpopulation of susceptible hosts grows according to birth B and recovery, $\theta(\alpha)I$, just as in Equation (a) of Box 2.1, but here this is diminished by infections acquired by direct contact, $\beta(\alpha)SI$, as well as from the water supply, $\beta_W WS$. According to Equation (2.2b), the subpopulation of infected hosts I is augmented by infection through direct contact, $\beta(\alpha)SI$, and through the water supply, $\beta_W WS$, and is reduced by the same sources of mortality and recovery as before, $[\alpha + d + \theta(\alpha)]I$, see Box 2.1. We assume that the rate of release of pathogens from each infected individual into the water supply is a function of pathogen-induced mortality, denoted by $\rho(\alpha)$. Equation (2.2c) specifies that the pathogen density W in the water increases by such a release, $\rho(\alpha)I$, and decreases through pathogen death, mW; the parameter m denotes the rate at which a propagule loses its viability in the water. We assume that the rate at which pathogens are lost from the external environment through ingestion is negligible relative to m, and that the pathogen does not replicate in the external environment.

Our goal at this stage is not to simulate competition between two variant strains of the pathogen, so the model does not explicitly incorporate competition between pathogens. Instead, we want to assess whether alterations in the potential for waterborne transmission alter virulence. We therefore compare the virulence expected (in the sense of an ESS as introduced in Box 2.2) if no waterborne transmission occurs in an area with the virulence expected if some waterborne transmission occurs in that area.

Reproduction ratios for waterborne and direct transmission

We define two reproduction ratios (see Box 2.2), one for pathogens that are directly transmitted and one for pathogens that are waterborne. The reproduction ratio for direct transmission, denoted by R_0, is obtained as described by Equation (a) in Box 2.2

$$R_0(\alpha) = \frac{\beta(\alpha)}{\alpha + d + \theta(\alpha)} .\qquad(2.3)$$

To analyze the consequences of this functional dependence of R_0 on α, numerical simulations were performed. The following values were chosen to generate coefficients that correspond at least roughly to real infectious diseases in humans: $c_0 = 1\,000$ and $c_1 = 100$ year^{-2}. The constant c_2 in the recovery rate $\theta(\alpha) = c_2/\alpha$ is set to 500 year^{-2}. The normal life span of hosts is set to 60 years so that $d = 1/60$ year^{-1}.

To find the value of α that maximizes R_0, denoted by α_{max}, we solve for $dR_0/d\alpha = 0$

$$\frac{dR_0}{d\alpha} = -c_0\alpha\frac{2\alpha^4 + d\alpha^3 - c_1 d\alpha - 2c_1 c_2}{(\alpha^2 + c_1)^2(\alpha^2 + d\alpha + c_2)^2} = 0 .\qquad(2.4)$$

This shows that α_{max} is the positive real root of Equation (2.5),

$$2\alpha^4 + d\alpha^3 - c_1 d\alpha - 2c_1 c_2 = 0 .\qquad(2.5)$$

For the given values of c_1, c_2, and d, this root was numerically computed to have the value $\alpha_{max} = 47.3$ year^{-1}. This value is less than half the pathogenicity that maximizes $\beta(\alpha)$ (see the dashed line in Figure 2.4). The resultant maximal value of R_0 is given by max $R_0 = 2.0$.

To determine the analogous values for waterborne transmission, denoted by α'_{max} and max R'_0, we first assume that the per capita rate of propagule release into water ρ is a linearly increasing function of α, $\rho(\alpha) = c\alpha$, with $c = 10$ propagules per infected host per unit pathogen-induced mortality rate. These "propagules" are best considered as infective units and therefore might comprise thousands or millions of individual pathogens, depending on the dosage required per infection. We also assume $\beta_W = 10$ year^{-1}: corresponding to the infections generated per year per propagule in the water, and $m = 500d$ (which corresponds to a life expectancy of propagules in the water of 44 days). To make the notation more compact we introduce the abbreviation $\chi = \beta_W c/m$. This notation also facilitates interpreting the effects of waterborne transmission, because χ is an indicator of the potential for waterborne transmission. The reproduction ratio in the presence of waterborne transmission, $R'_0(\alpha)$, is given by

$$R'_0(\alpha) = \frac{\beta(\alpha) + \frac{\beta_W \rho(\alpha)}{m}}{\alpha + d + \theta(\alpha)} = \frac{\frac{c_0\alpha}{c_1 + \alpha^2} + \chi\alpha}{\alpha + d + \frac{c_2}{\alpha}} .\qquad(2.6)$$

Comparison of optimal virulence levels

The virulence that maximizes the reproduction ratio R'_0 is denoted by α'_{max}, and is found, as in the case of direct transmission, by setting $dR'_0/d\alpha = 0$, and finding the real positive root. The optimal level of virulence thus determined is $\alpha'_{max} = 92.9$ year^{-1} i.e., about twice as large as α_{max}, the optimal level of virulence

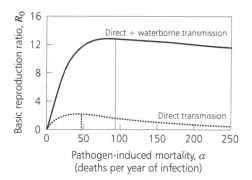

Figure 2.5 Basic reproduction ratio R_0 as a function of pathogen-induced mortality α for direct transmission and waterborne transmission. The continuous vertical line designates the virulence that maximizes R_0 for direct and waterborne transmission ($\alpha'_{max} = 92.9$ year^{-1}; max $R'_0 = 12.9$); the dashed vertical line designates the analogous value for direct transmission ($\alpha_{max} = 47.3$ year^{-1}; max $R_0 = 2.0$).

for direct transmission only. This result supports the hypothesis that waterborne transmission can increase virulence (Figure 2.5). Moreover, the maximum reproduction ratio in the presence of waterborne transmission, max $R'_0 = 12.9$, is over six times higher than the maximum reproduction ratio for direct transmission only, R_0, supporting the idea that the presence of waterborne transmission can greatly increase the potential for the spread of the infection (Figure 2.5).

This analysis therefore illustrates how the presence of opportunities for waterborne transmission may affect the outcome of virulence evolution, even though models that consider overall levels of hygiene (rather than differential effects of hygiene on different modes of transmission) do not show analogous effects. If the rate ρ of propagule release per host is considered to be an increasing saturating function of α (in accordance with assumptions made by Bonhoeffer *et al.* 1996), the overall results are similar.

How waterborne transmission rates affect virulence

To further investigate the effects of waterborne transmission on virulence, sensitivity analyses were performed with respect to $\chi = \beta_W c/m$. Figure 2.6 shows how pathogen-induced mortality changes with the percentage of waterborne transmission. Optimal levels of virulence increase at a greater than linear rate (Figure 2.6). A similar acceleration occurs whether ρ is assumed to be a linear function of α or a saturating function of α. It turns out that the acceleration is also relatively unaffected by changes in the degree of direct transmission c_0: changing the value of c_0 by over an order of magnitude does not change the association between virulence and percentage of waterborne transmission sufficiently to allow them to be distinguished in Figure 2.6. If, however, propagule production is not positively associated with virulence (i.e., if ρ is not an increasing function of α), the positive correlation between percentage of waterborne transmission and virulence vanishes.

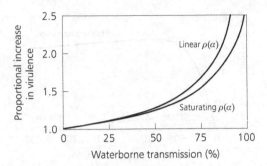

Figure 2.6 Pathogen-induced mortality as a function of the degree of waterborne transmission. Mortality is expressed as a proportional increase relative to that with no waterborne transmission. Waterborne transmission is expressed as a percentage of the total number of infections that occur via water as opposed to direct contact.

The theoretical trends in Figure 2.6 resemble the empirical trend in Figure 2.2; both show an accelerating rise in mortality as a function of increasing levels of waterborne transmission. Although the nonlinear rise was not predicted from the verbal theory that prompted the gathering of the data, the positive correlation between waterborne transmission and virulence was expected on the assumption that increased virulence would be positively correlated with the increased generation of propagules (see Section 2.1 and Ewald 1991a). The acceleration of the rise is more pronounced in the actual data than in the model's results (compare Figures 2.2 and 2.6), but considering the many simplifying assumptions made for the model and the uncertainties regarding the compatibility of measurements, even a qualitative agreement is noteworthy. The empirical trend, for example, is observed in terms of the percentage of outbreaks involving water, whereas for our model we have measured the percentage of infections involving water. To the extent that waterborne transmission generates far larger outbreaks, the conversion from outbreaks involving water to infections involving water would shift the curve in Figure 2.2 to the right along the horizontal axis, increasing the concordance between Figures 2.2 and 2.6. However, outbreaks that involve water also often involve nonwaterborne transmission; in response to this effect, the curve in Figure 2.2 would shift to the left, decreasing the concordance between Figures 2.2 and 2.6. Both the qualitative concordance and the quantitative uncertainties associated with a comparison of Figures 2.2 and 2.6 therefore draw attention to the potential value of developing the model and obtaining more refined epidemiological data.

2.5 Discussion: Applications and Implications

The model presented in Section 2.4 is cast in the context of waterborne transmission, yet its general form is applicable to other modes of transmission that are predicted to favor increased virulence. The model can, for example, be adjusted to sit-and-wait transmission simply by redefining some terms. In this case, ρ is the rate at which propagules are released into the terrestrial environment rather than into the water, β_W is the rate at which such pathogens are picked up by a susceptible host moving through that environment, and W is the density of pathogens in the terrestrial environment. Our results for the model of waterborne transmission suggest that conclusions as to the influences of durable pathogens on virulence (Bonhoeffer *et al.* 1996) may also change as the different modes of transmission are incorporated into the model.

Similarly, the model presented here should be adaptable to attendant-borne transmission in institutions such as hospitals. In this case the transmission is separable into direct transmission between patients, and attendant-borne transmission, which may involve the hands as the sole intermediary environment, or it may involve hands as part of a circuitous route that also involves objects in the environment (Ewald 1988, 1994a). The selection for increased virulence may differ substantially among different institutions, because different institutional settings may result in dramatically different shapes for the transmission rate as a function of virulence. In institutions for the retarded, patients may be relatively mobile, and the resultant dependence of transmission rate on virulence may have a shape similar to that shown in Figure 2.4. At the other extreme are nursery wards, in which even a healthy baby would not move around the ward contacting susceptible hosts. In this case, host illness may restrict transmission little if at all, and virulence could evolve to very high levels. Mortality associated with *E. coli* infections in such settings has been as high as 25% (Ewald 1988).

Our model may serve as a starting point for considering the integration of multiple modes of transmission with other influences on the evolution of virulence. Within-host competition, for example, is not incorporated into the model, yet it may contribute to increased virulence as pathogen durability in the external environment increases (Chapter 11). Immune mediation of competition may also have important effects.

As a simple optimization model, our approach does not incorporate frequency-dependent selection, which could maintain heterogeneity of pathogen virulence. This heterogeneity could be very important in applications to virulence management, for example, by providing variation on which natural selection can act. Such heterogeneity could increase the potential for rapid alteration of virulence through manipulation of the transmission mode, which thus enhances opportunities for virulence management.

Still, the model supports the idea that the evolution of virulence may depend on the presence of alternative transmission modes, particularly when one of the transmission modes is less dependent on host mobility. This outcome thus suggests specific options for virulence management: elimination of the transmission modes that are less dependent on host mobility should reduce virulence. In the case of diarrheal diseases this intervention would involve the provision of safe water supplies. As is discussed in Chapter 28, empirical evidence accords with this expectation.

Acknowledgments   Paul Ewald was supported by grants from Leonard X. Bosack and Bette M. Kruger Charitable Foundation and an Amherst College Faculty Research Award.

3

Wildlife Perspectives on the Evolution of Virulence

Giulio De Leo, Andy Dobson, and Andy Goodman

3.1 Introduction

The interaction between pathogens and their hosts is the most intimate of interspecific interactions. The pathogen is entirely dependent upon the host for resources and transmission to the next susceptible host in its life cycle. In contrast, the presence of pathogens usually leads to a reduction in host fitness through reductions in survival, fecundity, or opportunities to locate a mate. However, only a proportion of the host population is ever exposed to any particular parasite species, while all parasite populations are exposed to their hosts. These asymmetries in association and in the costs and benefits accrued to parasites and hosts are further compounded by asymmetries in the generation time of the two species: the generation time of the host often exceeds that of the parasite by several orders of magnitude. Consequently, when we examine the evolution of virulence and other components of parasite fitness, we usually focus on changes in parasite phenotype in response to constraints placed by the host's life history.

In this chapter, we analyze different aspects of the evolution of virulence in systems of free-living hosts and their parasites. First we establish the difference between micro- and macroparasite dynamics. Several documented population dynamic studies in which parasites have been shown to dramatically affect the abundance of host populations and the structure of biological communities are then discussed. We describe the spread of rinderpest epidemics in sub-Saharan Africa at the end of 19th century as one of the most striking examples of the impact of introduced pathogens on novel hosts. We then argue that host genetic diversity may have an important role in modifying epidemiological patterns in wildlife populations that are usually ascribed to other causes, and we outline the importance of diversity-generating mechanisms, such as sexual reproduction, as a way of escaping parasite attack. Next, we report a description of myxomatosis epidemics in the European rabbit in Australia, probably the best-documented case of evolution of virulence in wildlife. We then summarize the main aspects of the evolutionary race between the host and the parasite. Subsequently, we show how the competition among different strains of a pathogen can foster the selection for increased virulence.

We then briefly discuss the implication of interspecific transmission in terms of changes in virulence and transmissibility. We briefly explore this issue by referring to the specific case of *Pasteurella* in bighorn sheep in the western USA. We conclude the chapter by discussing potential impacts of wildlife disease on human health, with particular reference to some "new" or "resurgent" plagues that have recently afflicted human populations.

3.2   Microparasites versus Macroparasites

Theoretical arguments and empirical evidence for the evolution of virulence in wildlife are presented through this chapter for both micro- and macroparasitic infections. The distinction between micro- and macroparasites is common in epidemiology, and is based on the following line of reasoning. Microparasites are usually unicellular; they reproduce within the host (typically in host's cells) and do not necessarily require free-living stages or propagules during the course of their life cycle. Transmission usually occurs through direct contact between an infected and a healthy susceptible host, or via a vector such as ticks or mosquitoes. Quite often, microparasites can trigger a host immune response. Viruses and bacteria are typical microparasites. Macroparasites, on the contrary, are multicellular organisms that, in some cases, can achieve a considerable size. They grow within the host, but, in general, do not reproduce within the host; instead they require a free-living infective stage to complete their life cycle. The high antigenic diversity of the parasites means the host immune response takes longer to develop and may never occur. Macroparasitic life cycles may be monoxenic (just one host species) or heteroxenic (two or more host species). Moreover, macroparasites generally show a clumped distribution in their host population, with most of the hosts harboring few or no parasites, and few hosts harboring a large number of parasites. Macroparasites can be further classified as ectoparasites (such as ticks, fleas, mites, leeches, and several fungi) if they live on the host's skin, hairs, ears, or other cavities, or endoparasites (typically helminth worms, such as platyhelminths, nematodes, and acanthocephalans) when they live inside the host (in the gut or lungs, for example).

In the classic epidemiological analysis of microparasitic diseases, it is usually difficult to quantify the actual number of parasites within the host, because they are very small and reproduce quickly. As a consequence, emphasis is placed on the qualitative aspect of infection, namely the state of the host as healthy and susceptible, exposed (but not yet infective), infective, or recovered and immune. A simple model of host–microparasite dynamics is presented in Box 2.1 in Chapter 2. A crucial parameter in these kinds of systems is the basic reproduction ratio R_0, namely the expected number of secondary infections produced by a single infected host introduced in a population of susceptible individuals. The infective agent can establish in the host population only if $R_0 > 1$.

When analyzing macroparasite dynamics, the quantitative aspects of infection are important because the number of parasites actually harbored by a host may greatly affect its survival or reproductive success. As a consequence, the mathematical description of macroparasitic diseases is slightly more complicated than in the microparasitic case. In fact, the dynamics of the host–parasite system cannot be described in terms of few classes of individuals (susceptible/infected, etc.);

Box 3.1 The basic reproduction ratio in host–macroparasite systems

Macroparasites spend part of their life-cycle outside their host as free-living stages. In addition, the distribution of parasites over hosts is usually highly aggregated, with the majority of hosts harboring no or few, and few hosts harboring the largest fraction of parasites. Thus, not only the qualitative aspects of infection [such as presence/absence, as in susceptible-and-infected (SI) models for microparasites], but also its quantitative aspects (the number of parasites per host and their distribution) have to be taken into account.

The general host–macroparasite model consists of an infinite number of differential equations for variable $n_j(t)$, the number of hosts at time t harboring j parasites (Pugliese and Rosà 1995). Experimental data show that the distribution of parasites in their host can often be approximated by a negative binomial with a constant clumping parameter k (the smaller the value of k, the more aggregated are the worms in their hosts). May and Anderson (1978) have heuristically reduced the infinite number of differential equations, on the assumption that the system conserves the negative binomial form of parasite distribution, to

$$\frac{dN}{dt} = r(N)N - \alpha P , \tag{a}$$

$$\frac{dP}{dt} = \beta PN/(H_0 + N) - [\delta + \alpha + b(N)]P - \alpha(P^2/N)(k+1)/k , \tag{b}$$

where N and P are the overall densities of hosts and parasites, $r(N) = b(N) - d(N)$, in which b is the per capita birth rate and d the per capita death rate, is the per capita rate of increase of the host population in the absence of parasites (a decreasing function that becomes zero at $N = K$, the carrying capacity), α is the parasite-induced host mortality rate per parasite (on the assumption that this mortality increases linearly with the number of parasites per host), β is the rate of production of infective free-living stages, δ is the within-host worm mortality rate, and H_0 is the ratio of the natural mortality rate of the free-living stages to the transmission rate of the free-living infective stages to hosts. The term $\beta N/(H_0+N)$ in Equation (b) is derived by assuming that the dynamics of the short-lived, infective stages is so fast that it can be considered to be in a pseudo-steady state.

Consider a parasite-free host population at equilibrium K, where $r(N) = 0$ and $bK = dK$. Following the introduction of a single parasite, the parasite population can spread if its growth rate $[\beta K/(H_0 + K)] - [\delta + \alpha + d(K)]$ is positive, or, alternatively, if

$$R_0 = \frac{\beta K}{(H_0 + K)[\delta + \alpha + d(K)]} > 1 . \tag{c}$$

R_0 is the basic reproduction ratio and can be interpreted as the expected number of adult parasites produced, in the absence of density-dependent constraints, by a typical parasite during its entire period of reproductive maturity. R_0 is the product of three factors: the rate of production of infective free-living stages β, the life expectancy of a mature parasite $1/[\delta + \alpha + d(K)]$, and the probability of a free-living stage surviving to sexual maturity $K/(H_0 + K)$.

continued

Box 3.1 *continued*

Note that the R_0 concept presupposes clonal reproduction. Many macroparasites reproduce only sexually, thus making the R_0 concept moot. However, if low infections are sufficiently clumped, simplified clonal models may suffice. For more details see Nåsell (1985) and May (1977).

---

instead, the full distribution of parasites across the host population has to be considered (see Box 3.1). Here, the basic reproduction ratio, R_0, can be interpreted as the expected number of adult parasites produced (in the absence of density-dependent constraints acting on the parasite) by a typical adult during its entire period of reproductive maturity (Scott and Smith 1994).

In microparasitic diseases, attention is usually focused on the dynamics of either a single pathogen (simple infection) or several related strains of the same pathogen (multiple infection). The majority of cross-sectional surveys of macroparasites in wildlife, however, show that, in general, more than one parasite species is present in any given host (Bush and Holmes 1986; Goater *et al.* 1987; Goater and Bush 1988; Dobson and Keymer 1990). The combination of demographic and epidemiological parameters conferring the highest competitive advantage to a particular macroparasitic species is discussed in Dobson and Roberts (1994) and Gatto and De Leo (1998).

As a result of the complications inherent in macroparasite infection dynamics, empirical research on virulence evolution traditionally has been directed toward microparasitic infections. Most of the examples provided in this chapter refer to viruses and bacteria: because of their short generation time, microparasites are more likely to show evolution of resistance or virulence than the longer living macroparasites.

3.3 Impact of Parasitism on Community Structure

The need to develop a quantitative understanding of the evolution of virulence in wildlife stems from the pervasiveness of the parasitic mode of life: most of the species on our planet are parasitic (Price 1980), and this may have profound effects on patterns of genetic diversity, population dynamics, and community structure. Host–parasite interactions in wildlife differ from human or agroecosystem dynamics in many ways. First, the infective agents can alter the survival and/or reproductive ability of the host, and, in some cases, even their behavior, with potentially dramatic consequences on host density. In human epidemiology and agroecosystems, on the contrary, factors other than disease regulate population density, and thus the host population is generally assumed constant. Second, a host species is usually not isolated, but embedded in a complex web of ecological interactions. Since parasitism ultimately reduces host fitness, it is also likely to affect, and interact with, the interspecific and prey–predator relationships of the target host with other species in the community. Finally, the existence of a diverse community of hosts offers a wealth of opportunities for an infective pathogen to jump from one species to another, with often unpredictable effects at the population and community level. For these reasons, when we analyze the evolution of virulence in wildlife, it is generally convenient to account explicitly for the population dynamics of the potential hosts and of other species that can interact directly and indirectly with them, as unexpected connections and feedbacks among the different species of the community may well occur (Price *et al.* 1988).

Case studies of disease epidemics of wildlife illustrate this issue (Dobson and Hudson 1986; Minchella and Scott 1991; Dobson and Crawley 1994; McCallum and Dobson 1995). For example, parasites can indirectly tip the balance of competition and allow one host species to exclude another from a potentially sympatric range. Schmitz and Nudds (1993) have suggested that the meningeal helminth parasite, *Parelaphostrongylus tenuis*, has prevented moose and caribou from establishing in larger areas of the eastern USA, easing competition for white-tailed deer. In other cases, mass mortality resulting from cyclic epidemics may be able to regulate host population dynamics more effectively than predation or intra- and interspecific competition.

The evolution and spread of virulent pathogens is becoming a cause of great concern in the protection of threatened wildlife communities and ecosystems. Endangered species have small and potentially sparse populations, and, therefore, have a reduced ability to sustain continuous infections by virulent pathogens. However, endangered species can acquire these pathogens upon contact with more common and widespread species. Thus, pathogens that infect a range of host species cause great problems to endangered species (McCallum and Dobson 1995). Moreover, as most of the individuals in an endangered population have never been exposed to foreign pathogens, they have very little acquired immunity, and, thus, can suffer high levels of mortality. Canine distemper virus, for example, killed over 70% of the last remaining free-living colony of black-footed ferrets (Thorne and Williams 1988).

Striking evidence of the impact of infectious diseases on wildlife populations comes from outbreaks that have occurred following the introduction of a pathogen into a new area. The extinction of nearly half the endemic bird fauna of the Hawaiian Islands resulted from the combined effects of habitat alteration and the introduction of bird pathogens such as malaria and bird pox (Van Riper *et al.* 1986; Cann and Douglas 1999; Freed 1999).

In 1987–1988, over 18 000 harbor seals (*Phoca vitulina*) in northern Europe and several thousand Lake Baikal seals (*Ph. sibirica*) in Siberia died in two isolated morbillivirus epizootics. A similar outbreak caused severe mortality in striped dolphins (*Stenella coeruleoalba*) in 1990. The cause of the harbor seal outbreak was identified as phocine distemper virus, which might have been transmitted by asymptomatic harp seals (*Pagophilus grownlandicus*) in a large-scale migration from the Arctic to northern Europe in 1986–1987. The Siberian epizootic of Lake Baikal seals might have been transmitted from dogs or other terrestrial carnivores, demonstrating the potential for transfer of morbillivirus between terrestrial and aquatic hosts. As discussed later in this chapter, host-species transfer may be associated with selection for changes in virulence or the expanded host range of the virus.

Several theoretical studies show that macroparasites can also regulate the host population (Anderson and May 1978; May and Anderson 1978; Grenfell 1988; Scott 1990; Grenfell and Gulland 1995; Heesterbeek and Roberts 1995; Jaenike 1998). Empirical studies are less common and invariably reflect the complexity of the real world (Grenfell *et al.* 1995; Dwyer *et al.* 1997). The best-documented case is probably represented by the nematode *Trichostrongylus tenuis* in red grouse *Lagopus lagopus* in northern England (Hudson *et al.* 1992; Hudson and Dobson 1995). These studies show that clutch size, chick survival, and risk of predation by foxes were all related to parasite density. Unfortunately, further evidence from field studies is still sparse, and there is a great need for the development of field experiments to test predictions and assumptions of theoretical models on the evolution of virulence in wildlife and on its effect on population dynamics and community structure (Thompson and Lymbery 1996).

3.4 Example: The pan-African Rinderpest Epidemic

One of the most striking examples of the impact of introduced pathogens on novel hosts comes from the rinderpest epidemics that spread in sub-Saharan Africa at the end of the 19th century (Scott 1964). Rinderpest is a member of the *Morbillivirus* genus in the order *Paramyxoviridae*. This genus contains three other pathogens of great importance to humans and their domestic livestock: canine distemper, measles, and peste des petits ruminants. Although the virus has been present in Europe since the domestication of ungulates and canids, it probably did not spread south of the Saharan belt because of the low density of ungulate species that live in this deserted area (Dobson 1988; Plowright 1985). When a few infected cattle were accidentally introduced by Italian colonists to the African Horn in 1880, a massive pandemic swept through sub-Saharan Africa from Somalia to Cape Town, South Africa, in just ten years. The disease caused massive mortality among most ungulate species (kudu, eland, bushbuck, reedbuck, buffalo, wildebeest, impala, oryx, and giraffe). The population density of most ungulate species was strongly depressed, which led in turn to the decline of their carnivore predators. The impressive level of mortality of infected animals (up to 95%) suggested that the sub-Saharan ungulate populations had not previously been exposed to the pathogen (Plowright 1985). Cattle were the primary hosts of the virus, and vaccination of livestock resulted in the successful control of rinderpest. These inoculation programs, initiated in the 1950s, led in a few years to the eruption of the wildebeest population from approximately 300 000 to about 1 500 000. The increase of wildebeest, buffaloes, and other herbivores, in turn, stimulated population growth in a number of top predators, particularly lions and hyenas.

This example indicates a need to revise the common view that Savannah ecosystems are dominated by competitive and predator–prey interactions among large herbivores and the carnivores that prey upon them. Parasites and infective agents, which certainly comprise a negligible fraction of the Serengeti biomass, are likely to have an impressive impact on community structure at all trophic levels. Rinderpest has been able to keep herbivore populations far below carrying capacity, creating an ecosystem in which predator abundance is dependent upon prey density rather than determining it. Furthermore, the widespread decline in herbivore abundance allowed a massive pulse of recruitment to occur in plant species whose numbers were limited by browsers.

A similar series of ecological interactions has been observed in other East African game parks following anthrax outbreaks in impala populations (Prins and van der Jeugd 1993). Oak woodlands in southern England exhibited a parallel pulse of recruitment when myxomatosis massively reduced the rabbit population in the 1950s (Dobson and Crawley 1994).

These examples show that parasites, at least in some cases, can be more virulent when infecting exotic hosts (which have very little acquired immunity) than they are in their normal host (which may share a long history of coevolution and have a background level of herd immunity). In the long term, the selective pressure exerted by the parasite on the genetic pool of the host may favor rare resistant genotypes, limited neither by parasitic infections nor by resources or other ecological interactions. Diversity-generating mechanisms can thus play an important role in host–parasite interactions, as described in Section 3.5.

3.5   Role of Genetic Diversity

The interplay between parasite virulence and the costs of host resistance is an essential mechanism responsible for maintaining polymorphism in many ecological systems. Any mechanism capable of generating or increasing host genetic diversity may provide a way to escape parasite attack, because parasites generally focus on the most common genotype (Anderson and May 1991). The link between sexual reproduction of hosts and parasitism is thus worth attention, because sexual reproduction and the reassortment of genetic material can have a dramatic impact on the long-term evolutionary dynamics of host–parasite interactions (Ebert and Hamilton 1996; Ebert and Herre 1996). In fact, an essential role of sexual reproduction is to provide the host with a mechanism to produce and maintain genetic diversity. If a parthenogenetic individual is introduced into a population of individuals that reproduce sexually, its genotype is expected to spread in the population because it produces two reproductive offspring for every one (female) reproducer produced by a sexual individual (because the male produces no offspring). As the abundance of this genotype increases, however, it becomes more vulnerable to attack by parasitic agents that could specialize on its homogeneous, now common, genotype. A large number of studies (most theoretical, some empirical) have examined how sexual reproduction could confer an advantage through the production of offspring with higher levels of genetic diversity. This diversity effectively blunts the ability of pathogens to rapidly exploit any common genetic variety of host. Two field studies on snails suggest that parasites can indeed influence the level of sexual versus parthenogenetic reproduction (Anderson and Crombie 1985;

Lively 1992): parthenogenetic reproduction is employed at low parasitic loads, while sexual reproduction is favored when there is a high risk of parasitism. A further study on parthenogenetic and sexual geckos by Moritz *et al.* (1991) shows that parthenogenetic animals are indeed more likely than their sexually reproducing conspecifics to be infected with ectoparasitic mites. Sexual reproduction may also be important in increasing diversity in the parasite population, allowing sexual pathogens to continually challenge the host's immune response.

On the other hand, host genetic diversity may have an important role in modifying epidemiological patterns in wildlife populations that are usually ascribed to other causes (Read 1995). For instance, observations of age-dependent variation in the force of infection are usually ascribed to age-specific changes in the degree of mixing and contact within and among age classes. However, genetic heterogeneity in response to infection may also play a significant role in producing the observed age–prevalence curve: Anderson and May (1991, chapter 10) show that the age of first infection is dependent on susceptibility, as more-resistant hosts tend (on average) to acquire infection later in life.

Another debated case is provided by the aggregated distributions typical of parasitic helminth infections. The mechanisms that produce this distribution are not fully understood yet, even though the aggregation can be reasonably ascribed to the clumped spatial distribution of the free-living, infective stages and to age-dependent differences in immunity. On the other hand, Grenfell *et al.* (1995) have shown that the aggregated distribution ubiquitously observed in parasitic helminths may also be a consequence of host genetic heterogeneity, even though this undoubtedly interacts with other heterogeneities to produce any observed distribution.

To summarize, parasitism can affect the level of genetic diversity in its host, which, in turn, can affect the epidemiological pattern of host–parasite dynamics and potentially foster the selection for a more- or less-virulent strain. It is obvious that only through a full understanding of the mechanisms that generate genetic diversity and those that create opportunities for selection will it be possible to grasp and manage the evolutionary consequences of the host–parasite interaction (Burdon and Jarosz 1991).

3.6 Myxomatosis and the Coevolution of Virulence Traits

The myxoma virus, which was used to control populations of the European rabbit (*Oryctolagus cuniculus*) in Australia (Fenner and Ratcliffe 1965; Fenner 1994), is, without doubt, the best-documented example of the evolution of virulence in wildlife. Within a few years after their introduction in 1859, the rabbits spread impressively over the southern half of Australia, quickly becoming a major agricultural pest and an important cause of erosion in the semiarid interior. It was not until 1950 that control of the rabbit populations was achieved by the introduction of the myxoma virus as an epizootic disease. Myxoma virus, a member of the genus *Leporipoxvirus*, is a poxvirus that produces generalized diseases and rapid death in the European rabbit. The primary mechanism of transmission in

Australia was through mosquitoes that had bitten through skin lesions of infected rabbits. While the myxoma virus is usually only mildly virulent in its native host (the South American rabbit *Sylvilagus brasiliensis*), it is lethal when it infects the "exotic" European rabbit. Within 1 year of its first introduction, the disease had spread 1 100 miles from east to west and 1 000 miles from south to north. Between 1950 and 1953, thanks to extensive inoculation campaigns and seasonal conditions favoring mosquito breeding, myxomatosis was reported in all the Australian territories inhabited by the European rabbit.

In the beginning of the epidemic, the host population showed 99.8% mortality, but field observations and serological tests indicated that soon after the introduction, a somewhat less-virulent strain emerged in the population (Fenner and Ratcliffe 1965). During winter months, when mosquitoes are rare, strong selection may exist for less-virulent strains that cause reduced host mortality and thus increase the duration of infection. Hundreds of myxoma virus strains have been isolated in the 50 years following the initial introduction. By determining the survival time of rabbits inoculated intradermally with small doses of the virus, it was possible to grade these strains into six classes according to their virulence (class I the most virulent and class VI the least virulent). The strains with intermediate virulence (Grade III) became prevalent almost immediately and remained dominant for the next 30 years.

The emergence and dominance of a Grade III virus (which causes 90% mortality of infected rabbits) should conceivably provide selection for rabbits with greater innate resistance. Field observations showed that case fatality rate did indeed decrease over the series of planned epizootics, thus pointing toward selection for resistance to the myxoma virus in the rabbits. This resistance may well have a genetic component, although laboratory tests indicated an unknown "parental immunity" factor involved in the selection of more resistant rabbits (Willimans and Moore 1991). As rabbits with higher innate resistance became more common, more virulent strains of myxoma virus were selected (Fenner and Ratcliffe 1965). Therefore, even though the pattern of pathogen transmission in the European rabbit in Australia has apparently been stable for many years following the first introduction of myxomatosis, there is still the potential for changes in response to new selective forces. Historical data show that violent oscillations in the density of infected rabbits were more or less regularly followed by apparently stable situations in which the virus appeared to have settled at an intermediate level of virulence. In 1988, the rabbit population in Australia again experienced an enormous increase, thus showing that the "stable equilibrium" between rabbits and Myxoma may have been only an interlude in a longer coevolutionary saga.

3.7 Evolutionary Race Between Host and Parasite

The myxoma example illustrates how rapidly selection may occur in a host–parasite system. The continuous adaptation of the parasite and its host (known also as the "Red Queen hypothesis") is a consequence of the frequency-dependent nature of the selective forces exerted by the parasites. Anderson and May (1991,

p. 642) lucidly summarized this issue: "If different strains or genotypes of the parasite are present, and if different host genotypes respond differently to the various parasite genotypes, then, in general, the host genotype that is more abundant at any time will be differentially exposed to the adverse effects of infection. As the abundance of the most common host genotype declines, the abundance of some other genotype will correspondingly increase, and this host genotype will in turn suffer increasing depredation from the pathogen strains that most afflict it, and so on." The fact that in some host–parasite systems relatively few genes are involved in determining host resistance, while many are involved in parasite virulence, makes it possible that evolutionary change in the host–parasite relationship can occur in a wide variety of complex ways, with parasite selection much faster than that on the host (Frank 1996c).

3.8 Multiple Infection Alters the Evolution of Virulence

Current thinking has revised the traditional assumption that parasites continually evolve toward a state of symbiosis by maximizing transmission (see, for instance, Lipsitch and Moxon 1997). In the case of single infections, a benign, slow-reproducing strain may indeed be preferentially transmitted over fast-reproducing (and consequently harmful) competitor strains that rapidly kill their host before infecting new individuals. In this view, virulence is a costly side effect of within-host reproduction, as parasite-induced host mortality truncates parasite transmission from one living host to another. On the other hand, the benefits of reducing damage to the host through reduction of within-host growth needs to be balanced against the costs of producing few transmittable infective stages or propagules. Theoretical models (Anderson and May 1982) show that an intermediate level of virulence is selected so as to maximize the overall fitness of the infective pathogen (see also Chapter 2).

The situation changes in favor of an increase in virulence in the case of multiple infections, when more than one strain competes for the same host. In fact, the benefit of intermediate virulence of the resident strain may be overcompensated by the costs of slow reproduction, as faster reproducing strains will take over the host before the mild resident strain is able to transmit to another living host (Levin and Pimentel 1981; Van Baalen and Sabelis 1995a, 1995b). As a consequence, within-host competition fosters the selection of strains more virulent than expected under single infection.

3.9 Interspecific Transmission Influences Virulence

This balance between virulence and transmissibility is unique to the strain of a specific pathogen and a particular host. However, if a strain jumps to a new host species, the optimal virulence is generally shifted because of ecological, behavioral, and physiological differences between the two host species, as well as through disturbed molecular interactions (which include, but are not limited to, clues for localization and toxin production) between microparasite and the foreign host and the immune response of the host. In addition, the differences in population densities, and, therefore, in transmission rates will change; this effect is particularly appreciable when the jump is between domestic and wild populations. In some cases, parasites show devastating effects after accidental introductions into new populations (e.g., Dutch elm disease and chestnut blight, rinderpest in Africa, and *Pasteurella*-induced pneumonia in bighorn sheep in the western USA). However, these may be more the exception than the rule (Ebert 1994), as probably many failed introductions have passed unnoticed. In general, a parasite that infects a novel host should exhibit a reduction of fitness compared with its original host, with which it shares a long coevolutionary history (Ebert and Hamilton 1996). Furthermore, the more the novel host differs genetically from the host to which the parasite was adapted before the introduction, the stronger the reduction in virulence and transmissibility, as shown for viruses, fungi, helminths, and protozoans (Ebert and Hamilton 1996). Serial passage experiments (Ebert 1998a) and some field observations (Ebert and Hamilton 1996) show that the virulence of a pathogen introduced in a novel host, while initially mild, may substantially increase as a consequence of local adaptation.

3.10   Example: *Pasteurella* Outbreaks in Bighorn Sheep

Disease resulting from interactions between domestic and natural populations is a growing concern for conservationists and agriculturalists alike; an almost inevitable result of the national park system is that wildlife migrate beyond park boundaries and contact domestic animals on neighboring agricultural lands (Callan *et al.* 1991; Dobson and Meagher 1996). What effect will such contact have on these populations? A case study of a *Pasteurella haemolytica* bacterial outbreak in wild bighorn sheep in Hells Canyon, Oregon, USA, may present a case of massive die-off following contact with a domestic animal. Historically, *Pasteurella*-induced pneumonia has been a major killer of bighorns in the western USA (Hobbs and Miller 1991). These epidemics are often caused by interspecific transfer of *P. haemolytica* between domestic animals and bighorn sheep. Endemic bighorns disappeared from Hells Canyon by 1945, probably through over-hunting, competition with domestic livestock, and introduced disease (Cassirer *et al.* 1997b). Despite the fact that 329 bighorns have been transplanted into the canyon since 1971, the population seems to be limited by disease epidemics. Available habitat does not seem to be a limiting factor, as many suitable habitats (in terms of slope, available grassland, and proximity to water) are unoccupied (Cassirer *et al.* 1997a). Since the beginning of the transplantation projects (and careful record-keeping), significant die-offs have occurred seven times: five linked to contact with domestic sheep, one to contact with a domestic goat, and one to ectoparasites and drought (Cassirer *et al.* 1997b). Pneumonia seems to be the most likely cause of death for the contact-initiated epidemics.

Several empirical studies (Silflow *et al.* 1993; Silflow and Foreyt 1994; Sweeney *et al.* 1994; Cassirer *et al.* 1997b) suggest that *P. haemolytica* from domestic sheep are more virulent in bighorns than in their native host. Healthy bighorns inoculated with *P. haemolytica* from infected domestic sheep died within 48 hours of inoculation, while domestic sheep given the same inoculation remained healthy (Foreyt *et al.* 1994).

P. haemolytica has particularly high mutation rates (Lo and Macdonald 1991; Saadati *et al.* 1997), which can potentially induce a variation in virulence (Petras *et al.* 1995). As a consequence, there is the potential for selection for reduced virulence in the bighorns (Taddei *et al.* 1997a), just as observed for myxomatosis in the European rabbit. We thus expect that a careful analysis of recent outbreaks will provide an important opportunity to analyze the phenomenon of interspecific pathogen transfer in nature and to test whether attenuation of virulence will actually occur.

3.11   Potential Impact of Wildlife Diseases on Human Health

In the field of virulence management, it is perhaps inevitable that much of the concern will eventually focus on the implications of the evolution of virulence for human populations. More recently, however, the importance of virulence management of wild populations has been recognized, as the evolution and persistence of disease in free-living populations may have a profound influence on human health. Many diseases regarded as "new" or "emergent" in human populations likely existed for a long time at the endemic level, but were confined geographically in restricted areas or confined biologically to specific reservoir species. Some pathogens may convert from innocuous forms into lethal ones, or move from an animal species to human hosts (Gibbons 1993). The genetic makeup of pathogens is only one of many factors that contributes to the emergence of novel infectious and parasitic diseases. Ecological and social changes caused by human activity can inadvertently provide the appropriate conditions for infective agents and vectors to spread into new geographical ranges or to transfer into a human host population (Wilson *et al.* 1994; Schrag and Wiener 1995). Marburg and yellow fever viruses, for example, were originally only endemic in wild monkey. The hantavirus epidemics that struck the "Four Corners" region of the southwestern USA in 1992 at the end of a 6-year drought suggest that changes in ecological conditions can trigger the transmission of pathogens from a reservoir population to humans, as described below. The first rains led to an abundance of pinon nuts and grasshoppers (which are food for mice) and allowed a mouse population explosion unchecked by predators (which had been virtually eliminated by the drought). The increase in deer mouse density created the opportunity for the virus to be transmitted to humans. By February 1995, 102 cases of hantavirus pulmonary syndrome had been reported from the region, 53 of which were fatal (the mean age of death was 35 years). Similarly, in Zimbabwe and western Mozambique, periods of drought have regularly led to major infestations of rats that serve as carriers for a number of pathogens. The warmer climates in India and Colombia have fostered the spread of *Aedes aegypti* (a vector for dengue and yellow fever) to altitudes over 2 000 meters. Temperature restrictions had previously limited the mosquitoes to altitudes below 1 000 meters.

In Belem, Brazil, Oropouche fever was transmitted to humans from biting midges, which experienced a similar population explosion when settlers started clearing the forest. The open land neighboring the cacao plantations provided ideal breeding grounds for the midge population. The spread of Lyme disease in North America also originated in a series of human-induced ecological changes. In the 19th century, forest clearance drastically depressed the deer populations and caused the virtual extinction of their predators. Subsequent regrowth of forest during the 1900s allowed the deer population, now unregulated by predators, to rebound. The high density attained by the deer led to an increase in the densities of deer ticks (the carrier of *Borrelia burgdorferi*, the pathogen responsible for Lyme disease). At the same time, many more homes were built in forested sites, causing a great increase in the number of people bitten by deer ticks that had acquired *B. burgdorferi* from local rodents (Wilson *et al.* 1994). Since 1942, 40 000 cases of Lyme diseases have been reported in the USA, making it the most common vectorborne disease in North America.

This sampling of empirical data makes it apparent that sound management of wildlife diseases, along with an understanding of the genetic and ecological factors triggering the transmission of pathogens from animals to humans, is a prerequisite to avoiding similar unpleasant surprises in the future.

3.12 Discussion

Although there are many anecdotal cases of the evolution of virulence in wildlife, the paucity of data for wildlife populations and our poor knowledge of parasites and their biology and pathogenicity do not allow a quantitative risk assessment or prediction of the consequences of introducing a new pathogen in a novel host. Despite the complexity of the problem, wildlife managers should be aware that virulence and transmission may change, or evolve, in direct response to traditional management attempts to control pathogens and their hosts. Small-scale experiments should be carried out before treatment for a pathogen is applied on a larger scale, and the potential for more widespread consequences of pathogen removal and translocations on wildlife management should be carefully analyzed.

4

Adaptive Dynamics of Pathogen–Host Interactions

Ulf Dieckmann

4.1 Introduction

Over the past few decades, the expectations of scientists regarding stable patterns of pathogen–host interaction have undergone major transformations. During an initial phase it was widely agreed that pathogens and their hosts evolve in ways that would render benign the consequences of infection (May 1983). These predictions, fostered by the idea that evolution tends to act "for the benefit of the species", are challenged by the conspicuous existence of highly virulent, yet apparently rather stable, human and animal diseases. Within the paradigm of species-level selection, such examples could only be interpreted as transitory cases in which a pathogen has jumped to a new host species so recently that the predicted evolutionary loss of virulence has not yet progressed far enough.

To explain stable intermediate levels of pathogen virulence therefore required a paradigm shift in evolutionary theory: the seemingly conclusive (and, from today's perspective, almost too enthusiastic) demolition of scientific credibility for selection above the level of individuals (Williams 1966). This change in perspective was accompanied by the insight that, although a benign form of infection might benefit a pathogen population as a whole, individuals of a more aggressive pathogen strain might nevertheless invade to reap their harvest. The decisive criterion for the success or failure of such pathogens is their rate of spread through a given host population: if the new pathogen spreads faster than its predecessor does, it may invade and replace that predecessor. It is easily shown that this transmissibility of a pathogen can be highest at intermediate levels of virulence (Anderson and May 1982, 1991). If virulence is too low, symptoms may be absent or harmless and the pathogen may therefore have little opportunity to multiply massively and/or to leave its host. By contrast, if virulence is too high, the resultant symptoms are so severe that the host is likely to perish before it has spread much of the harbored pathogen population. It therefore appeared that evolution would tend to maximize the transmissibility of pathogens, rather than minimize their virulence.

This idea can be made precise. The so-called basic reproduction ratio of a pathogen, denoted by R_0, is defined as the expected number of infections produced by a single infected host individual *in an otherwise uninfected host population* (see Box 2.2). Analyses of relatively simple epidemiological models led to the conclusion that it is the value of R_0 that is raised by the successfully invading pathogens and that is therefore maximized by the evolutionarily stable strain. Since R_0 is a measure of effective transmissibility, maximizing a pathogen's R_0 is equivalent to maximizing its transmissibility.

This chapter explores how far the technique of R_0 maximization can take us when studying evolution in more complex epidemiological models. Section 4.2 reviews the conceptual limitations of the conventional R_0-based approach, and Section 4.3 introduces adaptive dynamics theory to overcome these limitations. Sections 4.4 and 4.5 focus on two different settings – pathogen evolution in a constant host population and pathogen–host coevolution – and illustrate how the results obtained by application of the new toolbox differ in interesting ways from those of traditional analyses.

4.2 Limitations of R_0 Maximization

The notion of R_0 maximization is plausible in general, applies rigorously to many well-studied models, and undoubtedly helps us to understand some major features of observed pathogen–host interactions. Yet it is not the full story – four crucial problems are not addressed by this approach.

First is the realization that it is not always R_0 that is maximized by evolution. Consider pathogen strains A and B, for which the R_0 of A exceeds that of B. The argument above leads us to expect that, among these strains, A will win the evolutionary race. This expectation is based on the infection's rate of spread in an uninfected host population, as specified in the definition of R_0. What we really should ask, however, is what happens once pathogen A has spread and substantial parts of the host population have thus become infected? In this situation the success or failure of a new strain is no longer determined by its performance in the initial environment, which comprised uninfected hosts only. Instead, we have to consider the strain's rate of spread in the current environment of hosts already infested by strain A. It may well be that in this case pathogen B is better adapted than A to the actual challenge of spreading in a partially infected host population. Under such circumstances, strain B, and not A, will be evolutionarily stable. In general, whenever the resident strains change the actual epidemiological environment in such a way that the performance of different strains in the uninfected environment is no longer indicative of their invasion success in the actual environment, R_0 maximization does not apply. This option raises the possibility of alternative optimization principles. It turns out that in some models it is indeed possible to find quantities other than R_0 that are maximized by evolution. In particular, it can be shown that the type of density regulation that operates in the system critically influences which quantity is maximized (Mylius and Diekmann 1995; Metz *et al.* 1996b).

Unfortunately, it is by no means clear that for a given system such an optimization principle exists at all. This is a second reason why the assumption of R_0 maximization often misleads. The well-known rock–scissors–paper game (rock beats scissors by crushing, scissors beats paper by cutting, paper beats rock by wrapping) is a very simple example of a situation in which no single quantity can be construed as being maximized by evolution. Likewise, it can happen that pathogen strain B outcompetes strain A in the environment that results from the prevalence of A, while strain C wins against B in the environment set by B, and A beats C in the C environment. The salient feature of such a scenario is frequency-dependent selection: selective pressures and the resultant invasion success depend on the composition of the established, or resident, pathogen population against which a variant strain is competing. Since frequency-dependent selection is ubiquitous in nature and also naturally arises in epidemiological models (unless the modeler explicitly tries to avoid it), the absence of an optimization principle is the rule, rather than the exception, in realistic pathogen–host interactions. It is important to stress that this does not imply our understanding cannot be furthered through modeling efforts. It merely shows that – instead of always having available the convenient shortcut of maximizing a certain quantity – we often have to evaluate which sequences of invasions are possible and to which evolutionary outcome they lead.

So far we have restricted attention to the evolution of a pathogen in a nonevolving population of hosts. Since pathogens often have much shorter generation times than their hosts, they may be expected to evolve faster than the hosts and therefore to experience essentially a nonevolving host population in the course of their adaptation. This situation appears to apply to acquired immunodeficiency syndrome (AIDS), in which evolutionary change on the part of the human immunodeficiency virus is so unusually rapid that it not only overwhelms the evolutionary potential of the host population, but it even tends to beat the immune system of individual hosts. However, even for the AIDS pandemic, which (in evolutionary terms) is still very recent, some genes that confer host resistance have been reported. Other examples show that evolutionary change in pathogens and their hosts can occur on similar time scales. A case in point is the swift coevolutionary race between the European rabbit and the myxoma virus in Australia, which commenced with the virus's introduction to the Fifth Continent in 1950 (see Figure 4.2a). As in this case, sexual recombination often allows hosts to match effectively the evolutionary pace of their asexual pathogens. It must therefore be concluded that pathogen and host evolutions do not always have different time scales. To conceive the adaptation of pathogen–host interactions in terms of coevolutionary dynamics makes it plain that no general optimization principle can predict adequately the evolutionary outcome of all possible arms races. Instead, we have to consider the potential for the invasion of a variant pathogen or host type into the environment jointly brought about by the prevalent pathogen and host types. This highlights the importance of the environmental feedback loop (Metz *et al.* 1996b; Heino *et al.* 1997) that operates in evolving pathogen–host systems: the current environment determines current selection pressures and, in turn, these selection pressures determine the future environments that result from the invasion of selectively favored types. In such a context, the rates at which new types are generated by mutation or recombination may be critical (Dieckmann and Law 1996) and dynamic descriptions therefore become essential – static optimization principles simply cannot account for such complexity.

Box 4.1 Pairwise invasibility plots

The invasion fitness of an evolving species (see Section 4.3) defines pairwise invasibility plots for resident and mutant phenotypes (Van Tienderen and de Jong 1986;
Metz *et al.* 1992, 1996a; Kisdi and Meszéna 1993; Geritz *et al.* 1997; see also
Taylor 1989). In the simplest case, these phenotypes are described by a single met-
ric character or quantitative trait. Plotting the sign of the invasion fitness f for each
of the possible combinations of mutant phenotypes x' and resident phenotypes x
reveals the shapes of the zero contour lines at which $f(x', x) = 0$. As shown
in the left panel below, these lines separate regions of potential invasion success
($f > 0$) from those of invasion failure ($f < 0$). The resident population precisely
renews itself when it is at equilibrium, so the resident trait value is neutral in its own
environment and the set of zero contour lines therefore always includes the main
diagonal.

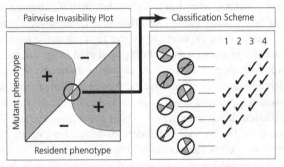

The shape of the other zero contour lines carries important information about the
evolutionary process. In particular, possible evolutionary endpoints are located at
the resident phenotypes for which a zero contour line intersects with the main diagonal. In characterizing these so-called evolutionarily singular points, adaptive
dynamics theory uses an extended classification scheme in which four different
questions are tackled simultaneously:

1. *Evolutionary stability.* Is a singular phenotype immune to invasions by neighboring phenotypes? This criterion amounts to a local version of the classic evolutionarily stable strategy (ESS) condition that lies at the heart of evolutionary
   game theory (Maynard Smith 1982).
2. *Convergence stability.* When starting from neighboring phenotypes, do successful invaders lie closer to the singular phenotype? Here the attainability of a
   singular point is addressed, an issue that is separate from its invasibility (Eshel
   and Motro 1981; Eshel 1983).
3. *Invasion potential.* Is the singular phenotype capable of invading populations of
   its neighboring types (Kisdi and Meszéna 1993)?
4. *Mutual invasibility.* If a pair of neighboring phenotypes lies either side of a
   singular phenotype, can they invade each other? Assessment of this possibility is
   essential to predict coexisting phenotypes and the emergence of polymorphisms
   (Van Tienderen and de Jong 1986; Metz *et al.* 1992, 1996a).

continued

Box 4.1 *continued*

All four questions are important to understand the nature of potential evolutionary endpoints. It is therefore remarkable how the four answers are obtained simply by examining the pairwise invasibility plot and reading off the slope of the zero contour line at the singular phenotype (Metz *et al.* 1996a; Geritz *et al.* 1997), as illustrated in the right panel above.

Three particularly interesting types of evolutionarily singular points are illustrated below. In each case, the staircase-shaped curve depicts a possible trait substitution sequence during which populations of resident phenotypes are repeatedly replaced by advantageous mutant phenotypes that invade successfully.

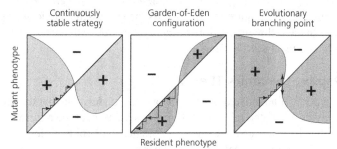

The left panel shows a singular point that is both evolutionarily stable and convergence stable. Such an outcome is called a continuously stable strategy (CSS; Eshel 1983). In the middle panel, the singular point is evolutionarily stable but not convergence stable. This means that, although the singular phenotype is protected against invasion from all nearby phenotypes, it cannot be attained by small mutational steps – a situation aptly referred to as a Garden-of-Eden configuration by Nowak and Sigmund (1989). The right panel shows an evolutionary branching point: here the singular point is convergence stable but evolutionarily unstable. This implies convergence to disruptive selection and thus permits the phenotypic divergence of two subpopulations that straddle the branching point (Metz *et al.* 1992, 1996a).

There is a fourth reason that necessitates a departure from classic concepts of evolutionary epidemiology. The principle of R_0 maximization is based on the notion that we should expect to see as evolutionary outcomes those types of pathogen or host that are unbeatable or evolutionarily stable against all possible other types that can, in principle, arise in their species. However, hopeful monsters are not frequently encountered in the biological world and substantial changes in morphology or physiology tend to be lethal. For this reason, adaptation can usually explore only the small range of variation that is accessible by gradual change. It is therefore not always meaningful to seek out those types of pathogens or hosts that cannot be beaten by any potential variant, including those that require major evolutionary reconstruction. In the presence of frequency dependence, this simple observation has substantial consequences. First, some evolutionary outcomes predicted by the analysis of evolutionary stability alone cannot actually be reached by a sequence of small adaptive steps, and, second, some outcomes actually attained in the course of evolution turn out not to be evolutionarily stable at all (for an illustration of these points, see Box 4.1). Consequently, evolutionary stability and attainability must always be considered in conjunction; it is only in the simple case of evolutionary processes governed by an optimization principle that the two notions coincide of necessity (Meszéna *et al.* 2000).

The conventional approach to maximize R_0 for a pathogen therefore has some fundamental limitations as a tool to describe the complex processes that arise from the evolution of general pathogen–host interactions. To overcome this obstacle, an extended framework is required to encompass the successful classic approach as a special case. In the following section the theory of adaptive dynamics is introduced as a candidate to meet this challenge.

4.3 Adaptive Dynamics Theory

The starting point of adaptive dynamics theory is to understand that the fitness of a type can only be evaluated relative to the environment that type experiences. This implies that we have to know the current ecological and epidemiological status of a host population before we can assess whether a given pathogen can spread within that population or not. A characterization of this status includes, *inter alia*, information about the types and abundances of other pathogen strains that are present in the host population. Likewise, we have to specify the resident host type, as well as the endemic strain or strains of the pathogen, to predict which variant host types excel at the evolutionary play staged in the given ecological theater.

These considerations naturally lead to the concept of invasion fitness (Metz *et al.* 1992). The invasion fitness of a type x is the expected long-term per capita growth rate f of that type in a given environment E, $f = f(x, E)$. If the invasion fitness of a type is positive it may invade in that environment, otherwise not.

As discussed above, those types $x_1, x_2, \ldots$ that are present in a given system in general affect the environment, $E = E(x_1, x_2, \ldots)$. One possible complication here is that the environment may not yet have fully settled to reflect the present set of types. This can happen, for instance, in the wake of an ecological perturbation or shortly after new types, very different from their predecessors, have started to invade the system. Often, however, evolution is slow enough for ecological processes to respond swiftly in comparison, in particular since gradual evolutionary change usually does not even require much ecological response for a population to stay at its ecological equilibrium or, more generally, its ecological attractor. To simplify matters, it is therefore convenient to assume that the state of the environment has come close to the attractor determined by the resident types. Under such conditions, the dependence of the invasion fitness f of a type x on the current environment E can be replaced by a dependence on the resident types $x_1, x_2, \ldots$, $f = f(x, x_1, x_2, \ldots)$. These types can belong to the same species as type x does,

or involve other coevolving species. For simplicity, it is often sufficient to characterize a population by its prevalent or average type (Abrams *et al.* 1993). Although
strictly monomorphic populations are rarely found in nature, it turns out that the
dynamics of polymorphic populations (which harbor, at the same time, many similar types per species) can often be well described and understood in terms of the
simpler monomorphic cases.

For pathogen–host systems that allow the coevolution of virulence x and resistance y, we thus arrive at the notation $f(x', x, y)$ for the invasion fitness of a
variant pathogen of virulence x' in a host population of resistance y that is infected by resident pathogens of virulence x. Analogously, in this infected host
population $f_h(y', x, y)$ is the invasion fitness of a variant host of type y'. Notice
that the variant types can arise from mutation, as well as from recombination and
immigration. In the absence of host evolution, pathogen fitness is simply denoted
by $f(x', x)$. (Throughout this chapter a prime denotes variant types, whereas no
prime refers to resident types; this keeps the notation shorter than using the more
explicit notation x_{mut} and x_{res}.) Based on these fitness functions so-called pairwise
invasibility plots can be constructed to explore which variant pathogens can successfully invade which resident pathogens, and the same analysis can be carried
out for evolution in the host (Box 4.1). Moreover, one of the explicitly dynamic
models of adaptive dynamics theory can be used to investigate the time course of
evolutionary or coevolutionary change in such systems (Box 4.2).

4.4 Pathogen Evolution

To illustrate how the theory of adaptive dynamics can elucidate the evolution of virulence, consider a generalized susceptible-and-infected (SI) model (see Box 2.1),

$$\frac{dS}{dt} = + b_S(x, S, I)S + b_I(x, S, I)I - d_S(x, S, I)S$$
$$- \beta(x, S, I)SI + \theta(x, S, I)I , \qquad (4.1a)$$

$$\frac{dI}{dt} = -d_I(x, S, I)I + \beta(x, S, I)SI - \theta(x, S, I)I , \qquad (4.1b)$$

which describes the dynamics of the density S of susceptible hosts and of the
density I of hosts infected by a single pathogen strain with virulence x. The per
capita birth and death rates, b and d, as well as the transmission rate β and the
recovery rate θ, can all depend on the virulence of the resident strain x and on the
current composition of the host population, in terms of densities S and I. The birth
rates of susceptible and infected hosts, b_S and b_I, can differ, as can their death rates
d_S and d_I; in particular, the pathogen-induced death rate is $\alpha = d_I - d_S$. Hosts are
born uninfected and the host population is assumed to be spatially homogeneous.

Evolutionary invasion analysis

A variant strain of the pathogen is now introduced into the resident population
described by Equations (4.1). The variant strain has virulence x' and the density

Box 4.2 Models of adaptive dynamics

Adaptive dynamics theory derives from considering ecological interactions and phenotypic variation at the level of individuals. Extending classic birth and death processes, as well as ecological descriptions of structured populations, adaptive dynamics models allow offspring phenotypes to differ from those of their parents, and thus enable studies of the interplay between population dynamics (changes in the abundance of individuals) and adaptive dynamics (changes in their heritable traits). Four types of dynamic model are used to investigate the resultant eco-evolutionary processes at different levels of resolution and generality:

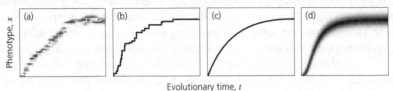

Evolutionary time, t

- With an individual corresponding to a single point in a population's trait space, situated at the individual's combination of trait values, populations can be envisaged as clouds of such points. These stochastically drift and diffuse through trait space as a result of selection and mutation (Dieckmann 1994; Metz *et al.* 1996a); see panel (a) above.
- If populations are large and mutation rates are sufficiently low, evolutionary change in clonal populations proceeds through sequences of trait substitutions (Metz *et al.* 1992; Dieckmann 1994; Dieckmann and Law 1996). During each such step, an advantageous mutant quickly invades a resident population, ousting the former resident. These steps are analyzed through the pairwise invasibility plots introduced in Box 4.1 and used in Figure 4.1. Concatenation of such substitutions results in a description of evolutionary change as a directed random walk in trait space; see panel (b) above.
- If, in addition, the mutation steps are sufficiently small, the staircase-like dynamics of trait substitutions are well approximated by smooth deterministic trajectories; see panel (c) above. It can be shown that these trajectories follow the canonical equation of adaptive dynamics (Dieckmann 1994; Dieckmann and Law 1996),

$$\frac{d}{dt}x_{jk} = \frac{1}{2}\mu_j n_j^*(x) \sum_l \sigma_{j,kl}^2 \frac{\partial}{\partial x_{jl}'} f_j(x_j', x)\bigg|_{x_j'=x_j}, \tag{a}$$

where x_{jk} is the value of trait k in species j, x_j is the resultant trait vector in species j, and x collects these trait vectors for all species in the considered ecological community. For species j, μ_j is the probability for mutant offspring, $n_j^*(x)$ is the equilibrium population size, σ_j^2 is the variance–covariance matrix of mutational steps, and f_j is the invasion fitness. The partial derivatives of f_j in Equation (a) are the components of the selection gradient g_j. Evolution in x_j comes to a halt where g_j vanishes, and the curves on which this happens are therefore known as evolutionary isoclines.

continued

Box 4.2 *continued*

- If, by contrast, mutation rates are high while populations are large, stochastic elements in the dynamics of phenotypic distributions become negligible; this enables mathematical descriptions of the reaction–diffusion type; see panel (d) above. However, the infinitely extended tails that phenotypic distributions acquire in this framework easily give rise to artifactual dynamics that have no correspondence to processes in any finite population.

At the expense of ignoring genetic complexity, models of adaptive dynamics are geared to analyze the evolutionary implications of ecological settings. This allows the study of all types of density- and frequency-dependent selection mechanisms within a single framework, into which coevolutionary dynamics driven by interspecific interactions are also readily incorporated.

---

of hosts thus infected is denoted by I'. Assuming that the resident population is at its demographic equilibrium $[S^*(x), I^*(x)]$, the mutant is rare, and super- or coinfections are negligible, we obtain

$$\frac{dI'}{dt} = f(x', x)I',$$ (4.1c)

where $f(x', x)$ denotes the mutant's invasion fitness,

$$f(x', x) = -d_I(x', S^*(x), I^*(x)) + \beta(x', S^*(x), I^*(x))S^*(x)$$
$$- \theta(x', S^*(x), I^*(x)).$$ (4.2a)

The lifetime reproductive success of the mutant in the resident population at equilibrium can also be determined,

$$R(x', x) = \frac{\beta(x', S^*(x), I^*(x))S^*(x)}{d_I(x', S^*(x), I^*(x)) + \theta(x', S^*(x), I^*(x))}.$$ (4.2b)

Analogously, the lifetime reproductive success of the mutant in an infection-free resident population that comprises S_0 susceptible hosts can be obtained,

$$R_0(x') = \frac{\beta(x', S_0, 0)S_0}{d_I(x', S_0, 0) + \theta(x', S_0, 0)},$$ (4.2c)

and is known as the mutant's basic reproduction ratio R_0 (see Box 2.2).

From Equations (4.2a) and (4.2b) it can immediately be seen that the invasion fitness of the mutant is positive – which indicates that the mutant can invade the resident population – if and only if its lifetime reproductive success exceeds one: $f(x', x) > 0 \Leftrightarrow R(x', x) > 1$. This is expected biologically and can be regarded as a trivial correspondence.

What is much less straightforward, however, is to formally link f and R to the widely used basic reproduction ratio R_0. For this link to become more transparent, we can exploit the relation $R(x, x) = 1$, which implies that, by definition, the density of infected hosts accurately replenishes itself once the disease has reached its endemic equilibrium. Applying this consistency condition to Equation (4.2b), an expression for $S^*(x)$ is obtained. This, in turn, yields

$$R(x', x) = \frac{\beta(x', S^*(x), I^*(x))/[d_I(x', S^*(x), I^*(x)) + \theta(x', S^*(x), I^*(x))]}{\beta(x, S^*(x), I^*(x))/[d_I(x, S^*(x), I^*(x)) + \theta(x, S^*(x), I^*(x))]} .$$

(4.2d)

This equation can be rewritten as $R(x', x) = R_0(x')/R_0(x)$ if the epidemiological rates β, d_I, and θ are density independent, that is, if the corresponding functions do not depend on their second and third arguments. It is therefore only under this condition that the convenient equivalence $R(x', x) > 1 \Leftrightarrow R_0(x') > R_0(x)$ can be taken for granted. Whether this equivalence also holds for some restricted types of density-dependent rates remains an open research question; to date no results on this have been obtained.

Next, these general considerations are illustrated with a suite of specific examples.

Virulence evolution toward benignity

Example I.   Let us start by investigating the most simplistic version of Equations (4.1). The rates for birth, transmission, recovery, and natural mortality are assumed to be constant: $b_S(x, S, I) = b_I(x, S, I) = b$, $\beta(x, S, I) = \beta$, $\theta(x, S, I) = \theta$, and $d_S(x, S, I) = d$, while the death rate of infected hosts increases with the virulence of the infecting strain, $d_I(x, S, I) = d + x$. The last relation sets the scale of virulence x in terms of disease-induced host mortality α.

With $S^*(x) = (x + b + \theta)/\beta$ the invasion fitness $f(x', x) = x - x'$ is obtained. A corresponding pairwise invasibility plot (Box 4.1) and evolutionary trajectory (Box 4.2) are shown in Figure 4.1a. Mutant strains x', with lower virulence than the resident strain x, can always invade and, therefore, the system will evolve toward the most benign strain. The same conclusion can be obtained by maximizing $R_0(x') = \beta S_0/(x' + d + \theta)$ with regard to x' – pathogen strains that harm their host as little as possible are always favored by natural selection.

Virulence evolution under transmission trade-offs

Under the simplistic assumptions made above, pathogens do not benefit from harming their hosts. However, many pathogens are more readily transmitted during individual contacts if they have a higher virulence: this introduces a trade-off for the pathogen between transmission probability and host longevity.

Example II.   It can be assumed, for instance, that transmission rates increase proportionally with virulence, $\beta = cx$. This results in $f(x', x) = (x' - x)(d + \theta)/x$, and $R_0(x') = cx'S_0/(x' + d + \theta)$. Thus an ever-increasing virulence (Figure 4.1b) would be expected, which is clearly unrealistic.

Example III.   Following the seminal work by Anderson and May (1982, 1991) a diminishing return for increased virulence is often considered by choosing, for instance, $\beta = x/(x + c)$; see also Equation (2.1). While maintaining a trade-off between transmission efficiency and host longevity, more emphasis is thus put on the latter. The resultant invasion fitness $f(x', x) = (x - x')[xx' - c(d + \theta)]/[x(x' + c)]$ has a vanishing selection gradient $g(x) = \frac{\partial}{\partial x'} f(x', x)\big|_{x'=x}$ (Box 4.2) at the intermediate virulence $x^* = \sqrt{c(d + \theta)}$, where also the basic reproduction ratio $R_0(x') = x'S_0/[(x' + d + \theta)(x' + c)]$ is maximized. A corresponding pairwise invasibility plot is shown in Figure 4.1c.

The ubiquity of density-dependent rates

Now consider situations in which the rates in the SI model depend on the densities of susceptible and/or infected hosts. Such density dependence can apply to the basic demographic rates d_S and b_S, as well as to the epidemiological rates. The latter include the disease-induced mortality $d_I - d_S = \alpha$, the disease-induced loss in fecundity $b_S - b_I$, the transmission rate β, and the recovery rate θ.

It is actually very implausible that all of these rates are density independent. Density dependence of demographic rates is already assumed in all simple non-epidemiological population models and is needed to prevent the density of susceptible hosts from diverging without bounds in the absence of the disease. The only justification for neglecting such dependence in simple versions of Equations (4.1) is to assume that the disease itself is fully responsible for regulating the host population density. However, even for the severest of diseases this must remain an approximation, whereas for most other infections the assumption is plainly wrong. A second way to avoid considering density-dependent demographic rates is to assume that the total host population size, $N = S + I$, stays strictly constant – independent of the virulence of the resident strain. Obviously, this is also an approximation at best and is likely to apply to very benign diseases only. As usual, reality lies between these mathematical extremes and density regulation in an infected population occurs partially through disease-independent factors and partially through the disease itself (May 1983).

The case for density-dependent rates becomes even stronger when the epidemiological rates, which are directly affected by the disease, are considered. An almost endless variety of mechanisms can cause such dependence; hence the following list is certainly not exhaustive:

- The number of patients an average doctor must treat may rise with the density of infected hosts. This can affect disease-induced mortality and loss of fertility, as well as recovery rates.
- The nutritional status of hosts, and thus their resistance against disease symptoms, may deteriorate with increases in total population density or in the population's morbidity level.
- The quality of medical services in terms of diagnostic and therapeutic options may improve with the wealth of a population. Such wealth may either increase or decrease with total population density and is likely to deteriorate with an increase in the density of infected hosts.

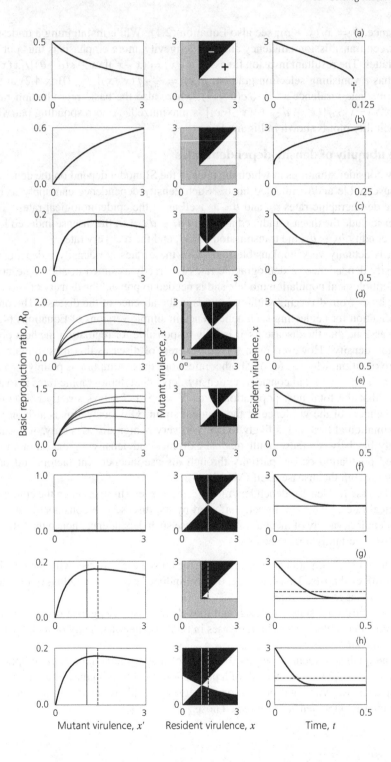

Basic reproduction ratio, R_0 — Mutant virulence, x' — Resident virulence, x

Mutant virulence, x' — Resident virulence, x — Time, t

Figure 4.1 Evolution of pathogen virulence as described by R_0 maximization (left column), pairwise invasibility plots (middle column), and evolutionary trajectories (right column, based on the canonical equation of adaptive dynamics). Rows (a) to (h) correspond to Examples I to VIII in the text. In the middle column, outcomes of virulence evolution are indicated by continuous lines and false predictions that result from R_0 maximization by discontinuous lines. Virulence ranges that do not allow the pathogen to remain endemic are depicted as gray areas. Whereas evolution in cases (a) and (b) leads to ever-increasing or -decreasing virulence, respectively, case (c) shows how a trade-off between transmission probability and host longevity induces evolution toward intermediate virulence. For these first three cases the outcome of virulence evolution can be predicted by R_0 maximization. Evolution in cases (d) and (e) also leads to intermediate virulence, but does not allow R_0 maximization, since the optimal virulence depends on which density of susceptible hosts is assumed. For these examples the left column shows several curves, corresponding to different assumptions about this density; the thick curves describe the self-consistent solutions. Rows (f) to (h) show cases for which R_0 maximization results in seriously misleading conclusions. Parameters: $b = 2$ (a–h), $d = 1$ (a–h), $\theta = 1$ (a–f, h), $\theta_0 = 1$ (g), $\beta = 1$ (a–b), $c = 1$ (c–d, f–h), $c = 0.5$ (e), $K = 10$ (d–h), $\mu\sigma^2 = 1$ (a–h, this scales the evolutionary time t).

- Awareness about potential transmission routes is expected to grow under conditions of high incidence. Transmission rates are then predicted to decrease when the density of infected hosts increases.
- The density of infected hosts changes the ambient density of infectious propagules to which susceptible hosts are exposed. Through the operation of the host's immune system, this propagule density may not translate linearly into the rate at which susceptible hosts acquire infections, and transmission rates then become dependent on the density of infected hosts.
- Changes in total population density are known to reshape social contact networks and thereby to affect the chances of disease transmission.

The last three mechanisms imply that the population-level rate of disease transmission is not proportional to the densities of susceptible and infected hosts and therefore cannot be described by the simplifying assumption of mass action (see Box 2.1). All six mechanisms together illustrate how far-fetched the assumption of fully density-independent rates really is. This conclusion, however, only has major consequences for virulence evolution if evolutionary outcomes in models with density-dependent rates can differ significantly from those in their simpler, density-independent counterparts. We therefore examine below how robust the method of R_0 maximization and the specific predictions thus obtained are for epidemiological models with density-dependent rates. To address this question, five further examples are studied.

Virulence evolution with rates dependent on susceptible host density

Example IV. This example originates from a slight modification of Example III by considering a density-dependent natural mortality of logistic type, $d_S =$

$d + S/K$, with carrying capacity K. The disease-induced mortality and the transmission rate remain density independent, $\alpha = d_I - d_S = x$ and $\beta = x/(x + c)$. This means that, in this example, density dependence extends only to the basic demographic rates, but not to the epidemiological rates. Examining the resultant invasion fitness $f(x', x)$ reveals that under the given conditions evolution converges toward the intermediate virulence $x^* = [c + \sqrt{c^2 K + cK(K-1)(d+\theta)}]/(K - 1)$ (Figure 4.1d).

This conclusion cannot be reached directly by maximizing the basic reproduction ratio $R_0(x') = x' S_0/[(x'+d+\theta+S_0/K)(x'+c)]$, since the resultant optimal virulence depends on the density of susceptible hosts in the absence of the disease, S_0. It is therefore clear that simple R_0 maximization ceases to work for examples like this. The reason is obvious: the optimal level of virulence depends on the density of susceptible hosts available for infection by a new strain, and this density in turn is affected by the resident strain. In other words, the existence of such an environmental feedback renders selection frequency dependent and usually precludes predicting the outcome of evolution through R_0 maximization.

It is therefore quite remarkable that this example nevertheless allows an optimization principle other than R_0 maximization. It can be shown that the optimal virulence x^* in this example can also be predicted by maximization of the function $\Phi(x') = [x'(K-1)-c]/[K(x'+d+\theta)(x'+c)]$ (J.A.J. Metz, personal communication). In agreement with the findings of Mylius and Diekmann (1995) and Metz *et al.* (1996b), the form of such alternative optimization functions is very sensitive to the way in which density dependence affects the rates of the epidemiological model, which implies that the generality of this particular choice of Φ is very limited. Notice that an analogous conclusion holds for R_0 itself: it primarily applies as an optimization principle for models with density-independent rates. Yet, such models have prevailed in the literature so far, which might have fostered a rather different impression.

Example V. As a second example for density-dependent rates, we return to density-independent mortalities, but now let the density of susceptible hosts affect the transmission rate, $d_S = d$, $\alpha = d_I - d_S = x$, and $\beta = x/(x + c/S)$. This means that the gain in transmission that results from a rise in virulence increases with the density of susceptible hosts. Analysis of the invasion fitness $f(x', x)$ shows that evolution again converges toward an intermediate virulence, this time given by $x^* = \sqrt{d+\theta}[\sqrt{d+\theta+4\sqrt{c}} - \sqrt{d+\theta}]/2$ (Figure 4.1e). Also, this example allows an alternative optimization principle, $\Phi(x') = x'/[z + \sqrt{z(z+4c)}]$ with $z = x'(x'+d+\theta)$. Since the form of density dependence has changed relative to that in Example IV, the two corresponding optimization principles also look very different.

While, for the previous two examples, the approach of R_0 maximization may be inconclusive, at least it does not turn out to be misleading. This is because, in these examples, the existence of the environmental feedback loop is unmistakably signaled by the dependence of R_0 on S_0. Subsequent to conventional R_0

maximization, the feedback loop can therefore be respected by choosing S_0 self-consistently. This is achieved by solving for a pair (x^*, S_0) such that, first, x^* maximizes R_0 given S_0 and that, second, S_0 is the equilibrium density of susceptible hosts for a resident virulence x^*. By adhering to such an extended R_0-based framework, it is thus sometimes possible to bypass the explicit analysis of invasion fitness. While evolutionary invasion analysis is applicable much more widely, the described alternative (but of course fully equivalent) route might appeal to those already familiar with conventional R_0 maximization.

Virulence evolution with rates dependent on infected host density

Example VI. Bypassing evolutionary invasion analysis is no longer an option when demographic or epidemiological rates that depend on the density of infected hosts are considered. Such a situation arises, for example, when the infection rate of susceptible hosts is assumed to change nonlinearly with the density of infected hosts. The relation $\beta = xI/(x+c)$ describes a setting in which the host's immune system is more likely to succumb to the onslaught of a disease if the ambient density of pathogens is high. Keeping the other rates as simple as in Example V, exactly the same expression is obtained for invasion fitness as when $\beta = x/(x+c)$, which predicts convergence toward the intermediate virulence $x^* = \sqrt{c(d+\theta)}$ (Figure 4.1f). Notice, however, that in this example R_0 for pathogens with any level of virulence x' vanishes, $R_0(x') = 0$ – erroneously suggesting that virulence is an evolutionarily neutral trait. The same conclusion pertains to any SI model in which the standard mass action term βSI is replaced by βSI^q with $q > 1$. For all these examples, an alternative optimization principle applies, $\Phi(x') = x'/[(x' + d + \theta)(x' + c)]$, and application of R_0 maximization is seriously misleading.

Example VII. Unfortunately, the error incurred by adhering to R_0 maximization can be even less conspicuous. Now consider an example in which the rate of recovery from the disease decreases with the number of infected hosts, $\theta = \theta_0/(1 + I/K)$. As mentioned above, such a situation could arise, for instance, when the care extended to individual infected hosts declines with their overall density. Here θ_0 is the recovery rate at very low disease incidence and K is the density of infected hosts at which that rate is halved. All other rates are assumed to be density independent, as in the previous examples; for the transmission rate we again revert to the classic trade-off relation $\beta = x/(x + c)$. As in Example III, R_0 for this setting is given by $R_0(x') = x'S_0/[(x' + d + \theta_0)(x' + c)]$ and it is immediately obvious that the density dependence of the recovery rate leaves this expression unchanged. This means that the parameter K cannot influence the optimal virulence $\tilde{x}^* = \sqrt{c(d+\theta_0)}$, predicted from maximizing R_0. Also, the birth rate b does not show in this result. For a particular choice of parameters ($b = 2$, $d = 1$, $c = 1$, $\theta_0 = 1$, and $K = 10$) R_0 maximization thus leads us to believe that, independent of b and K, evolution converges toward the intermediate virulence $\tilde{x}^* = \sqrt{2} \approx 1.414$. By contrast, a proper analysis of invasion fitness reveals that the selection gradient for this example actually vanishes at a significantly lower virulence, $x^* = 1.061$ (Figure 4.1g). Moreover, this evolutionarily stable outcome changes to $x^* = 1.253$ for $b = 1.75$ and to $x^* = 1.367$ for $K = 100$, qualitative effects that are altogether missed by the erroneous application of R_0 maximization.

Example VIII. The same conclusion applies when the density of infected hosts influences the disease-induced mortality. Here, consider an example described by $d_S = d$ and $d_I = d + x(1 + I/K)$. When disease incidence is low, disease-induced mortality $\alpha = d_I - d_S$ is given by x, just as in the preceding examples. Now, however, α increases with the density of infected hosts. As already mentioned, this could result, for instance, from the diminished care available to each infected host. For the other rates the same choices are made as in Example VII, except for the recovery rate θ, which is again simply kept density independent. Maximization of $R_0(x') = x' S_0 / [(x' + d + \theta)(x' + c)]$ yields the by now familiar expression $\tilde{x}^* = \sqrt{c(d + \theta)}$, which (for $b = 10$, $d = 1$, $c = 1$, $\theta_0 = 1$, and $K = 10$) gives $\tilde{x}^* = 1.414$. This prediction for the outcome of virulence evolution dramatically differs from $x^* = 0.194$, the accurate value derived from evolutionary invasion analysis. As in the previous example, R_0 maximization also fails to capture the dependence of x^* on b and K: $b = 2$ gives $x^* = 1.043$ (Figure 4.1h), and $K = 100$ gives $x^* = 0.219$.

Notice that the pairwise invasibility plots in all but the last two examples are skew-symmetric, that is, invariant under reflection along the main diagonal and simultaneous sign inversion (Figures 4.1a to 4.1f). The symmetry applies to the invasion fitness itself, $\operatorname{sgn} f(x', x) = -\operatorname{sgn} f(x, x')$, and hence is independent of the particular parameters chosen for the figures. According to the theory laid out by Metz *et al.* (1996b), this implies that the feedback loop in these examples acts through a one-dimensional environmental characteristic. If, in addition, the dependence of f on this characteristic is monotone, an optimization principle Φ can always be found – although the correct one often differs from R_0. By contrast, pairwise invasibility plots in Figures 4.1g and 4.1h are not skew-symmetric. As Metz *et al.* (1996b) have demonstrated, this means that the dimension of the environmental feedback loop exceeds one and no optimization principle can exist.

4.5 Pathogen–Host Coevolution

Evolution of pathogen virulence does not occur in isolation from other adaptive processes and is often accompanied by hosts changing their resistance toward infection. At first sight, the short life cycles of most pathogens suggest that pathogen adaptation greatly outpaces evolutionary responses on the part of the host. However, sexual reproduction in hosts often compensates for the pronounced asymmetries in demographic rates, and thus helps host populations to survive arms races with their pathogens.

This section briefly illustrates how models of adaptive dynamics are used to describe pathogen–host coevolution. Keeping in mind that R_0 maximization can be safely employed to predict virulence evolution only when demographic and epidemiological rates are density independent, the focus here is on the correspondence (or lack thereof) between processes of pathogen–host coevolution under

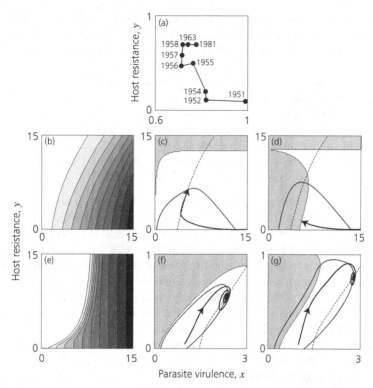

Figure 4.2 Coevolution of pathogen virulence and host resistance. (a) Coevolutionary trajectory observed after the introduction of the myxoma virus into the Australian rabbit population in 1950. Based on the trajectory's shape, a slight "viral backlash" can be conjectured, potentially resulting from the evolution of host resistance. *Data source*: Fenner and Ross (1994). (b) to (g) Coevolutionary trajectories that result from Examples IX to XII. Left column: dependences of disease-induced mortality on virulence and resistance (white: zero mortality, black: maximal mortality). Middle column: phase portraits of density-dependent models. Right column: phase portraits of corresponding density-independent models. Ranges of virulence and resistance that do not allow the pathogen to remain endemic are depicted as gray areas. Thin curves show the evolutionary isoclines of host (continuous) and parasite (discontinuous). Parameters: $b = 5$, $d = 1$, $\theta = 1$, $c = 2$, $K = 100$, $c_x = 4$, $y_0 = 10$, $c_y = 2$ (b–d); $b = 1.5$, $d = 1$, $\theta = 1$, $c = 1$, $K = 100$, $c_x = 0.4$, $y_0 = 1.75$, $c_y = 1$, $y_{max} = 10$, $c_{max} = 2$ (e–g); $(\mu\sigma^2)/(\mu_h\sigma_h^2) = 1$ (b–g).

density-dependent and density-independent conditions. To this end, host resistance y is introduced as a second trait in addition to pathogen virulence x, by slightly extending the SI model of Equations (4.1): all rates may now depend on (x, y, S, I), instead of on (x, S, I) as assumed in Section 4.4.

As a rough motivation for the examples considered below, Figure 4.2a shows the well-documented coevolutionary trajectory that resulted from the "escape" of the myxoma virus into the Australian wild rabbit population in 1950 (Fenner and

Ratcliffe 1965; Fenner and Ross 1994; Fenner and Fantini 1999). The data seem to indicate a slight gradual increase in pathogen virulence after about 1958, potentially in response to the substantial increase in host resistance between 1950 and 1958.

Below four models are considered to illustrate the evolutionary implications of density regulation. It must be emphasized that these simple models are by no means intended to capture the biological and dynamic complexity of myxoma–rabbit coevolution (for work in this direction see, e.g., Dwyer *et al.* 1990). For more details on the myxomatosis epidemic see Chapter 3, Section 3.6; the actual complexity of the involved evolution is neatly highlighted by the discussion of alternative selection pressures in Chapter 27, Section 27.2.

Example IX. The first example assumes that disease-induced host mortality decreases with increased resistance, $\alpha = x/[1 + e^{-(x-y)/c_x}]$ (Figure 4.2b). This function implies that, in the absence of resistance, disease-induced mortality is essentially proportional to virulence. If, however, resistance exceeds virulence, this mortality is greatly reduced (with the sharpness of the reduction determined by c_x). Also accounted for is that resistance is costly for the host, $b_S = b/[1 + e^{(y-y_0)/c_y}]$: while low levels of resistance are relatively cheap, resistance that approaches y_0 greatly reduces fertility (the sharpness of the cost increase is determined by c_y). Host mortality is assumed to be density dependent, $d_S = d + (S + I)/K$. Such density dependence is required to prevent the host population from diverging when the pathogen is not endemic. The other rates are given by $b_I = b_S$, $d_I = d_S + \alpha$, $\beta = \alpha/(\alpha + c)$, and θ. Evolutionary isoclines are those curves on which the selection pressure on virulence or resistance vanishes, $dx/dt = 0$ or $dy/dt = 0$. These isoclines are shown in Figure 4.2c, together with a coevolutionary trajectory that has a shape vaguely reminiscent of the empirical one in Figure 4.2a.

Example X. Example IX is now simplified by removing the density-dependent component of host mortality, $d_S = d$. A corresponding coevolutionary trajectory is shown in Figure 4.2d. Compared with Figure 4.2c, it is immediately obvious that the range of combinations of virulence and resistance for which the disease is endemic is greatly reduced. In particular, the coevolutionary attractor is now situated such that the coevolutionary process results in pathogen extinction. This is an example of evolutionary suicide, a process during which adaptation in a species is responsible for the extinction of that species (Matsuda and Abrams 1994; Ferrière 2000; Parvinen *et al.* 2000). Notice that, relative to Figure 4.2c, the shapes of the evolutionary isoclines, and therefore the position of the coevolutionary attractor, also change. The conclusion is therefore that to remove the density dependence of host mortality has serious implications for the expected coevolutionary outcome.

Example XI. Returning to density-regulated host mortality, $d_S = d + (S + I)/K$, now consider a slightly different dependence of that mortality on virulence and resistance, $\alpha = x/[1 + e^{-(x-\tilde{y})/c_x}]$ with $\tilde{y} = y_{max}y/(y + c_{max})$ (Figure 4.2e). This function describes a "resistance-is-futile" scenario. With investment in resistance exhibiting a diminishing return, effective resistance $\tilde{y}$ cannot increase beyond a maximum y_{max}, which is approached for large values of y (with the sharpness of the approach determined by c_{max}). This means that, in contrast to the two previous examples, it is now impossible for the host to fend off arbitrarily high virulence levels by increasing its resistance. Evolutionary isoclines and a coevolutionary trajectory that result from this scenario are given in Figure 4.2f, and show that the model gives rise to damped oscillations in virulence and resistance levels.

Example XII. The density-independent model that directly corresponds to Example XI can be considered by setting $d_S = d$. Comparing the results in Figure 4.2f with those in Figure 4.2g demonstrates that, for this case also, the shape of the evolutionary isoclines, the position of the coevolutionary attractor, and the domain over which the disease is endemic alter significantly. Coevolution now results in higher levels of virulence as well as resistance, and the coevolutionary oscillations become less pronounced. Notice in particular that the boundary of disease viability and the evolutionary isocline of the host essentially exchange their relative position. Thus, evolutionary suicide can again occur in the density-independent model, whereas such evolution-driven extinction of the disease is excluded in the density-dependent counterpart.

4.6 Discussion

This chapter evaluates the extent to which the traditional technique of R_0 maximization can be relied upon when studying the evolution of virulence traits. It is shown that R_0 maximization must be applied with great care to avoid erroneous conclusions. When demographic and epidemiological rates are density independent, R_0 maximization works well – unfortunately, however, such cases are quite simplistic. Once density regulation in these rates is accounted for, R_0 maximization may fail. Such failures may be conspicuous, as when the necessity to close the environmental feedback loop is signaled explicitly in the prediction derived from R_0 maximization, or they may go unnoticed and lead to serious mistakes. With such dangers lurking, the benefits of evolutionary invasion analysis are evident.

This conclusion is accentuated by comparison of models that describe the coevolutionary dynamics of parasite virulence and host resistance as resulting from density-dependent and density-independent rates. Although there are some rough similarities between the corresponding evolutionary scenarios, the shapes of the coevolutionary trajectories, as well as the positions of the evolutionary isoclines and attractors, turn out to be greatly affected by density regulation. A particularly intriguing finding in this context is that the conditions under which the evolution of virulence and resistance is expected to result in the extinction of the disease can differ greatly between these contrasting scenarios.

As pointed out in Section 4.4, density-dependent demographic and epidemiological rates appear to be virtually ubiquitous, so it is difficult to justify their omission from disease models. It may be argued that in industrialized nations human population densities are regulated by factors other than diseases; while the impact of population density on pathogen evolution must then still be considered, the feedback from disease evolution on population density may be negligible. This situation, however, is clearly different for the developing world, in which the prevalence of human diseases is highest and their evolution takes place. The same is true for many animal and plant populations, the demographies of which are greatly affected by endemic viral strains.

Although providing a convenient starting point, it is clear that the class of SI models studied in this chapter cannot capture the great variety of ecological stages on which processes of virulence evolution unfold in nature. Incorporating density regulation and the resultant mechanisms of frequency-dependent selection into more complex epidemiological models is therefore an exciting challenge. Such theoretical extensions have to address, in particular, the evolutionary implications of coinfection and metapopulation structure (Chapters 9, 10, and 11), spatially heterogeneous host populations (Chapters 7 and 8), and tritrophic interactions (Chapters 21 and 22).

As far as measures of virulence management are concerned, accurate predictions of the qualitative and quantitative effects of managerial interference on virulence evolution are indispensable. The theoretical consideration laid out in this chapter may foster this goal in several regards:

- First, it is not only asymptotic evolutionary outcomes that count in assessing strategies of virulence management: evolutionary transients toward such states may last long and must hence receive equal, if not primary, attention. Describing evolutionary transients requires dynamic models of adaptation and cannot be accomplished through consideration of optimization principles. Oscillatory transients, like that illustrated in Example XI, might actually be relatively widespread. A manager must be aware of such intrinsic instabilities, lest turning points in the dynamics are misinterpreted as indicators of faltering containment strategies.

- Second, R_0 maximization and adherence to models with density-independent rates can lead to grossly false predictions when mechanisms of density regulation are not negligible. As illustrated by Examples VI–VII and IX–XII, the resultant errors vary between quantitative inaccuracies and qualitative blunders. If simple models predict that interference with a demographic or epidemiological rate reduces the virulence of pathogens, while in actual fact such interference, properly analyzed, is expected to be inconsequential or even to result in more aggressive strains, efforts of virulence management can be seriously jeopardized.

- Third, the strength of density dependence may determine whether processes of evolutionary suicide can be utilized for the purposes of virulence management. Moving an evolutionary attractor out of the viability domain of the target pathogen by influencing the density dependence of demographic or epidemiological rates may sometimes result in runaway processes toward viral self-extinction, as illustrated by Examples IX–X and XI–XII. Such convenient opportunities may not arise too frequently, but, if an evolutionary attractor is situated in the vicinity of a viability boundary, limited managerial interference may well suffice to push it over the brink.

We must thus conclude that, as much as we would prefer evolutionary models of greater simplicity, continuing to overlook the adaptational repercussions of density-dependent demographic and epidemiological rates carries a high risk.

Acknowledgments   The author gratefully acknowledges insightful support offered by Hans Metz through many inspiring discussions on the role of optimization principles in evolution.

5

Dilemmas in Virulence Management

Minus van Baalen

5.1 Introduction

Both the patient who is infected with a communicable disease and the doctor treating the patient share a common interest: the eradication of the infection. That the treatment chosen by the doctor may have detrimental consequences for the population at large is not the primary concern of the doctor or the patient. Such matters are the concern of the larger-scale medical and political organizations that deal with the development of public health policies such as vaccination programs and possibly, as investigated in this book, "virulence management" strategies. Development of such policies is not only a complicated issue because of the intricacies of host–parasite interactions themselves, but also because the common aims of the public health authority and the population do not always overlap very well (Anderson *et al.* 1997). Of course, the community benefits when an individual ceases to be infective. However, parasites are not inert players in the game, and will adapt to any measures that are taken on a sufficiently large scale. Therefore, the development of some public health policies may not be beneficial to the community as a whole. The global resurgence of tuberculosis (TB) and the fact that many malaria parasites have become resistant against most preventive treatments are just two examples of the detrimental consequences of the large-scale application of individually beneficial medical treatment.

The insight that strategies to fight parasites should be based not only on short-term effects, but also on evolutionary considerations, is gaining ground (Ewald 1993, 1994a). For example, measures could be taken to counteract the development of resistance to antibiotics or other chemotherapeutic treatments (Baquero and Blázquez 1997; Bonhoeffer *et al.* 1997; Levy 1998; see Chapter 23). But other parasite traits evolve too. By working out how virulence may change in response to changes in the parasite's transmission cycle (Ewald 1994a; Van Baalen and Sabelis 1995b; see Chapter 2) one obtains an insight into the scope for such virulence management.

It has already been pointed out that measures taken to reduce the impact of a particular disease may involve ethical dilemmas. For instance, Anderson and May (1991) note that when a population is vaccinated against the poliomyelitis virus, the force of infection of this virus decreases. This means that fewer people will become infected, which is the desired beneficial effect. However, it also means that those who do become infected are likely to become so at a later age (polio was more commonly a childhood disease before vaccination); in the case of polio, as with some other childhood diseases, an infection at a later age may have more serious consequences. Thus, vaccination effectively means sacrificing the interests of a few individuals for the benefit of the population. A similar ethical issue arises when the degree of infection varies and treatment can be directed to either the (few) heavily infected individuals or the lightly infected majority.

The ethical dilemmas associated with public health measures are further intensified when the evolutionary response of the parasites is taken into account. It is increasingly recognized that the evolution of resistance against antibiotics is becoming a serious problem, and that antibiotics should be used sparingly and carefully to restrain this development (Baquero and Blázquez 1997; Bonhoeffer *et al.* 1997; Levy 1998; see Chapter 23). However, less attention has been given to the associated ethical dilemma: to what extent should an individual's interest be sacrificed for the good of the community? Analyses tend to predict that vaccination campaigns should select for decreased virulence [through the decrease in multiple infection and, hence, within-host competition (Van Baalen and Sabelis 1995b; Chapter 11)]; but what if vaccination favors more virulent parasites? In this chapter, I discuss a very simple model that suggests adequate treatment may indeed be a mechanism that selects for increased virulence.

The original model was formulated to study the question of how much of its resources a host should invest to create an immune system that eradicates infections (Van Baalen 1998). This chapter is based on the insight that, on an elementary level, visiting a doctor and receiving medical treatment is exactly analogous to the effect of the immune system. The corpus of medical knowledge and the availability of doctors and health insurance all work toward the eradication of infection. Of course, the relation between benefits and costs is less straightforward than the model assumes, but other than this shortcoming, the analogy can be carried quite far.

The analysis yields some results that may not be intuitively apparent. For example, individually optimal antiparasite measures may not lead to extinction of the parasites, but rather the opposite. Combining optimum defense with optimum counterstrategies on the side of the parasites suggests the possibility of even more worrisome outcomes. That is, when medical treatment becomes too effective, an "arms race" may be triggered, during which more and more resources need to be invested into developing more effective treatments against increasingly rare but increasingly virulent parasites. Any high-level community body (a "public health authority") may then have to "decide" which outcome is more desirable: a mild disease that affects many people, or a virulent disease that affects only a few.

In this chapter, I discuss to what extent this outcome depends on who pays the cost of medical treatment and at what level choices are made. That is, I compare the outcome for two possibilities: one in which all costs are paid by the individual on a case-per-case basis, and another in which all costs are paid by the community (so that every individual is required to pay an average fixed "health tax"). Decisions as to the effectiveness of the treatment are made by the individual (or doctor, assuming he or she does not balance the patient's interests against those of the community) or by the community.

This chapter is entirely speculative, and I make no attempt to analyze the results in terms of any real infectious disease. In fact, all the numerical examples have been chosen to demonstrate an effect rather than to give an indication of its likelihood or size.

5.2 Optimal Antiparasite Strategies

In this section, I compare the consequences of actions taken at different levels (i.e., at that of the individual or that of the community), while keeping parasite virulence constant. In Section 5.3, I allow the parasites to coevolve and respond to the antiparasite policies.

The basic points are illustrated by analyzing a simplistic susceptible–infected–susceptible (SIS) model for host–parasite dynamics. In the epidemiological literature, it serves as a reference base (e.g., see Anderson and May 1991) with which to contrast the consequences of more realistic extensions. This model also served as a framework to investigate coevolution of recovery rate and parasite virulence (Van Baalen 1998); here I present a reinterpretation of these results explicitly in terms of virulence management in which recovery is due to medical treatment.

The most important assumptions that underlie this model are that:

■ The host population grows logistically in the absence of disease;
■ The population is well-mixed so that overall transmission is a mass-action process;
■ Treated hosts become immediately susceptible again (no period of immunity).

This set of assumptions leads to

$$\frac{dS}{dt} = b(N)N - dS - \beta SI + \theta I \,, \tag{5.1a}$$

$$\frac{dI}{dt} = \beta SI - (d + \alpha + \theta)I \,, \tag{5.1b}$$

with $N = S + I$. Here, S and I represent healthy and infected hosts, respectively; d is the background per capita mortality rate, β is the per capita transmission parameter of the disease, α is the disease-induced per capita mortality rate (virulence), θ is the per capita recovery rate, and $b(N)$ represents the inflow of susceptible hosts due to births, with

$$b(N) = b_0(1 - \kappa N) \,, \tag{5.2}$$

where b_0 is the per capita birth rate and κ measures the density-dependent reduction of the recruitment rate.

In an SIS model with recovery, individual hosts switch back and forth between the susceptible and the infected states. Usually it is assumed that recovery occurs because the immune system clears the parasite, but here I assume that recovery is the result of medical treatment. The value of θ then embodies the efficiency of

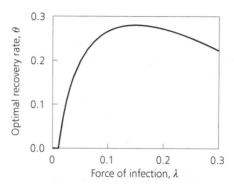

Figure 5.1 The relationship between individual optimal investment in recovery θ as a function of the force of infection λ (the risk per unit time of becoming infected). Parameters: $d = 0.02$, $\alpha = 0.3$, $c = 1$.

the entire public health system (i.e., the entire complex consisting of doctors, the availability of antibiotics, health insurance, etc.) in eradicating an infection.

Medical assistance is not free, of course. It is important to realize that the costs may be incurred at many levels, from the individual who pays a consultation fee to the community that finances the public health system (to train doctors, maintain hospitals, carry out research, etc.). There are two extreme cases: in the first individual hosts can pay for medical insurance – the quality of which determines their individual rate of recovery through treatment; in the second the rate of recovery is determined entirely by the community (through investment in a public health system).

Suppose an individual can increase his/her rate of recovery θ at the expense of a reduction in his/her rate of reproduction $b_0 = b_0(\theta)$, for example

$$b_0(\theta) = b_{max}e^{-c\theta} ,\tag{5.3}$$

where b_{max} is the maximum rate of reproduction and c is a measure (the cost) of how quickly the rate of reproduction decreases with a unit increase in θ. Of course, in reality this is more complicated, but, within the present simple framework, this is the most straightforward relationship. What is important to realize is that, in whatever way the costs are paid, the host population is involved in what is technically a "game." That is, the optimum strategy for the individual depends on the strategies that are adopted by the rest of the population (Maynard Smith and Price 1973; Maynard Smith 1982). In Van Baalen (1998), it is shown how the optimum investment in recovery rate depends on the risk of infection (see Figure 5.1). As can be seen, the optimal investment increases once the force of infection is greater than a threshold value, but decreases again for very high values of the force of infection. The reason for this is that if the force of infection is very high, hosts tend to become reinfected very quickly after they have recovered. No matter how quickly the infection is cleared, hosts spend most of their time in the infected state

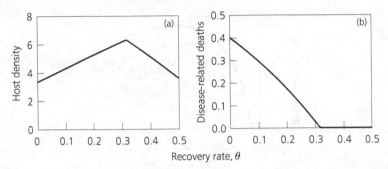

Figure 5.2. The relationship between tax-mediated investment in health care (leading to a recovery rate of θ) and total host population density (a) and proportion of disease-related deaths $p = \alpha I / (dN + \alpha I)$ (b). Note that host density decreases after the parasites have become extinct because the hosts keep paying their tax without accruing any additional benefit. Parameters: $b_{max} = 0.04$, $d = 0.02$, $\kappa = 0.05$, $c = 1$, $\alpha = 0.3$, $\beta = 0.1$.

anyway. Under such conditions, a host could just as well economize on health care and invest its resources otherwise (Van Baalen 1998).

Without a doubt, this model is far too simplistic to describe human population dynamics in any detail, let alone account for the complicated political decisions and the micro- and macroeconomic processes that govern the quality of public health. Having stated this, the model captures at least two ubiquitous relationships. First, parasites suppress host fitness (and hence population growth). Second, resistance to parasites imposes a cost (whether it is borne by individuals or averaged out over larger communities).

When the entire population of hosts tries to adopt the individually optimal strategy, the results may appear counterintuitive: the force of infection does not decrease, but rather is *maximized*. If the population invests little in health, then the parasites are given free reign – under which conditions it pays to invest in health care. If the host population invests heavily in health care, the parasite population will decrease. As a consequence, hosts can individually afford to "cheat" and economize on health care. Thus, one wonders to what extent the population as a whole benefits when investment in health care is based on individual decisions.

Contrast this with the case in which the cost of health care is uniformly distributed over the entire population. This would require a public organization that levies some sort of health care tax and ensures that every individual is treated once that individual is infected. What would be the optimal strategy for such a public organization? Taking the same cost–benefit function as defined before, the optimal strategy for the community seems obvious. Whether the aim is to maximize population density (Figure 5.2a) or to minimize disease-related deaths (Figure 5.2b), the best strategy for the community is to invest just enough to render the parasite extinct.

Some remarks are appropriate here. Once the parasites are extinct, it no longer makes sense to fight them. In principle, therefore, investments can then be re-allocated. However, this leaves the population susceptible to reinvasion by the parasite; thus, to protect the population against reinvasion, investments may have to continue. A second point is that for more realistic models (or with different cost–benefit relationships) the two criteria – maximizing mean wealth and mini-mizing parasite incidence – do not necessarily coincide. In that case, it must be decided what the most desirable outcome is – which may pose ethical dilemmas (see also Medley 1994).

5.3 Parasite Evolutionary Responses

Above, it was assumed that the parasites are evolutionarily inert. This, of course, is very unlikely. If health care becomes more efficient, then an elementary aspect of the parasites' environment changes – to which the parasites are expected to adapt. What will be the consequences?

A parasite's fitness is proportional to the product of its infectivity and the dura-tion of the infection (Anderson and May 1982; Bremermann and Pickering 1983). It is very likely that it cannot maximize both at the same time. An increase in infectivity is detrimental to the host, who is likely to die sooner, thus reducing the duration of the infectious period. Conversely, prolonging the infectious period may require a reduction in infectivity. Thus, the parasite's "host-exploitation strategy" should strike the optimal balance between the intensity and duration of infectivity (Anderson and May 1982; Bremermann and Pickering 1983; see Box 5.1).

To a parasite, it is irrelevant whether it stops transmitting because its host dies or because it is knocked out by antibiotic treatment. Therefore, if the host is likely to seek antibiotic treatment, the parasite should respond by shifting its policy toward quicker exploitation of the host. Thus, the availability of effective antibiotics is likely to favor more virulent parasites.

Often, it is argued that the best strategy for the application of antibiotics is to use them such that all parasites are killed. Then, it is claimed, even those parasites that are less sensitive to the antibiotic leave no descendants, and, hence, no resistance against the antibiotic can develop (Baquero and Blázquez 1997; Bonhoeffer et al. 1997; Levy 1998). This may be true, but it should not be forgotten that resistance is not the only parasite trait that evolves. The present analysis suggests that parasites respond evolutionarily even to perfect "magic bullet" types of antibiotics. In fact, the more effective the drug is, and the more likely a host is to seek treatment (resulting in a greater recovery rate θ), the stronger the evolutionary response. And it is worth noting that the direction of this evolutionary response is not at all desirable. I am not aware of any studies that show that the use of antibiotics has led to increased virulence, but the analysis in this chapter serves as a warning that there are reasons to expect such an evolutionary response!

If the parasites respond to increased treatment efficacy by becoming more viru-lent, then the ethical dilemmas associated with public health become more intense.

Box 5.1 Evolutionary optimization under infectivity–virulence trade-offs

A parasite needs to balance the short-term benefit of increased transmission and the longer-term benefit of host preservation. Suppose, as explained in Box 2.2, that the parasite experiences a trade-off between its infectivity (measured by its transmission coefficient β) and its virulence (measured by its disease-induced morality rate α). We can describe such a trade-off by a constraint that links these two parameters

$$\beta = \beta(\alpha) . \tag{a}$$

Under these conditions, what is the optimal virulence, that is, that level of virulence favored by natural selection? Ignoring the possibility of multiple infection (see Box 7.1), we can consider the dynamics of the density of hosts J that are infected by a mutant parasite with virulence α_{mut}

$$\frac{dJ}{dt} = \beta(\alpha_{mut}) S^*(\alpha_{res}) J - (d + \alpha_{mut}) J , \tag{b}$$

where d is the natural host mortality rate and $S^*(\alpha_{res})$ is the density of susceptible hosts, which, in turn, is determined by the resident parasite strain with virulence α_{res}. Whether or not the mutant invades depends on the sign of the right-hand side of Equation (b). This invasion condition is conveniently expressed in terms of the mutant's basic reproduction ratio

$$R_0(\alpha_{mut}, \alpha_{res}) = \frac{\beta(\alpha_{mut})}{d + \alpha_{mut}} S^*(\alpha_{res}) = Q(\alpha_{mut}) S^*(\alpha_{res}) , \tag{c}$$

where $Q(\alpha_{mut})$ is the "per-host exploitation factor." Notice that here the mutant's R_0 is a function of both its own virulence and the resident's, because the latter determines the density of susceptible hosts. (The relation with the R_0 introduced in Box 2.2 is explained below.) Since at equilibrium the resident's R_0 must exactly equal one, $R_0(\alpha_{mut}, \alpha_{res}) = 1$, we have

$$S^*(\alpha_{res}) = \frac{d + \alpha_{res}}{\beta(\alpha_{res})} . \tag{d}$$

If the resident strain adopts a virulence α_{res} such that

$$Q(\alpha_{res}) > Q(\alpha_{mut}) \tag{e}$$

for all levels of virulence α_{mut}, it is evolutionarily stable.

The evolutionarily stable level of virulence therefore maximizes the per-host exploitation factor Q (Van Baalen and Sabelis 1995a). Note that Q is expressed entirely in terms of individual-level rate coefficients and does not involve any population-level quantities, such as the number of susceptible hosts that appear in R_0. The evolutionarily stable level of virulence can be found graphically by determining for which α_{mut} the tangent on the curve $[\alpha_{mut}, \beta(\alpha_{mut})]$ passes through the point $(-d, 0)$ (see figure).

Boxes 2.2 and 9.1 explain how the evolutionarily stable virulence can be calculated by "maximizing R_0." Importantly, the R_0 introduced there is a slightly different quantity from the R_0 introduced here, although the two quantities are closely

continued

Box 5.1 *continued*

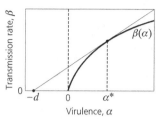

Graphical method for finding the evolutionarily stable level of virulence α^*. The evolutionarily stable virulence maximizes the ratio $\beta(\alpha)/(d + \alpha)$ and thus occurs for the α at which the tangent of the curve $\alpha, \beta(\alpha)$ passes through the point $(-d, 0)$.

related. In general, the R_0 of a certain type of individual is defined as the lifetime offspring production (in the case of a parasite, offspring are freshly infected hosts) in a certain reference environment. In Box 2.2 this reference environment is the parasite-free host population. By contrast, here the reference environment is a host population that is already infected with the resident parasite strain. Notice that, in the models under consideration in this box as well as in Boxes 2.2 and 9.1, the basic reproduction ratio R_0 in any reference environment with susceptible density S_0 is simply proportional to the "per-host exploitation factor" Q, $R_0 = QS_0$, but note that this no longer holds true if multiple infections occur. We can therefore choose any such reference environment to compare the basic reproduction ratios R_0 of a resident and mutant strain: this comparison gives the same result as one based on their per-host exploitation factors Q. The standard convention in the literature is to choose the disease-free environment to determine S_0. Yet, for models in which R_0 is always proportional to the density of susceptible hosts, environments with different S_0 can be chosen just as well. Nevertheless, it must be realized that models for which the evolutionarily stable virulence can be calculated through an optimization argument and for which the quantity to be optimized by a disease can be simply related to R_0 are special ones; unfortunately these two simplifying features do not apply to other, more general models (see Mylius and Diekmann 1995; Metz *et al.* 1996b; and Mylius and Metz, in press).

Consider again the case in which hosts individually decide on their health insurance. Now the game aspect involves not only the risk of infection, but also the consequences of being infected. For example, if the population is well-insured (resulting in a large population-wide value of the recovery rate θ), then the parasites may become rare but also very virulent. In fact, they may become so virulent that it pays an individual host to increase its own recovery rate even more. Thus, an arms race is triggered in which the hosts are forced to invest more and more resources in their defense, and the parasites become more and more virulent to counter this defense. Eventually a stable end result (i.e., a coevolutionarily stable strategy, or CoESS) may be reached, in which hosts pay heavily to defend themselves against a rare but serious disease.

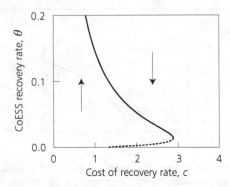

Figure 5.3 CoESS recovery rate θ as a function of relative cost c. For intermediate costs, there are two simultaneous CoESSs (one given by the full curve and one by $\theta = 0$, separated by the dashed curve). Arrows indicate the direction of selection. These results are based upon the assumption that parasite infectivity and disease-induced mortality are related through the constraint $\beta = \beta_{max}\alpha/(\delta + \alpha)$. Parameters: $b_{max} = 0.04$, $d = 0.02$, $\kappa = 0$, $\beta_{max} = 0.1$, $\delta = 0.02$.

This is not always an inevitable outcome as for some parameter combinations a second CoESS is possible: hosts tolerate the parasite, while parasites respond by staying relatively benign (Van Baalen 1998). Van Baalen (1998) argued that if such bistability occurs naturally (i.e., as a consequence of immune system and parasite coevolution), reinforcement of the immune system with an external medical component might destabilize the tolerance–avirulence CoESS and trigger an arms race that escalates to the defense–virulence CoESS. Presumably, when antibiotics become available, the cost of increasing the recovery rate will be reduced (antibiotics are likely to be much less expensive than gearing up the immune system to obtain a similar result). Again, whether such bistable outcomes are a reality remains to be confirmed; but if they are, it raises worrying questions. As can be seen in Figure 5.3, if the cost c is reduced below a certain threshold, an arms race is triggered that may be difficult to undo due to the hysteresis effect.

Note that the present model is too simplistic to assess the likelihood that such bistability occurs. But if such bistability is a reality, "virulence management" acquires a whole new aspect. Which of the two outcomes is preferable? Once again, this cannot be answered without addressing ethical issues. The question then really is whether "we" (i.e., presumably some governmental organization) should strive for a common avirulent disease or for a rare but virulent disease. This is not an easy question to answer, and certainly falls outside of the scope of pure science.

5.4 Discussion

There exists a very basic conflict of interest among the individuals of a population who are infected by parasites. Taking into account the evolutionary response of parasites against measures to fight them (whether on the level of individual treatment or of large-scale public health measures like vaccination) only intensifies this conflict of interest. Individuals profit from antibiotic treatment, but the community suffers from the evolution of resistance or increased virulence that follows.

In whatever form, defense against parasites is costly. Among the hosts there is an incentive to reduce these expenses. Moreover, there is a game-theoretical aspect to such defense. If the host population strongly defends itself, herd immunity creates opportunities for "cheats" to economize on defense. The end result (evolutionarily stable strategy, or ESS) is not the strategy that minimizes parasite load on the community – on the contrary. Rather, parasites effectively mediate competition among the hosts; the strategy that creates the highest parasite load while maintaining itself will outcompete any other (Mylius and Diekmann 1995). This scenario would create a bleak world. It is clear that under these conditions, a communal defense strategy may pay off for the community as a whole. That is, every host profits from the efforts of a public health authority that provides general health insurance. (An associated moral system, and possibly a judicial system to impose it, may be necessary to prevent cheats.)

Assuming that all hosts have ceded the most important decisions to such a public health authority, the problems are still far from over. The highest priority of such an authority would be, of course, to fight the parasites in the short term, such as by implementing public health measures, vaccination campaigns, provision of adequate medical care, etc. The decisions that must be taken at this level are complicated and must take into account all the effects of age structure, temporary or life-long immunity, multiple infection, cross-immunity, social structure, etc. (see Anderson and May 1991).

The purpose of this book is to discuss the possibilities of virulence management – that is, that set of public health measures that takes into account not only the short-term effects, but also the long-term evolutionary effects. The point of this chapter is that the design of such virulence management strategies may have to be developed in light of the partially conflicting interests between the individual and society, and, therefore, such strategies may require Machiavellian choices about whom to protect and whom to sacrifice. This may not be a welcome message, but turning a blind eye to it may present us with dire consequences. To end on a more positive note, virulence management allows us to exploit the forces that keep society together to improve the conditions for all. As such, virulence management may help the human society in its ongoing struggle to escape from its parasites (McNeill 1976).

Acknowledgments The author thanks John Edmunds for his valuable comments on a draft version of this chapter.

Part B

Host Population Structure

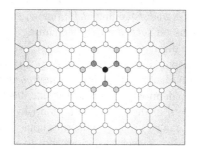

Introduction to Part B

Part B explores the impact of host population structure on the evolution of infectious diseases. While simple models of disease ecology and evolution conveniently ignore this complication, the following three chapters underline its importance. It is shown that host population structure can qualitatively alter expectations for the course and outcome of virulence evolution.

By linking individual-based mechanisms of transmission to the demographic consequences of epidemics in host populations, simple mathematical models offer an essential prerequisite for understanding and influencing the virulence evolution of a disease. Elaborations on such models, accounting for three different types of host heterogeneity, are discussed in this part. First, even in the absence of any spatial structure, a host population may be physiologically structured with respect to certain features of individual hosts. Relevant features could be age and size or could directly relate to epidemiological processes like disease-induced mortality, recovery from an infection, or disease transmission (investigated in Chapter 6). Second, host populations can be viscous in the sense that individual hosts are connected, by spatial proximity or social relations, not to the host population as a whole but to a relatively small number of neighbors. Implications of such connectivity structures are analyzed in Chapter 7. Third, connections between hosts may be organized in a hierarchical way such that infections spread more easily within host groups than between groups. A special case of such a metapopulation structure comprises just two groups of hosts, a large and viable host population (a "source") and a small host population (a "sink") that is prevented from extinction only by the continuous supply of immigrants from the source. As Chapter 8 shows, evolution of virulence or resistance in the sink population can only be understood by considering the impact of the source.

One key implication of host structure may be singled out for special emphasis: such structures often expose virulent pathogens to the detrimental consequences of aggressive host exploitation. Selection in structured host populations can favor pathogens of reduced virulence because those pathogens that exploit their victims excessively may soon run out of susceptible hosts. "Burn-out" phenomena of this sort are much more likely to occur in spatially structured populations; they offer important management opportunities to deliberately select for intermediate levels of virulence.

In Chapter 6, Dwyer, Dushoff, Elkinton, Burand, and Levin improve on basic epidemiological models by taking into account host heterogeneity in susceptibility and host seasonality in reproduction, key features of many insect–pathogen interactions. Their model is calibrated with experimental data on wild-type and genetically modified virus strains that can attack the gypsy moth, a polyphagous forestry pest. To assess the options for the modified virus to act as a biological control agent of the moth, the authors predict the rate of epidemic spread of both viral types in natural moth populations. They suggest that pathogens that are genetically engineered to have higher virulence may tend to be at a selective disadvantage.

Spatially or socially structured host populations are ubiquitous in nature. Chapter 7 describes how heterogeneity can arise from the local interactions among healthy and infected host individuals. Van Baalen explains why the resulting self-organized patterns of host abundance can lead to levels of pathogen virulence that qualitatively differ from those predicted for spatially unstructured populations. It is shown that increased regularity in the host's social structure selects for diminished virulence and that the same effect results when contacts between hosts become scarce. In general, any management strategy that keeps intact or even strengthens patterns of relatedness among infecting pathogens can be expected to favor the emergence of less virulent strains.

Considering examples of crop and livestock diseases and of hospital infections, Holt and Hochberg illustrate in Chapter 8 that source–sink structures are widespread in epidemiologically important situations. The authors show that virulent pathogens are less likely to conquer a sink habitat if host abundance in the sink is low, mutations have only a small effect, and invasions of benign pathogens (followed by local adaptation toward increased virulence) are rare. Conversely, resistant hosts, having reduced transmission rates for an infection, are more likely to evolve in a sink if host productivity is high, rates of pathogen transmission are low, and infected individuals are short-lived. In both cases, supply of novel genetic material from the source can be both detrimental (by swamping local adaptation) and beneficial (by providing the genetic variation needed to respond to local selection pressures).

Incorporating into a single model all possible aspects of host population structure evidently is impossible. The models considered in this part therefore separately focus on the main different types of host heterogeneity. Investigating interactions between the diverse evolutionary consequences discussed here is a challenge for future research.

6

Variation in Susceptibility:
Lessons from an Insect Virus

Greg Dwyer, Jonathan Dushoff, Joseph S. Elkinton,
John P. Burand, and Simon A. Levin

6.1 Introduction

A basic result of Anderson and May's (1982) early work on models of disease in
natural (nonhuman) populations is that pathogen fitness is $R_0 = \beta N/(\alpha + \gamma + d)$, where β is the horizontal transmission rate of the disease, α is the disease-
induced mortality rate, d is the background mortality rate, γ is the recovery rate
to the immune state, and N is host population density without the disease (see
Boxes 2.1 and 2.2). In this model, pathogen strains that maximize $\beta/(\alpha + \gamma + d)$
competitively exclude all others. A key insight, however, is that trade-offs among
fitness components prevent selection from driving horizontal transmission β to
infinity, and mortality α and recovery γ to zero. For the mosquito-vectored rabbit
disease myxomatosis, for example, virus strains that kill too rapidly have little
chance of being transmitted, because mosquitoes do not bite dead rabbits. On
the other hand, strains that kill too slowly produce such low concentrations of
virus that they are also unlikely to be transmitted (Fenner 1983). Assuming that
the rabbits evolve over a much longer time scale than does the virus, for such a
constraint an evolutionarily stable strategy (ESS) exists at the maximum of $\beta/(\alpha + \gamma + d)$; compare Boxes 2.1, 2.2, and 5.1.

Although later approaches to this problem have concentrated on this qualitative ESS approach, here we focus on a more quantitative feature of Anderson and
May's work: Anderson and May parameterized the trade-off that occurs in myxomatosis between recovery rate γ and mortality rate α (defined as virulence) by
fitting the function $\gamma = -c_0 - c_1 \ln \alpha$ to data from infections in rabbits in the lab.
When this parameterized function is inserted in the expression for R_0, it gives a
fitness maximum at a value that is not too far from the levels of virulence observed
in the field. Anderson and May were thus able to extrapolate from the characteristics of the disease in individual rabbits to the evolution of the disease in natural
populations. In this chapter, we emulate this approach by using parameterized
epidemic models to predict the fitness of insect pathogens, and by extrapolating
from small-scale measurements of transmission to large-scale epidemics. In doing
so, we extend Anderson and May's models by allowing for host heterogeneity in
susceptibility, and host seasonality in reproduction, which are key features of the
biology of many insect–pathogen interactions (Dwyer *et al.* 1997, 2000). Finally,

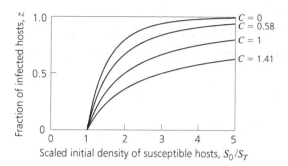

Figure 6.1 Fraction of hosts becoming infected during an epidemic, calculated by solving Equation (f) in Box 6.2 for z for $I_0 = 0$. C is the coefficient of variation of the distribution of host susceptibility.

we apply our understanding of insect–pathogen dynamics to see how genetic engineering for higher virulence affects pathogen fitness. An important caveat, however, is that we quantify fitness solely for the case in which one pathogen strain at a time infects the host population. Although, in so doing, we do not allow for pathogen coexistence, our hope is that this approach nevertheless allows at least preliminary insights into pathogen evolution in the face of host heterogeneity in susceptibility.

6.2 Theory of Multigenerational Epidemics

Anderson and May's work was based on earlier models of human epidemics (Mollison 1995), especially the work of Kermack and McKendrick (1927). As Kermack–McKendrick models (Box 6.1) are intended to represent single epidemics of human diseases, during which host population densities often change only very slightly, they typically assume constant host populations. Anderson and May's innovation in creating models of disease in natural populations was to allow for host reproduction (Anderson and May 1978; May and Anderson 1978). By considering only continuously reproducing hosts, however, Anderson and May effectively allowed endemic diseases only, yet insect pathogens and many other diseases are epidemic rather than endemic. More precisely, in many insects there can be only one epidemic each year, because only larvae can become infected, no new hosts are produced during the epidemic, and reproduction occurs only among the survivors (Dwyer *et al.* 2000). Kermack–McKendrick models allow us to incorporate this kind of seasonality into long-term host–pathogen models in a natural way, according to

$$S_{n+1} = g S_n \left[1 - z(S_n, P_n) \right] , \tag{6.1a}$$

$$P_{n+1} = p_{\text{survival},1} S_n z(S_n, P_n) + p_{\text{survival},>1} P_n . \tag{6.1b}$$

Here g is net fecundity, S_n and P_n are the densities of hosts and pathogens at the beginning of the epidemic in generation n, $p_{\text{survival},1}$ is the probability that

Box 6.1 SIR models in demographically closed populations

Kermack and McKendrick's disease model in its simplest form, also known as the SIR model, is given by

$$\frac{dS}{dt} = -\beta SI , \tag{a}$$

$$\frac{dI}{dt} = \beta SI - (\alpha + \gamma)I , \tag{b}$$

$$\frac{dR}{dt} = \gamma I , \tag{c}$$

where S is the density of susceptible hosts, I is the density of infected and also infectious hosts, R is the density of removed hosts, β is the rate at which the disease is transmitted horizontally, and $\alpha + \gamma$ is the rate at which infected hosts are "removed" from the infection process by death α and immunity γ (Kermack and McKendrick 1927). It is assumed that $R(0) = 0$. Note that $dN/dt = 0$, where $N = S + I + R$, so that $N = N(0) = S(0) + I(0)$ for all t. The more general version of Kermack's and McKendrick's model can distinguish between "infected, but not yet infectious" hosts (so-called latent infected) and "infected and also infectious" hosts, and allows for distributed delays between infection and infectiousness, just as between infection and death or recovery. R_0 for this model equals $[\beta/(\alpha + \gamma)]S_0$, where $S_0 = S(0)$.

Kermack and McKendrick (1927) showed that $z(S_0, I_0)$, the fraction of hosts that will become infected after a long epidemic, can be expressed implicitly as

$$1 - z = e^{-\frac{\beta}{(\alpha+\gamma)}(S_0+I_0)z} . \tag{d}$$

To derive Equation (d), one assumes that the epidemic continues until $t \to \infty$. This is loosely equivalent to continuing the epidemic until the density of susceptible hosts is too low to allow transmission, which is a more reasonable assumption than might first be imagined (Dwyer *et al.* 2000). The top curve in Figure 6.1 gives z as a function of the scaled density S_0/S_T, where $S_T = (\alpha + \gamma)/\beta$ is the threshold density and the value of I_0 is taken as negligibly small, so that Equation (d) reduces to

$$1 - z = e^{-\frac{\beta}{(\alpha+\gamma)}S_0 z} = e^{-R_0 z} , \tag{e}$$

with the basic reproduction ratio $R_0 = [\beta/(\alpha+\gamma)]S_0$ (compare Boxes 2.2 and 9.1). Figure 6.1 shows that no epidemic occurs for host densities below the threshold S_T, while above the threshold epidemic intensity climbs steeply with initial host density. Moreover, it shows that there is always a nonvanishing fraction of susceptible hosts that escape infection at the end of an epidemic, thus alerting us that the escapees after an epidemic in the real world do not necessarily represent resistant hosts.

pathogens produced during the epidemic survive to be infectious next season, and $p_{survival,>1}$ is the probability that pathogens surviving from previous epidemics survive to be infectious in the following season. Most importantly, $z(S_n, P_n)$ is the fraction of hosts that become infected during the epidemic, which is determined by a single-epidemic Kermack–McKendrick model such as Equations (a) and (b) in Box 6.1 [although, in practice Equations (d) and (e) in Box 6.2 describe epidemics of insect pathogens more realistically, as we discuss in Section 6.3]. For an appropriate choice of parameter values, Equations (6.1) show long-term cycles much like the cycles seen in many insect–pathogen interactions in nature (Varley *et al.* 1973; Dwyer *et al.* 2000). The single-epidemic Kermack–McKendrick approach can thus be as useful in understanding epidemics in natural populations as the continuous-time Anderson and May approach.

Both types of models, however, assume that host individuals are identical, whereas, for many diseases, heterogeneity among individuals has important epidemiological effects. In Box 6.2, we extend the Kermack–McKendrick approach to allow for variability among individuals' susceptibility to the pathogen, and we show that this complication can have a strong effect on the outcome of epidemics.

6.3 Controlling Gypsy Moths by Genetically Engineered Viruses

The theory that we outlined in Sections 6.1 and 6.2 incorporates models of single epidemics into models of long-term, host–pathogen population dynamics. In this section, we show that these models can explain data collected for a virus of gypsy moth (*Lymantria dispar*, a lymantriid lepidopteran) at a variety of spatial and temporal scales. This explanatory ability suggests that we can use these models to quantify the fitness of pathogen strains of different virulence, indicating, in turn, that we can predict the outcome of competition between wild-type and genetically engineered viruses. Although we focus on the nuclear polyhedrosis virus (NPV) of the gypsy moth, virus diseases have been found in a huge number of insect species (Martignoni and Iwai 1986), and often have an enormous impact on insect dynamics (Fuxa and Tanada 1987). We therefore expect our results to be of general usefulness in understanding the ecology and evolution of these pathogens.

Interest in genetically engineering NPVs has arisen because NPVs are often host-specific and are usually fatal, and, therefore, they can be useful in agriculture and forestry as environmentally benign insecticides (Black *et al.* 1997). A significant problem with using NPVs as insecticides, however, is that they often take 7 to 14 days to kill, which is much slower than conventional insecticides. Consequently, efforts to genetically engineer NPVs have usually been focused on increasing the speed of kill (here taken to be virulence). It remains to be seen, however, whether such genetically engineered strains are able to out-compete wild-type viruses, and thereby change the ecology of the insect–pathogen interaction. Part of our intent in what follows is to predict the environmental impact of releasing engineered virus strains into the environment.

Box 6.2 SIR models accounting for host diversity in susceptibility

We assume that any variability in the transmission results from differences in host susceptibility. To account for this variability we subdivide the single susceptible class from Box 6.1 into a distribution of susceptibilities, again called S, and change Equations (a) to (c) from that Box into

$$\frac{\partial S}{\partial t} = -\beta S I \,, \tag{a}$$

$$\frac{dI}{dt} = I \int_0^\infty \beta S(\beta, t) \, d\beta - (\alpha + \gamma) I \,, \tag{b}$$

$$\frac{dR}{dt} = \gamma R \,. \tag{c}$$

This model is used in this chapter to describe variability in the dose required to infect an insect (Dwyer *et al.* 1997), but a similar model has been used to describe heterogeneity in human immunodeficiency virus (HIV) transmission due to differences in sexual behavior (Anderson *et al.* 1986). The model framework is thus general enough to allow for different underlying mechanisms, and similar models can be constructed to include heterogeneity that depends on the infected host.

As with the Kermack–McKendrick model, allowing infectiousness and recovery to vary with time since infection is not difficult (Dwyer *et al.* 2000). For example, to describe epidemics of insect pathogens, the following model applies

$$\frac{\partial S}{\partial t} = -\beta S I \,, \tag{d}$$

$$\frac{dI}{dt} = I(t - \tau) \int_0^\infty \beta S(\beta, t - \tau) \, d\beta - \alpha I \,, \tag{e}$$

which accounts for most insect pathogens being fatal, and that transmission occurs only after the host dies; I now represents the density of infectious cadavers rather than infected hosts, and τ is the delay that occurs between infection and death. For the insect pathogens considered in this chapter (viruses of gypsy moths), epidemics typically end after about 70 days (or about 5 to 7 times τ) because larvae pupate and, therefore, can no longer become infected.

To see the effects of heterogeneity in susceptibility, for either Equations (a) through (c) or Equations (d) and (e), we can again calculate $z(S_0, I_0)$, the fraction of hosts infected at the end of a long epidemic (Dwyer *et al.* 2000), using

$$1 - z = \left[1 + \frac{\overline{\beta} C^2}{\alpha + \gamma} (S_0 z + I_0) \right]^{-1/C^2} \,. \tag{f}$$

Here $\overline{\beta}$ is the average transmission rate in the population, C is the coefficient of variation of the distribution of transmission rates in the population, and $S_0 = \int_0^\infty S(\beta, 0) \, d\beta$ [with $\gamma = 0$ in Equations (d) and (e)]. In addition to letting $t \to \infty$, the derivation of Equation (f) also assumes that transmission rates follow a gamma distribution. This is clearly only an approximation of reality, but possibly a good one.

continued

Box 6.2 *continued*

Figure 6.1 shows the fraction of hosts infected z as a function of the scaled host density S_0/S_T, where S_T is again the threshold population density, $S_T = \alpha/\bar{\beta}$. The figure demonstrates that increasing heterogeneity in susceptibility C strongly reduces the fraction of hosts that become infected during the epidemic z, but has no effect on the threshold density at which epidemics first appear. This reduction in the fraction infected z occurs even though the mean transmission rate $\bar{\beta}$ is unchanged, showing that highly resistant individuals have a disproportionately large effect on transmission.

Predicting virus epidemics in gypsy moth populations

First, we demonstrate that a single-epidemic model [Equations (d) and (e) in Box 6.2] can provide a useful description of the biology of the gypsy moth virus and the many similar viruses of other herbivorous Lepidoptera. The disease is transmitted horizontally when larvae consume the virus on contaminated foliage, and larvae that consume enough virus usually die (Cory *et al.* 1997; but see Rothman and Myers 1996). Near the end of the virus's lifecycle inside the insect, virally encoded chitinases and proteases break down the insect's integument, so that, shortly after death, the integument breaks open, releasing virus onto the foliage where it is available to cause new infections (O'Reilly 1997). At high enough densities of hosts and pathogens, epidemics occur that can annihilate gypsy moth populations.

NPVs can survive outside of their hosts because the virions that contain their DNA are packaged inside a polyhedral protein matrix that provides protection against environmental hazards such as dehydration and sunlight (Evans and Entwistle 1987). As is typical of Lepidopteran hosts of NPVs, adult gypsy moths cannot become infected, so virus epidemics must occur during the larval season. In gypsy moths, epidemics are begun when larvae hatch from contaminated egg masses (Murray and Elkinton 1989), and, in the years between epidemics, the virus apparently survives in the leaf litter on the forest floor (Elkinton and Liebhold 1990).

To understand the dynamics of the gypsy moth virus, we began with a standard Kermack–McKendrick model [Equations (a) and (b) in Box 6.1], into which we incorporated a delay between infection and death (Dwyer and Elkinton 1993). The delay is important because it takes about 10–14 days for the virus to kill an infected insect, which is a substantial fraction of the 9–10-week larval season (at the end of which larvae pupate, ending the epidemic). To test the usefulness of this model, we used it to make predictions of the intensity of epidemics, which required estimates of all of the parameters. Most of the parameters can be estimated easily from the literature, with the notable exception of the transmission rate β. To estimate β, we created small epidemics in experimental populations of gypsy moths. We confined healthy and virus-killed cadavers in mesh bags on red

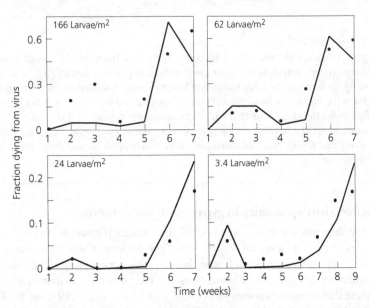

Figure 6.2 Model prediction of virus epidemics versus data from natural populations (Woods and Elkinton 1987, Woods *et al.* 1991). The model comprises Equations (d) and (e) in Box 6.2. All of the model's parameters were estimated independently of time-series data [mean transmission $\bar{\beta}$ and coefficient of variation C are averages of experimental values for feral larvae in Dwyer *et al.* (1997)]. Larvae/m^2 indicates initial host density. *Data source*: Woods *et al.* (1991).

oak (*Quercus rubra*) branches on trees in the field for a week, and then reared the healthy larvae in individual cups of artificial diet in the laboratory until pupation. Since the virus takes more than a week to kill, the experiment allows for only one round of transmission.

The resulting parameterized model gives an excellent prediction of the course of virus epidemics in natural populations at high density, but consistently underestimates virus mortality in populations at low density. The missing detail in the model appears to be host heterogeneity in susceptibility (Dwyer *et al.* 1997), specifically variability in the dose of virus that it takes to infect an individual. Allowing for this kind of heterogeneity in our model [Equations (d) and (e) in Box 6.2, with an epidemic of length 70 days], showed first that heterogeneity causes transmission to be a nonlinear function of virus density [where transmission is measured by $-\ln(S_T/S_0)$, and S_T/S_0 is the fraction of uninfected larvae at the end of the experiment]. Further experiments confirmed this effect (see Figure 6.3 for another example), and parameterizing the new model with the transmission data gave a much better ability to predict virus epidemics in the field (Figure 6.2).

Fitness and genetically engineered virulence: A case study

As Figure 6.2 demonstrates, an important feature of our model is that it allows us to extrapolate from the small scale of experiments to the large scale of natural epidemics. Since the epidemiology of a pathogen essentially determines its fitness, we have a practical method for estimating the fitness of gypsy moth viruses of different virulence. An important point, however, is that here we quantify absolute fitness somewhat differently than Anderson and May (or most disease modelers, see Diekmann *et al.* 1990). That is, the usual expression for absolute pathogen fitness R_0 comes from continuous-time disease models, so that R_0 is the number of new infections per old infection. Since many insect pathogens have only one epidemic per year, which is followed by a period of no transmission, a more natural measure of fitness is the number of new hosts infected in the current generation per infection in the previous generation – we use this alternative definition.

Given this definition, we can use the combination of theory and experiment outlined above to estimate the fitnesses of wild-type and genetically engineered gypsy moth virus strains. The engineered strain was produced by deleting the ecdysteroid-UDP-glucosyl-transferase (*egt*) gene, and is thus known as *egt-*. *egt* glycosylates ecdysteroids, which are molting hormones, so that insects infected with the wild-type virus ordinarily do not molt to the next larval stage. Insects infected with the *egt-* strain, however, do molt, unless their infection has advanced to a stage where they are close to death at the time of molting (Park *et al.* 1996). The *egt-* mutant kills about 25% faster than does the wild-type virus (Slavicek *et al.* 1999), suggesting that the *egt* gene is an adaptation that allows the virus to produce larger quantities of virus particles (although the molecular mechanism underlying the more rapid rate of kill is as yet unknown). Indeed, measurements of the amount of virus produced by *egt-* mutants in viruses of other insects (notably *Autographa californica*) show substantial reductions in the amount of virus produced, which is presumably translated into a reduction in transmission (O'Reilly 1997). More practically, because of their faster speed of kill, *egt-* mutants could be a more effective insecticide for controlling insect populations (Black *et al.* 1997). In fact, *egt* has been found in many different NPVs (O'Reilly 1997). However, for diseases like NPVs that must kill to be transmitted, higher virulence means a shorter generation time, and thus higher fitness. Therefore, strains that are engineered for higher virulence may be able to out-compete naturally occurring strains. On the other hand, if the wild-type gypsy moth virus does, indeed, produce greater amounts of virus, its slower speed of kill may be compensated by a higher transmission rate, leading to higher fitness overall. The kind of trade-offs seen in myxomatosis may therefore reduce the risk of releasing genetically engineered insect viruses.

To assess the risks of such releases, we compared the fitnesses of wild-type and genetically engineered *egt-* strains of gypsy moth NPV by experimentally measuring their transmission rates, following our standard protocol (Dwyer *et al.* 1997; except that we used oak branches in water jugs in the laboratory). Figure 6.3 shows that the transmission rate of *egt-* is indeed lower than that of the wild-type virus.

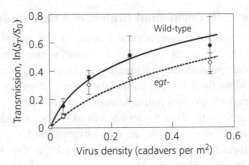

Figure 6.3   Results of a transmission experiment using wild-type and genetically-engineered (*egt-*) virus strains. Points are data, and curves show best-fit to the relation given by Equation (6.2).

From these data, we can estimate the mean transmission parameter $\bar{\beta}$ in Equations (d) and (e) in Box 6.2, noting first that the bags do not allow breakdown of the virus, so that we can set $I(t)$ constant. If we then assume that the initial distribution of susceptibility has a gamma distribution with mean $\bar{\beta}$ and coefficient of variation C, from Equations (d) and (e) in Box 6.2 we can derive an expression for transmission as a function of pathogen density (Dwyer *et al.* 1997),

$$-\ln\left(\frac{S_T}{S_0}\right) = \frac{1}{C^2} \ln\left[1 + \bar{\beta}C^2 I(0)T\right] , \qquad (6.2)$$

where T is the length of the experiment and $\hat{S}(t)$ is total host density at t, $\hat{S}(t) = \int_0^\infty S(\beta, t)\, d\beta$, so that S_0 and S_T are the densities of healthy hosts at the beginning and at the end of the experiment respectively.

To estimate $\bar{\beta}$ and C in Equations (d) and (e) in Box 6.2, we fit Equation (6.2) to the data in Figure 6.3. Before making use of these estimates, however, we first evaluate the usefulness of Equation (6.2), which gives a nonlinear relationship between transmission and pathogen density, for explaining the data in Figure 6.3, especially compared to the same model without heterogeneity, which is a linear function of pathogen density [for the linear model, $C \rightarrow 0$, so that on the logarithmic scale we have $-\ln(S_0/S_T) = \beta I(0)T$]. A lack-of-fit test rejects the linear model for *egt-* at $p < 0.05$ and at $p < 0.09$ for the wild-type, but does not reject Equation (6.2) for either strain ($p > 0.8$ in both cases), and the Akaike Information Criterion (AIC) in both cases chooses the nonlinear model. These statistical tests thus affirm that Equation (6.2) provides a useful explanation of the data, suggesting that there is a significant effect of host heterogeneity on each strain (see Dwyer *et al.* 1997 for further evidence for the wild-type). Above all, the two strains do not differ in heterogeneity C ($p > 0.5$, bootstrapped differences of C), but differ significantly in mean transmission rate, β ($p < 0.018$).

This statistical analysis also gives us estimates of the transmission parameters $\bar{\beta}$ and C, which we can use in Equations (d) and (e) in Box 6.2 to ask whether

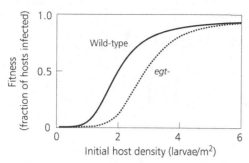

Figure 6.4 Fitnesses of wild-type and genetically-engineered *egt-* virus strains, calculated using Equations (d) and (e) in Box 6.2, for an epidemic that ends at 10 weeks due to host pupation. Parameters for wild-type: $\bar{\beta} = 1.77$ m^2/day, $C = 1.29$, $\tau = 12$ days. Parameters for *egt-*: $\bar{\beta} = 0.70$ m^2/day, $C = 1.14$, $\tau = 9$ days. $\bar{\beta}$ and C were calculated from the data in Figure 6.3.

the transmission rate of *egt-* is sufficiently reduced relative to the wild-type to outweigh its faster speed of kill [assuming for now that the breakdown rates of the two strains, α in Equation (e) in Box 6.2, are the same]. With *egt-* killing about 25% faster than the wild-type virus (Slavicek *et al.* 1999) – for example, 9 days instead of 12 – we can use the model Equations (d) and (e) in Box 6.2 to calculate the absolute fitness of each strain, for an epidemic that lasts 70 days. Figure 6.4 shows that, over the short-term at least, the wild-type virus's higher transmission rate strongly outweighs the fitness disadvantage of its slower speed of kill.

It thus appears that, for the gypsy moth virus, there may be a trade-off between virulence (speed of kill) and transmission rate (see also Cory *et al.* 1994). Moreover, this trade-off leads to a significant fitness advantage for the wild-type virus, so that the engineered strain apparently will not out-compete the wild-type. Nevertheless, we emphasize that our estimates of the relative fitnesses of the wild-type and *egt-* strains are preliminary; not only do we not yet know whether the two strains differ in the rates at which they break down in the environment, but also we have only considered monomorphic pathogen populations. In particular, the *egt-* strain is likely to have an added advantage in direct competition by virtue of killing faster, and thus reaching uninfected insects before the wild-type strain reaches them. The details of competition between the two strains within individual hosts, which are as yet unknown, may also affect their competitive balance (May and Nowak 1995).

These uncertainties mean that there is substantial room for doubt. Such uncertainties are not trivial, because if the engineered strain is able to out-compete the wild-type, then, as Figure 6.5 shows, the period of the cycles in the gypsy moth population would be reduced from about 9 years to about 7 years, the amplitude of the fluctuations would be reduced, and the mean density would be slightly higher. Substantial ecological change might therefore result.

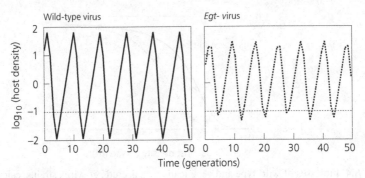

Figure 6.5 Effect on gypsy-population dynamics of wild-type and *egt-* virus strains, based on Equations (6.1), using model Equations (d) and (e) in Box 6.2 to calculate the fraction infected during the epidemic.

6.4 Discussion

In this chapter we have tried to show the usefulness of simple mathematical models for understanding the evolution of virulence. The general points that we emphasize are, first, that simple mathematical models can give important insights into the dynamics of disease in natural populations, and, second, that ecological models of disease dynamics can be useful in understanding the evolution of disease virulence. Finally, and most importantly, simple models can be used to understand the long-term and large-scale consequences of short-term, small-scale experimental measurements of pathogen fitness components. Simple mathematical models can thus be useful tools for understanding the complexities of natural populations.

For the purpose of managing the virulence of insect viruses in agriculture and forestry, our results tentatively suggest that viruses genetically engineered to have higher virulence may often have lower fitness than wild-type viruses. Before we can make this claim with confidence, however, we must develop a better understanding of how changes in virulence affect other pathogen fitness components besides transmission, such as survival in the environment. We hope to have also demonstrated that simple mathematical models can be a useful tool in these efforts.

Acknowledgments   Greg Dwyer, Jonathan Dushoff, and Simon A. Levin thank the National Science Foundation's Ecology Panel for financial support, through grant DEB-97-07610. Greg Dwyer also thanks Jeff Boettner, Jen Garrett, Gloria Witkus, and Don Wakoluk for their help with the experiment, Vince D'Amico and Raksha Malakar for helpful advice, and Juliette Langand for useful discussions. Finally, we thank Jim Slavicek of the US Forest Service, Delaware, OH, USA, for providing the *egt-* virus.

7

Contact Networks and the Evolution of Virulence

Minus van Baalen

7.1 Introduction

Virulence management can be defined as that set of policies that not only aims to minimize the short-term impact of parasites on their host population (e.g., incidence, mortality, and morbidity), but also to account for the longer-term consequences of the evolutionary responses of these parasites, for example by adopting measures that select for less virulent strains.

An important question pertaining to the scope of virulence management concerns the effect of contact structures in the host population. For successful transmission many parasites require close contact between the host they are infecting and new susceptible hosts. Consequently, the network of social contact in their host population is of paramount importance. It has already become clear that differently structured networks lead to different types of epidemiology (Keeling 1999). For example, a sparsely connected host population is more difficult to invade than a densely connected host population. But to what extent will the contact structure of their host population affect the evolution of the parasites, in particular of their virulence? Can we change the selective pressures on the parasites by modifying these contact structures? Claessen and de Roos (1995) and Rand *et al.* (1995) carried out computer simulations of evolving parasites in spatially structured host populations and concluded that less virulent (hypovirulent) parasites are favored with respect to well-mixed systems. Clearly, parasite evolution does depend on host population structure. Qualitative insight into the pertinent aspects of population structure, in the form of social networks, is still lacking, however (Wallinga *et al.* 1999).

Networks of social contacts may vary in a number of ways. First, the number of social contacts per host may vary (across the host population and in time). The relevance of whether a parasite's host interacts with a large or small number of other hosts is not immediately obvious, as explained below. Second, the overall structure of the social network may vary. Consider the contact structures depicted in Figures 7.1a and 7.1b. In both networks every host is connected to three other hosts, but in one the overall structure is laid out in a regular fashion (Figure 7.1a) whereas in the other it is completely random (Figure 7.1b). Watts and Strogatz (1998) and Keeling (1999) showed that such variations in network structure may have far-reaching consequences for, among other things, epidemiology. For example, a parasite can expand more rapidly in a random contact network than in a regular network, as suggested by the shaded nodes in Figures 7.1a and 7.1b. But

(a)   (b)

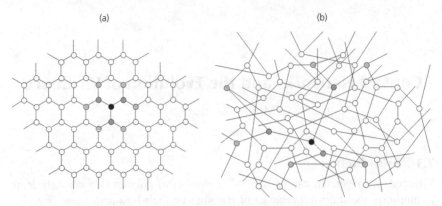

Figure 7.1 A regular network (a) and a random network (b). Both have a neighborhood size of three. In each structure, a focal host is indicated (black) with its neighbors up to two links away (dark and light gray).

how is parasite evolution determined by the number of contacts and the structure of the network? What are (if any) the evolutionary consequences of changes in a network structure?

It is not easy to find answers to these kinds of questions. In fact, a simple evolutionary analysis predicts *no* relationship between the number of contacts per host and the evolution of virulence. The reasoning is as follows. Whether a given mutant will increase (and hence invade a resident parasite population) is determined by its basic reproduction ratio R_0, that is, by the number of new infections produced by a host infected with the mutant parasite. In the simplest host–parasite models, this is given by the well-known expression

$$R_0 = \frac{\beta_{\text{mut}} S^*}{d + \alpha_{\text{mut}}},\tag{7.1}$$

where β_{mut} is the mutant's transmissibility, $S^* = S$ is the encounter rate with susceptible hosts (whose density S^* is set by the resident parasite), d the background host mortality rate, and α_{mut} the mutant's virulence (disease-induced mortality rate, Boxes 2.1 and 2.2; Anderson and May 1982; Bremermann and Pickering 1983; Lenski and May 1994; Van Baalen and Sabelis 1995a). [Note that this definition of R_0 is slightly different from the standard epidemiologic definition, in which it represents the number of secondary cases produced by a single infected individual in an entirely susceptible population. In an evolutionary setting, as in this chapter, the relevant fitness measure is the number of descendants of a *mutant* parasite introduced into a population in which a resident parasite is at equilibrium. See Box 5.1 and Mylius and Diekmann (1995) for a further discussion of the relationship between these R_0 concepts.]

A mutant parasite maximizes its fitness (i.e., its R_0) under all conditions if it strikes the optimal balance of infectivity and host longevity (as it cannot influence the density of susceptible hosts). Such optimal exploitation on a per-host basis depends only on the relationship between per-contact transmissibility (β) and disease-induced mortality (α). Since the optimum does not depend on population-level quantities, how many susceptible hosts there are, or how many of these an infected host will meet, is irrelevant (see Box 5.1 for more details). As a consequence, no change in the hosts' environment induces an evolutionary response in the parasite population. This implies that virulence management should focus on individual hosts and their current infections (for example, choosing among different medical treatments). Policies affecting the host–parasite interaction on a larger scale (vaccination, sanitary measures) may have desirable consequences on the epidemiological time scale, but may leave unchanged selection pressures on the parasites: lowering the density of infected hosts does not necessarily affect the optimal balance.

A number of observations suggest that parasite evolution *does* depend on such factors. Ewald (1993, 1994a, 1994b) discusses several examples in which the introduction of measures to hamper transmission is followed by a reduction in evolutionarily stable strategy (ESS) virulence. His explanation is that when transmission is more difficult the parasites are forced to deal more carefully with their host. For example, this explains the reduction in virulence of certain pathogenic bacteria infecting newborns in maternal wards after the introduction of measures to improve hygiene. Conversely, when transmission becomes "easier", more virulent strains have the advantage; this explains the emergence of more virulent parasites in response to the turmoil associated with war (e.g., the 1918 influenza pandemic), or to increased rates of global movement and partner change [as for the human immunodeficiency virus (HIV)].

As discussed, the standard "R_0-argument" cannot explain such evolutionary changes. However, since the argument is based on a number of simplifying assumptions, certain aspects of host–parasite relationships are not taken into account. For example, the standard argument assumes that hosts are exploited by single clones of parasites only. If this assumption is relaxed, then *within*-host competition among the parasites may drive an eco-evolutionary feedback (Van Baalen and Sabelis 1995a; Eshel 1977; Nowak and May 1994) that can explain Ewald's observations at least partially (Van Baalen and Sabelis 1995b; see Box 7.1). In this chapter, it is shown that Ewald's observations can also be explained if another assumption is relaxed, namely that of a "well-mixed" host population. The importance of this result is that contact structure becomes an essential aspect in explaining virulence.

Paraphrasing Tolstoy, it can be said that all well-mixed populations resemble one another, but that every structured population is structured in its own way. For example, host and parasite populations may be subdivided into discrete subpopulations, either because their habitat is patchy, or because the host forms different social groups that do not mix [see Anderson and May (1991) for a number of examples]. A common modeling approach for such cases assumes that subpopulations are well-mixed, while between subpopulations hosts and parasites disperse

Box 7.1 Models of virulence evolution accounting for within-host competition

Whenever multiple infections occur, a within-host conflict arises between the parasites. This conflict shifts the balance of virulence evolution toward the short-term advantage of faster host exploitation and away from host preservation. Multiple infection therefore favors increased virulence (Bremermann and Pickering 1983; Frank 1992a; Van Baalen and Sabelis 1995a).

If multiple infection is a factor determining the evolution of virulence, there will also be a feedback via epidemiology: the number of strains sharing a given host depends on the risk of infection, and this risk depends, in turn, on the strategies in the resident parasite population (Eshel 1977; Van Baalen and Sabelis 1995a; Van Baalen and Sabelis 1995b). Evolution will then depend on the small-scale interactions within hosts as well as on the large-scale interactions at the population level.

One of the earliest attempts to understand the evolutionary consequences of within-host competition is based on the assumption that more virulent parasites quickly replace less virulent clones. This process, called "superinfection," results in intermediate levels of virulence (Levin and Pimentel 1981) and increased levels of parasite polymorphism (Nowak and May 1994).

A problem with superinfection models is that the assumptions become highly artificial when applied to strains that differ very little in virulence: increasing virulence a tiny bit entails a huge fitness benefit since in these models the ancestral strain is assumed to be ousted immediately. Biologically, it is much more likely that strains that differ very little coexist within a host for a certain time. "Coinfection" models therefore make no assumptions about within-host competitive exclusion. However, unless alternative special assumptions are made, these models are more difficult to analyze, because the bookkeeping is more complex (hosts with one, two, three, etc., infections need to be tracked separately). Van Baalen and Sabelis (1995a) showed that, if the number of coinfections is limited to two, increased virulence results, but no polymorphism develops. Mosquera and Adler (1998) combined superinfection and coinfection models into a single framework. For more details on these models, see Chapters 9 and 10.

Another approach is to ignore the discrete character of infection events and focus instead on average relatedness among the parasites. This type of modeling, pioneered by Frank (1992a, 1994b, 1996c; see Box 11.1), allows analytic insight into the effects of within-host competition, but it is difficult to incorporate epidemiology into such models. Using this approach, Gandon (1998) argued that propagule survival affects the evolution of virulence through changes in average relatedness among the parasites (see also Chapter 11). Analysis of coinfection models suggests that within-host competition may include a component that favors *reduced* virulence, that is, parasites trade-in their capacity for within-host growth for an increased competitiveness (Chao *et al.* 2000).

Yet another approach ignores multiple infection altogether and focuses on the within-host diversity generated by mutations during within-host replication (Nowak *et al.* 1990; Nowak and May 1992). Such models are geared to take the immune system process into account, but are difficult to link to epidemiological models.

Notice that since within-host competition depends on multiple infection, the evolution of virulence depends on many epidemiological details that can be interfered with, thus greatly enhancing the scope for virulence management (Van Baalen and Sabelis 1995b).

Table 7.1 Events in a two-strain SIR model.

Event		Rate
Infection	$IS \rightarrow II$	β_I
	$JS \rightarrow JJ$	β_J
Recovery	$I \rightarrow R$	θ_I
	$J \rightarrow R$	θ_J
Loss of immunity	$R \rightarrow S$	ρ

"Mirror image" pair events always have the same rate.

much more slowly. Often, however, such a structure exists even when boundaries between subpopulations are less clear, or do not exist at all. Human populations, for example, tend to be highly structured, even when clear boundaries are absent. Such systems are much harder to analyze.

Populations without an imposed large-scale spatial structure, but with local dispersal, are called "viscous" (Hamilton 1964) or "mobility limited" (de Roos *et al.* 1991) populations. Such populations are much more "grainy" than well-mixed populations: instead of all members experiencing the same environment, individuals interact with their own local neighborhood, which consists of a finite number of other individuals, each of which may be infected or not.

This type of model is often studied by means of computer simulation of the so-called "probabilistic cellular automata" (PCA). In such simulations the state of a lattice of sites (a network of hosts, as is the case here) is changed by the occurrence of local events (birth, death, infection, and so on) that are governed by simple and local rules. Usually, the sites are arranged so as to form a regular, square lattice, in which every individual either interacts with its four or eight closest neighbors. This is, of course, a natural assumption if the system studied (plants, for example) inhabits a two-dimensional world. However, in many systems, in particular when interactions are determined by social relations rather than purely by geographical distance, other arrangements may be more appropriate (Keeling *et al.* 1997; Keeling 1999; Keeling 2000).

Spatial host–parasite dynamics have been studied for some time using the PCA approach (Satō *et al.* 1994; Rand *et al.* 1995; Rhodes and Anderson 1996; Jeltsch *et al.* 1997). More recently, the so-called pair dynamics (or correlation dynamics) approach has proved useful in explaining phenomena observed in these studies (Satō *et al.* 1994; Keeling *et al.* 1997; Boots and Sasaki 1999; Keeling 2000). This is a mathematical technique to analyze spatially extended systems (Matsuda *et al.* 1992; Satō *et al.* 1994; Van Baalen and Rand 1998; Van Baalen 2000; Iwasa 2000; Satō and Iwasa 2000). For example, a correlation dynamics model accurately predicted temporal patterns observed in childhood diseases in terms of contact structures (Keeling *et al.* 1997; Keeling 1999; Keeling 2000). Here, this technique is used to explore how social structure of the host population might affect the evolution of parasites, in terms of neighborhood size (the number of hosts a given host interacts with) and a parameter that describes the structure of the network.

Insight into the factors that determine selection pressures on parasites is vital for the development of virulence management strategies. If the shape of contact networks influences the evolution of virulence, then we know that virulence reflects not only small-scale (within-host) processes, but also larger-scale processes (comprising groups of hosts). Such knowledge may suggest ways to favor less virulent parasites by modifying the structure of contact networks or by changing the way parasites can spread through these.

7.2 Epidemics on Contact Networks

Hosts are assumed to form a fixed social network, in which every host is in contact with n other hosts, which here is called the host's interaction neighborhood. Any host is either susceptible S, infected with one of the two parasite strains I and J, or recovered and immune to infection by both parasite strains R. The events that change the state of the network are listed in Table 7.1. All these changes are stochastic; the associated rates are the probability per unit time for these events to occur. The model thus defines a PCA with asynchronous updating.

Assuming a fixed network means that there is no host dynamics in the model. This may be a reasonable assumption in many cases, but it renders the concept of "virulence" problematic. Usually, virulence is defined as the increase in host mortality [factor α in Equation (7.1)] or, more generally, as the reduction in host fitness. In the present model there is no host mortality, and thus no proper "virulence." For simplicity, it is assumed that there exists a trade-off between parasite virulence and clearance rate θ: the more infectious ("virulent") a parasite, the quicker it is eradicated by the host's immune system. The underlying idea is that such parasites are more detrimental to their hosts, which therefore put more effort into counteracting them. Of course, other relationships can be envisaged as well. For example, there could be a relationship between the parasite's host-exploitation strategy ("virulence") and the duration of the ensuing period of immunity. This yields a similar but more complex model (as different classes of immune hosts need to be tracked). A full analysis of the evolution of virulence requires that host dynamics be taken into account as well, but this is beyond the scope of this chapter.

Note that in Table 7.1 transmission events are characterized in terms of the per-contact transmission rate $\tilde{\beta}$. If this is fixed, the total transmission rate will be proportional to the number of contacts per host (i.e., to neighborhood size). Here, however, we are more interested in the consequences of the structure of the contact network than in the consequences of the absolute number of contacts. Therefore, it is assumed that per-contact transmission rates are inversely proportional to neighborhood size n,

$$\tilde{\beta}_x = \frac{\beta_x}{n} \, , \tag{7.2}$$

with $x = I$ or J. Consequently, the total infectivity β_x of an infected host is constant, but spread out over more hosts if neighborhood size increases (i.e., the per-contact transmission efficiency $\tilde{\beta}_x$ decreases, but this is counterbalanced by the larger number of contacts).

At the lowest level the model is thus defined exclusively in terms of local and discrete events. The question now is how the model behaves at a larger scale, that is, how the numbers of host in the various states change over time in the network.

7.3 Mean-field Dynamics

Before analyzing the viscous system, it is insightful to consider the equivalent nonstructured ("mean-field") model. This tells us what to expect in the standard case of no social structure, that is, when every host can potentially infect every other. The mean-field model is therefore obtained by letting the neighborhood size n go to infinity: every infected host can potentially infect every susceptible host. This yields

$$[S]' = -\beta_I[S][I] - \beta_J[S][J] + \rho[R] , \tag{7.3a}$$

$$[I]' = \beta_I[S][I] - \theta_I[I] , \tag{7.3b}$$

$$[J]' = \beta_J[S][J] - \theta_J[J] , \tag{7.3c}$$

$$[R]' = \theta_I[I] + \theta_J[J] - \rho[R] , \tag{7.3d}$$

where $[S]$, $[I]$, $[J]$, and $[R]$ are the densities of susceptible, infected, and recovered hosts, respectively. The total density of hosts does not change, and is scaled to one. This set of equations is a very basic model that has been studied extensively (Anderson and May 1991; see Boxes 2.1, 2.2, and 9.1). The main difference to the usual formulation is that host mortality is not included; normally, it is assumed that when a host dies, it is replaced instantaneously by a susceptible host, but this presupposes extremely tight control of the host population. In the present model, there is essentially no host population dynamics: from the viewpoint of the parasites, the host population (and its social structure) is "frozen" in time.

If one of the parasite strains (say strain J) is rare, the other strain (strain I) settles at a stable equilibrium ($[S]^*$, $[I]^*$, $[R]^*$). Parasite strain J is able to invade this equilibrium if $[J]'$ is positive when $[J]$ is small, which is the case if

$$\beta_J[S]^* - \theta_J > 0 . \tag{7.4}$$

The invasion condition can also be expressed in terms of the mutant's basic reproduction ratio $R_0(J)$,

$$R_0(J) = \frac{\beta_J[S]^*}{\theta_J} > 1 , \tag{7.5}$$

which gives the number of secondary infections caused by a host infected with the mutant J in a population infected by the resident parasite I. Since in a well-mixed population the mutant does not influence the density of susceptible hosts it

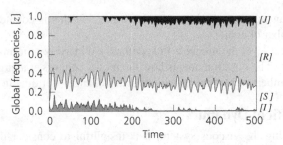

Figure 7.2 Simulations of the dynamics across a network. The hypovirulent strain J (black area) of parasites can invade and replace the strain adapted for well-mixed host populations I (dark gray shade; white, susceptible hosts; light gray shade, immune hosts). A total of 3 600 hosts are arranged in a triangular lattice (with periodic boundary conditions), in which every host is connected to its six nearest neighbors (i.e., $n = 6$). Parameters: $\beta_I = 30$, $\theta_I = 1$, $\beta_J = 25$, $\theta_J = 0.9$, $\rho = 0.1$. In addition to the events listed in Table 7.1, a small mutation rate is included: a fraction (0.001) of the infection events produces a host infected with the other type. The simulation was started with only parasite I present, close to the equilibrium for well-mixed populations.

encounters ($[S]^*$ is set by the resident), its evolutionary success is entirely determined by its "per-host transmission factor" β_J/θ_J, that is, by infectivity times the mean duration of the infective period.

Since equilibrium conditions imply

$$[S]^* = \frac{\theta_I}{\beta_I} , \qquad\qquad (7.6)$$

it can be concluded that the strategy with the largest per-host transmission factor β/θ is the ESS. Note that this maximum is independent of the density of susceptible hosts (the role of the per-host transmission factor is explained in more detail in Box 5.1; see also Bremermann and Pickering 1983; Lenski and May 1994; Van Baalen and Sabelis 1995a).

7.4   Across-network Dynamics

In a well-mixed population, all that matters to a parasite is to maximize the per-host transmission factor by striking the optimum balance of intensity and duration of infectiousness. In a structured host population, however, the situation is different. A first indication is provided by the simulation presented in Figure 7.2, which shows a parasite with a per-host transmission rate almost 8% lower than the strain that maximizes per-host transmission (the ESS in well-mixed populations can invade and replace the latter). Hence, compared to well-mixed populations, spreading through contact networks favors reduced virulence. Our task is now to determine why this is the case and to determine ESS levels of virulence in contact networks.

The change in selection pressure on virulence turns out to be tightly coupled to the distribution of susceptible, infected, and immune hosts across the contact

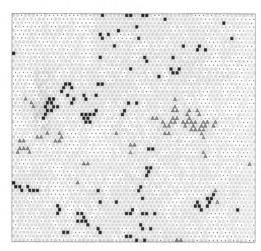

Figure 7.3 Clusters of the hypovirulent strain J (squares) competing with the strain adapted for well-mixed populations I (triangles) on a triangular lattice ($n = 6$). Susceptible hosts are represented by small points, immune hosts by gray circles. The snapshot is taken at $t = 200$ from the simulation presented in Figure 7.2.

network. Figure 7.3 is a snapshot of the network when the mutant of Figure 7.2 is invading. As can be seen, the parasites' distributions are far from homogeneous. Though there are no clear boundaries separating patches, patches of infected hosts tend to be surrounded by regions of immune hosts. This, of course, blocks the parasites' transmission into regions with many susceptible hosts. The distribution is highly dynamic. Patches of immune hosts lose their immunity and at some point in time a parasite breaks through and infects the hosts. The peaks in the time series (Figure 7.2) represent such episodes of parasites bursting into patches of susceptible hosts.

Since it is able to invade, mutant J is somehow better adapted to spread through the network. The snapshot presented in Figure 7.3 contains a clue about what may be happening. If we calculate the global densities in the network as well as the local densities that surround the two strains of parasites (results are shown in Figure 7.4), we observe that a host infected with a hypovirulent parasite J has on average more susceptible hosts in its immediate neighborhood than does the more virulent strain I (about twice as many). Reducing virulence therefore seems to allow the hypovirulent strain to exploit the fact that immune hosts lose their immunity after a while. Strain I cannot easily profit from this because the strains tend to be segregated across the network. Figure 7.4 shows that parasites of strain J have fewer parasites of strain I in their neighborhood and vice versa.

It can be shown (Matsuda *et al.* 1992; Van Baalen and Rand 1998; Van Baalen 2000; Dieckmann and Law 2000) that local densities equilibrate faster than global densities. In particular, a mutant parasite strain that is globally rare experiences a characteristic environment that includes related mutants, because an invading mutant tends to form clusters if infection is local. The characteristics of such

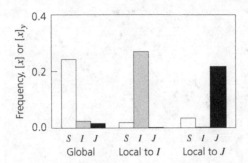

Figure 7.4 Global and local densities of susceptible and infected hosts and of parasites. The local densities experienced by parasites I and J correspond to the distribution shown in Figure 7.3.

clusters, viewed as more or less coherent units, determine the invasion success (Van Baalen and Rand 1998).

Thus, to understand the epidemiology and evolution in contact networks, we have to account for the heterogeneous distributions of the parasites. This is particularly important when considering the fate of (globally) rare mutants. One way to study evolutionary outcomes is to run simulations in which many strains of parasites compete over a range of aspects like neighborhood size and the geometrical structure of the contact network. An initial disadvantage of such an approach is that it is very computationally intensive (the single simulation shown in Figure 7.2 took several hours on a desktop computer). More importantly, even though the network is fairly large (3 600 hosts), the resultant dynamics are characterized by much demographic stochasticity. In particular, this is a major drawback if the aim is to study evolution, as numerous invasions have to be "tried" before one can decide that a certain mutant is likely to invade (Claessen and de Roos 1995). And even after numerous simulations, it may still be very difficult to gain insight into exactly which aspects of the interaction are important, as the simulations have to be repeated for many different combinations of parameters. It is here that the correlation dynamics approach can lead to greater insight.

7.5 Pair Dynamics

The differential Equations (7.3a) to (7.3d) keep track of the densities of susceptible, infected, and immune hosts. Such densities are nothing more than the probabilities that a randomly picked host is in a given state. In a similar fashion, we can define the densities of *pairs* (or "doublets") of neighboring hosts: these represent the probability that a pair of connected hosts is in a given combination of states (for example, one susceptible and the other infected). As for the "singlets," differential equations can be derived for the changes in the densities of doublets. This gives rise to an increased number of differential equations (one for every combination of states). These differential equations are rather complex because they keep track of all possible transitions that create and destroy pairs. For comparison: the only singlet "events" are $S \to I$, $I \to R$, and $R \to S$; the set of pair events is much larger: $SS \to IS$, $IS \to RS$, $SS \to RS$, $SR \to IR$, etc. The full set of equations taking into account all pair events is given in Box 7.2.

The major advantage of knowing pair densities is that local densities (the quantities shown in Figure 7.4) can be calculated directly. A local density of x experienced by y is simply the conditional probability that a given neighbor of a site in state y will be in state x, and is given by

$$[x]_y = \frac{[xy]}{[y]} \, , \tag{7.7}$$

where $[xy]$ is the density of xy pairs of hosts ($x, y = S, I, R$) and $[y]$ is the density of y-state hosts.

From the pair equations (given in Box 7.2) it follows that the global dynamics of both parasite strains are given by

$$[I]' = (\beta_I [S]_I - \theta_I)[I] \, , \tag{7.8a}$$

$$[J]' = (\beta_J [S]_J - \theta_J)[J] \, . \tag{7.8b}$$

Rewriting in terms of reproduction ratios, the condition for invasion of strain J is

$$R_0(J) = \frac{\beta_J [S]_J}{\theta_J} > 1 \, . \tag{7.9}$$

The important aspect is that the expected rate of increase of parasite strain J does not depend on the global density of susceptible hosts $[S]$, but on their local density $[S]_J$. A hypovirulent strain can then invade if it can offset its less efficient host-use (indicated by its reduced per-host transmission ratio β_J / θ_J) by surrounding itself with a higher density of susceptible hosts $[S]_J$. Hypovirulent parasites can then be said to exploit their local host population more prudently. Such prudent exploitation cannot evolve in well-mixed systems, in which both strains exploit the same global population of susceptible hosts. In viscous systems, however, both strains are segregated to a certain extent (see Figures 7.3 and 7.4), which makes it difficult for more virulent strains to profit from the increased density of susceptible hosts that surrounds the hypovirulent parasites.

The local density of susceptible hosts rises if the parasites shorten the immune period (at a cost of reduced infectivity). The parasites "wait," as it were, for immune hosts to become available again. If they are too "impatient" they surround themselves quickly by immune hosts, and so their spread is blocked. Their problem is that increasing the density of susceptible hosts may be exploited by competing parasite strains; they must not give away too much if there are too many of these in their neighborhood. It is therefore the clustering of the parasites in the network that favors the reduced virulence. In this system, reduced virulence is essentially an altruistic trait, disadvantageous for the parasite itself but of benefit to the parasites in their environment. In viscous populations these neighboring parasites tend to be related; hence the evolution of reduced virulence is an example of kin selection (Hamilton 1964; Maynard Smith 1964). Any parasite unit (a

Box 7.2 Pair approximation for incompletely mixed host populations

Since the pioneering work of Kermack and McKendrick (1927) the standard framework for epidemiological models is based on the assumption that the host population is well-mixed. That is, every host is equally likely to meet (and transmit any infection to) every other host in the population. This is obviously not true in many cases, and mathematical techniques have been developed to deal with the epidemiological consequences of spatial and/or social structures. One of these techniques is the correlation dynamics or pair approximation approach.

This method is based on three ingredients:

- The network that represents space or social structure (such as shown in Figure 7.1);
- The set of states that any site may be in (here, the susceptible state S, the infected states I and J, and the recovered state R);
- The set of rules for how sites may change state (listed in Table 7.1).

The simplest correlation dynamics equations keep track of the states of neighboring pairs of hosts on the lattice (Matsuda *et al.* 1992). If $[xy]$ denotes the proportion of pairs in states x and y, either the x or the y site can change through one of the events listed in Table 7.1. Notice that both individuals in such a pair also form pairs with other individuals for which events may occur. For the model summarized in Table 7.1, the differential equations for the resident SIR system therefore have to take into account all the transitions shown below.

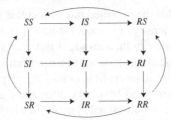

In deriving the differential equations for the changes in densities of pairs, use is made of the fact that the densities in the network of symmetrical pairs (for example, the densities of SI and IS pairs) are identical.

The resultant differential equations are:

$$[SS]' = 2\rho[RS] - 2\beta_I(1 - n^{-1})[I]_{SS}[SS],$$

$$[SI]' = \beta_I(1 - n^{-1})[I]_{SS}[SS] - \{\beta_I(n^{-1} + (1 - n^{-1})[I]_{SI}) + \nu_I\}[SI]$$
$$+ \rho[RI],$$

$$[SR]' = \theta_I[SI] - \{\beta_I(1 - n^{-1})[I]_{SR} + \rho\}[SR] + \rho[RR],$$

$$[II]' = 2\beta_I(n^{-1} + (1 - n^{-1})[I]_{SI})[SI] - 2\theta_I[II],$$

$$[IR]' = \beta_I(1 - n^{-1})[I]_{SR}[SR] - [\rho + \theta_I][IR] + \theta_I[II],$$

$$[RR]' = 2\theta_I[IR] - 2\rho[RR].$$

continued

Box 7.2 *continued*

$[x]_{yz}$ denotes the proportion of sites neighboring yz pairs that are in state x, i.e., $[x]_{yz} = [xyz]/[yz]$. The equations involving J are analogous to those involving I. Van Baalen (2000) explains in detail how these differential equations are derived from the set of transitions. Keeling *et al.* (1997) have analyzed a more complex version of this model.

Since $[x] = [xS] + [xI] + [xJ] + [xR]$, the dynamics of singlets follows from that of pairs,

$$[x]' = [xS]' + [xI]' + [xJ]' + [xR]'.$$

This results in Equations (7.8), but only if every host has the same number of contacts. If the number of contacts varies, differential equations that describe singlet dynamics have to be derived separately (Morris 1997; Van Baalen 2000).

The set of equations that describe pair dynamics is exact but it is not yet closed. The problem is that some of the equations depend on quantities of the type $[x]_{yz}$, which depend on the densities of xyz-triplets in the network. Van Baalen (2000) showed that the local density $[x]_{yz}$ can be approximated by

$$[x]_{yz} = [x]_y \{(1 - e) + eC_{xz}\}\tau_{xyz},$$

where e is the proportion of triplets that are in a closed triangular configuration and $C_{xz} = [xz]/([x][z])$ is the correlation between neighboring sites in state x and z (see also Keeling 1999). The remaining problem is to find correction factors τ_{xyz} that preserve the consistency of the system. One cannot simply assume that all $\tau_{xyz} = 1$ because then the $[x]_{yz}$ will not, as they must, add up to 1 when summed over all x. There are several alternative choices for the τ_{xyz} that preserve consistency but it is as yet unknown which one leads to the best approximation. The assumption adopted here is

$$\tau_{xyz} = 1 \quad \text{if} \quad x \neq z,$$

and τ_{xyz} is chosen such that

$$[z]_{yz} = 1 - \sum_{x \neq z} [x]_{yz}.$$

This is the simplest choice and has been demonstrated to work quite well in other cases (see Van Baalen 2000).

clone infecting a given host) does best by optimizing its per-host transmission ratio. Reducing virulence therefore does not benefit the individual itself, but rather the cluster of related individuals to which it belongs (Van Baalen and Rand 1998).

7.6 Implications of Network Structure

Let us now start to vary the structure of the contact network. Consider first the consequences of a finite neighborhood *per se*. Figure 7.5a shows that the hypovirulent parasite can only invade and replace strain I if the neighborhood size is fairly

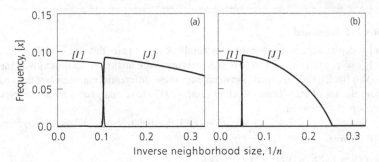

Figure 7.5 Equilibrium densities of strain I (adapted to well-mixed populations) and hypovirulent strain J for different neighborhood sizes n in (a) random and (b) regular ($e = 2/5$) contact networks.

small (i.e., if $1/n$ is larger than a threshold value). This is no surprise because if n becomes large the system approaches a well-mixed system to which strain I is adapted (as it maximizes the per-host transmission factor).

Working out the effects over all the network structure is more complicated. For this we need a parameter that describes whether the network is regular (as in Figure 7.1a), random (as in Figure 7.1b), or in between. Such a parameter emerges from a close consideration of the assumptions that underlie the pair equations (Van Baalen 2000).

The differential equations for singlets depend on the densities of pairs (for example, $[I]'$ depends on the density of SI pairs). In a similar fashion, the differential equations for the pairs may depend on triplet densities. More precisely, the differential equation for a given pair may depend on local densities of the form

$$[x]_{yz} = \frac{[xyz]}{[yz]} . \tag{7.10}$$

In effect, it is necessary to approximate $[x]_{yz}$ in terms of pairs. In technical terms, a "closure assumption" must be made. The simplest possible approximation is the "pair approximation" given by

$$[x]_{yz} \approx [x]_y , \tag{7.11}$$

which assumes that the probability of finding a neighbor of y in state x is independent of the fact that one of y's other neighbors is in state z (Matsuda *et al.* 1992).

Pair approximation is only one of the many possible ways to "close" the set of differential equations (see Van Baalen 2000; Dieckmann and Law 2000; Bolker *et al.* 2000), and consideration of the network structure becomes important in choosing a way to "close" the equations. The pair approximation can be shown to represent best random contact networks, such as depicted in Figure 7.1b. In more regular networks, however, a pair of neighbors share part of their neighborhood. For example, in the triangular lattice depicted in Figure 7.3, a pair always has two common neighbors.

For such triangular lattices we can use the fact that $2/5$ of all randomly picked triplets will be triangular (i.e., with a connection between the far ends). In Box 7.2 an expression is given for the conditional probability $[x]_{yz}$ that takes into account the correlation between x and y's z-neighbor based on the assumption that a proportion e of triplets are triangles.

This approximation allows the consequences of different degrees of network regularity to be assessed: $e = 0$ for random lattices, $e = 2/5$ for triangular lattices, and semi-regular networks have intermediate e values. In more everyday terms, e is a measure of the likelihood that a relation of one neighbor is also a relation of the other: if e is large, the friends of my friend are probably also my friends, whereas if e is zero, my friend and I have no common acquaintances. By varying e we can investigate some of the consequences of changes in overall social structure while keeping the number of contacts per host constant.

Figure 7.5b shows that for regular networks (with a fixed proportion $e = 2/5$ of triangles) the hypovirulent strain J can invade and replace strain I at a larger neighborhood size than for random networks. This suggests that regularity favors reduced virulence. Conversely, disruption of a social network from regular to random benefits the more virulent parasites.

Note that for the regular network, even the hypovirulent strain cannot maintain itself at low neighborhood sizes (approximately $n > 4$; see also Satō *et al.* 1994). It may be possible that strains with even lower virulence are able to maintain themselves. The information gained by letting only two strains compete is thus limited. More strains must be considered, but this poses some problems. Which strains are possible? Which of these will natural selection weed out? Does the parasite population become monomorphic? How will the ensemble of strains respond to changes in their host population? These and related question are the domain of adaptive dynamics models (Dieckmann and Law 1996; Metz *et al.* 1996a; Geritz *et al.* 1997). The focus here is on potential evolutionary endpoints only, at which the resident population comprises a single strain of parasites.

7.7 Evolutionary Stability

When considering a multitude of parasite strains, we have to specify a constraint that links all possible transmissibilities β to a recovery rate θ. Ideally, this constraint should be derived from a submodel of how the parasites interact with their host's immune system, but this is quite an ambitious undertaking. Here, we analyze the example

$$\theta(\beta) = \theta_0 + c\beta^2 , \tag{7.12}$$

which assumes that the recovery rate increases more than linearly with the parasite's infectivity. The idea behind the monotonic shape of this relationship is again that more transmissible parasites are likely to be more detrimental to their hosts, which will consequently put more effort into combatting them. Notice that we have to assume that the relationship between θ and β is nonlinear with a coefficient c, for otherwise there is no intermediate level of virulence that optimizes per-host exploitation (see the figure in Box 5.1). Equation (7.12) then is the simplest choice.

Maximization of the per-host transmission factor $\beta/\theta(\beta)$ gives the ESS virulence β^* in a well-mixed system,

$$\beta^* = \sqrt{\theta_0/c} \,. \tag{7.13}$$

The question now is how the ESS changes when the host population is not well-mixed, but socially structured. Whether or not a mutant J will invade a given resident population is, in general, determined by its invasion fitness, denoted by f_J and defined as its expected rate of increase when globally rare (Metz *et al.* 1992; Rand *et al.* 1994). For the socially structured SIR model, f_J cannot be calculated analytically, but for the present purpose a numerical analysis is sufficient.

The dynamics of the resident parasite I can be simply established by numerically integrating the differential equations for $[SS]$, $[SI]$, $[SR]$, $[IR]$, and $[RR]$. In a well-mixed model, the system always results in a stable endemic equilibrium; the same is true for the network-structured model (in fact, the analysis would be much more difficult if the resident parasite gave rise to cycles or chaotic dynamics). From the values of the pair densities we can infer the mean density of a resident parasite and the degree of clustering that it causes.

Subsequently, we can numerically calculate the selection pressure from the dynamics of a globally rare mutant parasite strain J with a strategy close to that of the resident (i.e., $\beta_J = \beta_I + \Delta\beta$, with $\Delta\beta \ll \beta_I$). The mutant dynamics are described by four more differential equations (we need differential equations for $[SJ]$, $[JJ]$, $[RJ]$, and $[IJ]$, which are derived as outlined in Box 7.2). When the mutant is rare, its effect on the resident system can be ignored and the pair densities $[SS]$, $[SI]$, $[SR]$, $[IR]$, and $[RR]$ remain at their equilibrium values. From the mutant dynamics, the mutant's invasion fitness can be calculated and from this, in turn, the selection pressure, which is proportional to $(f_J - f_I)/\Delta\beta$. By continually adjusting the resident strategy β_I in the direction indicated by the selection pressure until the selection pressure becomes zero, the evolutionary endpoint is eventually found numerically. (The method ensures that the point thus found is convergence stable. As only one mutant is tested at a time, it does not identify branching points, at which the parasite strains diverge because of disruption selection; see Metz *et al.* 1996a; Geritz *et al.* 1997.) Notice, however, that for this particular model such divergence is unlikely; the construction of the model means that the point found by the procedure used here will always be an ESS, that is, correspond to a fitness maximum.

Figure 7.6 shows the results of an extensive parameter survey relating ESS infectiousness β^* to the network parameters. It confirms that, in general, a reduction of the hosts' neighborhood size is followed by a reduction in ESS virulence. The same is true if the regularity of the network (the proportion e of triangular connections) increases. Note, however, that Figure 7.6 predicts that if e is large (many triangles) *and* n is low (few connections), ESS virulence may *increase* again; however, these are very extreme cases and it may actually be impossible to construct

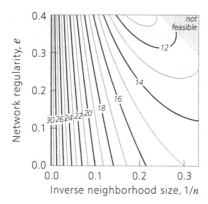

Figure 7.6 ESS transmissibility β^* [under the constraint $\theta(\beta) = \theta_0 + c\beta^2$], as a function of neighborhood size n and network regularity e. Parameters: $\theta_0 = 1/2$, $c = 1/1800$ (this parameter combination implies that $\beta^* = 30$ in a well-mixed host population), and $\rho = 0.1$. For combinations of small n and high e_{xyz} (gray area) the parasites always become extinct and hence an ESS is not feasible.

networks with such a combination of parameters. In most of the parameter region the pattern is remarkably clear: departures from well-mixedness favor reduced virulence.

7.8 Discussion

In socially structured host populations, less virulent parasites are favored compared to well-mixed host populations. To understand why one must focus on how a cluster of related parasites can expand through the social network. Clusters of virulent parasites tend to overexploit their local host population, which blocks their spread, whereas clusters of hypovirulent parasites exploit their hosts more prudently and can more easily expand. The effect is quite sensitive to the structure of contacts in the host population. The greater the neighborhood size of the hosts (the number of hosts a given host interacts with), the more the system approaches the dynamics of a well-mixed system, which benefits the more virulent strains. Figure 7.6 suggests that the same holds true when the network becomes more irregular.

Thus relatedness among parasites, whether it occurs *within* hosts (see Box 7.1) or *between* hosts, tends to favor reduced virulence. Anything that disrupts the pattern of relatedness among the parasites will favor increased virulence. It is shown here that increasing contact number or network randomness favors increased virulence; Boots and Sasaki (1999) showed that the same holds when infection is increasingly long-range as opposed to local. The same result can be expected for increases in host mobility, background host demographic rates, partner change (for sexually transmitted diseases), etc., since all these processes tend to disrupt patterns of relatedness.

An important aspect that remains to be studied is the relationship between the hosts' (physical) density and its contact structure. Neighborhood sizes may increase if host density increases, but also they might not. For example, in human populations one might suppose that the number of (sufficiently intense) social contacts would be almost independent of population density. Edmunds *et al.* (1997) recently carried out a survey to assess the number of contacts that are (presumably) suitable for the transmission of respiratory diseases. They found that the number of such contacts was remarkably constant across period of the week and age; subjects talked on average to about 30 persons per day (with the exception of Sundays). It would be very interesting to carry out similar surveys in different areas with different population densities, while also trying to assess network structure. The expectation is that the number of contacts will appear to be roughly the same, but that different densities will lead to different degrees of structure in the contact network. In rural communities an individual's contacts are also very likely to know each other, which is not necessarily the case in high-density areas (i.e., cities). If increasing (physical) density thus renders the contact structure less regular, we can expect parasites to evolve an increased virulence in response.

The model studied in this chapter is not very satisfactory as a model for the evolution of HIV: the assumption that every host has exactly *n* concurrent sexual relationships is rather extreme. (Contact network models for the spread of HIV are more complicated than those studied here and include processes like partnership formation and breakup, etc.; see Kretzschmar 1996; Kretzschmar and Morris 1996; Morris and Kretzschmar 1997.) Nonetheless, assuming that HIV infecting social networks exhibits similar relatedness patterns can give some insight into how HIV is likely to respond to changes in population structure. In particular, it is predicted that more virulent strains of HIV will emerge when the contact network becomes less regular. That is, evolutionary responses will ensue even if individual behavior does not change (the number of sexual partners remains constant and so forth).

The analysis, as reported in this chapter, confirms Ewald's hypothesis to a certain extent. That is, more "sparse" and more regular connection networks favor less virulent parasites. However, it should be realized that kin selection is at the heart of this phenomenon. The explanation is not that reduced virulence results because "the" parasite has to be more careful with "its" host, but rather that it allows a cluster of related parasites to exploit the local supply of hosts more efficiently. Changes in the pattern of relatedness thus entail an evolutionary response that results in a change in virulence.

Insight into the evolutionary consequences of contact structure is necessary to develop adequate "virulence management" strategies. In the first place, we need to know when (and how) social structure must be taken into account. Furthermore, such an insight might suggest opportunities to change the selection pressure acting on parasites. Of course, possible modifications to the contact network structure are limited and, in human populations, often plainly unethical. However, in some cases we are able to influence some social structures, such as classrooms in schools etc. (Keeling *et al.* 1997).

Changing contact patterns may lie behind the phenomenon of many "emerging diseases" (Morse 1993). For example, McNeill (1976) hypothesized that syphilis emerged in the Middle Ages when the parasite that causes a leprosy-like disease called yaws changed its transmission strategy in response to altered patterns of social interaction. This particular hypothesis is, of course, difficult to test; nonetheless it underscores that social behavior may affect parasite evolution and hence should be part and parcel of virulence management.

Acknowledgments   John Edmunds, Matthew Keeling, Kees Nagelkerke, and two anonymous referees are thanked for their valuable comments on a previous draft of this chapter.

8

Virulence on the Edge: A Source–Sink Perspective

Robert D. Holt and Michael E. Hochberg

8.1 Introduction

A recognition of spatial processes can be found even in the earliest glimmerings of intellectual understanding of the parasitic origin of infectious disease. As described in Ewald (1994a, p. 184), the Renaissance thinker Girolamo Fracastoro hypothesized that disease-specific germs could multiply within a person's body and be transmitted either directly over short distances, or over long distances (e.g., via contaminated objects). In recent years, a number of authors have emphasized how many epidemiological phenomena cannot be understood without explicitly considering infectious processes in a spatial context (e.g., Holmes 1997). There are several general issues that arise automatically when spatial aspects of the dynamics of infectious disease are considered. For instance, if infections are localized, spatial separation increases the degrees of freedom of a host–parasite system, permitting a rich array of dynamical behaviors to arise even in a spatially homogeneous world (e.g., Hassell et al. 1994). Moreover, spatial heterogeneity is the norm rather than the exception in ecological systems (Williamson 1981). Dispersal often couples habitats that differ strongly in local population parameters (e.g., carrying capacity), or involve anisotropic spatial flows. This leads to the potential for asymmetries among habitats in the degree of the impact of spatial coupling on local ecological and evolutionary dynamics.

In population ecology, an example of such asymmetries that has received considerable attention in recent years is "source–sink" dynamics. In these, in some habitats ("sinks") a species may persist despite a demographic deficit (with local births less than local deaths), because of immigration from "source" habitats (Box 8.1; see also Holt 1985; Pulliam 1996; Dias 1996). Sink populations may readily arise at the edges of species' ranges, or where habitats that differ greatly in productivity are juxtaposed. Given that genetic variation is present, natural selection might be expected to improve the ability of a species to utilize the sink habitat. Other chapters in this volume point out the potential for a rapid evolution of virulence in host–pathogen systems. However, recent theoretical studies suggest that there can be substantial constraints on adaptive evolution to conditions in sink habitats, leading to a kind of evolutionary conservatism in spatially heterogeneous environments (Bradshaw 1991; Brown and Pavolvic 1992; Kawecki 1995; Holt 1996; Holt and Gomulkiewicz 1997; Kirkpatrick and Barton 1997; Gomulkiewicz et al. 2000). Management practices that tend to foster such conservatism (for the

Box 8.1 Source and sink habitats in population biology

All naturalists know that species tend to be variable in abundance through space, being common in some places, rarer in others, and totally absent in yet others. Spatial variation in abundance can arise in part from chance, but most often reflects real spatial variation in habitat quality (including the abundance of other species such as competitors and predators). Such variation persists over time and can be quantified by ecologists (for example, variation in soil nutrient supply can underlie variation in plant seed production). Movement of individuals reshuffles abundances among different habitats and can obscure the influence of local demographic rates (births and deaths) on local abundances. This is particularly the case when there are sources and sinks, which have been the focus of much recent attention in population biology (Pulliam 1988; Holt 1993; Dias 1996).

A "sink" in common parlance is a "place where things are swallowed up or lost" (*Oxford English Dictionary*). As the coin of biological success is to leave successful offspring in future generations, a sink habitat is one in which residents on average do not quite replace themselves, because local deaths exceed local births. What is "lost" in a sink is the ability of individuals to have descendants into the indefinite future in that local environment. If a population is to persist at equilibrium in a sink, local losses must therefore be replenished by immigration from elsewhere and, in particular, from source habitats, where local births exceed local deaths.

Two general mechanisms can readily lead to a source–sink structure in population dynamics: passive dispersal or diffusion in heterogeneous landscapes (Holt 1985), and interference competition (e.g., territoriality) in high-quality habitats (Pulliam 1988). In general, movement that has a random component or is positively density-dependent (i.e., greater movement rates at higher densities) tends to move individuals down abundance gradients, increasing population size in low-quality or marginal habitats. In some situations ("true sinks"), births do not match deaths at any density, and therefore extinction is inevitable in the absence of immigration. A convincing example of a true habitat sink is provided by Keddy (1981), who found that interior dune populations of the seaside annual plant *Cakile edentula* would have become extinct in the absence of the wind-deposited seeds produced on the seaward edges of the dunes; at all densities, local seed production did not permit replacement of annual losses. In other situations, populations can persist without immigration, but only at a low carrying capacity; immigration from habitats with higher carrying capacity tends to push population size above these low numbers, and because of density-dependence local deaths then exceed local births at the higher equilibrium abundance induced by immigration. Watkinson and Sutherland (1995) refer to such habitats as "pseudosinks," because immigration is not absolutely required for population persistence. Thomas *et al.* (1996) and Boughton (1999) describe a complex spatial system for Edith's checkerspot butterfly, *Euphydryas editha*, in the Sierra Nevada of California, USA, including pseudosinks. In the pseudosinks, sufficient host plants are present to permit population persistence, but immigration from source populations inflates

continued

Box 8.1 *continued*

the local abundance of butterflies above carrying capacity. This leads to intense competition for host plants, such that individuals do not tend to replace themselves. (In addition to the pseudosinks, the system also contains true sinks, in which host plants are too rare or ephemeral to support a butterfly population in the absence of immigration.)

Ascertaining whether or not a given habitat is a sink is also important in evolutionary analyses as exemplified by the models discussed in the main text. More general theoretical studies (e.g., Gomulkiewicz *et al.* 1999, Holt 1996, Holt and Gaines 1992, Holt and Gomulkiewicz 1997, Kawecki 1995) further highlight how demographic constraints can hamper or even prevent natural selection from improving adaptation to sink environments. This is a phenomenon of general importance in evolutionary biology, for instance in understanding evolutionary dynamics at the edges of species' ranges, or understanding switches between host species (which can be viewed as distinct "habitats") by herbivores or pathogens.

---

pathogen) or weaken it (for the host) may help mitigate the long-term potential for highly virulent infectious diseases to evolve.

There is a rich and growing theoretical and empirical literature on the evolution of virulence (e.g., Anderson and May 1982; Bull 1994; Lenski and May 1994; Frank 1996c; Lipsitch *et al.* 1995a; Van Baalen and Sabelis 1995a), which focuses largely on how virulence reflects the balance of selective forces operating at different levels (within-host competition, and between-host transmission; e.g., Mosquera and Adler 1998; Koella and Doebeli 1999). The study of the interplay between gene flow and selection as determinants of local adaptation is, of course, a classic problem in evolutionary genetics (e.g., Antonovics 1976; Endler 1977; Nagylaki 1979). Yet few studies focus explicitly on the potential implications of source–sink dynamics for our understanding of the evolution of virulence and resistance (for an analysis of comparable issues in predator–prey coevolution along a gradient, see Hochberg and Van Baalen 1998). In a recent review Kaltz and Shykoff (1998) suggest that local adaptation by parasites to their hosts is often not observed, and they suggest this might arise from asymmetric gene flow in heterogeneous environments. This suggestion has particular force in systems with sources and sinks, which automatically contain asymmetric flows of individuals among habitats.

The organization of this chapter is as follows. First, as the motivation for theoretical studies we sketch several hypothetical examples that illustrate how, in principle, host–pathogen interactions of practical interest in human, animal, and plant epidemiology could match qualitatively the ecological assumptions of asymmetrical spatial flows that generate source–sink systems. We then present several models of evolution in which a source habitat is linked to a sink habitat, either for a host or pathogen, and discuss the initial stages of adaptation for a host–pathogen interaction in a sink habitat. The models are deliberately quite simple, but their qualitative conclusions illuminate a much broader range of source–sink systems. Finally, we point out some potential conclusions of our results for applied evolutionary epidemiology.

8.2   Sources and Sinks: Pervasive in Host–Pathogen Systems?

Source–sink dynamics may be common in many important applied epidemiological situations. The three situations described next are hypothetical, but we believe quite plausible.

1. An organic farmer is attempting to grow corn (an annual plant) in an environmentally responsible manner, and so uses no pesticides or fungicides. The crop is generated from retention of some seeds from the previous year's production, supplemented by purchases from a commercial seed company. A fungal blight is present in the field and is reducing crop yield. Ideally, the farmer would like to develop a local strain that could be resistant to the blight. This goal implicitly involves the evolution of resistance in the host to a resident blight; the pathogen could either be a specialist on corn and so dynamically responsive to the corn crop itself, or a generalist that inflicts many local species, and so less tightly coupled to the corn. The farmer would like to know how many seeds she should retain, so as to balance the long-term goal of fostering local adaptation by her corn population to the blight, against the shorter-term economic goal of maximizing seed yield. What should a population biologist tell her?

2. A group of ranchers husband cattle on ranches, where the cattle usually range at low densities. The livestock carry a pathogen, which is usually benign as measured by its effect on mortality and morbidity. However, in recent years the practice has arisen to ship the cattle from different ranches to a common feedlot, to be fattened before being sent off for slaughter. Should these ranchers be concerned at all about the emergence of a more serious disease from the historically benign pathogenic infection, arising because of the admixture of different herds?

3. Doctors managing a large nursery ward are concerned with the potential for outbreaks of serious neonatal diarrheal diseases. These doctors are aware that in human epidemiology many bacterial species that are usually maintained in human populations in a relatively benign form can develop virulent forms in hospitals or other institutional settings (Ewald 1994a). For instance, *Escherichia coli* can lead to diarrheal diseases in hospital nurseries, despite being an innocuous component of the gut microflora in most people. Ewald (1994a) has argued that this localized evolution of virulence in institutional settings reflects the evolution of specific virulent strains of the bacterium in hospital wards. Presumably, attendants, parents, and visitors to wards all carry the benign community strain, out of which the virulent strain has evolved. What general conditions characterize the evolution of locally adapted bacterial strains in these situations? Should the doctors minimize visits by parents to their babies, or focus on other management procedures in the hospital environment?

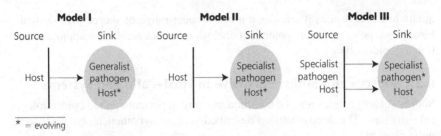

* = evolving

Figure 8.1 Three evolutionary models of source–sink populations discussed in this chapter. In Models I and II, the host in the sink adapts, respectively, to a generalist or to a specialist pathogen. In Model III, the pathogen in the sink adapts to its host.

In the first of these situations, the practical issue is to develop management practices that foster the evolution of reduced susceptibility to infection in the host. In the second and third situations, the focus is on how to prevent evolution toward greater virulence in a local population of the pathogen. What unifies these three situations is that they all involve spatial dynamics (in the broad sense of mixing together individuals drawn from distinct populations); depending upon the quantitative details, these scenarios could involve a source–sink structure for either the host or pathogen. Recent theoretical studies on adaptive evolution in sink environments (e.g., Kawecki 1995; Holt 1996; Gomulkiewicz *et al.* 2000) suggest management practices that could reduce the likelihood of the evolution of a virulent, locally adapted strain of pathogen, or enhance the evolution of resistance in the host.

8.3   A Limiting Case: Two Coupled Patches

Imagine that a host–pathogen interaction exists in a landscape with two distinct habitat patches. Spatial flows of individuals and heterogeneity in local demographic properties are assumed to generate a strong asymmetry, such that ecological and evolutionary dynamics in one habitat are strongly influenced by coupling with the other habitat, but without a marked reciprocal effect. We consider, in turn, three models that correspond to the three situations schematically depicted in Figure 8.1. In the first model, the source contains the host alone (effectively, in a refuge from the parasite), whereas the sink has both the host and pathogen. The pathogen is a generalist, so its dynamics are decoupled from the focal host species. Only the host disperses from the source into sinks. In the second model, we also assume the host flows from source to sink, but now the pathogen is a specialist with a dynamical response, such that the realized level of infection depends upon local host dynamics. For these two models, we examine the evolution of resistance of the host to infection. In the third model, we assume that both habitats contain the host and pathogen. However, there is cross-habitat infection, with infected individuals in the source infecting healthy individuals in the sink. For this model, we examine evolution of virulence of the pathogen in the sink habitat.

These models are not meant to duplicate faithfully the detailed dynamics of the hypothetical examples sketched above, but instead to illustrate more broadly how source–sink dynamics can lead to constraints on the evolution of virulence and resistance. We believe the simple models explored below capture some essential features of the above hypothetical situations, and are limiting cases of potentially much more complex models. Models I and II pertain to the first situation above, whereas model III is relevant to the other two situations. However, we stress that specific management suggestions for the evolution of resistance require detailed, empirically validated demographic models for the specific systems in which the evolution is occurring. The models below are strategic tools to help highlight broader issues, rather than tactical models directly useful in the development of policy decisions.

Simplifying assumptions

We make a number of simplifying assumptions at the outset. With respect to genetics, we assume haploid or clonal variation, and that the source population is fixed and is not itself evolving (relaxing these genetic assumptions does not fundamentally alter our basic conclusions; Gomulkiewicz *et al.* 2000). With respect to basic ecology, we assume that the host exhibits continuous generations, and that any direct density-dependence in the sink is dominated by the effects of the pathogen. Further, to simplify the many potential ramifications of virulence, we make a number of assumptions about basic epidemiology: once a host is successfully infected so that it itself is infectious, it cannot be super- or multiply-infected; and infected hosts do not recover, do not give birth, and remain infectious until removed from the population. In future work, it will be important to relax these simplifications.

Evolution of the pathogen can, of course, influence virulence, for instance if a higher transmission rate from infected hosts leads to a higher death rate (Frank 1996c). At first glance, because we assume infected hosts do not recover, it might seem that host evolution has no impact on virulence evolution. We suggest that a more subtle view may be appropriate. With no potential for host recovery, host evolution in response to the pathogen is related to the likelihood of successful infection in the first place and to the production of infected host individuals who themselves are infectious. The transmission parameter β at the heart of a standard epidemiological model defines the rate at which susceptible hosts themselves become infective. As a result of the deleterious effects of infection upon fitness, it is reasonable to assume that selection on the host tends toward a lower β value, all else being equal. This could happen either through effective avoidance of initial infection, so the host individual is not penetrated by the pathogen at all, or because of rapid host defenses that (when successful) reduce the pathogen titer quickly to trivial levels within the host body. By contrast, hosts whose defenses fail may continue to carry a high pathogen load, and so be infectious to other hosts (viz., be counted in the infectious class). Selection through host defenses could increase the frequency with which some hosts recover so rapidly that they, for all practical purposes, remain uninfected. If virulence is measured by assessing the average fitness across all hosts carrying the pathogen, including those whose successful defenses are reducing the pathogen titer toward extinction, selection on hosts can clearly influence the mean realized virulence experienced in the host population.

We first consider the evolution of a focal host species, immigrating into a habitat where it faces a genetically fixed pathogen. Let S be the density of the immigrant (= ancestral) host type in the sink habitat, S_{mut} the density of a novel type, and I the density of infected hosts. We consider in turn two distinct kinds of pathogen dynamics, and for each derive conditions for the initial increase of host alleles favored in the sink. We then turn to a situation in which infections of hosts in one habitat arise because of pathogens maintained in a source habitat, and examine evolution in the pathogen.

Model I: Host evolution in a sink with generalist pathogen

Here, we assume that the pathogen is maintained by alternative hosts, and that the density of infected hosts is, to a reasonable approximation, fixed at I. In our corn example, the pathogen might be a fungal blight sustained by grass species in pasturelands surrounding the field. In this case, we are not concerned with pathogen persistence. Let b be the intrinsic birth rate of the host, and d its death rate in the absence of the pathogen, so that $r = b - d$ is the focal host species' intrinsic growth rate. We assume that healthy hosts immigrate at a constant rate H, and that infection from the resident infected alternative hosts is described by a mass action law, parameterized by β, a transmission rate. The evolution of virulence is governed by evolution in β.

The dynamics of the immigrant susceptible host in the sink habitat are described by

$$\frac{dS}{dt} = rS - \beta SI + H .\tag{8.1}$$

When r is sufficiently greater than 0, the habitat is not a sink at all. In this case, the focal host species should increase in abundance when rare, and eventually Equation (8.1) no longer characterizes its dynamics adequately (e.g., direct density-dependence should become more important). We assume here that this is not the case, but instead that the habitat is a demographic sink.

There are two basic ecological situations that can lead to a sink for the host. First, one might have $r < 0$, that is, the habitat is an *intrinsic* sink, regardless of the presence of the pathogen. Second, one could have $0 < r < \beta I$. In this case, it is the presence of the pathogen itself, at sufficient abundance, that creates the sink habitat for the host; the sink condition is *induced* by the infectious disease agent. As we show, the potential for adaptive evolution of the resistance by hosts to infection is profoundly different in intrinsic than in induced sinks.

We now consider the fate of a novel allele arising in the sink, which is *not* part of the immigrant stream. Assume that the novel mutation experiences a lower infection rate, $\beta_{\text{mut}} < \beta$. Moreover, assume that $r_{\text{mut}} = r$, so there is no cost associated with this lowered rate of infection. Clearly, the novel type has a higher relative fitness than the immigrant type in the sink environment. But will it be retained by evolution?

The dynamics of the novel type are described by

$$\frac{dS_{\text{mut}}}{dt} = r_{\text{mut}} S_{\text{mut}} - \beta_{\text{mut}} S_{\text{mut}} I \, . \tag{8.2}$$

In an intrinsic sink, we have $r_{\text{mut}} = r < 0$. It is immediately clear that, regardless of the magnitude of β_{mut}, S_{mut} tends toward 0. Thus, if the local habitat is a sink for the host even in the absence of the natural enemy, natural selection cannot retain alleles that increase host resistance (reduced rate of infection) to the pathogen, even if these alleles are cost-free. This implies that the likelihood of local adaptation by hosts to pathogens via the accumulation of locally favored mutations is greatly reduced in intrinsic sink habitats.

Alternatively, the sink may be induced by the pathogenic infection itself, such that $0 < r < \beta I$. A novel allele in the host can increase in abundance (and thus frequency) provided $r > \beta_{\text{mut}} I$. Hence, an allele with a sufficiently low susceptibility can successfully invade and convert the induced sink into a source. The minimum magnitude of the mutational effect (measured by the quantity $\Delta\beta = \beta - \beta_{\text{mut}}$) required for a decrease in transmission to be deterministically retained by evolution is $\Delta\beta = \beta - r/I$, where β is the transmission rate experienced by immigrant hosts. One can imagine that there is a distribution of mutational effects on β, centered on the transmission rate of the immigrant type. Mutants that arise with a higher β than the immigrant are, of course, selectively disfavored and should rapidly disappear. Mutants that have a very small effect upon transmission are not retained in the local host population, even though they have a higher relative fitness than immigrant hosts. If there is a cost to reduced susceptibility, so that $r_{\text{mut}} < r$, the threshold mutational effect required for a mutant with lowered susceptibility to be selected will exceed $\Delta\beta$.

We can therefore draw the following conclusions:

- Local adaptation of the host to the pathogen is less likely if productivity in the sink (r) is low, or if pressure from the pathogen (I) is high. This means, for example, that if $r_{\text{mut}} < r$ due to a cost associated with higher resistance in the mutant host, an even larger decrease in β is required, compared to the resident type, than without such a cost.
- At larger I, or smaller r, evolution occurs only if mutants with a sufficiently large effect arise.
- In intrinsic host sinks, evolution in the hosts toward reduced transmission rates is not expected at all.

What is the role of host immigration in this sink model? First, immigration from the source defines an ancestral condition, against which the effect of each new mutation is measured. Second, in this particular model, the rate of immigration does not directly influence conditions for the retention of novel, favorable alleles.

(Recall that the model assumes hosts do not directly compete; with direct density-dependence, high immigration rates can depress local fitnesses by increasing local population size, and thereby hamper local adaptation; Holt and Gomulkiewicz 1997.) A final effect not directly addressed in the above model (which focuses on the fate of single mutations) is that of immigration on genetic variation in sink populations (Gomulkiewicz *et al.* 2000). Higher rates of immigration can increase local population size of the host in the sink habitat, making it more likely that favorable mutants will arise. A larger immigration rate also provides a larger sample drawn from the variation available in the source. For these genetic reasons, larger rates of immigration might indirectly facilitate local adaptation to the sink habitat, by increasing the total amount of variation available for selection in the sink. Combining these distinct effects, the greatest scope for local adaptation to a sink habitat may be provided by intermediate rates of immigration from source habitats (Gomulkiewicz *et al.* 2000).

Model II: Host evolution in a sink with specialist pathogen

In model I, for simplicity we assumed that the abundance of infected hosts was fixed by alternative host species. We now assume that the pathogen is a specialist, maintained solely by the focal host, so that the magnitude of infection is a dependent dynamical variable of the interaction. A canonical model for basic host–pathogen dynamics (Anderson and May 1981) is

$$\frac{dS}{dt} = rS - \beta SI + H \,, \tag{8.3a}$$

$$\frac{dI}{dt} = \beta SI - \mu I \,. \tag{8.3b}$$

This is the usual SI model with an additional term for immigration by healthy hosts.

The parameter μ equals the sum of intrinsic host deaths d plus additional deaths due to the infection α, so $\mu = d + \alpha$. At equilibrium, $S^* = \mu/\beta$, and $I^* = r/\beta + H/\mu$. If $r < 0$, then we must have $H > |r|d/\beta$ for the pathogen to persist. In other words, a specialist pathogen must be sufficiently transmissible to persist in an environment that is an intrinsic sink for its host. Given that the pathogen persists, at equilibrium the habitat is always a sink (intrinsic or induced) for the host; the abundance of infected individuals rises until the negative growth rate of the host population just matches the rate of input from outside.

As before, the initial dynamics of a rare, novel mutation in the host are described by

$$\frac{dS_{\text{mut}}}{dt} = r_{\text{mut}} S_{\text{mut}} - \beta_{\text{mut}} S_{\text{mut}} I \,. \tag{8.4}$$

Here, I is the abundance of infected hosts, exerting a force of infection on the invading host type, but determined indirectly by the dynamics of the resident host. As before, consider cost-free mutations lowering disease transmission such that

$r_{mut} = r$, but $\beta_{mut} < \beta$. If the habitat is an intrinsic sink for the host ($r_{mut} < 0$), then S_{mut} tends to 0. Hence, as in model I, evolution does not promote local adaptation by the host to the pathogen, even if the pathogen is dynamically dependent upon that host. If, by contrast, the habitat is an induced sink for the host ($r_{mut} > 0$), then $dS_{mut}/dt > 0$ if

$$\beta_{mut} < \beta / \left(1 + \frac{H\beta}{r\mu}\right). \tag{8.5}$$

Host immigration H and local production r have similar ecological effects on the incidence of infection in the local population, as shown by the expression for equilibrium incidence, $I^* = r/\beta + H/\mu$. However, Equation (8.5) shows that these two parameters have diametrically *opposing effects* on local adaptation to the pathogen. Increasing the host intrinsic growth rate r increases the range of mutational effects on β that can be captured by selection within the local host population; by contrast, increasing the immigration rate H makes local adaptation more difficult.

This model leads to several interesting predictions regarding local adaptation by hosts to parasites in sink habitats. Local adaptation of hosts toward lower transmission rates for an infection is more likely for:

- Habitats with high host productivity (and is conversely particularly unlikely in intrinsic sinks);
- Habitats in which the pathogen initially has a low rate of transmission (low β);
- Habitats in which infected individuals are short-lived [high μ, which can arise from either high intrinsic death rates (high d) or a highly virulent pathogen (high α)];
- Unproductive source habitats (low H).

Host immigration indirectly increases pathogen abundance, and thus increases the magnitude of the mutational effect required to retain a novel host mutant with lower β. These predictions hold even if mutations that affect the rate of infection are cost-free for the host. Including such costs (i.e., assuming $r_{mut} < r$) makes local adaptation by the host to the pathogen more difficult. [Comparable results arise in predator–prey coevolution along gradients (Hochberg and Van Baalen 1998).]

Model III: Pathogen evolution in a sink

In models I and II, the host evolves, but the pathogen does not. We now look at a counterpart model in which the host–pathogen interaction occurs in both habitats, and evolution of pathogen transmission occurs in the pathogen population given unidirectional movement of the pathogen (or infected hosts) from a source to a sink habitat. This model schematically matches the feedlot and neonatal ward situations discussed earlier. The source habitat in these models is comparable to the notion of disease reservoirs in epidemiology (Anderson and May 1991). In the hypothetical example of the cattle feedlots sketched above, for instance, the source or reservoir could be a benign infection maintained in low-density, free-ranging cattle herds. In the sink, the dynamics of the infection itself are described by

$$\frac{dI}{dt} = \beta_{\text{source}} I_{\text{source}} S + \beta I S - \mu I \ . \tag{8.6}$$

Here the term $\beta_{\text{source}} I_{\text{source}} S$ denotes the force of infection on healthy, susceptible hosts in the sink caused by infected individuals in the *source* (e.g., long-distance dispersal of infective propagules; alternatively, infected individuals could immigrate from the source at a fixed rate, and die or emigrate at a constant rate, in which case I_{source} denotes the equilibrium abundance of such individuals). For simplicity, we assume I_{source} is constant and that $S = K$, a constant (e.g., K could be host-carrying capacity, which is reasonable if the infection is initially very rare). By definition, the habitat is an intrinsic sink for the pathogen if $\beta K < \mu$, or $\beta < \mu/K$. We assume that the habitat is initially a sink for the pathogen (otherwise, the assumption that $S = K$ would be unreasonable). Biologically, three factors are likely to make a host population an intrinsic sink for an immigrating pathogen strain:

- The host is scarce (low K);
- Transmission rates are low (low β);
- Infected hosts have high death rates, either because hosts have intrinsically high death rates (high d), or the pathogen is highly virulent (high α).

As a result of external inputs, a pathogen can persist in a local host population that is intrinsically a sink with respect to local pathogen dynamics.

Imagine that a novel pathogen strain arises with a different rate of transmission and death rate (e.g., because of a correlation between virulence and transmissibility) than the immigrant type. When rare, this strain has dynamics described by

$$\frac{dI_{\text{mut}}}{dt} = \beta_{\text{mut}} I_{\text{mut}} K - \mu_{\text{mut}} I_{\text{mut}} \ . \tag{8.7}$$

The novel strain can increase provided $\beta_{\text{mut}} > \mu_{\text{mut}}/K$. Unlike the host models I and II presented above, in which there was no evolution on host resistance to transmission in an intrinsic sink, a sufficient increase in transmission in the parasite is always selected if it has no pronounced effects on virulence. If $\mu = \mu_{\text{mut}}$ (so no virulence costs are associated with increased transmission), a novel mutation will be successful only if it increases the transmission rate from β to β_{mut} by at least an amount $\mu/K - \beta$. All else being equal, novel strains in the parasite that provide a small increase in transmission are more likely to be favored if the host population is abundant than if it is scarce. Small increases in the transmission rate can be favored if infected hosts have low μ. If the host is an intrinsic sink for the pathogen, then either K is low, or μ is high. Evolution of the pathogen in a host habitat that is a sink for the pathogen thus occurs only if mutants that have a large effect upon transmission arise.

Now, consider a host–pathogen interaction in which there is a trade-off between transmission and virulence, such that increasing transmission translates into

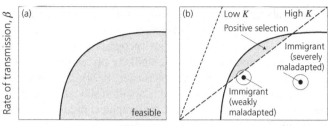

Figure 8.2 (a) A fitness set for a pathogen. Increased transmission to healthy hosts incurs the cost of higher mortality of infected hosts. (b) Evolutionary potential in a pathogen sink. For any pathogen allele to be favored, it must increase when rare (i.e., $\beta K > \mu$, where K is the abundance of the host). The dashed lines show the minimal configurations of β and μ that permit pathogen persistence at low host K and high host K. Given the available variation (which lies within the fitness set), no feasible pathogen genotype can persist (without immigration) in the low host K environment, so evolution there is impossible. In the high host K environment, some feasible genotypes permit persistence. If most mutants have a small effect, we can represent the immigrant genotype and available mutational variation as a dot at the center of a small cloud of available variation. If immigrants are severely maladapted, most mutants are likewise maladapted, and evolution of the pathogen does not occur. By contrast, if immigrants are only weakly maladapted, it is likely that mutants of modest effect will arise with positive growth, and hence will be retained by evolution.

greater host mortality. The shaded zone in Figure 8.2a represents all feasible phenotypes for the pathogen. In Figure 8.2b, the straight lines denote, for two different values of host K, combinations of β and μ that permit demographic persistence of a rare allele; for each case, pathogens with values of β and μ falling below the line are expected to go extinct. Immigrant types have a particular value of β and $\mu = d + \alpha$ (denoted by large dots in Figure 8.2b for two different possible cases corresponding to weakly and strongly maladapted immigrants). Mutants with a small effect may be more likely to arise than mutants that have a large effect (the circles around the dots indicate zones of likely mutational input in this phenotype space). At low K, given the array of possible pathogen phenotypes in the sink, no pathogen strain can increase when rare. In this circumstance, no evolutionary change in pathogen transmission (or virulence) would occur in the sink habitat. At higher K, it is feasible for some pathogen strains to invade, if the immigrant strains are not too badly maladapted to the host in the sink habitat.

This simple model suggests that if a pathogen is sustained by immigration into a given host population, which demographically is a sink for the pathogen, then local adaptation in rates of transmission and virulence are more likely if:

- The host population is abundant;
- The immigrant parasite strain is not too badly maladapted to the local host in the first place.

The effect of host abundance on the retention of parasite alleles leading to higher transmission suggests that pathogen adaptation is more likely in habitats that are favorable for hosts. Across a host species' geographical range, if host Ks are low near the range margin, these sites are unlikely places for local adaptation of the pathogen (see also Hochberg and Van Baalen 1998). If host abundances increase toward the range interior, pathogen adaptation should become increasingly feasible. (We caution that making precise geographical predictions depends upon knowing the detailed spatial texture of abundances, relative flux rates among habitats, and so forth, across the range.) Finally, if the host population in the sink is dynamically responsive to parasitism, a fuller analysis shows that increasing pathogen immigration from a source tends to depress the abundance of available, susceptible hosts in the sink, a demographic effect that reduces pathogen fitness. This suggests that increased pathogen immigration from a source can indirectly hamper local adaptation in the parasite in a sink, as measured in an adaptive balance between transmission and virulence. As this effect operates by a reduction in the availability of healthy hosts, it is unlikely to be of practical utility in managing the evolution of virulence.

Concretely, a more virulent pathogen strain is most likely to invade a sink habitat when:

- Mutations occur that confer a substantial difference between the immigrant and mutant pathogen strains. If most mutations tend to be small in effect, pathogens are not likely to be become locally adapted to hosts in sink habitats.
- Host density in the sink, K, is high. (With increasing K, the slope of a dashed line in Figure 8.2b decreases, increasing the feasible invasion space for more virulent pathogens.)
- The pathogen in the immigrant stream is low in virulence to begin with. (Again, invasion requires that the dynamic properties of the mutant be in the hatched parameter space of Figure 8.2a.)

The final message, therefore, is that evolution of virulence in habitats receiving inputs of pathogens depends on the initial transmissibility and virulence of pathogen streams from not-too-distant habitats. If a given pathogen is not approximately adapted to the sink habitat in the first place, so that for the pathogen the host in the sink habitat is a severe intrinsic sink, local adaptation is not likely to occur. The particular trajectory of evolution in a given focal habitat and, indeed, whether pathogen evolution is likely at all, depends on the intrinsic quality of local host populations and on the initial virulence of the immigrant pathogen strain.

8.4 On to Praxis

What practical advice do these theoretical ruminations suggest? Let us return to the three hypothetical situations sketched in the introduction, given that the simple, abstract models described above surely miss crucial details of concrete real-world situations. In particular, the assumptions made in the models about basic host population dynamics (e.g., continuous clonal growth, and no age or stage structure) would have to be modified to match the complexities of the actual host species dynamics. However, these strategic models do help to highlight some general issues that practitioners should think about in managing the evolution of virulence and resistance.

Consider first the crop yield problem, in which the organic farmer wishes to develop a variety of an annual crop adapted to a blight resident in her field. One simple "rule-of-thumb" that the host evolution models (I and II) suggest is that local adaptation to a pathogen should not occur in a host that inhabits an intrinsic sink. What makes a habitat an intrinsic sink for an annual plant is simply the number of expected successful offspring an average individual leaves. The farmer can influence the "sink" quality of her crop by the magnitude of seed retention. In years of bad harvest it might be tempting to sell all or most of an entire meager crop, and to purchase seeds from a seed company for the following year. This is good economic sense, but removes any possibility of local adaptation to particular strains of the blight, because it makes the local habitat an intrinsic sink for the crop plant. Local adaptation obviously requires retention of locally recruited plants.

The model also suggests that if the blight is sufficiently serious for the crop not to be self-sustaining, any action taken to make it more self-sustaining (e.g., changes in cultivation practices that reduce the impact of the blight by lowering β, or increasing local fecundity, r) indirectly may facilitate adaptive evolution to the blight, by making it more feasible for mutants that have a smallish effect on fitness to invade the host population. One interesting issue, which goes beyond the particular model discussed above, is the magnitude of foreign (nonlocal) seed that should be introduced, and its genetic character. If the seed source is itself genetically variable, introduced seeds can provide a valuable source of novel genetic variation. However, if instead the seed source is genetically homogeneous (a far more likely scenario in this era of hybrid seeds marketed by giant agrobusiness firms), little evolutionary traction is provided by supplementation from external host sources, and external seeds may vitiate local adaptation by competition with resident, better-adapted plants (Gomulkiewicz *et al.* 2000).

The problem faced by our hypothetical cattle ranchers and neonatal pediatricians, by contrast, is to prevent the evolution of novel, virulent infectious diseases. The ranchers have a difficult problem. By pooling their livestock together in feedlots they in effect increase the size of the potential host populations, and because interindividual distances are shorter, pathogen transmission is likely to be easier. Conditions here seem ripe for the evolution of a more virulent form of the benign pathogen brought in from the range, along with the cattle. The only model parameter open for manipulation appears to be d, the removal rate of infected hosts. The intrinsic "death" rate of an animal in the feedlot is likely to be determined by the amount of time required for sufficient fattening before being sent off for slaughter; this is governed by market requirements and other factors, largely out of the ranchers' control. The one remaining parameter is α, the additional rate of death or removal an animal incurs upon infection. The ranchers already have an obvious incentive to remove infected animals whenever encountered, namely to reduce the rate of infection of still healthy individuals. In addition to the direct ecological benefit of quick removal of infected hosts, our theoretical results help highlight an additional evolutionary benefit of this practice – it may make it harder for the immigrant pathogen to evolve a more virulent strain in the first place. Thus an even higher premium is now placed upon earlier detection, and removal, of infected animals. It would behoove the ranchers to invest in diagnostic procedures to facilitate this task.

There is a similar problem in the hospital neonatal ward – the enhanced risk of infection posed by parental visits to their babies and by other hospital procedures. With respect to the evolution of self-sustaining hospital strains, parental visits provide a source of variation for pathogen evolution. By contrast, hospital management practices can determine the selective fate of novel pathogen strains. Minimizing contact among infants (including contacts via indirect channels such as hospital personnel who come into contact with numerous infants) reduces transmission rates. The economies of scale that lead to the creation of large homogeneous wards for a given class of patients, such as suites for neonates, automatically creates a long-term evolutionary risk. Likewise, managing care so as to reduce the length of hospital stays in effect increases μ, the depletion rate of infected infants. Managed care plans in the USA today, for crass economic reasons, often reduce patient stays to the barest minimum. Though bad for any individual patient, unwittingly this practice may eventually benefit patients as a class, by reducing the chances of evolution of virulent pathogen strains adapted to the hospital environment. Both reducing transmission rates among patients (e.g., reducing β by creating multiple small wards with few patients, or attempting to reduce assiduously routes of contact among patients) and decreasing hospital stays (e.g., increasing the parameter μ) in effect turn hospital wards into sink habitats for invasive pathogens, and make the evolution of locally adapted, possibly more virulent strains, a more difficult evolutionary hurdle for the pathogen.

Concern is growing about emerging diseases for which the evolutionary origins are in species other than humans and domesticated species (Ewald 1994a). A simple message of the above models is that the demographic context of the initial stages of contact with novel hosts may be crucial in predicting emergence. If the demographic context is that the novel hosts are sinks for pathogens invading from another species, then adaptation by the pathogen to the host requires mutants of large effect; if such genes are rare, the emergence of the disease as a serious problem may be unlikely. The epidemiological goal of preventing a self-sustaining infectious disease in a novel host thus has the useful side-effect of precluding adaptive transformation in the disease agent.

8.5 Discussion

Many of these suggestions about management are commonsensical, and involve practices that are useful in reducing disease incidence even without considering host–pathogen evolution. In general, the emergent infectious diseases of greatest public health concern may not be those that are initially the most devastating to their individual hosts, and acquired mainly by recurring infection from alternative reservoir hosts, but instead any novel disease that can, with relatively little modification, become self-sustaining in the focal host species of interest.

If managing the evolution of virulence in pathogens is to become seriously integrated into public policy in agriculture, husbandry, and medical practice, it must be recognized that the direction of evolution in host–pathogen systems can be profoundly influenced by the direction and magnitude of spatial flows in heterogeneous environments, and that such flows often involve sources and sinks. Sophisticated management of the evolution of virulence can exploit the evolutionary impact of asymmetric spatial flows. As John Donne once famously remarked "No man is an island"; doubtless, our pathogens would agree.

Acknowledgments   Robert Holt would like to acknowledge National Science Foundation support, and thanks Susan Mopper for inviting me to participate in the National Science Foundation–Centre National de la Recherche Scientifique workshop where these ideas were first presented, and Jean Clobert for his hospitality in Paris. This chapter is also a contribution to the National Center for Ecological Analysis and Synthesis Working Group on "The Ecological and Evolutionary Dynamics of Species' Borders."

Part C

Within-Host Interactions

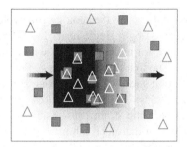

Introduction to Part C

For a long time, epidemiology essentially dealt with the spread of diseases within a population of susceptible, infected, or recovered hosts. Progress in microbiology and molecular biology has allowed us to study the full life cycle of pathogens: this comprises not only their transmission from one host to the next, but also their population dynamics within individual hosts. It has become clear that within-host interactions between pathogen strains can profoundly influence selection on virulence.

This part therefore concentrates on the ecology and evolution of microparasites within the biosphere presented by a single host. In particular, it focuses on two aspects of utmost importance for understanding the combined effects of mutation and within-host selection. The first aspect is of an ecological nature and relates to competition between different strains for the ecological niche offered by the host. In particular, competitive exclusion among strains can lead to the takeover of the host by the most virulent parasitic strain: this is the case of superinfection. Co-existence of several strains, on the other hand, leads to coinfection. The second aspect of within-host interaction that is crucial for virulence evolution is kin selection; it is based on genetic considerations, in particular on the genetic relatedness of parasites to each other. Very roughly speaking, the two effects pull in opposite directions: competitive exclusion tends to increase the virulence level, whereas kin selection tends to decrease it.

In Chapter 9, Nowak and Sigmund investigate simplified models of multiple infection. The first part of the chapter deals with superinfection: the more virulent strain quickly outcompetes its rivals. The other part deals with coinfection: the rate of new infections produced by one strain is unaffected by the presence of other strains. The two cases differ in expectations for the resultant range of strains within the host population; they are similar in that both predict a considerable increase in virulence. This underscores that mathematical arguments for the evolution of virulence based on optimizing the basic reproduction ratio of the pathogen do not work if several strains of pathogens compete within the host.

Adler and Mosquera Losada in Chapter 10 offer a considerably more detailed picture of multiple-infection processes. These authors investigate the full range of infection patterns possible for two strains of pathogen, ranging from coinfection to superinfection. In particular, they take into consideration the order of infection. Their mathematical analysis highlights some usually neglected subtleties of super- and coinfection processes that depend on the relation between virulence of strains, their ability to infect a susceptible or singly infected host, and their impact on the coexistence patterns of competing strains.

When pathogens replicate inside their hosts, their relatedness tends to be high and kin selection prevails. In Chapter 11, Gandon and Michalakis analyze how the coefficient of pathogen relatedness is influenced by four ecological parameters:

population size, pathogen dispersal rate, cost of dispersal, and transmission mode. On this basis, they investigate how the separate and joint evolution of pathogen virulence and dispersal rate is affected by these parameters. Applying these findings to identify options for virulence management, the authors conclude that in the presence of multiple infections long-term benefits arise from sanitation and vaccination that would otherwise be absent.

In Chapter 12, Read, Mackinnon, Anwar, and Taylor evaluate the relevance of kin-selection models for malaria epidemiology, and critically assess data on the influence of genetic relatedness among parasites on the outcome of the disease. Correlations observed in the field and laboratory experiments support the conclusion that plausible mathematical models may rely on wrong assumptions about the effects of within-host competition on between-host transmission. This strikes a cautionary note and stresses that, at present, the models serve to suggest further experiments.

Resolving the many open questions that surround within-host interactions may be the most important milestone on the road toward consolidating existent models of disease ecology and evolution. Much empirical testing has to be carried out before the current thicket of within-host models that has sprung up in recent years gives way to harvestable cultures – an intermediate slash-and-burn stage seems inevitable.

9

Super- and Coinfection: The Two Extremes

Martin A. Nowak and Karl Sigmund

9.1 Introduction

As is well known, the "conventional wisdom" that successful parasites have to become benign is not based on exact evolutionary analysis. Rather than minimizing virulence, selection works to maximize a parasite's reproduction ratio (see Box 9.1). If the rate of transmission is linked to virulence (defined here as increased mortality due to infection), then selection may in some circumstances lead to intermediate levels of virulence, or even to ever-increasing virulence (see Anderson and May 1991; Diekmann *et al.* 1990, and the references cited there).

A variety of mathematical models has been developed to explore theoretical aspects of the evolution of virulence (see, for instance, Chapters 2, 3, 11, and 16). Most of these models exclude the possibility that an already infected host can be infected by another parasite strain. They assume that infection by a given strain entails immunity against competing strains. However, many pathogens allow for multiple infections, as shown in Chapters 6, 12, and 25. The (by now classic) results on optimization of the basic reproduction ratio cannot be applied in these cases.

The mathematical modeling of multiple infections is of recent origin, and currently booming. Levin and Pimentel (1981) and Levin (1983a, 1983b) analyzed two-strain models in which the more virulent strain can take over a host infected by the less virulent strain. They found conditions for coexistence between the two strains. Bremermann and Pickering (1983) looked at competition between parasite strains within a host, and concluded that selection always favors the most virulent strain. Frank (1992a) analyzed a model for the evolutionarily stable level of virulence if there is a trade-off between virulence and infectivity, and if infection occurs with an ensemble of related parasite strains. In Adler and Brunet (1991), Van Baalen and Sabelis (1995a), Andreasen and Pugliese (1995), Lipsitch *et al.* (1995a), and Claessen and de Roos (1995), further aspects of multiple infection are discussed.

In this chapter, following Nowak and May (1994) and May and Nowak (1994, 1995), we deal with two opposite extreme instances of multiple infection by several strains of a parasite. These simplified extreme cases, which are at least partly amenable to analytical understanding, seem to "bracket" the more general situation. The first case deals with *superinfection*. This approach assumes a competitive hierarchy among the different parasite strains, such that a more virulent parasite can infect and take over a host already infected by a less virulent strain. Multiply infected hosts transmit only the most virulent of their strains. The opposite scenario is that of *coinfection*. In this case, there is no competition among the different strains within the same host: each produces new infections at a rate that is unaffected by the presence of other strains in the host.

Both these extremes are amenable to analytical understanding, at least in some simplified cases. Mosquera and Adler (1998) produced a unified model for multiple infections (by two strains), which yields both superinfection and coinfection (as well as single infection) as special cases (see also Chapter 10). The long-term goal is, of course, to combine the full scenario of multiple infections in a single host with the adaptive dynamics for evolution within and among hosts. Such studies will mostly rely on computer simulations, but it is important to understand the basics first.

What happens when many different strains are steadily produced by mutation? Both for superinfection and for coinfection, the virulence will become much larger than the optimal value for the basic reproduction ratio. There are interesting differences, however, in the packing of the strains and in the increase of their diversity, depending on whether superinfection or coinfection holds. Furthermore, in the case of superinfection, removal of a fraction of the hosts implies a lasting reduction of the average virulence. This last fact has obvious implications for virulence management: it is quite conceivable that even an incomplete vaccination campaign will have a decisive impact on population health, not by eradicating the pathogen but by making it harmless.

9.2 Superinfection

In this section we expand the basic model for single infections (Box 9.1) to allow for superinfection. We consider a heterogeneous parasite population with a range of different strains j (with $1 \leq j \leq n$) having virulence α_j, with $\alpha_1 < \alpha_2 < \ldots < \alpha_n$. Furthermore, we assume that more virulent strains outcompete less virulent strains on the level of intra-host competition. For simplicity we assume that the infection of a single host is always dominated by a single parasite strain, namely that with maximal virulence. In our framework, therefore, superinfection means that a more virulent strain takes over a host infected by a less virulent strain. Only the more virulent strain is passed on to other hosts. The translation of these assumptions into mathematical terms is given in Box 9.2.

To arrive at an analytic understanding, we consider the special case that all parasite strains have the same infectivity, β, and differ only in their degree of virulence, α_j. For the relative frequencies i_j of hosts infected by strain j we obtain from Equation (c) in Box 9.1 the Lotka–Volterra equation

$$i'_j = i_j \left(r_j + \sum_{k=1}^{n} a_{jk} i_k \right) , \tag{9.1}$$

Box 9.1 Population dynamics of pathogen diversity in SI models

We consider the model of Box 2.1 with the recovery rate γ set equal to zero,

$$\frac{dS}{dt} = B - dS - \beta SI ,$$

$$\frac{dI}{dt} = I(\beta S - d - \alpha) . \tag{a}$$

The basic reproduction ratio of the parasite for this model is

$$R_0 = \frac{\beta}{d + \alpha} \frac{B}{d} . \tag{b}$$

If R_0 is larger than one, then the parasite will spread in an initially uninfected population, and damped oscillations lead to the stable equilibrium

$$S^* = \frac{d + \alpha}{\beta} , \qquad I^* = \frac{\beta B - d(d + \alpha)}{\beta(d + \alpha)} . \tag{c}$$

To understand parasite evolution, consider a number of parasite strains competing for the same host. The strains differ in their infectivity β_j and their degree of virulence α_j. If I_j denotes the density of hosts infected by strain j, and excluding the possibility of infection by two strains at once, then

$$\frac{dS}{dt} = B - dS - S \sum_j \beta_j I_j ,$$

$$\frac{dI_j}{dt} = I_j(\beta_j S - d - \alpha_j) . \tag{d}$$

For a generic choice of parameters there is no interior equilibrium, and coexistence between any two strains in the population is not possible. To see this, consider two strains, which, without loss of generality, are called 1 and 2. Now $h_{1,2} = \beta_1^{-1} \ln I_1 - \beta_2^{-1} \ln I_2$ is introduced, which gives

$$\frac{dh_{1,2}}{dt} = \frac{d + \alpha_2}{\beta_2} - \frac{d + \alpha_1}{\beta_1} . \tag{e}$$

So $h_{1,2}$ goes to $-\infty$ or $+\infty$ depending on which of the two terms is the larger. Since the model does not allow I_j to go to infinity, the conclusion is that strain 2 always outcompetes strain 1 if

$$\frac{\beta_2}{d + \alpha_2} > \frac{\beta_1}{d + \alpha_1} . \tag{f}$$

This is exactly the condition that the transversal eigenvalue $\lambda_2 = \partial I_2'/\partial I_2$ at the two-species equilibrium $E_1 = (S^*, I_1^*, I_2 = 0)$ is positive, while the transversal eigenvalue $\lambda_1 = \partial I_1'/\partial I_1$ at the two-species equilibrium $E_2 = (S^*, I_1 = 0, I_2^*)$ is negative; that is, strain 2 can invade 1, but 1 cannot invade 2. Applying

continued

Box 9.1 *continued*

Condition (f) to any pair of two strains shows that ultimately, out of the full diversity, only one strain remains, which is the one with the highest value of R_0.

If there is no relation between infectivity and virulence, then the evolutionary dynamics will increase β and reduce α. In general, however, there is some relationship between α and β, see Box 5.1. This can lead to an intermediate degree of virulence prevailing, corresponding to the maximum value of R_0. Other situations allow evolution toward ever higher or lower virulences. The detailed dynamics depends on the shape of β as a function of α.

---

on the positive orthant R_+^n, with $r_j = \beta - \alpha_j - d$ (here, d is the background mortality of uninfected hosts) and $A = (a_{jk})$, given by

$$
A = -\beta \begin{pmatrix}
1 & 1+\sigma & 1+\sigma & \cdots & 1+\sigma \\
1-\sigma & 1 & 1+\sigma & \cdots & 1+\sigma \\
1-\sigma & 1-\sigma & 1 & \cdots & 1+\sigma \\
\vdots & \vdots & \vdots & \ddots & \vdots \\
1-\sigma & 1-\sigma & 1-\sigma & \cdots & 1
\end{pmatrix},
\tag{9.2}
$$

where the parameter σ describes the vulnerability of an already infected host to infection by another strain (with higher virulence). In the extreme case $\sigma = 0$, infection confers complete immunity to all other strains (an effect similar to vaccination); for $\sigma = 1$, an infected individual is as vulnerable as an uninfected one; for $\sigma > 1$, infection weakens the immune system so that invasion by another strain becomes more likely.

In Nowak and May (1994) it is shown that Equation (9.1) has one globally stable fixed point, that is, one equilibrium that attracts all orbits from the interior of the positive orthant. If this equilibrium lies on a face of the positive orthant, then it also attracts all orbits from the interior of that face. In Nowak and May (1994) this equilibrium is computed.

The important special case $\sigma = 1$ offers a quick solution. The unique stable equilibrium is then given recursively in the following way,

$$
i_n^* = \max\{0, 1 - \frac{\alpha_n + d}{\beta}\},
\tag{9.3a}
$$

$$
i_{n-1}^* = \max\{0, 1 - \frac{\alpha_{n-1} + d}{\beta} - 2i_n^*\},
\tag{9.3b}
$$

$$
i_{n-2}^* = \max\{0, 1 - \frac{\alpha_{n-2} + d}{\beta} - 2(i_n^* + i_{n-1}^*)\},
\tag{9.3c}
$$

Box 9.2 SI models accounting for superinfection

In this box the simple model of Box 9.1 is modified to cope with superinfection. We now have to deal with a number of different strains of parasite, which will be labeled with the index j. If I_j denotes the density of hosts infected with strain j, then we obtain

$$\frac{dS}{dt} = B - dS - S \sum_{j=1}^{n} \beta_j I_j \,,$$

$$\frac{dI_j}{dt} = I_j \left(\beta_j S - d - \alpha_j + \sigma \beta_j \sum_{k=1}^{j-1} I_k - \sigma \sum_{k=j+1}^{n} \beta_k I_k \right), \quad j = 1, \ldots, n \,. \quad \text{(a)}$$

Here α_j denotes the virulence of strain j. Without restricting generality, we assume $\alpha_1 < \alpha_2 < \ldots < \alpha_n$. In our model a more virulent strain can superinfect a host already infected with a less virulent strain. The parameter σ describes the rate at which infection by a new strain occurs, relative to infection of uninfected hosts. If either the host or the parasite has evolved mechanisms to make superinfection more difficult, then σ would be smaller than one. If already-infected hosts are more susceptible to acquiring a second infection (with another strain), then $\sigma > 1$, that is, superinfection occurs at increased rates. The case $\sigma = 0$ corresponds to the single-infection model discussed in Box 9.1.

To arrive at an analytical understanding we make the simplifying assumption that the immigration of uninfected hosts exactly balances the death of uninfected or infected hosts, $B = dS + dI + \sum_{j=1}^{n} \alpha_j I_j$. In that case we can divide through by $N = S + \sum_{j=1}^{n} I_j$ to obtain an equation for the relative frequencies

$$\frac{di_j}{dt} = i_j [\beta_j (1 - i) - d - \alpha_j + \sigma (\beta_j \sum_{k=1}^{j-1} i_k - \sum_{k=j+1}^{n} \beta_k i_k)], \quad j = 1, \ldots, n \,,$$

$$\text{(b)}$$

where $i = \sum_{j=1}^{n} i_j$. This is a Lotka–Volterra system of equations,

$$\frac{di_j}{dt} = i_j \left(r_j + \sum_{k=1}^{n} a_{jk} i_k \right), \quad j = 1, \ldots, n \,, \quad \text{(c)}$$

with $r_j = \beta_j - \alpha_j - d$ and the matrix $A = (a_{jk})$ is given by

$$A = - \begin{pmatrix} \beta_1 & \beta_1 + \sigma\beta_2 & \beta_1 + \sigma\beta_3 & \cdots & \beta_1 + \sigma\beta_n \\ \beta_2(1-\sigma) & \beta_2 & \beta_2 + \sigma\beta_3 & \cdots & \beta_2 + \sigma\beta_n \\ \beta_3(1-\sigma) & \beta_3(1-\sigma) & \beta_3 & \cdots & \beta_3 + \sigma\beta_n \\ \vdots & \vdots & \vdots & \ddots & \vdots \\ \beta_n(1-\sigma) & \beta_n(1-\sigma) & \beta_n(1-\sigma) & \cdots & \beta_n \end{pmatrix} . \quad \text{(d)}$$

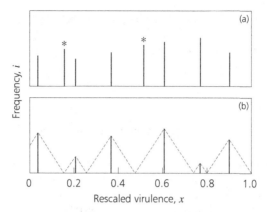

Figure 9.1 For $\sigma = 1$ there is a simple geometric method to construct the equilibrium configuration. Suppose there are n strains, given by their virulences $\alpha_1, \ldots, \alpha_n$, and let i_j^* be their relative frequencies. We set $x_j = (\alpha_j + d)/\beta$. (a) We only have to consider strains with $0 < x_1 < \ldots < x_n < 1$ and their corresponding frequencies. (b) Draw verticals with abscissae x_j and construct a polygonal line with $45°$ slopes, starting on the horizontal axis at abscissa 1, at first to the north-west until the first vertical is reached, from there to the south-west until the horizontal axis is reached, then to the north-west until the next vertical is reached, then south-west again, etc. The vertices on the verticals correspond to the i_j^* values that are positive. The strains with other virulences, marked by a star in (a), are eliminated. *Source*: Nowak and May (1994).

$$\vdots$$

$$i_1^* = \max\{0, 1 - \frac{\alpha_1 + d}{\beta} - 2(i_n^* + i_{n-1}^* + \ldots + i_2^*)\}, \qquad (9.3d)$$

This fixed point is saturated, that is, no missing species can grow if it is introduced in a small quantity. Indeed, for each parasite strain j with equilibrium frequency $i_j^* = 0$ we obtain $\partial i_j'/\partial i_k < 0$ for a generic choice of parameters, see Hofbauer and Sigmund (1998). Hence this fixed point is the only stable fixed point in the system.

Equations (9.3) correspond to a very simple and illuminating geometric method for constructing the equilibrium (see Figure 9.1).

For a given σ, one can estimate $\alpha_{\max}$, the maximum level of virulence present in an equilibrium distribution. Assuming equal spacing (on average), that is, $\alpha_j = j\alpha_1$, Nowak and May (1994) derive

$$\alpha_{\max} = \frac{2\sigma(\beta - d)}{1 + \sigma}. \qquad (9.4)$$

For $\sigma = 0$, we have $\alpha_{\max} = 0$, that is, only the strain with the lowest virulence survives, which for our scenario (with all transmission rates equal) is also the strain with the highest basic reproduction ratio [see Equation (c) in Box 9.1]. For $\sigma > 1$, strains can be maintained with virulences above $\beta - d$. These strains by themselves are unable to invade an uninfected host population, because their basic reproduction ratio is less than one.

From Equation (9.4) it can be deduced that the equilibrium frequency of infected hosts $\sum_{j=1}^{n} i_j$ is given by

$$i = \frac{\beta - d}{\beta(1 + \sigma)}. \tag{9.5}$$

Hence, with greater susceptibility to superinfection (larger σ) one obtains fewer infected hosts!

Let us now return to the model with different strains having different infectivities, β_j, as given by Equation (c) in Box 9.2. Here the solutions need not always converge to a stable equilibrium. For $n = 2$, either coexistence (i.e., a stable equilibrium between the two strains of parasites) or bistability (in which either one or the other strain vanishes, depending on the initial conditions) is possible. An interesting situation can occur if $\sigma > 1$, and strain 2 has a virulence that is too high to sustain itself in a population of uninfected hosts ($R_0 < 1$), whereas strain 1 has a lower virulence with $R_0 > 1$. Since $\sigma > 1$, infected hosts are more susceptible to superinfection, and thus the presence of strain 1 can effectively shift the reproduction ratio of strain 2 above one. In this way, superinfection allows the persistence of parasite strains with extremely high levels of virulence.

For three or more strains of parasite we may observe oscillations with increasing amplitude and period, tending toward a heteroclinic cycle on the boundary of R_+^n, that is, a cyclic arrangement of saddle equilibria and orbits connecting them (comparable to those discussed in May and Leonard 1975, and Hofbauer and Sigmund 1998). Accordingly, for long stretches of time the infection is dominated by one parasite strain (and hence only one level of virulence), until suddenly another strain takes over. This second strain is eventually displaced by the third, and the third, after a still longer time interval, by the first. Such dynamics can, for example, explain the sudden emergence and re-emergence of pathogen strains with dramatically altered levels of virulence.

To explore the case of nonconstant infectivities, Nowak and May (1994) assume a specific relation between virulence and infectivity, $\beta_j = c_1 \alpha_j / (c_2 + \alpha_j)$ for some constants c_1 and c_2. For low virulence, infectivity increases linearly with virulence; for high virulence the infectivity saturates. For the basic reproduction ratio this means that, for strain j

$$R_{0,j} = \frac{c_1 B \alpha_j}{d(c_2 + \alpha_j)(d + \alpha_j)}. \tag{9.6}$$

The virulence that maximizes R_0 is given by $\alpha_{opt} = \sqrt{dc_2}$. For $\sigma = 0$ (no multiple infection), the strain with largest R_0 is, indeed, selected. For $\sigma > 0$, selection leads to the coexistence of an ensemble of strains with a range of virulences between two boundaries α_{min} and α_{max}, with $\alpha_{min} > \alpha_{opt}$.

Thus superinfection has two important effects:

- It shifts parasite virulence to higher levels, beyond the level that would maximize the parasite reproduction ratio;
- It leads to the coexistence of a number of different parasite strains within a range of virulences.

We note from Figure 9.2 that strains have a higher equilibrium frequency if the strains with slightly larger virulences have low frequencies. Conversely, if a strain has a high frequency, strains with slightly lower virulence are extinct or occur at very low frequencies. This implies a "limit to similarity," that is, a spacing of the coexisting strains, which agrees well with the construction of the equilibrium in the special case of constant β and $\sigma = 1$, see Figure 9.1.

Limits to similarity are well-known in ecology and, indeed, the epidemiological model above turns out to be equivalent to a metapopulation model introduced independently, and in an altogether different context, by Tilman (1994). The different strains play the role of distinct species and the hosts play the role of ecological patches. This is further analyzed in Nowak and May (1994) and Tilman *et al.* (1994); also see Nee and May (1992) for a related analysis.

If mutation keeps generating new strains with altered levels of virulence, then there will be an ever-changing parasite population, in which the virulences are restrained by selection to a range between α_{min} and α_{max}. Indeed, there will always be new strains capable of invading the polymorphic population. Some of the old strains may then become extinct, and many of those surviving strains with lower virulence than the newcomer will have altered frequencies.

If this evolutionary dynamics is iterated for a very long time, then one can define a distribution function $i(\alpha)$ that describes the long-term equilibrium frequencies of strains as a function of their virulence, α. A semi-rigorous argument suggests that $i(\alpha)$ is given by a uniform distribution over the interval $[\alpha_{min}, \alpha_{max}]$. Extensive numerical experiments suggest that this distribution is globally stable for the mutation–selection process.

9.3 Coinfection

We now turn to the case of coinfection, and assume therefore that the infectivity of a strain is unaffected by the presence of other strains in the same host. Again, we derive a simple model and investigate it first analytically (after further simplifications) and then by means of numerical simulations.

As before, we denote by i_j the fraction of the host population infected by strain j, and assume that the strains are numbered in order of virulence: $\alpha_1 < \ldots < \alpha_n$. Several parasites can be present in the same host, and so $\sum_{j=1}^{n}$ can exceed the fraction of all hosts that are infected.

If we assume that the death rate is determined by the most virulent strain harbored by the host, we obtain a simple dynamic model presented in Box 9.3.

The equilibria of Equation (a) in Box 9.3 must satisfy, for all j, either

$$i_j = 0 , \qquad\qquad\qquad (9.7a)$$

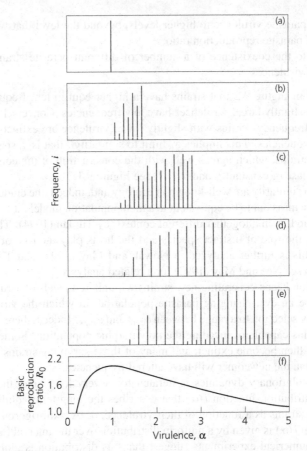

Figure 9.2 (a) to (e) Equilibrium distribution of parasite virulence for the superinfection model. The horizontal axis denotes virulence, and the vertical axis indicates equilibrium frequencies (always scaled to the same largest value). The simulation is performed according to Equation (b) in Box 9.2 with $B = 1$, $d = 1$, $n = 50$, $\beta_j = 8\alpha_j/(1 + \alpha_j)$ and $\sigma = 0, 0.1, 0.5, 1,$ or 2 [in (a) to (e)]. The individuals α_j are assumed to be regularly spaced between 0 and 5. Thus $\alpha_1 = 0.1, \alpha_2 = 0.2, \ldots, \alpha_{50} = 5$. For $\sigma = 0$ (the single-infection case) the strain with maximum basic reproduction ratio, R_0 [displayed in (f)], is selected. With $\sigma > 0$ we find coexistence of many different strains with different virulences, α_j, within a range $\alpha_{\min}$ and $\alpha_{\max}$, but the strain with the largest R_0 is not selected; superinfection does not maximize parasite reproduction. For increasing σ, the values of $\alpha_{\min}$ and $\alpha_{\max}$ also increase. *Source*: Nowak and May (1994).

or

$$i_j = 1 - (\overline{\alpha}_j + d)/\beta_j . \tag{9.7b}$$

Using Equations (b) and (c) in Box 9.3, the equilibrium values of i_j can be computed in a recursive way, starting from $i_n = 1 - (\alpha_n + d)/\beta_n$.

Box 9.3 SI models accounting for coinfection

With i_j denoting the fraction of individuals harboring strain j (possibly in addition to various other strains), a simple model for coinfection is

$$\frac{di_j}{dt} = i_j[\beta_j(1 - i_j) - d - \bar{\alpha}_j], \qquad j = 1, \ldots, n .\tag{a}$$

The total population size of hosts is assumed to be held constant, and is normalized to one. The infectivity (transmission rate) of strain j is denoted by β_j. Strain j can invade any host that is not already infected by strain j. Thus $\beta_j i_j (1 - i_j)$ is the rate at which new infections with strain j occur.

There is a natural death rate d and a disease induced death rate $\bar{\alpha}_j$ which denotes the average death rates of hosts infected by strain j, and is assumed to be given by the strain with the highest virulence in the host. We define p_j as the probability that a host is not infected with a strain *more* virulent than j. That is,

$$p_j = \prod_{k=j+1}^{n} (1 - i_k) .\tag{b}$$

Note that $p_n = 1$ and $p_i = (1 - i_{j+1}) p_{j+1}$. The fraction of hosts that are uninfected is given by $p_0 = \prod_{k=1}^{n}(1 - i_k)$. The probability that k is the most virulent strain found in a host is $i_k p_k$, and

$$\bar{\alpha}_j = \alpha_j p_j + \sum_{k=j+1}^{n} \alpha_k i_k p_k .\tag{c}$$

This coinfection model is completely defined by Equations (a) to (c). We note that infection and death rules are devised such that if the strains are randomly assorted relative to each other, this continues to be the case, so that Equation (a) remains correct.

If the transmission rates β_i are all equal to some value β, then, as shown in May and Nowak (1995), the following expressions for the average virulence $\bar{\alpha}$ and the fraction s^* of uninfected hosts are approximately valid (see Figure 9.3)

$$\bar{\alpha} = \beta - d - \sqrt{2\beta(\beta - d)/n} ,\tag{9.8a}$$

and

$$s^* = 4 \exp[-\sqrt{2n(\beta - d)/\beta}] .\tag{9.8b}$$

One can similarly investigate coinfection if the transmission rate is not constant, but an increasing function of virulence, for instance

$$\beta_j = c_1 \alpha_j / (c_2 + \alpha_j) ,\tag{9.9}$$

with constants c_1 and c_2. The basic reproduction ratio for strain j is given by

$$R_{0,j} = \frac{c_1 \alpha_j}{(c_2 + \alpha_j)(d + \alpha_j)} .\tag{9.10}$$

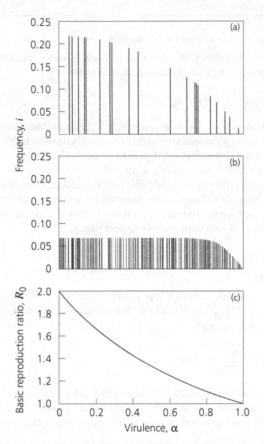

Figure 9.3 Equilibrium distribution of parasite virulence for the coinfection model given by Equations (a) to (c) in Box 9.3 with uniform transmission rate $\beta = 2$ and $d = 1$. The individual parasite strains have randomly assigned levels of virulence ranging from 0 to 1. For different numbers of strains n the equilibrium population structure is computed according to Equation (9.7b). (a) $n = 20$ parasite strains. (b) $n = 200$ parasite strains. For large n there is excellent agreement between the numerical calculations and the theoretical curve, given by Equation (9.8a). (c) The basic reproduction ratio R_0 as a function of virulence. *Source*: May and Nowak (1995).

R_0 is thus maximized by the strain with virulence $\alpha = \sqrt{dc_2}$, and takes the value $c_1/(\sqrt{d} + \sqrt{c_2})^2$. The minimum and maximum virulence values for strains that have the potential to maintain themselves within the host population, α_- and α_+, respectively, are given by

$$\alpha_\pm = \frac{1}{2}\left[c_1 - d - c_2 \pm \sqrt{(c_1 - d - c_2)^2 - 4dc_2}\right].\tag{9.11}$$

In Figure 9.4 the results for coinfection are illustrated for transmission rates that increase with virulence.

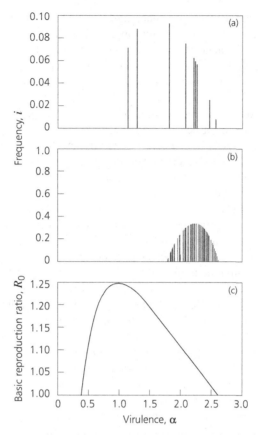

Figure 9.4 Equilibrium distribution of parasite virulence for the coinfection model with a trade-off between transmission rate β_j and virulence α_j given by $\beta_j = 5\alpha_j/(1 + \alpha_j)$. The natural death rate is again $d = 1$, and the parasites have levels of virulence uniformly distributed between 0 and 3. The virulences of the persisting strains are between $\alpha_{\min}$ and the maximum level of virulence that corresponds to $R_0 = 1$, i.e., $\alpha_+ = (3 + \sqrt{5})/2$. (a) $n = 20$ parasite strains. The average virulence is $\bar{\alpha} = 1.9246$ and the fraction of uninfected hosts is $s^* = 0.5716$. (b) $n = 200$ parasite strains. Here $\bar{\alpha} = 2.3039$ and $s^* = 0.1952$. (c) The basic reproduction ratio, R_0, as a function of virulence. *Source*: May and Nowak (1995).

9.4   Discussion

Multiple infections cause intra-host competition among strains and thus lead to an increase in the average level of virulence above the maximal growth rate for a single parasitic strain.

The simple models for superinfection (transmission only of the most virulent strain within a host) and for coinfection (all strains transmit independently of other strains present in the host) represent extremes that are likely to bracket the reality of polymorphic parasites. In both cases, we find the expected tendency toward the predominance of strains with a virulence significantly higher than that maximizing reproduction success of parasites in the single-infection case. The number of persisting strains and the range of their virulence, however, differ in the two cases of super- and coinfection. The latter allows for a larger number of coexisting strains, more closely grouped around the virulence level with the maximal reproduction ratio, than does the former.

The basic reproduction ratio is not maximized. With superinfection, the strain with highest R_0 may even become extinct, and strains with very high levels of virulence can be maintained (even strains so virulent that they could not persist on their own in an otherwise uninfected host population). Both superinfection and coinfection lead to polymorphisms of parasites with many different levels of virulence within a well-defined range.

Superinfection can lead to very complicated dynamics, with sudden and dramatic changes in the average level of virulence. The higher the rate σ of superinfection the smaller the number of infected hosts.

It is particularly interesting to investigate evolutionary chronicles. What happens if mutation, from time to time, introduces a new strain? In the case of superinfection, according to the "limit to similarity" principle, only those mutants sufficiently different from the resident strain with next-higher virulence can invade; they then affect the equilibrium frequencies of the resident strains with lower virulence, possibly eliminating some of them. The average total number of strains increases slowly (logarithmically in time). On the other hand, these limits to similarity result in a wide range of virulence values persisting in the system.

By contrast, coinfection models have no limits to similarity, and surviving strains are packed ever closer as time goes on, constrained to a narrow band of virulence values. If we assume again that mutants are produced at a constant rate, we find that, asymptotically, the total number of persisting strains increases with the square root of time.

In the superinfection case, removing a certain percentage of potential hosts (for instance by vaccination) results in a sharp drop in the number of strains, eliminating the most virulent strains. Indeed, if there are fewer hosts, then the overall incidence of infection is lower, and fewer hosts are superinfected; thus strains favored by their within-host advantage do less well than those favored by their between-host advantage. After the onset of vaccination, the total number of strains slowly recovers again, but not the average virulence (see Figure 9.5). Thus even if vaccination eliminates only a fraction of the potential hosts, and therefore has little long-term effect on the number of strains, it produces a lasting effect by reducing the average virulence.

At present, many instances of multiple infections are known, but there are disappointingly few data on the coinfection function (the actual rate of invasion by a more virulent strain). Mosquera and Adler (1998) make the point that many previous models are based on the assumption that this coinfection function is discontinuous: even a marginally more virulent strain will immediately, and certainly, displace its less virulent predecessor (see, e.g., May and Nowak 1994, 1995; Van

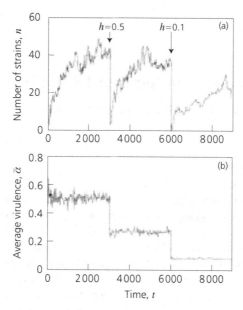

Figure 9.5 (a) The number n of pathogen strains present at time t, in the superinfection model, with mutations arising uniformly in the interval $0 \leq \alpha \leq 1$. At time $t = 3\,000$, the total number of hosts h is decreased by 50%. The number $n(t)$ subsequently increases again. At $t = 6\,000$ the number of hosts is reduced to 10% (since the rate of new mutants able to invade is 10% of the former value, the growth in n proceeds at a slower rate). (b) Corresponding average values of the virulence as a function of time. Removal of a fraction of the hosts permanently reduces the average virulence by that same fraction. *Source*: May and Nowak (1994).

Baalen and Sabelis 1995a). Continuous coinfection functions produce different results. Individual-based modeling and clinical research are needed to test the implications of the current superinfection models on the evolution and management of virulence.

10

Super- and Coinfection: Filling the Range

Frederick R. Adler and Julio Mosquera Losada

10.1 Introduction

How many different strains of a disease can coexist in a single population of hosts? What effect do different mechanisms of coexistence have on the properties of diseases? The principle of competitive exclusion (Armstrong and McGehee 1980; Levin 1970) states that no more species can coexist in a system than the number of resources or limiting factors allow, which can be thought of, somewhat imprecisely, as stating that a single trade-off can support only a single species – the one that deals best with that trade-off. Disease models describe a simple ecological interaction, with hosts acting as resources, to test the limits of competitive exclusion. Trade-offs for the disease often involve virulence, a trait of abiding interest to hosts.

In the absence of a trade-off between host mortality and transmission efficiency, the disease strain with the lowest virulence would always win out in competition, and diseases would be favored to evolve ever-reduced virulence. When such a trade-off between host mortality and transmission efficiency exists, the single strain that maximizes the basic reproduction ratio R_0 will persist (see Boxes 2.2, 5.1, and 9.1; Bremermann and Thieme 1989). Ecological factors that affect this trade-off, such as host density, might favor higher or lower virulence (Ewald 1994a). However, in the absence of spatial or temporal variation in these factors, only one strain persists [but see Andreasen and Pugliese (1995) for a case in which coexistence is due to density-dependence in the host].

As many authors have shown, including an additional trade-off between virulence and competitive ability (ability to take over from or share hosts with less virulent strains) may not only favor higher levels of virulence, but may also support coexistence of multiple strains (Hastings 1980; Levin and Pimentel 1981). In fact, this single additional trade-off has the potential to support an entire continuum of strains (May and Nowak 1994, 1995; Nowak and May 1994; Tilman 1994; see also Chapter 9). In addition, the pattern of coexistence has been shown to differ depending on the mode of interaction between strains, whether it is coinfection (two strains sharing the same host) or superinfection (one strain having the capability to quickly take over a host from another).

This chapter investigates two issues related to this coexistence based on some earlier work (Mosquera and Adler 1998; Adler and Mosquera 2000). Mosquera and Adler (1998) explicitly derive the superinfection model as a limit of the coinfection model, based on the argument that hosts are removed rapidly from the doubly infected class through either recovery or death. This sort of derivation points out an often neglected subtlety of the super- and coinfection processes: the existence of a discontinuous function relating virulence to the ability to coinfect or superinfect (Pugliese 2000).

Let the coinfection function $\phi(\alpha, \eta)$ describe the rate at which a strain with virulence α can coinfect a host infected with strain with virulence η, relative to its ability to infect an uninfected host (Mosquera and Adler 1998, Pugliese 2000). This function will be increasing in α and decreasing in η. Competition is asymmetric because more virulent strains have an advantage within hosts. Other models have used a step function for the coinfection function (see Figure 10.2a), such as

$$\phi(\alpha, \eta) = \begin{cases} \sigma & \text{if} & \alpha > \eta \\ 0 & \text{if} & \alpha \leq \eta \end{cases} \tag{10.1}$$

(Tilman 1994; May and Nowak 1994). Biologically, a slightly more virulent strain has the same advantage as a much more virulent strain, which is probably quite unrealistic. Mathematically, $\phi(\alpha, \eta)$ is discontinuous at $\alpha = \eta$.

Mosquera and Adler (1998) derived pairwise invasibility plots for smooth forms of the coinfection function, showing that the picture changes qualitatively when $\phi(\alpha, \eta)$ is continuous or differentiable at $\alpha = \eta$ (Figures 10.2b to 10.2d). Furthermore, the possibilities for coexistence in superinfection models differ depending on how the superinfection limit is approached (the mechanism by which doubly infected hosts are rapidly removed). Pairwise invasibility plots, however, address only two strains at a time. They show that every strain is invasible when the coinfection function is discontinuous, but cannot reveal the actual diversity of creatures that can coexist in the model.

In related work, Adler and Mosquera (2000) investigated diversity in a superinfection model, simplified by ignoring the trade-off between virulence and transmission efficiency. Smoothing the superinfection function (so that it has a continuous second derivative) eliminates the infinite number of strains that can coexist. Smooth superinfection functions can, when sufficiently steep at $\alpha = \eta$, support a large number of strains, but the number depends sensitively on the slope.

This chapter outlines an approach to diversity and virulence with two trade-offs: the trade-off between virulence and transmission efficiency, and that between virulence and coinfection ability. We present the coinfection model, and take the superinfection limit. The shape of the resultant superinfection function depends both on the underlying assumptions about the coinfection process and on how the superinfection limit is approached. We then examine the number and virulence of coexisting strains. Management of virulence through public health measures requires an understanding of the sensitivity of the results to the details of intrahost competition.

10.2 Coinfection and the Superinfection Limit

Complete models of coinfection can be crushed by the weight of their own notation. Authors have avoided this collapse by assuming that the interaction between strains is simple (May and Nowak 1995; see also Chapter 9), weak (Adler and Brunet 1991), or that hosts can harbor no more than two strains at one time (Van Baalen and Sabelis 1995a).

To illustrate some of the complexity and to introduce the notation, we present a model of a population of hosts beset by many strains of a disease. For simplicity, we assume that the population size is constant and that a single host can be infected by no more than two strains simultaneously. The variables α and η both represent the virulence level and index the disease strains. Let $i(\alpha)$ denote the fraction of hosts infected only by strain α, and $i(\alpha, \eta)$ denote doubly infected hosts who were first infected with strain α and later with η (Figure 10.1). If $\beta(\alpha)$ gives the rate of infection of susceptible hosts by strain α, the differential equations describing this system are

$$\frac{di(\alpha)}{dt} = \beta(\alpha)(1 - i_{++})i_+(\alpha) - \alpha i(\alpha)$$

$$+ \int \left[\theta_1(\eta, \alpha)i(\eta, \alpha) + \theta_2(\eta, \alpha)i(\alpha, \eta) \right] d\eta$$

$$- \int \beta(\eta)\phi(\eta, \alpha)i_+(\eta)i(\alpha) \, d\eta \,, \tag{10.2a}$$

$$\frac{di(\alpha, \eta)}{dt} = \beta(\eta)\phi(\eta, \alpha)i_+(\eta)i(\alpha)$$

$$- \left[(\theta_1(\alpha, \eta) + \theta_2(\eta, \alpha) + \delta(\alpha, \eta) \right] i(\alpha, \eta) \,. \tag{10.2b}$$

The shorthand i_{++} represents the total infected fraction, or

$$i_{++} = \int i(\alpha)d\alpha + \iint i(\alpha, \eta) \, d\eta \, d\alpha \,, \tag{10.2c}$$

while $i_+(\alpha)$ represents the total infectivity of strain α or

$$i_+(\alpha) = i(\alpha) + \int \left[i(\alpha, \eta) + i(\eta, \alpha) \right] d\eta \,. \tag{10.2d}$$

$\theta_1(\alpha, \eta)$ gives the rate of recovery from strain α if infected first by α, $\theta_2(\alpha, \eta)$ gives the rate of recovery from strain α if infected second by α, and $\delta(\alpha, \eta)$ gives the death rate when infected first by α and then by η. The notation is summarized in Figure 10.1.

There are four ways that the virulence of a strain affects its success. First, it has a direct effect on mortality. Second, the infectiousness $\beta(\alpha)$ generally is an increasing function of α. More virulent strains have an advantage in transmission. In the absence of coinfection or superinfection, the strain that maximizes the ratio of transmission efficiency $\beta(\alpha)$ to virulence α excludes all others in competition (Bremermann and Thieme 1989). Third, the coinfection function $\phi(\alpha, \eta)$ describes how virulence determines the ability of strain α to coinfect strain η.

Finally, the virulence might affect the rate of recovery from each strain and the host mortality in doubly infected hosts. If hosts infected with two strains tend

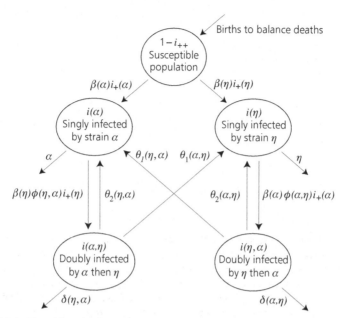

Figure 10.1 Transitions in the general coinfection model. Hosts can be infected with at most two strains. The arrows give the per capita rate at which hosts move from one category to another; arrows pointing into empty space represent deaths.

to recover from the less virulent strain, then the more virulent strain has another advantage in intra-host competition. If hosts tend to recover from the more virulent strain, then the less virulent strain acts as a vaccine, and might be using the host immune system to gain an advantage in intra-host competition. If doubly infected hosts die quickly, the model reduces to the single-infection case (Mosquera and Adler 1998).

Suppose that Equation (10.2b) has reached equilibrium. We can solve for $i(\alpha, \eta)$ and obtain

$$i(\alpha, \eta) = \frac{\beta(\eta)\phi(\eta, \alpha)i_+(\eta)i(\alpha)}{\theta_1(\alpha, \eta) + \theta_2(\eta, \alpha) + \delta(\alpha, \eta)} . \tag{10.3}$$

If the dynamics within hosts are fast relative to the infection dynamics, we can substitute Equation (10.3) into the differential Equation (10.2a) for $i(\alpha)$ to find

$$\frac{di(\alpha)}{dt} = \beta(\alpha)(1 - i_{++})i_+(\alpha) - \alpha i(\alpha)$$

$$+ \int \frac{\theta_1(\eta, \alpha)}{\theta_1(\eta, \alpha) + \theta_2(\alpha, \eta) + \delta(\eta, \alpha)} \beta(\alpha)\phi(\alpha, \eta)i_+(\alpha)i(\eta)\, d\eta$$

$$- \int \frac{\theta_1(\alpha, \eta) + \delta(\alpha, \eta)}{\theta_1(\alpha, \eta) + \theta_2(\eta, \alpha) + \delta(\alpha, \eta)} \beta(\eta)\phi(\eta, \alpha)i_+(\eta)i(\alpha)\, d\eta . \tag{10.4}$$

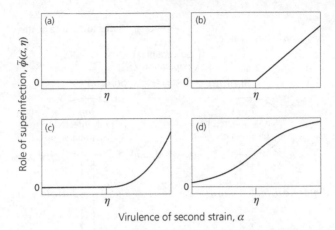

Figure 10.2 Four possible shapes of the superinfection function. Each shows the rate at which strain α takes over individuals already infected with strain η. In the first three panels, a less virulent strain cannot superinfect. (a) The discontinuous case, in which a slightly more virulent strain has a high ability to superinfect. (b) The piecewise differentiable case, in which the ability to superinfect begins increasing immediately. (c) The differentiable case, in which the ability to superinfect shows no sharp change at $\alpha = \eta$. (d) The differentiable case, in which less virulent strains superinfect more virulent strains at a low rate.

Although this might appear to be a one-dimensional superinfection model because doubly infected hosts are not tracked explicitly, they appear implicitly in $i_+(\alpha)$ and $i_+(\eta)$, the total infectivity of strains α and η. To write a closed system solely in terms of $i(\alpha)$ and $i(\eta)$, we generalize the approach of Mosquera and Adler (1998) by considering the limiting case in which hosts are removed rapidly from the doubly infected class. In this case, $i_+(\alpha) = i(\alpha)$ and Equation (10.4) becomes

$$\frac{di(\alpha)}{dt} = \left\{ \beta(\alpha)(1 - i_{++}) - \alpha \right.$$
$$\left. + \int \left[\beta(\alpha)\tilde{\phi}(\alpha, \eta) - \beta(\eta)\tilde{\phi}(\eta, \alpha) \right] i(\eta)\, d\eta \right\} i(\alpha)$$
$$+ \beta(\eta)\tilde{\delta}(\alpha, \eta) , \qquad (10.5a)$$

where

$$\tilde{\phi}(\alpha, \eta) = \frac{\theta_1(\eta, \alpha)}{\theta_1(\eta, \alpha) + \theta_2(\alpha, \eta) + \delta(\eta, \alpha)} \phi(\alpha, \eta) , \qquad (10.5b)$$

and

$$\tilde{\delta}(\alpha, \eta) = \frac{\delta(\alpha, \eta)}{\theta_1(\alpha, \eta) + \theta_2(\eta, \alpha) + \delta(\alpha, \eta)} . \qquad (10.5c)$$

The form of the superinfection function $\tilde{\phi}(\alpha, \eta)$ depends on the coinfection function $\phi(\alpha, \eta)$ and the mechanism that leads to quick removal of hosts from the doubly infected class. Most simply, hosts could be removed quickly if one of the three terms of the denominator of $\tilde{\phi}(\alpha, \eta)$ is large:

- Rapid mortality when doubly infected [large value of $\delta(\alpha, \eta)$]. In this case, $\tilde{\phi}(\alpha, \eta) = 0$ and the coinfection process is irrelevant (this rapid mortality does matter when host population size can change). We do not consider this case explicitly and henceforth assume that $\tilde{\delta}(\alpha, \eta) = 0$.
- Rapid recovery from the less virulent strain [large values of $\theta_1(\alpha, \eta)$ and $\theta_2(\alpha, \eta)$ if $\alpha < \eta$]. In this case, $\tilde{\phi}(\alpha, \eta) = 0$ when $\alpha < \eta$ because a less virulent strain cannot superinfect (Figures 10.2a to 10.2c).
- Rapid recovery from the more virulent strain [large values of $\theta_1(\alpha, \eta)$ and $\theta_2(\alpha, \eta)$ if $\alpha > \eta$]. In this case, $\tilde{\phi}(\alpha, \eta) = 0$ when $\alpha > \eta$. A less virulent strain acts as a vaccine, favoring evolution of a lower level of virulence (results not shown).

In addition, it is possible that individuals recover rapidly from strains as a function of absolute rather than relative virulence. For example, suppose that $\theta_1(\alpha, \eta) = \theta_2(\alpha, \eta) = \theta_0\alpha$ for some large value of θ_0, meaning that recovery is proportional to virulence. The superinfection function is

$$\tilde{\phi}(\alpha, \eta) = \frac{\eta}{\eta + \alpha}\phi(\alpha, \eta) . \tag{10.6}$$

This particular choice for the recovery rate reduces the advantage of more virulent strains in a superinfection model.

If the coinfection function is nonvanishing for $\alpha < \eta$, the superinfection function retains this property (Figure 10.2d). A less virulent strain can take over hosts from a more virulent strain, although probably at a low rate.

10.3 Coexistence and the Superinfection Function

The shape of the superinfection function affects several aspects of the evolutionary outcome in the system:

- the value and existence of an evolutionarily stable strategy (ESS);
- the number of strains in an evolutionarily stable coalition when there is no ESS;
- the abundances of strains in the evolutionarily stable coalition.

We illustrate results with pairwise invasibility plots to show the structural differences among the cases, and with numerical simulations of multiple strains in competition.

As a baseline, we compute the ESS in the single-infection case. A single strain α in isolation follows the equation

$$\frac{di(\alpha)}{dt} = \left[\beta(\alpha)(1 - i_{++}) - \alpha\right]i(\alpha) , \tag{10.7a}$$

with stable equilibrium at

$$i^*(\alpha) = i^*_{++} = 1 - \frac{\alpha}{\beta(\alpha)} \,. \tag{10.7b}$$

In the absence of superinfection, the ESS occurs where i^*_{++} is maximized. With the function

$$\beta(\alpha) = 5\frac{\alpha^2}{4+\alpha^2} \,, \tag{10.8}$$

the ESS virulence level in the single-infection case is equal to 2.

   With superinfection, the per capita growth rate of a strain α invading a population at equilibrium for strain η is

$$f(\alpha, \eta) = \beta(\alpha)(1 - i^*_{++}) - \alpha + \left[\beta(\alpha)\tilde{\phi}(\alpha, \eta) - \beta(\alpha)\tilde{\phi}(\eta, \alpha)\right]i^*_{++} \,, \tag{10.9}$$

where $i^*_{++} = i^*(\eta)$ obeys the equilibrium for strain η [Equation (10.7b)]. A critical point or evolutionary singularity (Geritz *et al.* 1998) occurs where

$$\left.\frac{\partial f(\alpha, \eta)}{\partial \alpha}\right|_{\alpha=\eta} = 0 \,. \tag{10.10}$$

The second derivative of $f(\alpha, \eta)$ with respect to α is negative if $\beta''(\alpha) < 0$, meaning that this critical point is a local maximum when transmission shows diminishing returns. When the critical point becomes invasible, therefore, the invading strain has virulence rather different from that of the former ESS. From Equation (10.9), the critical point occurs where

$$\left.\frac{\partial f(\alpha, \eta)}{\partial \alpha}\right|_{\alpha=\eta} = \beta'(\alpha)(1 - i^*_{++}) - 1$$

$$+ \left[\beta'(\alpha)\tilde{\phi}(\alpha, \alpha) + \beta(\alpha)\left.\frac{\partial \tilde{\phi}(\alpha, \eta)}{\partial \alpha}\right|_{\alpha=\eta}\right.$$

$$\left. - \beta(\alpha)\left.\frac{\partial \tilde{\phi}(\eta, \alpha)}{\partial \alpha}\right|_{\alpha=\eta}\right]i^*_{++} = 0 \,. \tag{10.11}$$

The critical point depends on both the value and the derivative of the superinfection function at the point $\alpha = \eta$. Positive values of $\tilde{\phi}(\alpha, \alpha)$ or of the derivative of $\tilde{\phi}(\alpha, \eta)$ increase the virulence at the critical point above that in the single-infection case.

   With a discontinuous superinfection function (Figure 10.2a), every strain can be invaded by a slightly more virulent strain (Figure 10.3a). There is thus no critical point or candidate ESS, although highly virulent strains can be invaded only by slightly more virulent strains. The resultant coalition (not shown) does not consist of one of the many pairs of mutually invasible strains that appear in black in the pairwise invasibility plot, but of a continuum of strains (Tilman 1994; May and Nowak 1994). Although any single strain could be invaded by a more virulent

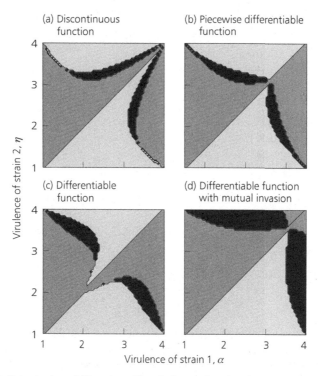

Figure 10.3 Pairwise invasibility plots. The virulence of strain 1 is plotted on the horizontal axis, and the virulence of strain 2 is plotted on the vertical axis. In the lightly shaded regions, strain 1 can invade strain 2, in the darker regions strain 2 can invade strain 1, and in the black regions both can invade each other. In each case, the transmission function is $\beta(\alpha) = 5\alpha^2/(4+\alpha^2)$, chosen to set the uninvasible virulence at $\alpha = 2$ in the absence of coinfection [Equation (10.8)]. (a) Discontinuous superinfection function (Figure 10.2a). (b) Piecewise differentiable superinfection function (Figure 10.2b). With $x = \alpha - \eta$, the functional form is $\tilde{\phi}(\alpha, \eta) = \sigma x/1 + \sigma x$ if $\alpha > \eta$ and $\tilde{\phi}(\alpha, \eta) = 0$ if $\alpha < \eta$. The parameter σ represents the slope at $x = 0$, and is set to $\sigma = 1$. (c) Differentiable superinfection function (Figure 10.2c). With $x = \alpha - \eta$, the functional form is $\tilde{\phi}(\alpha, \eta) = \sigma x^2/1 + \sigma x^2$ if $\alpha > \eta$ and $\tilde{\phi}(\alpha, \eta) = 0$ if $\alpha < \eta$. The parameter σ is set to $\sigma = 1$. (d) Differentiable superinfection function with nonvanishing invasion rate by a less virulent strain (Figure 10.2d). The functional form, with $x = \pi/2(\alpha - \eta)$, is $\tilde{\phi}(\alpha, \eta) = 2/\pi \tan^{-1}(\sigma x) + c$, where σ is the slope at $x = 0$. The slope at $x = 0$ is set to 1.

strain, the mix of more and less virulent strains in the coalition places an upper bound on the persisting virulence.

A piecewise differentiable superinfection function (Figure 10.2b) does include a critical point (Figure 10.3b). As the slope of the positive part of the curve increases, the virulence at the critical point increases and then undergoes evolutionary branching (Figure 10.4). With the parameter values in Figure 10.3b, the critical value is unstable, and thus can be invaded by a less virulent strain. With a yet steeper slope, the coalition includes several less virulent strains. Unlike the

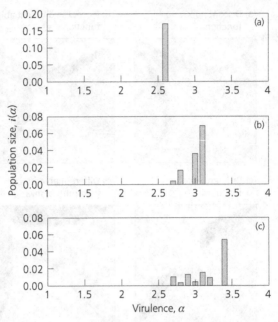

Figure 10.4 Evolutionary outcomes for the piecewise differentiable superinfection function (Figure 10.2b). Slopes σ at $\alpha = \eta$ are 0.5, 1.0, and 2.0 in panels (a), (b), and (c), respectively. The simulations are based on forty evenly spaced strains started from uniform initial conditions and were run until convergence occurred.

discontinuous case, the number of strains increases one by one as the slope of the superinfection function is increased.

A differentiable superinfection function with no invasion by less virulent strains (Figure 10.2c) maintains the critical point at $\alpha = 2$, even as the positive part of the curve increases more quickly (Figure 10.3c). This occurs because the slope at $\alpha = \eta$ is always 0 (Mosquera and Adler 1998). As the curve increases more quickly, the ESS value remains at 2 until it bifurcates, when the more common strain both begins to increase in virulence and is invaded by a less common strain of even higher virulence (Figure 10.5). With a more rapid increase in the curve, more strains join the coalition one by one. This case differs from the piecewise differentiable case in that virulence levels only increase after the critical point destabilizes, and in that the most common strain is the least rather than the most virulent.

A differentiable superinfection function that allows some invasion by less virulent strains (Figure 10.2d) produces results most similar to the piecewise differentiable case because the slope at $\alpha = \eta$ is neither infinite (as in the discontinuous case) nor 0 (as in the differentiable case). In both cases, the critical virulence increases as the slope increases, and the coalitions that arise are dominated by the most virulent strain (Figure 10.6).

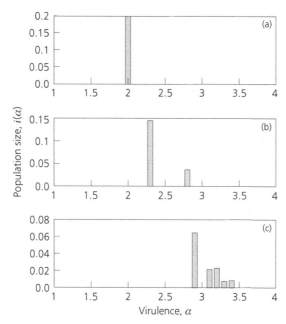

Figure 10.5 Evolutionary outcomes for the differentiable superinfection function (Figure 10.2c). The parameter σ takes on values 0.5, 1.0, and 2.0 in panels (a), (b), and (c), respectively.

However, the virulence levels supported in this case tend to be higher because of the additional mixing among strains. As a simple example, consider the invasion criterion in Equation (10.11) in the simplified case in which any strain can take over from any other at the same rate, or $\phi(\alpha, \eta) = \overline{\phi}$ for any α and η. Substituting into Equation (10.11) gives

$$\frac{\partial f(\alpha, \eta)}{\partial \alpha}\Big|_{\alpha=\eta} = \beta'(\alpha)(1 - i_{++} + \phi i_{++}) - 1 = 0 . \qquad (10.12)$$

Substituting for i_{++} [Equation (10.7b)] and solving for $\beta'(\alpha)$ gives the condition

$$\beta'(\alpha) = \frac{\beta(\alpha)/\alpha}{1 - \phi + \phi\beta(\alpha)/\alpha} . \qquad (10.13)$$

When $\phi = 0$, this reduces to the usual condition (Mosquera and Adler 1998). If we treat the critical value of α as a function of ϕ and differentiate both sides, it is not difficult to show that α is an increasing function of ϕ at $\phi = 0$. Superinfection, even in the absence of any competitive advantage for the more virulent strains, favors strains with higher virulence.

10.4 Discussion

The virulence and coexistence results obtained from models of coinfection depend on the assumptions underlying the coinfection process:

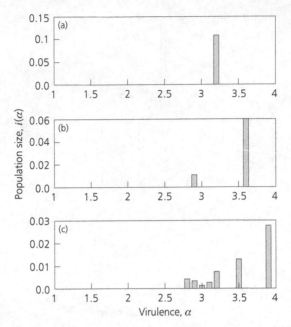

Figure 10.6 Evolutionary outcomes for the differentiable superinfection function with non-vanishing invasion rate by the less virulent strain (Figure 10.2d). The slope σ at $\alpha = \eta$ is 0.5, 1.0, and 2.0 in panels (a), (b), and (c), respectively.

- Which strains can coinfect;
- How coinfection depends on the virulence of the two strains;
- The fate of doubly infected hosts.

When the time scale of double infection is short, these assumptions manifest themselves in the shape of the superinfection function in the resultant model.

The evolutionarily stable coalition that results depends on the shape of the superinfection function for similar strains. If there is a discontinuity, so that slightly more virulent strains can superinfect at a high rate, a continuum of strains can coexist. This result vanishes with a continuous function. In general, as the slope becomes larger, the community adds strains one by one, with a concomitant increase in average virulence. Furthermore, a nonvanishing rate of superinfection by identical strains increases the virulence in any coalition. The virulence of the most abundant strain in the coalition also depends on the shape of the superinfection function. Only for the differentiable superinfection function (see Figure 10.5) is the least virulent strain the most common.

Different shapes of the superinfection function thus produce different patterns of coexistence, with more detailed differences including whether the most abundant strain is the most or least virulent. Ideally, these predictions could be compared with estimates of real superinfection functions to test whether the real systems exist in the region of parameter space where large numbers of strains can coexist, or whether this particular trade-off tends to support only a few strains with a characteristic pattern of abundance.

Is manipulation of the superinfection function a viable strategy for virulence management? In general, reducing the extent of superinfection reduces virulence. More specifically, our models show that decreasing the competitive advantage of more virulent strains reduces not only the mean virulence, but also the variance in virulence that provides the potential for further evolution. Any control measures that selectively harm more virulent strains could help control virulence in both the short and long term.

11

Multiple Infection and Its Consequences for Virulence Management

Sylvain Gandon and Yannis Michalakis

11.1 Introduction

Many parasites exploit their host in order to accomplish within-host reproduction and allow transmission to new hosts. However, an extreme exploitation strategy may incur a cost since it might decrease the life expectancy of the host and, as a consequence, the chances of the parasite being transmitted. In this respect, virulence (i.e., the deleterious effect induced by the parasite) can be considered a by-product of the parasite's host-exploitation strategy. Such a trade-off leads to the conclusion that parasites should evolve toward intermediate levels of virulence. This idea has been formalized by several authors (Anderson and May 1979; Ewald 1983; Van Baalen and Sabelis 1995a; Frank 1996c), who found that an evolutionarily stable level of virulence depends on several life-history parameters for both the host and parasite, as well as some constraints (such as the classic trade-off between transmission ability and virulence).

Moreover, it has been shown that multiple infections (i.e., the ability of the parasite to infect an already infected host) increase within-host competition and select for higher levels of virulence (Eshel 1977; Bremermann and Pickering 1983; Frank 1992a, 1994b, 1996c; Nowak and May 1994; May and Nowak 1995; Van Baalen and Sabelis 1995a). Several models have been developed around this idea (see Box 5.1), but we believe the kin-selection model proposed by Frank remains the simplest way to address this question (but see Box 11.1). Frank formalized the idea that there is a strong analogy between the evolution of altruism and that of parasite virulence. More intense host exploitation can be viewed as a selfish strategy selected for at the within-host level, but selected against at the between-host level since it induces a cost (higher virulence results in lower transmission) on the entire group of parasites that share the same infected host. Therefore, if the relatedness among parasites within infected hosts is very high, lower levels of virulence should be selected for (Frank 1992a, 1994b, 1996c).

In this chapter, we study the consequences of multiple infections on the evolution of parasite virulence. First, following the approach developed by Frank, we present a general kin-selection model for the evolution of virulence and, since transmission ability is tightly linked with multiple infections and parasite virulence, the evolution of parasite dispersal. Second, we use this model to analyze

Box 11.1 Kin-selection models of within-host competition

Since each infected host harbors a population of parasites, a population of hosts can be viewed as a metapopulation of parasites. Frank (1994b, 1996c) and Taylor and Frank (1996) have developed an approach that formalizes the notion of kin selection for the evolution of parasite virulence in a parasite metapopulation structure. This approach is based on a two-step argument: first, the derivation of the evolutionarily stable virulence and, second, the derivation of relatedness.

Derivation of evolutionarily stable virulence. The fitness w of an individual parasite depends on its own virulence α and on the average level of virulence, $\overline{\alpha}$, of the other parasites that share the same infected host, $w = w(\alpha, \overline{\alpha})$. The following simple expression for parasite fitness,

$$w(\alpha, \overline{\alpha}) = \frac{\alpha}{\overline{\alpha}} (1 - \overline{\alpha}) , \qquad (a)$$

has been suggested by Frank (1994b, 1996c). The first term, $\alpha/\overline{\alpha}$, describes within-host success (a high relative virulence within the host increases parasite fitness), whereas the second term, $1 - \overline{\alpha}$, describes between-host success (a high average level of virulence decreases the transmission rate of the parasite and, in consequence, parasite fitness). The model presented in this chapter is more general than this and is based on a more realistic expression for parasite fitness; here we use Frank's model for illustration.

We suppose that the virulence of a parasite is determined by its genotype and consider a monomorphic population with virulence α. If

$$\left. \frac{dw}{d\alpha} \right|_{\alpha=\overline{\alpha}=\alpha^*} = 0 , \qquad (b)$$

then α^* is an evolutionarily stable level of virulence. The total derivative $dw/d\alpha$ is understood here as the sensitivity of the fitness w to varying the virulence α of a focal individual together with that of all its kin, that is, of its identical-by-descent relatives. Since w then depends on α not only directly but also through $\overline{\alpha}$, this derivative is given by

$$\frac{dw}{d\alpha} = \frac{\partial w}{\partial \alpha} + \frac{\partial w}{\partial \overline{\alpha}} \frac{d\overline{\alpha}}{d\alpha} . \qquad (c)$$

The sensitivity of the average virulence $\overline{\alpha}$ of parasites within a host to varying the virulence α of a focal individual together with that of all its kin,

$$\rho = \frac{d\overline{\alpha}}{d\alpha} , \qquad (d)$$

is called the coefficient of relatedness. If all parasite individuals within an infected host are unrelated, then varying the virulence α of a single parasite has practically no effect on the average virulence $\overline{\alpha}$ within that host; ρ is therefore (very close to) zero. By contrast, if all parasites within a host are kin, then $\rho = d\overline{\alpha}/d\alpha = d\alpha/d\alpha = 1$. Using Equations (a) to (d), the evolutionarily stable level of virulence in Frank's model can be calculated; after some algebra this yields

continued

Box 11.1 *continued*

$$\alpha^* = 1 - \rho \,. \tag{e}$$

As shown by Gandon (1998), it turns out that this simple result also holds for a model in which the death rate of hosts is taken into account for determining parasite fitness.

Derivation of relatedness. The second step is to reveal the dependence of the coefficient of relatedness ρ on other model parameters. If one assumes that selection is weak (i.e., that the virulence of mutants differs only slightly from that of their ancestors), ρ can be derived from classic identity-by-descent coefficients (Taylor 1988; Taylor and Frank 1996). For example, in Frank's model (1994b) in which parasites are asexual and haploid and hosts do not die, the coefficient of relatedness is given by

$$\rho = 1/[P - (P-1)(1-\lambda)^2] \,, \tag{f}$$

where P is the number of parasites within each infected host and λ is the parasite immigration rate or force of infection. This, in turn, depends on the fraction p_{leave} of parasites that leave their host in each parasite generation, on the probability $1 - p_{\text{failure}}$ for these to enter a new host, and on the extinction rate μ of parasite populations (Frank 1994b),

$$\lambda = \frac{(1 - p_{\text{failure}})(1 - \mu)p_{\text{leave}}}{1 - p_{\text{leave}} + (1 - p_{\text{failure}})(1 - \mu)p_{\text{leave}}} \,. \tag{g}$$

The assumptions of asexuality and haploidy are relaxed in Taylor (1988) and the assumption of no host mortality is relaxed in Gandon (1998).

Strengths and limitations of the kin-selection approach. The approach outlined here offers a simple and powerful method to analyze the qualitative effects of multiple infections on the evolution of virulence in structured populations in which selection operates at two levels (local population versus metapopulation) and to separate the direct effects of selection from indirect ones that result from relatedness.

Limitations of the kin-selection approach arise because the phenotype of a mutant is assumed to be very close to that of the resident. This simplifies derivations of relatedness since one can use classic identity-by-descent coefficients. However, an important drawback of this assumption is that it does not include the potential occurrence of parasite polymorphism, a feature shown to emerge in other types of models (e.g., Bonhoeffer and Nowak 1994a; May and Nowak 1994, 1995; Nowak and May 1994). The kin-selection approach further assumes that the parasite–host system has reached a stable epidemiological equilibrium at which every single host is infected and that there is an infinite number of infected hosts so that parasites from two randomly chosen infected hosts are not related. How the epidemiological details of the parasite–host interaction affect the evolutionary outcome is still largely unclear (but see Box 7.1 and Frank 1992a, 1996c; May and Nowak 1994, 1995; Van Baalen and Sabelis 1995a; Gandon *et al.* 2001).

how parasite propagule survival may affect the evolution of virulence when multiple infections occur. This example leads us to a more general discussion on the effects that emerge only through kin selection of several other parameters (i.e., contact rate, host resistance, host clearance rate). Finally, in light of our analysis, we discuss the implications of several classic health policies (i.e., sanitation, medical treatment, vaccination, etc.) for virulence management.

11.2 Multiple Infection, Virulence, and Dispersal

In this section, we present the general kin-selection model that we use throughout this chapter to study the effects of multiple infection on the evolution of parasite virulence and parasite dispersal. Details of this general model are presented elsewhere (Gandon 1998), but in the following we present the main assumptions of the host and the parasite life cycles.

Let us first assume that the habitat is filled with an infinite number of hosts. For the sake of simplicity, we assume that host mortality is only due to parasite virulence α.

The model further assumes the following parasite life cycle:

1. Generations are discrete.
2. Every single host is infected by a constant number P of haploid and asexual parasites.
3. The average within-host relatedness among parasites is ρ, and throughout the chapter we assume that parasites from different hosts are unrelated.
4. Parasites compete against each other for resources provided by the host, and the level of parasite competitiveness is measured by α.
5. We also assume that within-host competition has a deleterious effect on hosts, since the parameter α measures the parasite-induced host mortality (i.e., parasite virulence). We further assume that the death of an infected host leads to the extinction of the whole parasite population before parasite dispersal. Hence, the extinction rate of parasite populations μ and the overall host mortality are linked: $\mu = \alpha$. Note that assumptions (4) and (5) induce a trade-off between parasite virulence and parasite transmission.
6. After reproduction, a proportion p_{leave} of the offspring leave their host and try to reach another.
7. During the transmission phase these dispersed propagules pay a cost of dispersal by failing to infect a susceptible host with a probability p_{failure}.
8. Effects of the mode of parasite transmission are characterized by the probability p_{common} of common origin of migrants (Whitlock and McCauley 1990; Gandon 1998). In particular, when $p_{\text{common}} = 0$ all immigrant parasites come from different hosts. For example, this might be the case for an airborne disease. At the other extreme, when $p_{\text{common}} = 1$, all immigrant parasites come from a single infected host. This situation might be closer to a vectorborne type of transmission or to a sexually transmitted disease.

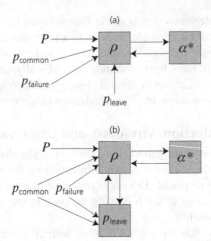

Figure 11.1 Indirect effects via relatedness ρ. Effects of various parameters (P, P_{leave}, P_{failure}, P_{common}) on the evolution of parasite virulence α^* when virulence and dispersal are unlinked traits. In (a) parasite dispersal P_{leave} is a passive trait and in (b) dispersal P_{leave} coevolves with parasite virulence.

9. Finally, we assume that the parasite fecundity is sufficiently large to allow the infection of each susceptible host. Therefore, all hosts are infected after the parasite dispersal phase.

Under these assumptions, the host population reaches a stable age structure distribution that depends only on the host mortality and on the age of the host (Olivieri *et al.* 1995)

$$n_a = \mu(1 - \mu)^a, \tag{11.1}$$

where n_a is the frequency of hosts of age a.

Note that our model relies on several oversimplified assumptions. In particular, we assume that all the hosts are infected. This greatly simplifies the algebra, removes the potential consequences of host–parasite epidemiological dynamics, and, as a consequence, allows us to focus on the effects of multiple infections in a simple case (i.e., when the parasite has reached an epidemiological equilibrium).

Evolution of parasite virulence

Under the assumptions presented above, it is possible to formulate explicitly the inclusive fitness of an individual parasite (Gandon 1998) and, following the approach developed by Taylor and Frank (1996), to search for the evolutionarily stable life-history strategies of the parasite. When dispersal is not correlated with parasite virulence, the evolutionarily stable virulence α^* is equal to $1 - \rho$ (Gandon 1998). This simple expression, first derived by Frank (1994b, 1996c) in a simpler model (see also Box 11.1), captures the kin-selection argument underlying the effects of multiple infections: higher relatedness among parasites decreases virulence.

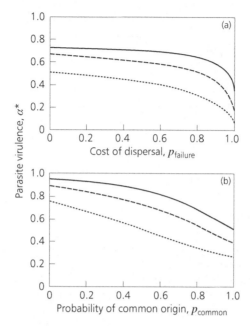

Figure 11.2 Evolution of parasite virulence. (a) Dependence of the evolutionarily stable level of virulence, α^*, on the cost of dispersal, $p_{failure}$, for three different parasite dispersal rates: $p_{leave} = 0.2$ (dotted curve), $p_{leave} = 0.5$ (dashed curve), and $p_{leave} = 0.8$ (continuous curve). Other parameter values: $P = 10$, $p_{failure} = 0.9$. (b) Dependence of the evolutionarily stable level of virulence on the probability of common origin, p_{common}, for three different parasite population sizes: $P = 5$ (dotted curve), $P = 10$ (dashed curve), and $P = 20$ (continuous curve). Other parameter values: $p_{leave} = 0.5$, $p_{failure} = 0.9$.

However, it can be argued that relatedness among parasites is not a fixed parameter but a dynamic variable that depends on several other parameters. Therefore, several parameters may affect the evolution of virulence indirectly through their effects on relatedness (see Figure 11.1). Lower costs of dispersal or higher dispersal rates increase the probability of a given host being infected several times. As a consequence, the relatedness among parasites decreases and the evolutionarily stable virulence increases (Figure 11.2a). Similarly, when the number of infecting parasites P increases or when the probability of common origin p_{common} decreases, then relatedness decreases and the evolutionarily stable parasite virulence increases (see Figure 11.2b).

Evolution of parasite dispersal

Using the approach used to derive the evolutionarily stable parasite virulence (see Box 11.1), it is possible to derive the evolutionarily stable dispersal rate of the parasite. For situations in which virulence and dispersal are not correlated traits, Gandon and Michalakis (1999) showed that the evolutionarily stable dispersal probability is given by the following analytical expression:

$$p^*_{\text{leave}} = \frac{1}{2B}\left[A - \sqrt{A^2 - 4\mu(1 - \rho p_{\text{common}})B}\right], \tag{11.2a}$$

where

$$A = p_{\text{failure}} + \mu^2(1 - p_{\text{failure}}) + \mu - \rho(1 - \mu)$$
$$- 2\mu p_{\text{common}}\rho[p_{\text{failure}} + \mu(1 - p_{\text{failure}})] \tag{11.2b}$$

and

$$B = \left[p_{\text{failure}} + \mu(1 - p_{\text{failure}})\right]^2 - \rho(1 - \mu) - p_{\text{common}}\rho\left[(1 - p_{\text{failure}})^2\right.$$
$$\left. - \mu(3 - 6p_{\text{failure}} + 2p^2_{\text{failure}}) + \mu^2(3 - 4p_{\text{failure}} + p^2_{\text{failure}})\right]. \tag{11.2c}$$

This expression formalizes the effects of p_{failure}, ρ, μ, and p_{common} on the evolution of dispersal. In general, lower p_{failure} or p_{common} and higher ρ or μ select for higher dispersal rates. Indeed, Figures 11.3a and 11.3b show that higher costs of dispersal, higher probability of common origin, or lower virulence tend to decrease the evolutionarily stable dispersal rate.

The effect of the mode of dispersal can be explained by a kin-selection argument. When p_{common} is large, immigrants have to compete against related individuals (i.e., other immigrants originating from the same population). This induces an extra cost of dispersal and selects for lower dispersal rates (Gandon and Michalakis 1999).

When ρ is used as a dynamic variable, some parameters may indirectly affect the evolution of dispersal. For example, a higher within-host population size decreases relatedness among parasites and, as a consequence, increases the evolutionarily stable parasite dispersal rate (Figure 11.3b).

Let us now assume that virulence and dispersal are two coevolving traits. In this case, there is no simple analytic expression for the evolutionarily stable strategies of virulence and dispersal. The derivation of such strategies, however, can be pursued with numerical simulations. Figure 11.4 presents the evolutionarily stable virulence α^* and dispersal p^*_{leave} versus the cost of dispersal p_{failure}. Not surprisingly, higher cost of dispersal decreases both virulence and dispersal. However, note that the effects of higher cost of dispersal seem to be more pronounced when virulence and dispersal coevolve (compare Figures 11.2a, 11.3a, and 11.4). This results from synergistic effects that emerge from the coevolution between virulence and dispersal. First, higher costs of dispersal result in decreases in both virulence and dispersal, independently (Figures 11.2a and 11.3a). Second, a drop in virulence selects for lower dispersal rates (see Figure 11.3a), and, reciprocally, lower dispersal selects for lower virulence (see Figure 11.2a). These interactions strengthen the effect of higher cost of dispersal on both the evolution of virulence and dispersal (Gandon 1998).

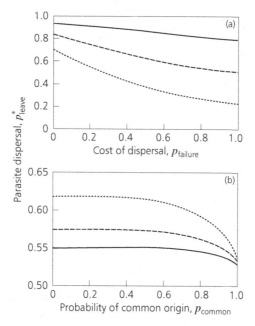

Figure 11.3 Evolution of parasite dispersal. (a) Dependence of the evolutionarily stable dispersal rate, p^*_{leave}, on the cost of dispersal, p_{failure}, for three different levels of parasite virulence: $\alpha = 0.1$ (dotted curve), $\alpha = 0.5$ (dashed curve), and $\alpha = 0.9$ (continuous curve). Other parameter values: $P = 10$, $p_{\text{failure}} = 0.9$. (b) Dependence of the evolutionarily stable dispersal rate on the probability of common origin, p_{common}, for three different parasite population sizes: $P = 5$ (dotted curve), $P = 10$ (dashed curve), and $P = 20$ (continuous curve). Other parameter values: $\alpha = 0.5$, $p_{\text{failure}} = 0.9$.

11.3 Indirect Effects

In Section 11.2, we show that several parameters may affect the evolution of virulence indirectly (via relatedness). In the following, we develop this argument and show how several classic environmental and life-history parameters may act indirectly on the evolution of virulence. First we analyze the indirect effects of propagule survival on parasite virulence. Second, we generalize this kin-selection argument to other parameters.

The curse of the pharaoh

Ewald proposed that higher propagule survival may promote evolution toward higher levels of parasite virulence (Ewald 1987a, 1994a). The basic argument in favor of this hypothesis is that the "cost of virulence" should decrease when propagule survival increases: even highly virulent strains find a susceptible host to infect if they can survive for a very long time in the environment. This hypothesis is also known as the "sit-and-wait" hypothesis (Ewald 1987a, 1994a) or as "the curse of the pharaoh" hypothesis in reference to the highly virulent and very long-lived pathogen that some have claimed was responsible for the mysterious death

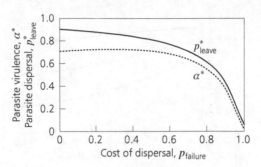

Figure 11.4 Coevolution of parasite virulence and parasite dispersal. Dependence of the evolutionarily stable level of virulence, α^* (dotted curve), and of dispersal, p_{leave}^* (continuous curve), on the cost of dispersal, $p_{failure}$, with $p_{common} = 1$ and $P = 10$.

of Lord Carnarvon after it lay dormant in the tomb of Tutankhamen (Bonhoeffer *et al.* 1996).

Bonhoeffer *et al.* (1996) formalized this argument to test the validity of the hypothesis. They found that if the host–parasite system has reached an ecological equilibrium, parasite propagule survival does not affect the evolution of virulence. However, in nonequilibrium situations, and, in particular, during an epidemic, higher propagule survival increases the evolutionarily stable parasite virulence.

In the following, we extend the investigation of Bonhoeffer *et al.* (1996) to the case in which multiple infections can occur. For the sake of simplicity, we focus on a host–parasite system that has reached a stable epidemiological equilibrium. Using a modified version of the general kin-selection model presented in Section 11.2, we assume that:

■ Dispersed propagules fail to infect a susceptible host with a probability $p_{failure}$ (the basic cost of dispersal).

■ Unsuccessful propagules have a probability $p_{survive}$ to survive until the next generation.

If parasite propagules reach the next generation, they have another chance to infect a host (see Figure 11.5). Under these assumptions, a propagule effectively infects a host with a probability p_{infect} (i.e., the transmission efficiency of the parasite) given by

$$p_{infect} = (1 - p_{failure}) \sum_{t=0}^{\infty} (p_{failure} p_{survive})^t = \frac{1 - p_{failure}}{1 - p_{failure} p_{survive}} . \tag{11.3}$$

It is worth returning to the proper definition of the effective cost of dispersal, which is

$$1 - p_{infect} = \frac{p_{failure}(1 - p_{survive})}{1 - p_{failure} p_{survive}} . \tag{11.4}$$

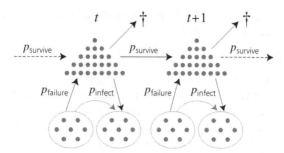

Figure 11.5 Schematic representation of parasite life cycle with parasite propagule. At time t, parasites from a given infected host disperse and, eventually, infect a new host. With a probability p_{failure}, parasite propagules fail to infect a host. In this case, parasite propagules have a probability p_{survive} to survive until the next generation $t + 1$ where they will have another chance to infect a host. Under this assumption the probability for a given parasite propagule to effectively infect a host is p_{infect} (see more explanations and the explicit formulation of p_{infect} in the text).

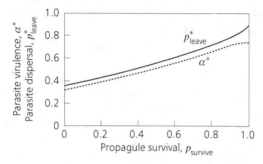

Figure 11.6 The curse of the pharaoh hypothesis. Dependence of the evolutionarily stable parasite virulence, α^* (dotted curve), and parasite dispersal, p_{leave}^* (continuous curve), on propagule survival, p_{survive}. Parameters: $P = 10$, $p_{\text{common}} = 1$, $p_{\text{failure}} = 0.9$.

This simple expression shows that higher survival rates decrease the effective cost of dispersal. From Section 11.2, we know that lower costs of dispersal increase the evolutionarily stable virulence (see Figure 11.2a), and, not surprisingly, we found that higher propagule survival tends to increase the evolutionarily stable parasite virulence in accordance with the curse of the pharaoh hypothesis (see Figure 11.6). Gandon (1998) showed that this qualitative effect seems to be quite robust for a wide range of parameter values. However, in some situations (and in particular when dispersal and virulence are negatively correlated traits), higher propagule survival may also select for lower evolutionarily stable virulence.

The effect of higher propagule survival is used here as a case study to illustrate the potential importance of indirect effects (see also the direct effects of propagule survival analyzed in Chapters 2 and 28). In the following subsection, we analyze several other hypotheses concerning the potential effect of some parameters in the light of our kin-selection approach.

Other indirect effects

Contact rate. Ewald (1994a) proposed that a higher number of contacts between hosts may select for larger parasite virulence. In particular, Ewald suggested that higher rates of transmission through sexual contacts or needleborne transmission would result in selection for more virulent strains of human immunodeficiency virus (HIV). The validity of this prediction has been studied formally by Lipsitch and Nowak (1995), who found that the contact rate affects the evolution of parasite virulence only in nonequilibrium situations. In particular, during an epidemic, higher contact rates increase parasite virulence. However, when the host–parasite system has reached an epidemiological equilibrium, the contact rate no longer affects parasite virulence. These results are similar to the conclusion of Bonhoeffer *et al.* (1996) regarding the effect of propagule survival. Indeed, both contact rate and propagule survival affect a more generic parameter of parasite life-cycle: the probability of transmission. In a broader perspective, it has been noted (e.g., see Frank 1996c) that the probability of transmission does not affect the evolution of parasite virulence at equilibrium, but may increase the evolutionarily stable virulence during an epidemic. However, this very general prediction relies on the assumption that multiple infections do not occur, and we know from our study on the effects of propagule survival that this general prediction may be altered when this hypothesis is relaxed. This strongly suggests that if multiple infections are allowed, then higher host contact rates may also increase parasite virulence when the whole system is at equilibrium because of indirect effects. Indeed, if we modify our model by simply assuming that the cost of dispersal is a decreasing function of the contact rate, then we find that higher contact rates increase virulence.

Van Baalen (Chapter 7) addressed a similar question by looking at the effects of contact rate on the evolution of parasite virulence in a viscous host population. He found that, in accordance with Ewald's prediction, higher contact rates increase virulence. Although multiple infections are not allowed in Van Baalen's model, kin-selection processes are also influential because of the viscosity of the host population. In this situation, kin selection does not operate at the individual host scale but at the scale of the cluster of infected hosts (Van Baalen and Rand 1998). There is, however, a strong analogy with our kin-selection model that results from the incorporation of some spatial structure. The indirect effects of higher contact rates that we study here emerge via this spatial structure.

So far, we have focused our attention on the effects of transmission rates; however, as we noticed in the previous subsection, the mode of transmission may also affect the evolution of parasite virulence. In our model, the probability of common origin p_{common} offers a simple way to account for different modes of transmission. Very generally, higher p_{common} increases within-host relatedness among parasites and tends to decrease parasite virulence (Figure 11.2b). Note the interesting analogies between this result and the effect of the mode of transmission in

Van Baalen's model (where ϕ describes the regularity of the network of hosts; see Chapter 7).

Vaccination and host heterogeneity. In the previous subsections, we assumed that all hosts are fully susceptible. Relaxing this assumption may also affect indirectly the evolution of parasite virulence if multiple infections occur. For example, if we assume that the host population is composed of a certain proportion r of fully resistant hosts, then the parasite successfully infects a host with a probability $(1 - p_{failure})(1 - r)$. A higher proportion of resistant hosts decreases the probability of infection (this effect is best known as "herd immunity") and, as a consequence, increases within-host relatedness among parasites. In this respect, our prediction is that a higher proportion of resistant hosts may decrease parasite virulence. This effect has actually been observed by May and Nowak (1994, see also Chapter 9). They found that when superinfection occurs, the removal of a certain proportion of the host population selects for lower parasite virulence. This result is consistent with the kin-selection argument we have proposed.

Host clearance rate. It has been shown that when only single infections occur, higher intrinsic host death rates select for higher parasite virulence. Ebert and Mangin (1997) pointed out that the occurrence of multiple infections may strongly alter this prediction. Indeed, higher host death rates may decrease the probability that a given host will be infected by multiple strains and, as a consequence, increase the average within-host relatedness among parasites. This process may result in a decrease of parasite virulence with higher host death rate, because such an indirect effect may be stronger than the classic direct effect that tends to increase parasite virulence when host mortality increases (Gandon *et al.* 2001).

11.4 Virulence Management

Virulence management aims at decreasing the deleterious effects of parasites on their hosts. There are two main components in the deleterious effect induced by the parasite:

- The risk of being infected (i.e., prevalence);
- Once infected by a particular parasite, the pathogenicity of the parasites (i.e., virulence).

Ideally, virulence management should promote policies that counter both. This may lead to short-term (epidemiological) and long-term (evolutionary) beneficial effects.

In the following, we try to partition the effects of three different types of classic health policies between these two components (Table 11.1). This analysis allows us to stress the potential conflict that may emerge in virulence management, since a particular intervention may be beneficial in the short term but could have negative consequences in the long term. Moreover, we aim to show the importance of the occurrence of multiple infections in understanding the effects of some classic

Table 11.1 Short- and long-term effects of some classic health policies when only single infections occur as opposed to situations when multiple infections occur. "+" and "−" indicate *positive* and *negative* effects, respectively, while a "0" indicates no effect and "?" indicates that the effect is not known. We only consider situations in which the host–parasite system has reached a stable epidemiological equilibrium. Note the differences between, first, the single and multiple infection cases and, second, the short- and long-term effects. See the text for more explanations and discussion.

Health policies	Consequences	Single infection		Multiple infection	
		Short-term effect	Long-term effect	Short-term effect	Long-term effect
Sanitation (lower contact rate)	Decrease in transmission	+	0	+	+
Medical treatment	Increase of clearance rate	+	−	+	?
Vaccination	Decrease in prevalence	+	0	+	+
Large-scale use of antibiotics	Emergence of resistance	+	−	+	−−

health policies on the evolution of virulence. Table 11.1 summarizes the effects of different interventions in two different types of models:

- Only single infections occur;
- Multiple infections are allowed.

We emphasize that we restrict our analysis to cases in which the host–parasite system is at epidemiological equilibrium, and we further assume that the host population is homogeneous and spatially unstructured (i.e., a parasite has an equal probability to reach any individual host). Relaxing one of these assumptions could greatly alter the predictions presented in Table 11.1.

First, some classic interventions (e.g., sanitation) aim to reduce contact rates. More generally, prophylactic interventions act directly through a reduction of parasite transmission rate. This, of course, has a straightforward beneficial effect in the short term since it lowers the risk of being infected by a parasite. However, there might also be a long-term effect of such interventions since, if multiple infection occurs, we expect lower transmission rates to decrease parasite virulence.

The development of medical treatments (e.g., antibiotics) reduces the time a single host is infected; in reference to classic epidemiological models (Anderson and May 1991; Van Baalen and Sabelis 1995a), the clearance rate increases. This is beneficial for the individual host that needs to be cured. However, in models with only single infections, one expects that these interventions should increase parasite virulence. A dilemma emerges here since the short-term effect may benefit a particular individual host, but have long-term deleterious effects on the entire host population. This is not necessarily the case when multiple infections occur, since higher clearance rates tend to decrease the transmission rate and, consequently, the rate of multiple infections (see Section 11.3 on the effect of clearance rate).

Finally, let us briefly examine the case of large-scale use of antibiotics and vaccination campaigns. First, in many intensive production units of cattle, farmers systematically include antibiotics in bovine nutrients. Such use of antibiotics significantly increases productivity because it both prevents parasitic infections and promotes the growth of animals. Second, in a similar way, vaccination campaigns are large-scale interventions that prevent infection by particular strains of parasites (here we do not consider the use of imperfect vaccines). Both of these interventions are beneficial in the short term because they lower the risk of being infected and decrease parasite prevalence. If multiple infections occur, one might also expect a long-term beneficial effect through a decrease in parasite virulence because of the drop in transmission rate. However, there might also be long-term detrimental effects if the ability of the parasite to evolve resistance against antibiotics or vaccines is taken into account (Baquero and Blázquez 1997). It has long been pointed out that the systematic use of antibiotics may promote the emergence of resistant strains of parasites. This may have important consequences on cattle and, hence, on human populations, since antibiotics used for the cattle are often similar to those used for human treatments. A similar argument holds for vaccination campaigns. Moreover, if multiple infections occur, there might be another deleterious effect of emerging resistant strains, and a particular resistant strain may be expected to reach high levels of prevalence in the host population. This, in turn, may increase the probability of multiple infections and, as a consequence, may select for high virulence. We indicate these two evolutionary consequences by putting two minus signs in the last cell of Table 11.1. We believe these effects could be even stronger if the epidemiological aspects of the emergence of resistant strains are considered.

These particular examples show:

▪ The importance of including the impact of multiple infections in this type of analysis;
▪ The difficulties that may occur when the effects of particular interventions are examined at different temporal scales (see also Chapter 5).

All the classic health policies used are beneficial in the short term. However, some of them (e.g., medical treatment, large-scale use of antibiotics) may also have negative consequences in the long term. This indicates that particular policies should be promoted or avoided according to the time scale of interest.

11.5 Discussion

In this chapter, we demonstrate some implications of multiple infections for the evolution of parasite virulence. It has long been suggested that multiple infections tend to increase parasite virulence. However, the interaction between the occurrence of multiple infections and other parameters has not attracted much attention. Our model clearly shows that some parameters may affect the evolution of parasite virulence only when multiple infections occur. For example, parasite transmission may indirectly act on parasite virulence, since higher transmission tends to decrease within-host relatedness among parasites and, as a consequence, increase virulence.

More generally, kin-selection processes and indirect effects emerge naturally from the host–parasite interaction when details of the spatial structure of the host–parasite interaction are accounted for. In our model, kin selection emerges from the parasite population structure through multiple infection. Van Baalen showed that kin-selection processes (and qualitatively similar mechanisms) could also emerge from spatial viscosity in the host population (see Chapter 7). In these situations, it is particularly relevant to study the inclusive effects of a given parameter on the evolution of virulence by considering the combination of direct and indirect effects.

These inclusive effects have important implications for virulence management. Indeed, we show that classic health policies may have long-term evolutionary consequences on the parasite. A better understanding of these consequences may help to identify particularly efficient policies with both short-term (epidemiological) and long-term (evolutionary) beneficial effects. Our analysis is only a first step toward this ultimate goal, since our model remains an oversimplification of parasite evolution. In particular, we did not include the epidemiological details that determine the dynamics of host and parasite populations. This greatly simplifies the algebra and allows us to identify explicitly the occurrence of genetical feedbacks. However, by doing so we definitely exclude ecological feedbacks (Van Baalen and Sabelis 1995a, 1995b; Gandon *et al.* 2001) that may occur via the dynamics of host–parasite interactions. The next step toward the design of strategies for virulence management is to test the robustness of our predictions under more realistic assumptions.

Acknowledgments We would like to thank Minus van Baalen for very useful discussions. John Jaenike and Michael Hochberg helped to clarify a first draft of this manuscript. We gratefully acknowledge support from the Wellcome Trust (grant 06429 SG), Foundation Singer Polignac, the British Council, and the Centre National de la Recherche Scientifique.

12

Kin-selection Models as Evolutionary Explanations of Malaria

Andrew F. Read, Margaret J. Mackinnon, M. Ali Anwar,
and Louise H. Taylor

12.1 Introduction

Malaria, a disease caused by protozoan parasites of the genus *Plasmodium*, can substantially reduce host fitness in wild animals (Atkinson and Van Riper 1991; Schall 1996). In humans, the major disease syndromes – severe anemia, coma, and organ failure, as well as general pathology such as respiratory distress, aches, and nausea – cause considerable mortality and morbidity (Marsh and Snow 1997).

Biomedical research attributes malaria to red cell destruction, infected cell sequestration in vital organs, and the parasite-induced release of cytokines (Marsh and Snow 1997). But mechanistic explanations are just one type of explanation for any biological phenomenon, and, in recent years, evolutionary biologists have become interested in offering evolutionary explanations of infectious disease virulence. This is entirely appropriate (Read 1994). In the context of malaria, for example, the clinical outcome of infection has an important impact on parasite and host fitness and is – at least in part – determined by heritable variation in host and parasite factors (Greenwood *et al.* 1991). Yet in the recent rush to provide evolutionary explanations of disease, there has been, in our view, too little interaction between the models built by evolutionary biologists and reality. There is unlikely to be a simple, general model of virulence: the causes of disease and the fitness consequences for host and parasite are too variable. Instead, different models, and even different frameworks, will be relevant in different contexts. Only by evaluating specific models in the context of specific diseases will sensible evolutionary explanations of virulence be realized. Such evaluations seem to us an essential step if one aim of an evolutionary explanation is to contribute to virulence management. An evolutionary explanation of malaria would answer the question "Why has natural selection not eliminated the disease?" and would perhaps contribute to answering the question "Why is the clinical outcome of infection so variable?"

One evolutionary explanation, for instance, postulates that malaria is maintained by natural selection because it enhances the fitness of the parasite that causes it, since sick hosts have reduced antivector behavior (Day and Edman 1983; Ewald 1994a). Rather than evaluate that idea, we instead examine an idea that has attracted more attention from theorists. Kin-selection models of virulence represent an important component in the evolution of virulence literature. They postulate that the genetic relatedness of parasites within hosts affects the outcome of virulence. In this chapter we attempt to evaluate the relevance of these models to malaria.

12.2 Kin-selection Models of Virulence

Most evolutionary models consider disease virulence to be a consequence of selection acting on parasite life history. The most frequently espoused view is that virulence is an incidental and unavoidable consequence of parasites extracting resources from hosts to maximize the production of effective transmission stages. Virulence *per se* is seen as detrimental to parasite fitness (it increases the risk of death of the hosts and, hence, of the parasites), but host damage is necessary for transmission. Thus, observed levels of virulence are said to represent schedules of host exploitation that optimize some measure of parasite fitness by balancing the risk of death with the need to maximize transmission-stage output. This idea has been much reviewed (Bull 1994; Read 1994, Frank 1996c; Ebert 1998b) and we hope to evaluate the relevance of it to malaria in due course. Here, we assume that this idea is applicable, and, therefore, we evaluate the relevance of an important development of the idea.

Many authors have pointed out that where mixed-genotype infections are common, levels of virulence greater than those optimal for single-genotype infections are favored by natural selection. This is because optimal rates of host exploitation are altered when unrelated parasite genotypes compete (Hamilton 1972; Eshel 1977; Axelrod and Hamilton 1981; Levin and Pimentel 1981; Bremermann and Pickering 1983; May and Anderson 1983a; Knolle 1989; Bremermann and Thieme 1989; Sasaki and Iwasa 1991; Frank 1992a, 1996c; Herre 1993, 1995; Nowak and May 1994; May and Nowak 1995; Van Baalen and Sabelis 1995a, 1995b; Ebert and Mangin 1997; Leung and Forbes 1998; and see Chapters 5 and 9). Parasites that slowly exploit hosts are outcompeted by those that exploit hosts more rapidly. Even if host life expectancy is reduced so that all parasites do worse, the prudent parasites do disproportionately badly, and are thus eliminated by natural selection. This "tragedy of the commons" appears in many areas of evolutionary biology (e.g., social evolution; Trivers 1985); the common link is relatedness. Here, prudent exploitation of hosts is favored when relatedness within an infection is high (e.g., all parasites are members of the same clone). But the kin-selective fitness benefits of prudence are reduced when within-host relatedness is lowered – that is, more selfish genotypes win.

It follows from these ideas that where mixed-genotype infections occur, levels of virulence favored by natural selection are greater. There are two mechanisms by which natural selection acting on parasites could match virulence to within-host relatedness. Schedules of host exploitation could have become genetically fixed at levels that are evolutionarily stable for the average frequency of mixed-genotype infections found in a population. Alternatively, conditional strategies might have evolved, whereby parasites alter their exploitation schedules, and hence virulence, according to the type of infection they find themselves in (Sasaki and Iwasa 1991;

Frank 1992a; Van Baalen and Sabelis 1995a). Facultative life-history strategies are a common feature in many taxa (e.g., Wrensch and Ebbert 1993; Godfray 1994; Via *et al.* 1995). If conditional virulence strategies exist, there should be an association between within-host genetic diversity and virulence within a host population; if only genetically fixed strategies are possible, there will be no such association within populations, but there should be across them.

Are these ideas applicable to malaria, as several evolutionary biologists have suggested (Pickering 1980; Bremermann and Pickering 1983; Frank 1992a; Ewald 1994a)? *Plasmodium* infections consist of asexually replicating genotypes, which transmit to mosquitoes by producing gametocytes – terminal forms that are incapable of further replication in the vertebrate host. Natural infections often consist of unrelated genotypes, acquired from either the same or different infectious bites. Multiplicity of infection (the frequency of mixed infections, or the number of clones per host) is variable within populations and on average higher in areas where transmission rates are high (Day *et al.* 1992; Babiker and Walliker 1997; Paul and Day 1998; Arnot 1999). The potential for kin selection to affect the outcome of virulence evolution thus exists.

But does it? We begin by asking whether the multiplicity of infection correlates with disease outcome within populations, as would be expected if there are conditional virulence strategies. We then consider the issue of genetically fixed strategies, before summarizing results from our experimental work, which address some assumptions implicit in the foregoing arguments. We end by discussing the management implications of these ideas and data.

12.3   Conditional Virulence Strategies

In this section, we discuss field correlations and data from laboratory experiments concerning conditional virulence strategies.

Field correlations

Direct measurements of the genetic composition of infections that differ in clinical status are increasingly available from human populations afflicted by malaria. Genetic diversity can be assayed using monoclonal antibody analysis, isoenzyme analysis, and, most recently, polymerase chain reaction (PCR) amplification of highly polymorphic loci. This has made it possible to ask whether infections that consist of more than one genotype are more virulent, as would be expected if parasites are facultatively increasing rates of host exploitation in the presence of coinfecting competitors.

Although such studies are in their infancy, available data are summarized in Table 12.1. Care is needed in the interpretation of such data. Many estimates of the multiplicity of infection are almost certainly underestimates (Arnot 1999), and comparisons across studies are of limited value because the loci under study and clinical definitions vary. Nevertheless, within-study comparisons probably are meaningful, and here the picture that emerges is, if anything, opposite to that expected from kin-selection models of virulence. In the majority of studies, the

Table 12.1 Multiplicity of infection and disease status in field studies of humans infected with *Plasmodium falciparum*. n = number of people who are PCR-positive for parasites in the respective groups; ns = between-group difference not significant; s = between-group difference significant (significance as presented by author or from appropriate tests based on data presented); nt = between-group significance not tested and not possible to test from presented data. Standard errors given where reported or could be calculated from presented data.

Location	Ref.	n	Average number of clones/person in people with			Proportion single clone infections in people with		
			Asymptomatic infections	Mild malaria[a]	Severe malaria[b]	Asymptomatic infections	Mild malaria[a]	Severe malaria[b]
Senegal	[1]	30 and 56		2.3 ← ns →	2.4		0.29 ← s →	0.65
Senegal	[2–4]	24 and 10	4.0 ← nt →	1.4				
Senegal	[5]	166 and 25	1.65 ← s →	2.3		0.52 ← s →	0.20	
Gabon	[6]	99 and 99		1.3±0.1 ← ns →	1.2±0.4		0.50 ← ns →	0.70
Tanzania	[7]	76 and 71	5.0±0.25 ← s →	3.4±0.3				
The Gambia	[8]	118 and 35		2.0±0.1 ← ns →	2.1±0.2		0.45 ← ns →	0.46
Kenya	[9]	c.172 and 25	2.0 ← ns →	2.2		0.33 ← ns →	0.25	
Papua New Guinea	[10]	116 and 111	1.3±0.1 ← ns →	1.2±.04		0.74 ← ns →	0.82	
Sudan	[11]	160 in longitudinal study	c.1.39 ← s →	c.1.59		0.62 ← s →	0.49	
Papua New Guinea	[12]	82 single and 53 multiple infections	Prospective study: children infected with multiple clones had significantly lower risk of subsequent clinical attack					

[a]Febrile and parasite positive.
[b]Parasite positive and severe anemia, altered consciousness, convulsions, or at least one other symptom of severe malaria (Warrell et al. 1990).
Sources: [1] Robert et al. (1996a), [2] Ntoumi et al. (1995), [3] Contamin et al. (1996), [4] Mercereau-Puijalon (1996), [5] Zwetyenga et al. (1998), [6] Kun et al. (1998), [7] Beck et al. (1997), [8] Conway et al. (1991), [9] Kyes et al. (1997), [10] Engelbrecht et al. (1995), [11] Roper et al. (1998), [12] Al-Yaman et al. (1997).

number of clones in an infection is unrelated to the severity of clinical symptoms. At least three studies provide evidence of an association between genetic diversity in an infection and disease severity (Robert *et al.* 1996a, Mercereau-Puijalon 1996, Beck *et al.* 1997, Al-Yaman *et al.* 1997), but it is the less diverse infections that are the more virulent. Only two studies show evidence that symptomatic infections – those detected when sick people report to clinics – contain more genotypes than infections discovered by random sampling of asymptomatic people (Roper *et al.* 1998, Zwetyenga *et al.* 1998).

A major problem in the interpretation of these studies is the (almost scandalous) lack of understanding of naturally acquired immunity against malaria. To the extent that there is a consensus view on the immunoepidemiology of malaria, it might be summarized as follows. Immunity is of two sorts: antiparasite and antidisease. The precise nature of either, or of the link between them, is unknown, but they are certainly not two sides of the same coin. For example, semi-immune people can often harbor high densities of parasites without any obvious effect on the host. Antiparasite immunity has a large strain-specific component. Effective protection may require multiple exposures to the same genotype and/or rapidly decay. Memory of recent or low-grade concurrent infections thus determines specificity of effective responses against new infections. Clinical disease is caused by antigenic types not previously seen by that individual. As children in malaria-endemic regions age, the repertoire of genotypes to which the immune system has been exposed increases, and they become protected against progressively more parasite genotypes. A variety of indirect immunological and epidemiological evidence is consistent with this view (Day and Marsh 1991; Gupta *et al.* 1994c; Mendis and Carter 1995; Mercereau-Puijalon 1996), but the evidence is far from definitive.

If this view is even approximately correct, an important implication is that the effects of previous exposure and genotype-specific immune responses are a major – perhaps *the* major – proximate factor to determine disease outcome. If so, any effect of conditional host exploitation strategies may be hard to detect. It may also explain why in some studies lower genetic diversity is associated with greater virulence. Genotypes not previously seen by a host may grow unchecked to high densities and trigger nonspecific effectors [tumor necrosis factor (TNF), fever, nitrous oxide, oxygen radicals] which eliminate other genotypes or suppress them below PCR-detection thresholds. Alternatively, high multiplicity of infection may indicate recent exposure to more genotypes, which reduces the chances of encountering a previously unseen genotype in the near future.

In light of these complexities, it may be possible to reconcile the data summarized in Table 12.1 with the existence of conditional host-exploitation strategies. Indeed, it is intriguing that both places where higher multiplicity of infection is associated with disease are areas with low year-round transmission (Roper *et al.* 1998, Zwetyenga *et al.* 1998), and so immunity against previously experienced genotypes may have time to wane. Ideally, what is required are comparisons of the severity of disease following infection with one or more previously unseen genotypes in hosts with identical exposure histories. In the uncontrolled world of field correlations, such data are unlikely to be forthcoming. In this respect, animal models can play an important role.

Laboratory experiments

Using the rodent malaria *Plasmodium chabaudi* in laboratory mice, we compared the virulence of mixed clone and single clone infections (Taylor *et al.* 1998b). We used anemia and weight loss as virulence measures, because these measures are correlated with mortality rates (Mackinnon and Read 1999a). All mice were infected with the same number of parasites; mixed clone infections were initiated with varying ratios of the two clones. We found that mixed clone infections were more virulent. Mice infected with two clones lost about 30% more weight than those infected with one; mice with mixed clone infections were also more anemic. These findings are certainly consistent with the theory that parasites conditionally alter host exploitation strategies in response to the presence of competing clones. However, parasite densities were no higher in mixed clone infections. The rate of parasite proliferation correlated with virulence across all mice, but for a given rate of proliferation, mixed clone infections were still more virulent. If the parasites employed conditional host-exploitation strategies, the effects were not detectable in terms of parasite replication, as is conventionally assumed in models of virulence.

We believe our data are most parsimoniously explained not by conditional virulence strategies, but instead by the additional costs to hosts of mounting a response against genetically diverse parasites, in terms of both consumption of host resources and immunopathology. Diverse parasite populations may, for example, stimulate a larger number of T- or B-cell clones or stimulate a greater immune cascade, causing the destruction of more red blood cells (RBCs), or trigger increased production of self-damaging effectors such as TNF and fever. Direct evidence for any of this is currently lacking, but the idea is amenable to experimental testing. What we do know is that infections with genetically diverse parasites take longer to clear (Taylor *et al.* 1997a, 1997b, 1998b; Read and Anwar, unpublished) and that prolonged infection results in prolonged anemia (Read and Anwar, unpublished). Longer clearance times do not, however, explain the greater weight loss induced by mixed clone infections: maximum weight loss occurs during "crisis" (well before clearance) when there is a rapid reduction in parasite numbers associated with low RBC densities and strong nonspecific immune activity (Jarra and Brown 1989).

In sum, then, field data from *P. falciparum* provide, with two exceptions, either no evidence of conditional virulence strategies, or evidence against them. Uncontrolled field correlations are hard to interpret, especially in the face of strain-specific immunity, but controlled experiments with *P. chabaudi* in mice also fail to show any evidence of facultative alterations in growth strategies in response to the presence of coinfecting genotypes. A suggestion of conditional virulence strategies in lizard malaria (Pickering *et al.* 2000) is based on a correlation between surrogate measures of virulence and genetic relatedness. There is no evidence of a correlation between the same surrogates in other lizard malarias (Schall 1989),

P. falciparum in humans (Robert *et al.* 1996b), *P. chabaudi* in rodents (Taylor 1997), or in *Haemoproteus*, a related genus of avian blood parasites (Shutler *et al.* 1995).

12.4   Genetically Fixed Virulence Strategies

Conditional virulence strategies require the ability of a clonal lineage to recognize the presence of nonkin and modify host exploitation strategies accordingly. It may be that such sophistication is beyond what is, after all, just a single-celled protozoan (however, this "simple" organism is sufficiently sophisticated to outwit a century of biomedical science). If so, kin-selection models of virulence predict that host exploitation strategies appropriate for some average level of within-host competition in a population should be favored by selection.

This idea requires heritable variation in the levels of virulence induced by malaria parasites on which selection acts. Moreover, this variation should be positively and genetically correlated with replication rates within hosts and, in the absence of host death, with transmission rates between hosts. Theoreticians have suggested that various epidemiological patterns are consistent with the existence of virulent genotypes or strains of *P. falciparum* circulating within human populations (Gupta *et al.* 1994c), but the issue is contentious (Marsh and Snow 1997). The only parasite phenotype that has been found to correlate consistently with disease outcome is rosetting, whereby uninfected erythrocytes become stuck to infected cells (Carlson *et al.* 1990). The ability to rosette is under parasite genetic control, being encoded by specific variant types of the *var* multigene family (Rowe *et al.* 1997; Chen *et al.* 1998). In the laboratory, rapid increases in the virulence of rodent malaria have been attributed to point mutations (Yoeli *et al.* 1975). In controlled laboratory infections of single clones of *P. chabaudi* in a single mouse genotype, we found substantial differences between clones in virulence. These differences were repeatable over successive passages. Moreover, the genetic architecture was as assumed by parasite-centered models of virulence: virulence and rates of within-host replication were genetically correlated, as were virulence and infectivity to mosquitoes (Mackinnon and Read 1999a). Insofar as these results are generalizable, there appears to be the necessary raw material for natural selection on virulence to act in accordance with the evolutionary models and generate genetically fixed virulence strategies.

Are these strategies fixed as we would expect from the kin-selection models? In areas with high transmission, where there is a high multiplicity of infection (e.g., *P. falciparum* in Tanzania; Babiker *et al.* 1994), levels of parasite virulence should be higher than in areas where the force of infection is lower, so that the majority of hosts are infected with a single clone (e.g., *P. falciparum* in Papua New Guinea; Paul *et al.* 1995). Testing that prediction is unfortunately fraught with difficulties. Levels of host immunity are also likely to vary with transmission rates and, hence, the multiplicity of infection, which confounds cross-community correlations between average levels of within-host diversity and morbidity and mortality measures. Genetic differences between host populations may also confound any such tests. Direct comparisons of virulence of isolates from different populations grown in a "common garden" would resolve that difficulty; the problem is to find an ethical or biologically realistic garden. Of the strains used for malaria therapy of neurosyphilis in nonimmune Europeans in the first half of the 20th century, a number of geographically distinct races were recognized that differed in their clinical virulence. Recent analysis of data gathered at that time reveals repeatable strain differences in within-host growth rates, but comparable data on virulence seems to lacking (Gravenor *et al.* 1995).

Once virulence factors and the parasite genes that encode them have been identified, informative field data should be forthcoming. It would be of considerable interest, for example, to determine whether mean rosetting rates correlate with multiplicity of infection across populations.

12.5 Within-host Competition and Between-host Fitness

If the predictions of the kin-selection models are currently hard to test in the malaria context, what of the models' assumptions? Two distinct sources of selection for increased virulence when mixed infections occur can be identified in the theoretical work to date. The first arises when the presence of "competing" parasites increases the likelihood of host death. Even if the transmission rates of individual clones are otherwise unaffected by coinfecting parasites, this situation favors higher levels of virulence (May and Nowak 1995; Leung and Forbes 1998). The other source of selection arises from exploitation or interference competition, in which the population sizes and/or transmission rates of clones that proliferate within a host are reduced by the presence of competitors (e.g., Frank 1992a; Van Baalen and Sabelis 1995a; Herre 1995). This could occur through conventional resource competition or through apparent competition (*sensu* Holt 1977), with the immune response triggered by one population having a detrimental effect on the other. The most mathematically tractable case (or at least the most frequently modeled), is of competition so severe that less virulent parasites do not transmit at all from mixed infections (e.g., Levin and Pimentel 1981; Bremermann and Pickering 1983; Knolle 1989; Bremermann and Thieme 1989; Nowak and May 1994; Leung and Forbes 1998) – what Van Baalen and Sabelis (1995a) term superseding infections.

We do not know if the presence of coinfecting malaria clones increases the probability of host death. As described above, one experimental study (Taylor *et al.* 1998b) and two field studies (Roper *et al.* 1998, Zwetyenga *et al.* 1998) suggest virulence increases with the multiplicity of infection; a number of other field studies suggest that it does not (Table 12.1). As well as the attendant ambiguities associated with this data, we do not know how disease levels translate into mortality rates, or even whether observed mortality rates are sufficiently high to impose selection on virulence; case fatality rates may be as low as 2.5 per 1 000 (Greenwood *et al.* 1991).

On the other hand, it seems highly likely that resource and/or apparent competition affect within-host population sizes. We are unaware of any direct evidence

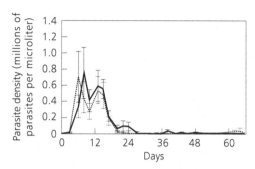

Figure 12.1 Parasite density during the course of infections in mice consisting of one (dotted curve) or two (continuous curve) clones of *Plasmodium chabaudi* (vertical lines are ± standard errors). $n = 9$ single clone infections; $n = 11$ two-clone infections. *Source*: Read and Anwar (unpublished).

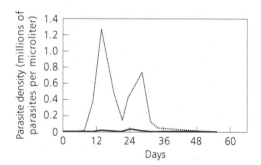

Figure 12.2 Density of two clones of *Plasmodium chabaudi*, AS (continuous curve) and CB (dotted curve), in a single mouse. Clone AS was inoculated 3 days after clone CB (day 0). Clones were distinguished by monoclonal antibody labeling. *Source*: Read and Anwar (unpublished).

from humans, but we have found in the rodent malaria *P. chabaudi* in laboratory mice that the total number of blood-stage parasites produced during an infection is unaffected by the genetic diversity of the inoculum, implying a cap on total densities (Figure 12.1). Depending on initial conditions, clonal populations can be reduced to <10% of that achieved in a single clone infection by the presence of coinfecting genotypes (Figures 12.2 and 12.3; Taylor *et al.* 1997a, 1997b, 1998b; Taylor and Read 1998; Read and Anwar, unpublished).

However, for within-host competition to have any long-term evolutionary consequences, it has to affect the transmission success of individual clones. None of our experimental data are consistent with the idea of superseding infections: in all the infections we examined, all the clones present successfully infected mosquitoes. Moreover, and quite unexpectedly, we found that, despite comparable parasite densities in infections consisting of one or two clones, mixed infections had higher gametocyte densities and were more infectious to mosquitoes

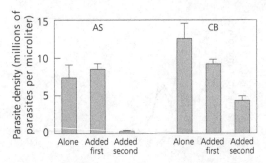

Figure 12.3 Total parasite densities in mice inoculated with either AS alone, CB alone, or with clones added sequentially. When added first, clone AS does as well as it does on its own; when added second (3 days after CB), it does substantially worse. CB always does better on its own, does somewhat worse when AS is added 3 days later, and does even worse if added second. Clones were distinguished by monoclonal antibody labeling; each bar is the mean of 4–6 infections. *Source*: Read and Anwar (unpublished).

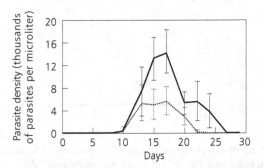

Figure 12.4 Gametocyte (transmission stage) densities in peripheral blood of mice infected with one (dotted curve) or two (continuous curve) clones (vertical lines are ± standard error). Same infections as for Figure 12.1; total gametocyte density is greater in mixed clone infections ($p = 0.034$). *Source*: Read and Anwar (unpublished).

(Figure 12.4; Taylor *et al.* 1997a; Read and Anwar, unpublished). Molecular genetic analysis of the parasites that successfully infected the mosquitoes showed that clones in mixed infections transmitted at least as well as they did from single clone infections, and often did substantially better (Figure 12.5; Taylor *et al.* 1997b). This is because, in some cases, competitively suppressed clones are able to achieve higher densities toward the end of the infection when the transmission stages are being produced (Figure 12.6, Taylor and Read 1998). We hypothesize this occurs because the clone that dominates the bulk of the infection also dominates the attention of the specific component of the host immune response, so that in effect the "successful" competitor shields the "suppressed" genotype from immune clearance. This theory is amenable to experimental testing; it would also

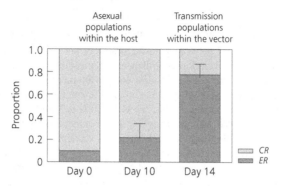

Figure 12.5 Relative frequencies of two *Plasmodium chabaudi* clones, ER and CR, in mixed infections in mice. On day 0, clones were inoculated at a 9:1 ratio, a difference that was maintained through the first 10 days of the infection, when the bulk of the parasites are present. Nevertheless, almost the opposite ratio was observed among parasites that successfully transmitted to mosquitoes. Clones in mice were distinguished by monoclonal antibody assays; genotype frequencies in mosquitoes were determined by PCR. *Source*: Taylor *et al.* (1997b).

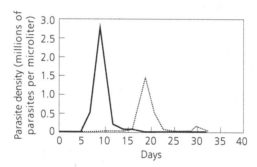

Figure 12.6 Density of two clones of *Plasmodium chabaudi*, AS (continuous curve) and CB (dotted curve), in a single mouse. Clone AS was inoculated 3 days before clone CB (day 0). Clones were distinguished by monoclonal antibody labeling. *Source*: Read and Anwar (unpublished).

benefit from theoretical work on within-host competition in the presence of strain-specific and nonspecific immunity. Such models are in their infancy and their very complexity may rule out simple generalizations (Box 12.1).

Whatever the mechanism, our data clearly demonstrate that despite often substantial competition within an infection, individual genotypes transmit at least as well from mixed infections. This is counter to the assumption of all current kin-selection models of virulence. Assuming that the patterns in mice generalize, it would be of substantial interest to understand the population level consequences of the *positive* feedback between the multiplicity of infection and infectiousness that we find, for both disease epidemiology and the evolution of virulence.

Box 12.1 Models of within-host competition between parasite strains

Within-host competition between parasite strains is the critical element of kin-selection models of the evolution of virulence. Yet, most theoretical work concerns population level (epidemiological) models with no attempt to model explicitly the within-host processes involved (e.g., Levin and Pimentel 1981; Nowak and May 1994; Van Baalen and Sabelis 1995a; see Boxes 5.1, 9.2, and 9.3). These models generally assume the outcome of within-host competition to be fixed in some mathematically tractable way (e.g., only the more virulent clone transmits from a mixed-clone infection, see Levin and Pimentel 1981). Models for the evolution of virulence that describe the outcome of competition between parasite strains as the emergent property of explicit within-host processes do not yet exist (but see Chapter 22 for such a model in a predator–prey metapopulation context).

Actually, explicit within-host models of competition in any context are relatively rare. A number of models of single genotype infections incorporate some sort of intra-clone competition, either with explicitly modeled limiting resources, or by including unspecified logistic constraints on growth (e.g., Anderson *et al.* 1989; Gravenor *et al.* 1995; Hetzel and Anderson 1996). While an important first step toward modeling the more-than-one strain case, single-strain models necessarily ignore parasite heterogeneity in competitive ability, immunogenicity, and susceptibility to immune clearance.

There are two published attempts to model explicitly within-host competition between strains. Smith and Holt (1996) argue that the machinery of mechanistic resource–consumer theory (Tilman 1982) provides a useful lens through which to view the internal struggle between pathogens. In their view, within-host dynamics can be seen as a consequence of competition for limited resources, such as glutamine or iron. The essential output of this approach is predictions about the ability of a pathogen to invade or exclude a competitor. The determinant of this is the critical resource concentration at which a strain's birth rate balances its death rate; that is, the strain with the lower critical resource concentration wins. This approach can be extended to incorporate competition for multiple resources and – by considering the effects of increasing pathogen mortality on critical resource concentration – also the immune pressure. However, we see two principal challenges. First, it is an equilibrium approach. In reality, equilibrium may not be achieved before the host clears the competitors. And even if competitive exclusion does occur, the excluded strain may achieve substantial transmission in the interim. Indeed, if there were a trade-off between persistence and rapid growth, it is possible to envisage situations whereby the excluded pathogen achieves higher total transmission stage densities than the eventual "winner." Second, the important complexities of strain-specific and strain-transcending immunity need to be incorporated. Resource limitation may be an important determinant of competition for only a minority of infections, if at all: host protective responses may halt population growth first.

Hellriegel (1992) explicitly incorporated those complexities. She extended the coupled ordinary differential equations of Anderson *et al.* (1989) to include two coinfecting malaria strains, competition for resources (erythrocytes), and specific and nonspecific immunity against different parasite stages. At its height, her

continued

Box 12.1 *continued*

model was not analytically tractable as it involved 15 equations with at least 21 parameters and variables. Assessment of equilibria provides some insight into long-term behavior, but does not necessarily reveal the most interesting dynamic features. Numerical simulations are the only way to explore nonequilibrium dynamics and these show that the population dynamics of a clone can be dramatically altered by the presence of a competitor, the order of infection of competitors, and the kind of immune response elicited.

Finally, models of within-host competition are somewhat analogous to models of within-host competition between antigenic variants and virulence mutants generated within an infection of a single strain (Bonhoeffer and Nowak 1994a, 1994b; Antia *et al.* 1996). Models such as those of Antia *et al.* (1996) incorporate variant-specific and cross-reactive immunity, as well as differences in growth and clearance rates of "competing" variants. Again, numerical simulation seems to be the only way forward: these models show that a hugely diverse range of outcomes is possible, and it is unclear what generalities might be revealed by further numerical exploration. And again, statements about who finally wins within a host may not be relevant; Bonhoeffer and Nowak (1994a) give an example in which strains that outperform their competitors in the long run nonetheless achieve small population sizes when summed over the whole infection.

12.6   Management Implications

Much of the motivation for thinking about the evolution of virulence (and the motivation for this volume) is that evolutionary models of virulence will contribute to disease management. It is certainly one of our interests. Yet we hope that the above summary cautions against the understandable urge to assume that elegant theory, even when relatively well-developed, is relevant to disease control in the field. We are not yet in a position to say even whether current kin-selection models are relevant to malaria. There is some evidence that the genetic architecture of malaria parasites is of the sort assumed by kin-selection models of virulence, but relatively little field evidence is in accord with the expectations of the models (indeed, the bulk of the evidence points to the reverse), and experimental evidence most likely to support the models has other explanations. Competition within hosts does occur, but our evidence to date suggests that within-host competition actually *enhances* the transmission success of individual clones. That it might actually be preferable to be competitively inferior within a host is an unexpected conclusion, and one that raises many new questions. If it proves to be a widespread phenomenon, it is difficult to see how kin-selection models of virulence, at least in their current form, can be profitably applied to malaria parasites. At the very least, these data demonstrate that unexpected phenomena may exist, which can confound theory based on intuitively appealing assumptions.

In light of this, we consider that the formulation of management advice from parasite-centered models of virulence currently is premature. The clinical outcome of a malaria infection is undoubtedly affected by many things, including ecological factors such as inoculation dose and prior exposure (Box 12.1) and genetic factors in hosts as well as in parasites. All of these probably vary with the epidemiological situation. This makes it a challenge to assess the impact, if any, of evolutionary arguments that place parasite genetics center stage. As things stand, it is entirely conceivable that parasite adaptation may play, at best, a trivial role. Even if it is important, confounding factors may alter or even reverse the outcome of intervention strategies derived from evolutionary theory.

12.7 Discussion

We believe the data summarized above point to the need to understand both the epidemiological and evolutionary consequences of variation in the multiplicity of infection (see also Van Baalen and Sabelis 1995b). Intervention strategies designed to reduce exposure to infectious mosquitoes presumably reduce the average number of clones per host (data on that would be very interesting). What does this mean for average levels of infectiousness? Do multiple infections select for *reductions* in rates of host exploitation? How could that be stable? Experimentally, there are many challenges. When clones (which, on their own, are relatively virulent or avirulent) are in the same host, what happens to total virulence? In the absence of host death, can clones ever *reduce* the transmission success of others in the same host? Is the intrinsic virulence of a clone a more important determinant of the outcome than the initial conditions, such as size of inoculum, infection sequences, and inter-infection intervals? We hope this chapter has demonstrated how such questions are brought into sharp focus by trying to view the theoretical models in the context of particular disease realities. None of the issues are intractable; in the next few years it may be possible to evaluate more successfully the relevance of evolutionary theory for malaria control.

Acknowledgments Our empirical work has been supported by the Leverhulme Trust, Biotechnology and Biological Sciences Research Council, Medical Research Council, University of Edinburgh, and the Royal Society. We are grateful to D. Arnot, A. Rowe, and R. Timms for useful discussion.

Part D
Pathogen–Host Coevolution

Introduction to Part D

Virulence is not a property of the parasite, but of the interaction between host and parasite. Accordingly, the evolution of virulence is the result of a coevolutionary process and to understand it we have to account for both sides. As a result of their generation time, which is usually much shorter than that of the host, microparasites seem to be at a huge advantage. However, sexual reproduction allows host organisms to present a moving target (while at the same time inevitably offering opportunities for parasites to use sexual contacts between hosts to infect new susceptibles). In particular, genetic recombination helps to preserve heterozygosity and leads to a wide diversity of immune responses. But there are many other examples of the intricate struggles between parasite and host and of the trade-offs imposed on them.

Part D explores how reciprocal selection between host and parasite populations influences the evolution of host resistance and parasite virulence. Chapters 13 and 14 deal with parasite–host interactions in which investments in resistance and/or virulence incur a cost. The question here is how the resultant trade-offs influence the coevolutionary process. In Chapters 15 to 17 trade-offs play no role. In these, coevolution acts on the ability of the host to recognize the parasite and discriminate it against cells and tissue of its own, while, at the same time, parasites attempt to bypass recognition by the host. In its simplest form this leads to gene-for-gene coevolution. The question posed in these chapters is whether this process can explain the great diversity in resistance and virulence genes observed in parasite–host systems. Chapter 18 focuses on the role of sexual selection for parasite-free or parasite-resistant mates and its consequences for the health of the offspring. The final chapter of Part D, Chapter 19, is devoted to phylogenetic techniques that help to glean coevolutionary trends from historical reconstructions of species-branching patterns.

In Chapter 13, Krakauer models the coevolution of pathogens and host cells to identify the conditions under which we should expect apoptosis, that is, programmed cell death, to be induced by the host, the virus, or by both. Apoptosis is commonly thought of as a host strategy to create "scorched earth" around a virus-infected cell, thereby hampering progress of the disease. The obvious response of the virus is to inhibit apoptosis and to shift an infection to a more persistent latent form while gaining net productivity. Some viruses, however, can even stimulate apoptosis to promote virus extrusion to surrounding cells. These intricate trade-off mechanisms suggest various routes for intervention to protect the host.

In Chapter 14, Hochberg and Holt explore patterns of virulence and resistance in coevolving parasite–host systems along a gradient of habitat suitability. Their model predicts the parasite's virulence and the hosts' resistance to rise with increasing host productivity along the gradient. However, this prediction critically hinges on the assumption that cost functions of attack and defense do not depend on the habitat. If this does not hold, as can easily be the case because of some inherent trade-off, the trend can even be reversed. Model predictions are thus extremely sensitive to the underlying assumptions. Hochberg and Holt discuss the consequences of their findings for selecting suitable natural enemies for biological pest control.

In Chapter 15, Beltman, Borghans, and de Boer critically assess the common belief that heterozygote advantage is sufficient to explain the widespread polymorphism in molecules of the major histocompatibility complex (MHC). They show that the evolutionary response of the pathogens involves a frequency-dependent selection, which leads to a much higher diversity in MHC molecules than results from selection for host heterozygosity alone. This illustrates that, if defense is subject to genetic constraints, parasite–host coevolution may well contribute to the diversity in host defense and parasite virulence genes. The implication for virulence management is that health in the host population may decrease whenever possibilities for host evolution are limited, as is the case, for instance, in breeding programs for endangered species and in livestock production.

Chapter 16 studies the genetic response of the host population to parasite onslaught. An understanding of this response is crucial to assess the long-term impact of measures of virulence management. Andreasen investigates classic one-locus, two-allele models, both in discrete and in continuous time. Fitness of host genotypes depends on differential susceptibilities and hence on the prevalence of the disease, which in turn depends on the genetic composition of the host population. This relation can be used to assess the consequences of virulence management measures on polymorphic equilibria in the host population, for example, in the context of malaria-induced sickle-cell polymorphism.

In Chapter 17, Sasaki analyzes the coevolution of virulence and resistance in plant–pathogen systems by using a class of mathematical models that incorporate the genetic composition of both the host and the parasite population. This leads to coevolutionary dynamics with a high degree of instability based on complex cycles in genotype frequencies and in genetic polymorphism, which reflects an endless arms race between the interacting populations. A consequence of potential relevance to virulence management is that the analysis allows an estimation of the number of resistant host varieties necessary to protect a host population from disease.

Gene-for-gene interactions may play an important role in the evolution of sexual reproduction. This is highlighted by the Red Queen hypothesis, which emphasizes that there is a continual arms race between parasite and host. In particular, hosts (and parasites too) can benefit from outbreeding because the mere random recombination of host genes already acts to forestall the optimal adaptation of parasites to their host. In Chapter 18, Wedekind explains how sexual selection may play a role in parasite–host arms races. Sexual selection in the host population can act in two ways – on the one hand through uniform preferences for healthy and vigorous mates, and on the other hand through active preferences for complementary genes, especially for loci of the MHC (a crucial component of host–parasite interactions).

Such sexual preferences for dissimilar types have been observed in mites, mice, and man. Since free natural mate choice may well be important for the health of host populations, Wedekind points out the dangers of assisted reproductive technology in humans and breeding programs for endangered species; in both cases, possibilities for mate choice are limited.

The models for the evolution of diseases presented in this book eventually have to be gauged against field data. These data can be observations of genetic changes in response to various selection pressures, but also historical reconstructions of the origin of various diseases, as well as comparative data. To assess the effect of different selective environments the latter have to be considered against the background of historical relatedness. In Chapter 19, Rannala presents techniques to reconstruct phylogenetic trees and applies these to a number of case studies to demonstrate the insights that phylogenetic analyses provide in virulence evolution.

Arms races between hosts and parasites offer some of the most dramatic and intriguing examples of coevolutionary dynamics. They not only add excitement to theoretical modeling, but also lead to testable predictions and, indeed, suggest promising opportunities for virulence management.

13

Coevolution of Virus and Host Cell-death Signals

David C. Krakauer

13.1 Introduction

The death of a cell is no longer thought of as something undesirable for the individual. It is understood that cell death is an essential complement to cell division, without which the building of complex multicellular organisms would be impossible. Programmed cell death, or apoptosis, is the mechanism by which cells are eliminated by proteins encoded by the host genome. The genes that code for these proteins have been found in all eukaryotic organisms investigated, and are recognized as homologous (Vaux *et al.* 1994). It is customary to distinguish apoptosis from necrosis, a series of irreversible changes to the cell following injury. During necrosis, a swelling of the cytoplasmic membranes culminates in rupture and the release of lysosomal enzymes. *In vivo*, necrosis is often accompanied by an inflammatory response. During apoptosis, compaction and segregation of nuclear chromatin is accompanied by a convolution of the plasma membranes. These membranous folds give rise to "apoptotic bodies" filled with densely packed organelles. Apoptosis is also associated with double-strand cleavage of nuclear DNA between nucleosomes, which results in a "ladder" of oligonucleosomal fragments on an electrophoretic gel (Wyllie 1987). Apoptosis has been further divided into heterophagic (type I) and autophagic (type II) mechanisms, the latter of which appears similar to necrosis. Conventionally, type I apoptosis is seen in highly mitotic lines or in the reticuloendothelial system, and involves nuclear collapse, condensation of chromatin, and cell fragmentation. Type II apoptosis is common in secretory cells, in which the majority of cells die and the bulk of the cytoplasm is consumed by expansion of the lysosomal system (Zakeri *et al.* 1995).

Viruses are obligate parasites of autonomously replicating organisms. They are able to maximize their efficiency of replication by dispensing with those proteins provided by their host cells. A conflict of interest arises between virus and host as the virus seeks to derive the maximum benefit from the cell's replicative machinery at the least cost to itself. Simultaneously, the host attempts to minimize the costs of virus replication and, hence, infection. Apoptosis plays an important part in the host cell's response to viral infection, whereby infected cells can commit suicide to reduce the host's total virus population (Thompson 1995). From a medical perspective, the ensuing tissue damage qualifies as virulence. From an evolutionary perspective, this damage can ensure the continued viability of the host and a reduced proliferation of the virus. Cell death neatly illustrates these two concepts of virulence: one based on an idea of health and the other on an idea of fitness. Preempting a cell's attempted suicide, many viruses are able to express genes that inhibit apoptosis. Inhibition of apoptosis is thought to provide a virus with an opportunity to shift infection from an acute lytic form, to a nonlytic, persistent, and latent infection, over which the net virus productivity will be higher.

The adenovirus *E1B* gene blocks apoptosis by binding to the cell's *p53* tumor suppressor protein (Tollefson *et al.* 1996); the Epstein–Barr virus LMP-1 protein induces the expression of the host's *bcl-2* protein (Gregory *et al.* 1991); baculoviruses express the gene *p35* and members of the "inhibitor of apoptosis" (*iap*) gene family (Clem *et al.* 1991); and the cowpox gene-product *CrmA* is reported to inhibit a suite of pathways, including Fas-, tumor necrosis factor, and cytotoxic lymphocyte-induced apoptosis by inhibiting the activity of the interleukin-1β-converting enzyme (Ray *et al.* 1992). In these examples, the cytopathic effects of infection are caused by the host immune response. Increasing virulence in the evolutionary sense (e.g., increasing virus load) is achieved by a reduction in virulence in the medical sense (fewer cells killed). We assume, of course, that host fitness is positively correlated with the number of cells, and that by killing infected cells there are fewer chances for new infection.

Virus can also stimulate cell death. The adenovirus *E1A* protein stabilizes *p53*, causing it to accumulate in the nucleus and thereby block the cell cycle (Lowe and Ruley 1993); and a virus-dependent stimulation of the TcR–CD3 complex on T-lymphocytes, can induce apoptosis in human immunodeficiency virus (HIV; Gougeon *et al.* 1993). When the alpha virus chimera over-expresses the anti-apoptotic gene *bcl-2*, the result is a significantly lower host mortality rate in infected mice by comparison with chimeric control mice (Levine *et al.* 1996). The lower mortality is attributed to reduced nerve cell death. In these examples, the virus is itself cytopathic and, hence, evolutionary, and medical definitions of virulence are in agreement. Increased virus load or replication rates are associated with greater cell death.

In this chapter, I present models applicable to predicting the evolutionary trajectory of apoptosis. These models review and extend the Krakauer and Payne apoptosis model (Krakauer and Payne 1997). The evolutionary outcome (inhibition or induction of apoptosis by the virus) is found to be a function of the host cell-death rate and the virus life cycle. It is also shown that within-host competition among virus strains can cause the virus to switch from inducing cell death to a strategy of inhibition of cell death, with an opposite reversal in the host.

13.2   Mathematics of Cell Death

In the model, we use three variables: the densities of free virus particles v, uninfected cells x, and infected cells y. We assume that the uninfected cells are produced and die naturally at a rate given by the function $g(x)$. Free virus interacts with the uninfected cells to produce infected cells at a rate βxv. This is an assumption of mass action. It is quite likely that there is a strong local component to infection, which would require a model with explicit spatial structure. Infected cells die at a rate μy, and are killed by the virus by lysis at a rate Ly. Free virus is produced by the infected cells by extrusion or secretion at a rate εy, and by lysis at a rate kLy. Virus dies at a rate δv. The model can thus be represented by the following system of ordinary differential equations

$$x' = g(x) - \beta v x ,\tag{13.1a}$$

$$y' = \beta x v - \mu y - Ly ,\tag{13.1b}$$

$$v' = (kL + \varepsilon)y - \beta x v - \delta v .\tag{13.1c}$$

Assuming that committed, uninfected cells are produced from precursor cells at a constant rate, and are themselves incapable of replication such that $g(x) = b - dx$ (where b is the rate at which susceptible cells are produced), the basic reproduction ratio of the virus (the number of secondary infections produced on average by each primary infection) is given by

$$R_0 = \frac{\varepsilon + kL}{\mu + L} \cdot \frac{\beta b}{\delta d + \beta b} .\tag{13.2}$$

If $R_0 > 1$, then the system converges to the stable equilibrium values

$$x^* = \frac{\delta(\mu + L)}{\beta(\varepsilon - \mu - L + kL)} ,\tag{13.3a}$$

$$y^* = \frac{b}{\mu + L} - \frac{\delta d}{\beta(\varepsilon - \mu - L + kL)} ,\tag{13.3b}$$

$$v^* = \frac{b(\varepsilon - \mu - L + kL)}{\delta(\mu + L)} - \frac{d}{\beta} .\tag{13.3c}$$

Cell death is controlled by modifying the value of the variable L, such that some measure of parasite fitness or host fitness is maximized. For example, we might assume that a virus seeks to establish a level of lysis that maximizes the free virus population v^*, and is thereby more likely to be transmitted between susceptible hosts; or that the host seeks to maximize the total number of cells that remain uninfected x^*, and thereby minimize tissue damage. First, we require an understanding of how the equilibria are influenced by the value of L. I explore the case in which $g(x) = b - dx$ in some detail to clarify the logic of the model. The behavior of the equilibrium values depends crucially on the value of the parameter

$$\Delta = \frac{\varepsilon}{\mu} - k .\tag{13.4}$$

The right side of the equation can be interpreted as the average number of virions produced per cell when there is no lysis (ε/μ), minus the average number of virions produced per cell assuming replication through lysis alone (k). The sign and magnitude of Δ therefore provides a measure of the relative contribution through these two modes of replication. We can express the most likely direction of evolution of the virus or host's cell-death program in terms of Δ:

1. If $\Delta > 0$ (i.e., extrusion and secretion are potentially more productive than lysis), then

$$\text{as } L \text{ increases} \begin{cases} x^* \text{ increases} \\ v^* \text{ decreases} \\ y^* \text{ decreases} \end{cases}.$$

2. If $\Delta < 0$ (i.e., lysis is potentially more productive than extrusion and secretion), then

$$\text{as } L \text{ increases} \begin{cases} x^* \text{ decreases} \\ v^* \text{ increases} \\ y^* \begin{cases} \text{decreases if } \Delta_c < \Delta < 0 \\ \text{has a maximum if } \Delta < \Delta_c \end{cases} \end{cases},$$

where

$$\Delta_c = (k-1)\left(\sqrt{\frac{\delta d}{\beta b (k-1)}} - 1 \right). \tag{13.5}$$

Whether a virus or host stimulates (increases the parameter L) or inhibits apoptosis (reduces the parameter L) depends on the efficiency of lytic and nonlytic replication and the natural death rate of infected cells. In all cases, I assume that these are options available to the virus and, hence, we are considering principally encapsulated viruses. The final value of L is likely to represent a biased outcome of virus and host pressures, and, therefore, depends on the relative contributions to fitness of these two mechanisms of replication. For the virus, it is reasonable to assume in most cases that it evolves to maximize fitness by maximizing the free virus level. Hence, when $\Delta > 0$, it is in the virus's interest to minimize L, and in any infection for which $\varepsilon/\mu > k$, we expect to find virus mechanisms that act to inhibit apoptosis. Whereas when $\Delta < 0$, it is in the virus's interest to maximize L, and thus in infections for which $\varepsilon/\mu < k$, we expect the virus to have evolved mechanisms to induce apoptosis.

The situation for the host is potentially more complex, as the host might benefit from maintaining a combination of both uninfected and infected cells. If infected cells are of little value to the host, then we assume that the host tries to maximize x^*. Hence, when $\Delta > 0$, the host cell has evolved mechanisms to inhibit apoptosis, whereas when $\Delta < 0$, the host cell attempts to undergo apoptosis when infected. If infected cells remain important to the host (e.g., as in the nervous system) and $\Delta > 0$, an intermediate level of L is best for the host – the precise value of which depends on the relative contribution of the uninfected and infected cells to host fitness. However, when $\Delta < 0$, with the exception of a small range of parameter values ($\Delta < \Delta_c$, see Appendix 13.A for derivation), the total number of host cells (both infected and uninfected) will be maximized by evolving mechanisms that inhibit apoptosis.

Choosing an alternative function, in which cells are able to replicate, but only up to some maximum number, we can let $g(x) = (b - dx)x$, where b/d is the carrying capacity of the host's uninfected cell population. This reflects some form of density-dependent limitation on the total number of cells within the host. In this case, as with $g(x) = b - dx$, v^* will always increase and x^* will always decrease with increasing L for $\Delta < 0$, so the overall conclusions stay the same.

13.3 Evolutionary Dynamics of Cell-death Signals

In Section 13.2, I assume that the virus evolves toward that level of lysis L that maximizes the virus load and, thereby, increases the likelihood of between-host transmission. The outcome of within-host competition among genetically different viruses should also be considered. We do this by establishing what effect a mutant virus has on the wild-type virus at equilibrium.

Consider the case in which there are two virus strains 0 and 1. This means that we add a few equations to our system to include a possible, coinfecting mutant strain

$$x' = g(x) - \beta x(v_0 + v_1) , \tag{13.6a}$$

$$y_0' = \beta x v_0 - \mu y_0 - L y_0 , \tag{13.6b}$$

$$y_1' = \beta x v_1 - \mu y_1 - L y_1 , \tag{13.6c}$$

$$v_0' = (k_0 L + \varepsilon_0) y_0 - \beta x v_0 - \delta_0 v_0 , \tag{13.6d}$$

$$v_1' = (k_1 L + \varepsilon_1) y_1 - \beta x v_1 - \delta_1 v_1 . \tag{13.6e}$$

These two strains are able to differ in all parasite-specific parameters: the number of virions produced during lysis (k_0, k_1), the efficiency of extrusion ($\varepsilon_0, \varepsilon_1$), and the death rate ($\delta_0, \delta_1$). If we assume that the wild-type virus v_0 is already at its equilibrium [as described by Equation (13.3c)], then strain 1 will invade and replace strain 0 within the host whenever

$$\frac{k_1 L + \varepsilon_1 - \mu - L}{\delta_1} > \frac{k_0 L + \varepsilon_0 - \mu - L}{\delta_0} . \tag{13.7}$$

The significant points are that when $g(x) = b - dx$ or $g(x) = (b - dx)x$, increasing k or ε through competition is always associated with an increase in virus

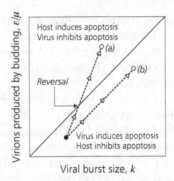

Figure 13.1 The outcome of within-host competition on virus and host cell-death strategies. Along the vertical axis is a measure of the number of virions produced by budding, and along the horizontal axis is the viral burst size. The diagonal line represents all points along which $\Delta = 0$. In the trajectory marked (a), within-host competition causes the virus to cross the $\Delta = 0$ line and switch from inducing apoptosis to inhibiting apoptosis. In the trajectory marked (b), the increment in the burst size is accompanied by an increase in budding, leaving the strategies of both virus and host unchanged.

load. Thus, within-host competition does not conflict with between-host selection toward maximizing the virus load. This is an important consideration, for there are examples in which the within-host phase leads to the emergence of a strategy that is at odds with the between-host requirement for increased transmissibility (Bonhoeffer and Nowak 1995). The inclusion of within-host competition does lead to some new possibilities discussed in Section 13.4.

13.4 Threshold Reversals

The evolution of the virus-specific parameters can cause the value of Δ to pass its threshold at 0, and thereby reverse the direction in which we predict the value of L to evolve in both the virus and the host (Figure 13.1). In other words, within-host evolution of the virus can lead to a situation in which the virus switches from inducing cell death to inhibiting cell death, with an opposite reversal in the host response.

Let us consider within-host competition in the case when $g(x) = b - dx$, and treat only the populations of susceptible cells and the free virus. When the character described by the parameter grouping ε/μ (the number of virions budding from the infected cell) is most easily modified during evolution (i.e., assuming it is a particularly labile or adaptable trait), then competition favors those strains of virus that lower the value of L. This is because an increase in the value of ε/μ makes it more likely for Δ to remain greater than 0. If the parameter k is most adaptable, then competition favors strains that increase the value of L. This makes it more likely for Δ to remain less than 0. When $g(x) = (b - dx)x$, the same result applies. As illustrated in Figure 13.1, these reversals only occur when the starting conditions permit the line $\Delta = 0$ to be crossed and when $\partial(\varepsilon/\mu)/\partial k > 1$.

13.5 Experimental Case Studies

Table 13.1 represents the results of a literature review of case studies in which cell death was observed to be the result of infection with a virus. Wherever possible, I have also noted the effects of cell death on virus load and on the number of uninfected cells. Of course, there is no evidence that these populations have reached equilibrium. There is also very little data on host fitness that might allow us to determine the optimal composition of infected and uninfected cells within the host. I chose to explore the following cases in more detail to determine when and where the virus or the host is likely to have evolved control of the parameter L.

Case 1: RNA virus Sindbis

The RNA virus Sindbis (SIN) is able to produce a fatal encephalitis that is persistent in neurones and lytic in the majority of vertebrate cell lines. The expression of *bcl-2* by host cells is able to block virus-induced lytic replication in postmitotic neurones (Levine *et al.* 1993). The host is thought to promote viral persistence by inhibiting virus-induced suicide and thereby mitigate viral lytic potential. The over-expression of *bcl-2* in a recombinant alpha virus chimera expressing the human *bcl-2* protein reduces host mortality and reduces the net viral titer (Levine *et al.* 1996). Host fitness is increased by reducing nerve-cell depletion. From the perspective of the virus, we appear to have a case in which $\Delta < 0$ reduced lysis results in a reduced virus load (lower triangle of Figure 13.1). This happens when extrusion is more productive than lysis. SIN virus, in common with all alphaviruses, employs nucleocapsids and virus-encoded transmembrane glycoproteins for effective extrusion through the cell plasma membrane. The lipid composition of the membrane must closely match that of the alphavirus, and this confers a degree of host specificity on SIN infection. Therefore, the alphaviruses have evolved elaborate mechanisms for budding from infected cells. From the perspective of the host, both the susceptible and infected cell populations contribute to fitness and, hence, both of these cells should be preserved. The models suggest that when susceptible cells replicate, the virus strategy should be the same as with nonreplicating, postmitotic cells. How might we explain the lytic strategy of this virus in non-nervous tissues? It has been shown that within-host evolution of a virus through competition can lead to such a reversal (Figure 13.1). Infection of long-lived nerve cells is often associated with a reduction in virus replication rates and a more local pattern of virus dispersal. Both of these factors reduce opportunities for competition. In the absence of competing strains there is no strong selection for increased rates of proliferation. In rapid turnover cells in which rates of replication are higher (accompanied by more mutations) and mixing is more frequent, competition becomes probable. In such a context, the virus evolves toward inhibiting cell death and the host toward inducing cell death (Figure 13.1).

Case 2: Herpes simplex virus

The $\gamma_1 34.5$ gene of herpes simplex virus 1 inhibits neuroblastoma cells from triggering the shut-off of protein synthesis characteristic of programmed cell death,

Table 13.1 A literature review of cases in which cell death results from virus infections.

Cell type	Virus species	Inhibition (−) or induction (+)	Cell-death signal	Host cells (+/−)	Virus load	Reference
Nerve cell	HSV-1	−	Protein kinase U-S3	−	+	Leopardi et al. (1997)
Peripheral blood T-lymphocyte	HSV-1	+	?	−	+	Ito et al. (1997)
Neuroblastoma cell	HSV-1	+	$\gamma_1 34.5$	−	+	Chou and Roizman (1992)
B-lymphocyte	EBV	+ (and) −	?	−	+	Gregory et al. (1991)
Bombyx mori B5 cell	AcMNPV	−	p35	+	+	Castro et al. (1997)
Microglia and macrophage	HIV-1	−	?	−	−	Badley et al. (1997)
Peripheral macrophage	HIV-1	+	Mitochondrial damage	−	+	Carbonari et al. (1997)
Activated PBMC	HIV-1	+	Impaired bcl-x	−	+	Blair et al. (1997)
Lymphocyte	MV	+	PMA and ionomycin	−	+	Ito et al. (1997)
Cervical epithelial cell	HPV	+	HPV-16 E6 acts through MPT	−	+	Brown et al. (1997)
Transformed fibroblast	BPV	−	?	+	−	Melchinger et al. (1996)
Lymphocyte	IBDV	+	?	−	−	Ojeda et al. (1997)
Macrophage	NDV	+	?	−	+	Lam (1996)
Nerve cell	SINV	+	?	−	+	Levine et al. (1993, 1996)
HL-60 leukemic cells	HGE	+	Inhibition of bcl-2	−	+	Hsieh et al. (1997)
Nerve cell	Prion protein	+	?	−	−	Kretzschmar et al. (1997)
HT-29 enterocyte-like cell	Rotavirus	+	?	−	+	Superti et al. (1996)
Endothelial cell	HTLV	−	Induction of bcl-2	?	?	Nicot et al. (1997)
Peripheral blood lymphocytes	FIV	+	?	−	+	Momoi et al. (1996)
BSC-1 cells	TMEV	+	Cytoplasmic event	−	+	Jelachich and Lipton (1996)
Nerve cell	VEE	+	?	−	+	Jackson and Rossiter (1997)
Nerve cell	Reovirus	+	?	−	+	Oberhaus et al. (1997)
Nerve cell	Ad	+	ADP	−	+	Tollefson et al. (1996)
Endothelial cell	AHV	+	gp85	−	+	Sela-Donnenfeld et al. (1996)

Abbreviations: AcMNPV = *Autographa californica* multiply embedded nuclear polyhedrosis virus; Ad = Adenovirus; ADP = Adenovirus death protein; AHV = Avian hemangioma virus; BPV = Bovine papilloma virus; EBV = Epstein–Barr virus; FIV = Feline immunodeficiency virus; HGE = Human granulocytic ertichiosis; HIV = Human immunodeficiency virus; HPV = Human papilloma virus; HSV = Herpes simplex virus; HTLV = Human T-cell lymphotrophic virus; IBDV = Infectious bursal disease virus; MV = Measles virus; MPT = Mitochondrial permeability transition; NDV = Newcastle disease virus; PBMC = Peripheral blood mononuclear cell; PMA = Phorbol 12-myristate 13-acetate; SINV = Sindbis virus; TMEV = Theiler's murine encephalomyelitis virus; VEE = Venezuelan equine encephalitis.

thereby allowing the virus to replicate. In contrast, viral mutants incapable of expressing $\gamma_1 34.5$ cause a shut-off of protein synthesis (Chou and Roizman 1992). Hence, cell death appears to be inhibited by the virus rather than by the host in the susceptible nerve cells. From the model, this implies that $\Delta > 0$, the reverse pattern of that observed with SIN. As μ is a host-dependent factor, and the cell tropism of these two viruses is similar, they must differ in the parameters, k or ε. The herpes virus is able to bud efficiently through the internal nuclear membrane and can be directed through the Golgi along pathways employed by soluble proteins. This suggests that the parameter ε is high. Nerve cells themselves have a relatively low death rate μ. The ratio ε/μ is therefore likely to be large, as predicted by $\Delta > 0$. However, to be certain that the virus is inhibiting apoptosis, and not the host cell that is somehow benefiting, a measure of k, the number of virions produced at lysis, is required.

Case 3: Epstein–Barr virus

The Epstein–Barr virus (EBV) is a human herpes virus able to establish a persistent asymptomatic infection in circulating B-lymphocytes by entering into the memory B-cell pool. Expression of the full set of eight virus-coded "latent proteins" protects B-cells from cell death and activates B-cell proliferation. Phenotypes expressing only one of the latent proteins, nuclear antigen EBNA 1, remain sensitive to apoptosis. When EBV-positive Burkitt's lymphoma (BL) cells expressing all eight EBNAs are placed in 10% (optimal) or 1% (suboptimal) fetal calf serum (FCS), cells grow to saturation density and no cells enter apoptosis. In contrast, cells expressing only EBNA 1 in 1% FCS rapidly enter apoptosis, while cells in 10% FCS grow to saturation (Gregory *et al.* 1991). Thus, an increased cell-death rate brought about by a suboptimal environment (1% FCS) induces apoptosis in EBNA 1 clones of BL cells. This corresponds to the *in vivo* properties of BL cells. BL cells are derived from within the germinal centers of lymphoid follicles, sites of rapid B-cell proliferation, and death. In early B-cell development, μ is likely to be high and virus load increases with increasing rates of virus-induced cell death ($\Delta < 0$). Following transit through the proliferative cell compartment, selection may induce the activation of the full repertoire of eight EBNAs, reduce the sensitivity of the cell to apoptosis, and allow it to enter into the long-lived B-cell pool (low value of μ) in which a low rate of lysis promotes viral fitness ($\Delta > 0$). Thus, EBV may evolve between different apoptotic strategies in response to a change in the death rate of its target cell. The alternative explanation for this switch is, of course, within-host competition. Competition among different strains could lead to the successive activation of each latent protein, resulting in full resistance by the time the virus enters into the memory pool.

Case 4: AcMNPV

The baculo-virus *Autographa californica* multiply-embedded nuclear polyhedrosis virus (AcMNPV) produces an acute disease with lysis within 72 hours of infection. The viral gene product *p35* is transcribed on entry into the cell, and it is able to block the apoptotic response by the cells of the host organism *Spodoptera frugiperda*. In the larvae of *S. frugiperda*, BV *p35* mutants have a medium lethal dose (LD_{50}) larger by a factor of 1 000 than that of the wild-type virus, and the titer of BV in mutants is reduced by a factor of 100; occluded virus production is eliminated completely (Clem *et al.* 1991). These results suggest that we are dealing with $\Delta > 0$. In the late phase of infection, virus egresses by extrusion, after interacting with gp64-rich sections of the plasma membrane. The rate of release increases exponentially between 10 and 20 hours postinfection. This is consistent with ε remaining relatively high and k low in the model.

Case 5: Avian hemangioma virus

The avian retrovirus avian hemangioma virus (AHV) is capable of inducing hemangiomas (vascular tumors composed of continuously dividing endothelial cells) in hens *in vivo*, while inducing a strong cytopathic effect in cultured endothelial cells (Sela-Donnenfeld *et al.* 1996). The AHV glycoprotein gp85 is responsible for killing the host cell by apoptosis, and its efficacy is dependent on the proliferative state of the cell: quiescent G_0/G_1 cells are more sensitive to AHV-induced apoptosis than actively dividing cells. Thus, in AHV, there is a relationship between the cell cycle and the apoptotic strategy. It is possible that apoptosis in tissue culture reflects the outcome of virus evolution in conditions of high death rates (high μ). Tumor cells with a typically protracted lifetime promote inhibition of cell death to obtain the most from budding. This could indicate a shift from the high-lysis strategy of inequality $\Delta < 0$ in culture, to the low-lysis strategy of inequality $\Delta > 0$ in cancer.

Case 6: Adenovirus death protein

The protein adenovirus death protein (ADP) is required for efficient lysis in adenovirus-infected (*Ad*) cells: mutations in the ADP that render it nonfunctional (denoted by *adp*) do not influence the replication rate of the virus, but cause the virus to lyse cells more slowly than does the wild type (Tollefson *et al.* 1996). *Ad*-infected cells of the *adp* type remain viable for much longer than those of the wild type ($\mu \gg d$), and become swollen with virus with little to no virus released by the cell into the surrounding cytoplasm. This describes a situation in which ε/μ is very low. Assuming k is sufficiently large, we might deduce that $\Delta < 0$, which suggests that the virus has evolved the ADP to increase the rate of lysis and thereby maximize viral load – the opposite mechanism of that employed by AcMNPV with the *p35* protein. Thus *Ad*, which seems incapable of efficient extrusion, may have evolved a more cytopathic mechanism of replication than AcMNPV.

13.6   Lessons from Case Studies

Assuming that cell death is an important component of virulence, I have presented simple models that might help us to understand when and where a virus is likely to behave cytopathically. Virulence is shown to depend on one predominantly host-dependent factor, the cell-death rate, and virus-dependent factors: the rates of virion extrusion and secretion, the efficiency of virus production during lysis, and the death rate of the infected cell. These parameters establish a threshold across which virus fitness is maximized by adopting one of two apoptotic strategies directed toward the host cell. These strategies differ among viruses and target cells, and reflect the outcome of a conflict of interest between the host and the virus in which each is evolving mechanisms related to cell death to maximize its own fitness.

For most of the examples discussed, the crucial parameter values are unavailable, and we are therefore unable to decide which party benefits most from apoptosis. In some cases, it is not at all clear if either party benefits. As a general rule of thumb, a virus evolves toward increasing cytopathicity through lysis when the mean lifetime of a cell is high and the rate of extrusion and secretion is low; a virus evolves toward reduced cytopathicity when the mean lifetime of the cell is low and the rate of extrusion and secretion is high.

Coinfection is important in a virus's apoptotic strategy. One simple way to address this in terms of the model is that coinfection might have the effect of raising the parameter μ for a coinfecting virus, and thereby cause it to switch from a latent, persistent strategy in which it inhibits apoptosis, to an acute one in which it induces apoptosis. This need not imply that the virus has some means of gauging cell-death rates, and does not require that the virus recognizes the presence of another virus species that shares the same cell. It merely states that evolution upon a background of coinfection (an elevated death rate) drives virus evolution toward an earlier induction of cell death.

13.7   Testing the Model

There are two approaches available for evaluating the model. The first involves estimating the parameters and then manipulating the rate of lysis, and recording the effects on the virus population and host phenotype. The second strategy requires manipulating the individual parameters and, by calculating Δ, predicting the course of virus evolution. Levine *et al.* (1993) adopted the first approach, in which *bcl-2* was over-expressed in an alpha virus chimera, leading to a reduction in viral titer and host mortality. Here apoptosis is clearly of value to the virus and not to the host. Unfortunately, even in this case, data on the longevity of the infected and uninfected cells, and on the burst size are unavailable. If they were, they would provide a critical test of the model, which predicts that $\Delta < 0$. In the second approach, the model predicts that the use of agents that reduce the efficiency of budding should favor the evolution of more cytopathic viruses, while agents that reduce the efficiency of lysis should favor the evolution of less cytopathic viruses. Thus, cell death is not manipulated directly as in the previous example, but through the parameters identified as important by the model.

Not only the viruses, but many pathogens have discovered through evolution that manipulating cell death is a fruitful mechanism. It has been noted that cell death is a feature of several single-celled eukaryotes, including the kinetoplastid parasite *Trypanosoma cruzi* (Ameisen *et al.* 1995). In this parasite, apoptosis could have evolved to allow selection of the fittest cell in a colony (Ameisen 1996), for inclusion of the best cells into a primordial germ line, or as a means of controlling their own parasite infestations. The bacterial pathogens, Shigella, Salmonella, and Pseudomonas, are able to influence host cell apoptosis by producing a diversity of proteins, including *IpaB* (in the case of Shigella), which binds to interleukin-1β-converting enzyme, initiating apoptosis (Finlay and Cossart 1997). In all of these cases, the replicative gains accruing from subverting or initiating cell death provide a significant selective advantage to the obligate parasite. It is worth stressing that viruses might represent something of a special case, as their genomes are particularly flexible and tend to acquire new host genes (or other virus genes) through frequent recombination events. The objective of this chapter is to provide an adaptive explanation for the many cases of virus-induced apoptosis, and, consequently, a means of deciding which party has gained the (temporary) advantage in the evolutionary arms race.

13.8 Medical Implications

How might knowledge of the coevolutionary dynamics of host and virus cell-death signals be put to use? The first thing we must do is identify exactly which parameter regime prevails in our system. It is then a matter of deciding whether modifications to the host or to the virus are more amenable to intervention, and which is more robust and less easily overcome during the rapid evolution of the virus.

One can envisage four categories of intervention, each with their respective strengths and weaknesses:

▪ *Transgenic modification of the virus genome to carry virally antagonistic death factors.* This strategy has the obvious advantage that it only affects cells infected by the virus. The problem is that it will not last long, as the modified virus will ensure its own demise. It will, however, behave somewhat like a vaccine, enabling the host to mount an immune response against a strain attenuated to cell death. A new and very promising use for engineered viruses is as vectors to carry apoptotic genes into cancerous tissues, many of which have lost cell-death genes during neoplastic progression (Tos *et al.* 1996). Thus, the *p53* tumor-suppressor gene has been delivered into lung cancer cells by an adenovirus/DNA complex (Nguyen *et al.* 1997). A consideration of cell death is also important when using viral vectors for the delivery of wild-type genes into mutated tissues. The introduction of beta-galactosidase into pancreatic, islet cells to ameliorate diabetic symptoms must take into consideration the possible risks of cell death if effective doses require high virus loads (Clouston and Kerr 1985).

▪ *Modification of the host cell lines using gene therapy to block the virus induction and inhibition signals.* This strategy would be effective against many different virus infections, but might also abrogate the beneficial host response to infection.

▨ *Synthesis of antisense oligonucleotides or anti-cell-death antibodies to the virus cell-death proteins.* Antisense techniques have been shown to be effective against growing myeloid leukemia cells, in which a BCR-ABL antisense oligodeoxynucleoside can induce apoptosis (Maekawa *et al.* 1995). Tumor necrosis factor alpha (TNF alpha) is correlated with an increase in cell death, and TNF-associated cell death has been blocked by constructing an anti-TNF antibody (Ebert *et al.* 1997). These methods have the advantage of efficient targeting, but interfere with host protein if the protein was produced by a xenologous gene acquired by the virus from the host at some time in their association.

▨ *Modification of the mitochondrial membrane potential using antioxidants.* This strategy has been applied to HIV-gp120 induced apoptosis, in which the antioxidants (ascorbic acid and glutathione) are able to inhibit cytopathic cell death (Radrizzani *et al.* 1997). A related strategy under investigation is caloric restriction, in which an energy source is made scarce or, more directly, inhibitors of mitochondrial electron transport or oxidative phosphorylation are administered at very low doses, thereby lowering mitochondrial free-radical production (Wachsman 1996). This approach might be therapeutic when dealing with long-lived cells such as neurones, but can compromise homeostatic cell-death processes in high turnover tissues.

13.9   Discussion

In conclusion, cell death is one of the proximate determinants of virulence and, in certain cases only, associated with the evolutionary optimum for a pathogen. In many cases, cytopathic effects such as cell death are induced by the host and are associated with a reduced fitness in the virus. Paradoxically, these might even lead to the death of the host. Evolutionary definitions of virulence can be positively or negatively correlated with tissue damage. The sign of this correlation depends on the life history of the virus and its cell tropism. When discussing virulence, we must therefore take care to distinguish fitness effects from viability effects. A more inclusive modeling of pathogen biology, treating both mechanisms and function, will hopefully help us toward a better understanding of disease management.

Acknowledgments   David C. Krakauer thanks the Alfred P. Sloan Foundation, The Ambrose Monell Foundation, The Florence Gould Foundation, and the J. Seward Johnson Trust for support. Thanks to Robert J.H. Payne, with whom the first apoptosis model was developed, and to Sebastian Bonhoeffer for his ideas and comments on the chapter.

Appendix 13.A   The Cell-death Model: Assessment of Extrema

How do the equilibrium values of the number of uninfected cells, infected cells, and virus particles, that is, x^*, y^*, and v^* as given in Equations (13.3), depend on the virus-induced cell-death rate L? Consider the derivatives with respect to L of the three state variables at equilibrium, which are given by

$$\frac{\partial x^*}{\partial L} = \frac{\delta \mu \Delta}{\beta (\varepsilon + kL - \mu - L)^2},$$

(13.8a)

$$\frac{\partial y^*}{\partial L} = \frac{\delta d(k-1)}{\beta(\varepsilon + kL - \mu - L)^2} - \frac{b}{(\mu + L)^2}, \quad \text{and}$$ (13.8b)

$$\frac{\partial v^*}{\partial L} = \frac{-b\mu\Delta}{\delta\,(\mu + L)^2},$$ (13.8c)

where $\Delta = \varepsilon/\mu - k$. Note that

$$\text{sign}(\frac{\partial x^*}{\partial L}) = \text{sign}(\Delta)$$ (13.9a)

and

$$\text{sign}(\frac{\partial v^*}{\partial L}) = -\text{sign}(\Delta).$$ (13.9b)

The behavior of $y^*(L)$ is less obvious. Extrema can exist if $\partial y^*/\partial L = 0$ has solutions, that is, if

$$b\beta(\varepsilon + kL - \mu - L)^2 = \delta d(k-1)(\mu + L)^2.$$ (13.10)

After rearranging, this yields

$$\left(\frac{\mu\Delta}{\mu + L} + k - 1\right)^2 = (k-1)\frac{\delta d}{\beta b},$$ (13.11)

and therefore

$$L = \mu\left(\frac{\Delta}{\Delta_c} - 1\right),$$ (13.12)

where

$$\Delta_c = (k-1)\left(\sqrt{\frac{\delta d}{\beta b(k-1)}} - 1\right).$$ (13.13)

Biologically we are only interested in those cases in which $y^* > 0$. This requires

$$\frac{\mu\Delta}{\mu + L} + k - 1 > \frac{\delta d}{\beta b}.$$ (13.14)

Substituting this result into Equation (13.11) gives

$$\sqrt{\frac{\delta d}{\beta b(k-1)}} < 1.$$ (13.15)

But we know that $k \gg 1$ (where k is the number of virions produced per lysis event) and hence $\Delta < 0$. Thus, the only way to obtain a solution of $\partial y^*/\partial L = 0$ for positive values of L is by fulfilling the condition $\Delta < \Delta_c < 0$; this solution must be a maximum.

14

Biogeographical Perspectives on Arms Races

Michael E. Hochberg and Robert D. Holt

14.1 Introduction

Natural enemies include parasites, pathogens, parasitoids, and predators (in the order of how we generally perceive their increasing impact on the survival of their individual victims). It has been increasingly recognized since the 1970s that the ecological dynamics of natural enemies and their victims can be diverse (Begon *et al.* 1996), and that understanding such dynamics has important implications for applied disciplines such as pest control (Chapter 32) and conservation biology (Dobson and McCallum 1997; Clarke *et al.* 1998; Hochberg 2000).

It is undeniably the case that natural enemies can be geographically widespread, yet most individuals spend their lives within the limited range of environments suitable for their species. Environmental differences over the geographical range of a natural enemy could, in turn, lead to spatial variation in population and adaptive dynamics. Large-scale environmental variation manifests itself in at least three ways.

First, all species have geographical boundaries, either abiotic barriers such as mountains or lakes, or biotic variables such as the abundance and quality of food, and the presence of competitors or predators (Brown *et al.* 1996; Holt *et al.* 1997). For many (but by no means all) species, geographical boundaries approximate those experienced by their resources. However, a more functional view of geographical boundaries of a species would include all of those habitats in which natural selection operates (Holt 1996). Such habitats could include vectors (for some parasites and pathogens), breeding grounds (for migrating predators), and nectar sources (for some species of parasitoid wasp).

Second, all species exhibit variations in community structure (Cornell and Lawton 1992). This notion combines:

- Spatial variation in the degree to which an enemy exploits each of its potential victim species;
- Spatial changes in community composition;
- Spatial variation in the types and strengths of indirect interactions between the natural enemy and other community members.

Such variation in community structure can lead to spatial variation in the population dynamic role of a specialized natural enemy on itself and other interacting species within the communities (Hochberg 1996).

Third, natural enemies and the interacting members of their communities exhibit spatial variation in the genetic structure of their populations. Such structure integrates gene flow, mutation, recombination, selection, and drift. Adaptation of natural enemies to their victims and vice versa may be rapid, especially if population sizes are large and generation times short. Given pronounced geographical differentiation in environmental factors, two-species interactions may vary considerably (and, as argued below, predictably) over geographical ranges. Despite an increasing appreciation of the spatial dynamics of natural enemies (e.g., Hassell *et al.* 1991; 1994), their importance in determining ecological and evolutionary patterns over geographical spatial scales has, thus far, received little attention (Hochberg and Van Baalen 1998).

In this chapter, we focus on a nearly neglected facet of natural enemy population biology: biogeography. As with single-species perspectives, examining multiple-species interactions should lend itself to biogeographical interpretations. MacArthur and Wilson's (1967) theory of island biogeography dealt with geographical patterns in species diversity fueled by chance historical events (colonization and extinction) and local adaptation. Much of their theory was inspired by terrestrial animals, especially birds and insects. Their basic ideas surely apply to the world of natural enemies, albeit with modifications because of the (often astonishing) array of local habitats available for persistence and evolution. Just how natural selection in local habitats interacts with migration over larger spatio-environmental scales should be, in our view, a major focus of research in the future (Thompson 1994).

Our aim is to elucidate how the impact of natural enemies on their victims could vary over geographical ranges. Causes for such variation may include:

- Ecology [i.e., no evolution occurs, but because of ecological factors (see Section 14.2) the impact varies among localities];
- Adaptive reasons independent of the species interaction (i.e., parameters that affect impact may evolve as a correlated response to other selected factors);
- Adaptive reasons arising from the interaction, but not reciprocally (i.e., selection is not tightly coupled between natural enemy and victim);
- Reasons of reciprocal evolution or coevolution [i.e., the interaction is sufficiently coupled such that selection and counter-responses to selection dominate evolution in both species (Thompson 1994)].

This chapter is divided into three sections. First, we briefly discuss why a biogeographical perspective may be necessary to understand natural enemy impacts. Second, a simple model of a predator–prey interaction is presented and its behavior discussed. Finally, we speculate on how our model framework may be of use in understanding applied issues such as biological control and population conservation.

14.2   Importance of Species and Space in Population Dynamics

Little is known about the extent to which spatiotemporal dynamics are dominated by the dynamic entities themselves and/or by underlying variation in the abiotic environmental template over which they play out their dynamics. Most models consider the dynamic entities in isolation, and are often referred to as "self-structuring" models (e.g., de Roos *et al.* 1991; Hassell *et al.* 1991, 1994; Rand *et al.* 1995). The presence of spatial time lags in population densities (or allelic frequencies) is a key aspect in the structure of these systems. In many respects, these models dominate the way we view the role of space in ecological systems (Tilman and Kareiva 1997).

Relatively few models include both species interactions and spatial variation in the environment extraneous to the interaction. These are "landscape-dynamic" models, and they have been increasingly applied in problems of population ecology (e.g., Oksanen *et al.* 1981; Holt 1984, 1985; McLaughlin and Roughgarden 1992; Leibold 1996; Clarke *et al.* 1997) and adaptive evolution (e.g., García-Ramos and Kirkpatrick 1997; Hochberg and Van Baalen 1998). Given that the self-structuring approach is a potential component of the landscape-dynamic approach, it will be an important challenge in the future to learn the conditions under which environmental templates are necessary to explain spatial variation in nature, or whether we can often rely on self-assembly rules. Our approach below includes both species interactions and landscape effects.

14.3 (Co)Evolution of Impact by Natural Enemies

A simple way to interpret how species demography may drive geographical variation in natural enemy impact is to make ecological parameters functions of spatial position x. Given that selection acts differently in different parts of a species' range, individual movement among sites can influence the realized spatial pattern of adaptation: even if dispersing individuals are maladapted to novel environments, once present, they may reproduce (producing more or less adapted offspring) and compete with residents (reducing the fitness of resident genotypes). Understanding how gene flow and selection combine to influence geographical patterns of variation is a major issue in the study of microevolution. Much of the relevant literature has concentrated on genetic dynamics alone (e.g., for host–parasite systems, see May and Anderson 1983a; Seger 1992). However, it is increasingly apparent that to analyze adaptive dynamics and population dynamics simultaneously can be useful, because different population dynamic scenarios can entail different conclusions about the importance of gene flow as a constraint on local adaptation (Holt and Gomulkiewicz 1997).

Several indices are relevant to how a natural enemy affects its victim; these include population density of the victim, both with and without the natural enemy (Beddington *et al.* 1975), selective impact of the enemy on the victim (e.g., Abrams and Matsuda 1997), and attack rate (e.g., Hochberg 1991). We focus on what we call "impact," which in the model given below is the per capita attack rate of enemies on their victims. Since we consider only nonpolymorphic, coevolutionarily stable strategies, impact translates readily into the notion of risk of attack (i.e., fraction of victims killed over a particular span of time).

Model assumptions and structure

The model is based on a recent study of predator–prey dynamics (for details, see Hochberg and Van Baalen 1998). It can apply to predators as well as to pathogens and parasitoids that kill their host rapidly after infection/attack.

A number of simplifying assumptions are made at the outset. With respect to genetics, we assume clonal variation. With respect to basic ecology, we assume that the victim exhibits continuous generations. Further, we assume random encounters between the two species. The two species interact over a network of patches. Species flows among patches (and mutation) are assumed to be sufficient to maintain all genetic variants in all patches, but insufficient to have density effects on dynamics (but see Discussion). Therefore, we do not consider in detail colonization–extinction dynamics or "swamping" effects of migration.

Let N and P be the densities of prey and predators, respectively. Within any given patch, the interaction is described by

$$\frac{dN_j}{dt} = (b_N - d_{N_j})N_j - \kappa N_j \sum_j N_j - N_j \sum_k \beta_{j,k} P_k . \tag{14.1a}$$

$$\frac{dP_k}{dt} = \phi P_k \sum_j \beta_{j,k} N_j - d_{P_k} P_k , \tag{14.1b}$$

where the indices j and k refer to genetically different clones. Any of the parameters b_N, d_N, ϕ, β, κ, or d_P could be patch-specific. The prey is limited by two forms of density dependence: logistic-type limitation (the κN term), resulting in a standing crop of $(b_N - d_N)/\kappa$ prey; and predator-driven limitation at a per capita rate of βP, with the production of ϕ predator offspring per prey consumed. We call βP the impact λ of the predator. In studies on host–parasite associations this is often called the "force of infection."

We assume that the capacities of predator attack and prey defense each incur costs according to a quantitative genetic model for the respective species [see Hochberg and Van Baalen (1998) for details], such that the predation constant between strains k of the parasite and j of the victim is

$$\beta_{j,k} = \beta_0 + \beta_1(k - j) . \tag{14.2}$$

Impact λ on victim strain j is therefore

$$\lambda = [\beta_0 + \beta_1(k - j)]P_k , \tag{14.3}$$

where β_0 and β_1 are constants. Thus, impact has components influenced by a combination of environment (β_0, β_1), evolution (k, j), and ecology (P).

The costs of the interaction are deducted from the natural survival rates of each species. Such costs could include reductions in life span because of increased mortality by generalist natural enemies, or reductions in reproductive rate through a shifting of resource allocation from reproduction to the interspecific interaction modeled here.

The natural mortality rate of a predator expressing level k attack is

$$d_{Pj} = d_{P0} + d_{P1}k^{c_P} ,$$ (14.4a)

whereas the mortality rate of a prey expressing level j defense is

$$d_{Nj} = d_{N0} + d_{N1}j^{c_N} .$$ (14.4b)

The quantities d_{P0}, d_{P1}, d_{N0}, and d_{N1} are constants; the mortality rate for the most efficient predator strain is $d_{P0} + d_{P1}$, and for the most defensive prey strain it is $d_{N0} + d_{N1}$. The constants c_P and c_N reflect nonlinearities in trade-offs involving impact (Frank 1994a). If $c > 1$ ($c < 1$), then costs increase at a greater than (less than) linear rate with marginal increases in character state j or k.

Geographically labile parameters

Numerous biological characters of a species are likely to vary over geographical ranges for many reasons peripheral to the predator–prey interaction. Environmental suitability for the prey is the propensity of the prey population to grow, and is measured by parameters such as b_N, d_N, and κ. Suitability may also influence impact-related parameters (i.e., β_0 and β_1).

Below, we briefly discuss how, in predator–prey systems, local environments influence the dynamic impact of enemies on their victims. Three classes of factors are likely to vary over the prey's geographical range: the prey's net rate of reproduction $r = b_N - d_N$, the intensity of density-dependence acting on the prey κ, and predator attack β:

■ Spatial variation in the net rate of increase, r, is fundamental to defining range limits (Holt and Kiett 2000) and spatial variation in abundance (Holt *et al.* 1997). Low values of b_N could result from unfavorable abiotic conditions or low quantity and quality of the prey's own resources (e.g., for a herbivore, its preferred plants may be rarer or of low nutritional quality). Higher values of d_N may also reflect scarce, low-quality resources, or the greater impact of other (e.g., generalist) natural enemies near the edges of the prey's distribution.

■ Variation in the intensity of prey density-dependence κ could result from variation in the impact of, for example, (1) other natural enemies, (2) intraspecific interference, and/or (3) levels of resources in the system. Areas with low prey densities could arise from higher impacts of natural enemies, competitors, or lower resource levels. Were κ to be decomposed into a different component for each of these factors, then certain components might increase from the center to the periphery of the prey's distribution (e.g., effects of lack of resources), whereas others decrease (e.g., effects of generalist natural enemies). We equate high κ with habitat marginality for the prey.

■ With regard to the negative effects that the predator has on its prey, β, it is reasonable to expect that as the prey's environment becomes less favorable, successful attacks are more frequent (i.e., higher β), although it could equally be argued that the conditions for predator attack decrease even more rapidly than those for prey defense (meaning lower β).

Population ecological results

In the case where neither species evolves (i.e., $j = k$ fixed at 1) it can be shown that the equilibrium impact

$$\lambda^* = r - \kappa d_P / \phi \beta , \tag{14.5}$$

which means that independent of evolution, impact varies over geographical ranges purely through population–environment interactions. Since habitat suitability is positively correlated with r and negatively correlated with κ and β, impact should decrease with decreases in habitat suitability, but only if components of suitability extrinsic to the interaction (i.e., r and κ) change more over the geographical range of the victim than does vulnerability to the natural enemy β.

Coevolutionary results

Now we consider what happens when both species evolve. Hochberg and Van Baalen (1998) presented a technique for finding the coevolutionarily stable strategy (CoESS) solution to this system for cases in which the parameters that control nonlinearities in trade-offs [see Equations (14.4)] are both greater than one (Box 14.1). Employing the CoESS approach, the system evolves toward a single observed equilibrium point for $c > 1$. When $c < 1$, numerical techniques are needed to find the polymorphic solutions. We present here only the CoESS solutions (for additional results, see Hochberg and Van Baalen 1998).

Figure 14.1 shows how impact and its components are expected to vary with four habitat suitability parameters. In all cases, declining habitat suitability is associated with lower natural enemy impact. Higher natural enemy impact λ is always associated with less evolved victim defense, $j^* - k^*$; or, in other words, the victim defends itself less in productive environments than in nonproductive ones! Note from these figures that components of λ may (Figures 14.1a, 14.1c, and 14.1d) or may not (Figure 14.1b) vary in the same fashion, meaning that focusing on single correlates of impact may belie other components of the index λ.

How does migration affect these results? By employing numerical simulations of Equations (14.1a) and (14.1b), Hochberg and Van Baalen (1998) showed that migration tended to expunge spatial patterns in local adaptation (e.g., patterns in impact explored in this chapter); this was especially true when there was no spatial pattern in habitat suitability (see their figures 3–5). Migration tends to differentially favor the global representation of adaptations to productive environments, leading to the expectation of overall heightened natural enemy impact in dispersive systems through a species' range.

Thus, increasing habitat suitability for the victim should be associated with higher impact by the natural enemy. However, this central result overlooks an important consideration: interactions between habitat and gene, which are encapsulated in parameters β_1, d_{N1}, and d_{P1}. Such interactions mean that marginal changes in genotypes (i.e., the ability of the natural enemy to attack, or the ability of the victim to defend itself) have different weightings in different habitat types.

Box 14.1 Determining coevolutionarily stable strategies

The strategy set for a natural enemy and victim that resists invasion from all possible mutants is called a coevolutionarily stable strategy (CoESS). The victim's strategy is denoted by x, and y is the enemy's strategy. Assume there is one resident strain of each species, x_{res} for the victim and y_{res} for its enemy. If x^* and y^* are a CoESS, then $x_{mut} = x^*$ and $y_{mut} = y^*$ should both be local fitness optima for mutants x_{mut} or y_{mut} in the environment created by the pair $(x_{res}, y_{res}) = (x^*, y^*)$ (Vincent and Brown 1989; Van Baalen and Sabelis 1993).

The fitness of a rare mutant victim strain with strategy x_{mut} is

$$f_{victim}(x_{mut}; x_{res}, y_{res}) = \left.\frac{dN_{mut}}{N_{mut}dt}\right|^*_{res},$$

where N is the population density of the victim (the density of the enemy will be denoted by P), and $|^*_{res}$ denotes "evaluated at the resident equilibrium (N^*_{res}, P^*_{res})." If $f_{victim}(x_{mut}; x_{res}, y_{res})$ is positive, then the mutant strain with strategy x_{mut} invades. To find the evolutionarily stable strategy (ESS) x^*, for a given fixed value of y, one must first find the optimum strategy x^0_{mut} for rare mutants arising in a population dominated by a resident population x_{res}, the so-called best reply, and then identify that resident strategy x^* that is its own best reply. To find the CoESS pair, this procedure has to be followed for the pair (x, y) simultaneously, that is, determine the optimal (x^0_{mut}, y^0_{mut}) for each possible (x_{res}, y_{res}) and then determine the $(x_{res}, y_{res}) = (x^*, y^*)$ such that (x^*, y^*) and the corresponding (x^0_{mut}, y^0_{mut}) coincide.

Setting the partial derivatives $\partial f_{victim}(x_{mut}; x_{res}, y_{res})/\partial x_{mut}$ and $\partial f_{enemy}(y_{mut}; x_{res}, y_{res})/\partial y_{mut}$ equal to zero to obtain the mutant optimum and at the same time setting mutant and resident equal to x^* for the victim and y^* for the enemy, gives two equations in two unknowns

$$\left.\frac{\partial f_{victim}(x_{mut}; x_{res}, y_{res})}{\partial x_{mut}}\right|_{x_{mut}=x_{res}=x^*, y_{res}=y^*} = 0,$$

$$\left.\frac{\partial f_{enemy}(y_{mut}; x_{res}, y_{res})}{\partial y_{mut}}\right|_{x_{mut}=x^*, y_{mut}=y_{res}=y^*} = 0.$$

from which CoESS can be determined.

For example, the marginal cost to the prey of evolving from a given strategy j to $j + 1$ would be expected to be higher in marginal environments than in productive ones (in which case d_{N1} would decrease with productivity). Although not discussed in detail here, larger d_{N1} and β_1 (predicted to be associated with poorer habitats for the prey) actually lead to higher impact. Therefore, strong habitat and gene interactions can produce the opposite trends predicted for variation in habitat alone, and their relative weighting as compared to habitat-based effects should be more relevant to the overall effect on natural enemy impact on its victim.

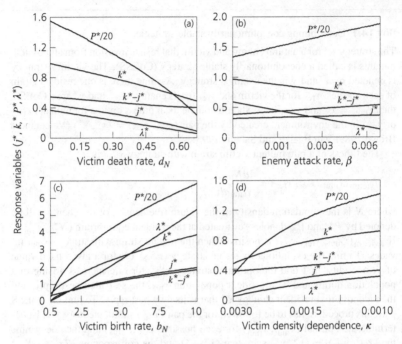

Figure 14.1 Numerical results of CoESSs as a function of habitat suitability for the victim: (a) victim death rate d_N, (b) enemy attack rate β, (c) victim birth b_N, and (d) victim density dependence κ. Vertical axis shows response variables j^* (investment by victim), k^* (investment by exploiter), $k^* - j^*$ (difference in investments by exploiter and victim), P^* (equilibrium density of exploiter), and $\lambda^* = \beta_{j,k} P^*$ (impact). Note that by assuming an effectively infinite number of strains in enemy and victim populations, the variables j and k can be expressed as continuous variables. Other parameter values: $c_N = c_P = 2, d_{P1} = 5, d_{N1} = 0.7, d_{P0} = 1, d_{N0} = 0.1, \beta_0 = 0.001, \beta_1 = 0.02, b_N = 1, \phi = 1$, and $\kappa = 0.001$.

14.4 Discussion

According to our analysis, increasing habitat suitability for a victim should be associated with higher natural enemy impact as long as there are no interactions between habitat and gene. Below, we discuss how other factors may impinge on these findings, some empirical support of the model predictions, and ways in which the results can be applied to real-world problems.

Main factors

We begin our discussion by identifying several major factors at work in determining patterns of natural enemy impact over geographical ranges:

- *Landscape.* Aspects of species biologies that evolve independently of the interaction set the template for the evolution of impact. It remains to be seen in real systems whether spatial pattern formation in impact reflects the underlying variation in habitat suitability for the victim, but preliminary empirical studies indicate this (see subsection on empirical support below). We expect the most detectable patterns to occur for a victim that experiences a wide range of habitat suitabilities over its geographical range.

▪ *Reciprocal selection.* Natural enemy monophagy (see Holt and Lawton 1993 for discussion) is not a prerequisite for selection to occur (Takasu 1998). Conversely, little or no evolution may occur in some tightly coupled systems (Hochberg and Holt 1995; Holt *et al.* 1998). We suggest that systems in which the natural enemy has the latitude to have a major impact on its victim are the most likely to yield spatial patterns.

▪ *Genetic systems and genetic structure.* The genetic mechanisms that control reconnaissance (e.g., distinguishing self from nonself) and response (e.g., defending self, or mounting an immune response) in natural enemy–victim interactions (Frank 1993a; 1996c), and the genetic diversity of these interactions (Frank 1993a; Hochberg 1997), can be pivotal determinants of evolutionary trajectories. For instance, if indeed the selection pressure exerted by natural enemies varies predictably across geographical gradients, then this leads to the predictions that (1) the diversity and amplitude of defenses should increase from habitats of low to high suitability (Hochberg 1997), and (2) defenses and countermeasures to these defenses should be less specific in high-quality as opposed to low-quality habitats (Hochberg and Van Baalen 1998). How genetic systems of reconnaissance and response impinge on geographical patterns of impact and its components requires further research, but preliminary analyses suggest that the relative constraints (e.g., trade-offs, allelic diversity, single or multilocus genetic systems, or metabolic costs) associated with each of these two broad categories have a major impact on their relative contributions to geographical patterning (Hochberg 1998; Hochberg and Strand, unpublished simulations).

▪ *Patch size.* Patch size influences the local extinction of genotypes and even entire populations. In an island biogeographical setting for our models, we expect that local adaptation and the maintenance of genetic diversity on small islands should be hampered compared with that on large islands because of demographic stochasticity and other factors (Frankham 1997; Holt 1997). Thus, the spatial patterns in impact we predict will be more obscure in systems of small patches.

▪ *Interpatch flows.* Dispersal may either promote or destroy local adaptation. It promotes local adaptation by introducing novel genotypes to areas where they may proliferate (e.g., Holt 1996); but if flows are too intense, it destroys local adaptation and overall diversity by shunting maladapted genotypes that compete with locally adapted genotypes (García-Ramos and Kirkpatrick 1997; Hochberg and Van Baalen 1998).

▪ *Landscape ruggedness.* Which-patch-is-next-to-which can have very important implications for geographical patterns in impact, whereby continuous variation in environments from patch to patch tends to conserve more genotypes of each species globally, and make geographical differentiation of impact more identifiable than a more rugged variation (Hochberg and Van Baalen 1998). In other words, when patch suitability varies irregularly through space, swamping effects from productive to neighboring unproductive patches are much more commonplace than in systems in which neighboring patches tend to have similar productivities.

■ *Temporal dynamics.* Holt *et al.* (1998) showed that unstable temporal population dynamics can hamper the evolution of resistance of a victim species to its exploiter. An important issue to examine in future studies is the influence of geographical-scale variation in population instability on the spatial patterning of impact.

Empirical support

Three studies go some way to explain how natural enemies and their victim may evolve in areas of different habitat suitability for the victim.

Work by Lenski and colleagues vividly illustrates how habitat productivity for a host influences the persistence of susceptible and resistant forms to a phage pathogen. Their basic approach was to vary the level of glucose input into chemostats and monitor the population dynamics of susceptible and resistant forms of *Escherichia coli* and various bacteriophages. Bohannan and Lenski (1997) demonstrated that both predation pressure increases and the rate of replacement of phage-sensitive clones by resistant clones increases with nutrient enrichment. As the phage did not evolve in this experimental set-up, invasion of the resistant bacteria resulted in the system being transformed from parasite-limitation to resource-limitation.

Another system that yields results consistent with our predictions about transmissibility is wild oats (*Avena*) and their rust parasite *Puccinia coronata* in New South Wales, Australia. Burdon *et al.* (1983) showed that northern populations of oats in more favorable (mesic) conditions were more resistant to the rust than populations in southern, arid environments. Oates *et al.* (1983) considered the flip-side of the interaction and showed a trend for increasing parasite virulence from arid to mesic sites. These studies suggest that both antagonists show spatially varying adaptations in their association, and that both are increasingly engaged in the interaction as habitat quality increases.

A third example involves fruit fly hosts (*Drosophila melanogaster*) and their insect parasitoids (*Asobara tabida* and *Leptopilina boulardi*). Mollema (1988) and later Kraaijeveld and van Alphen (1994, 1995) showed that the highest levels of virulence of *A. tabida* to a single reference strain of *D. melanogaster* tends to occur in the southernmost latitudes of Western Europe, and that the highest encapsulation abilities of the host to a single reference strain of *A. tabida* occur in the central latitudes of Western Europe. Why *D. melanogaster* does not exhibit a geographical pattern in encapsulation to another parasitoid (*L. boulardi*) is unknown.

Managing natural enemies

Our approach and the results we present here have applications in the reintroduction of endangered or locally extinct species and in the introduction of exotic natural enemies to biologically control pest species. However, one should be cautious in interpreting the discussion below, as it is designed to elucidate some potential ways to apply a biogeographical approach. More studies are necessary to evaluate our findings, and only specific models applied to particular systems of interest should ever be employed in real policy making.

Reintroduction of endangered natural enemies. The conservation of natural enemies has received relatively little attention in the applied literature, and theoretical models (which could be useful to understanding the important factors of their conservation) are rarely applied to this type of problem (Hochberg *et al.* 1998; Hochberg 2000). Natural enemies are generally candidates for conservation as long as they are perceived to have some kind of "value"; detailed discussion for the case of insect parasitoids can be found in Hochberg (2000). What is relevant to the present scenario is the conservation of a natural enemy that is part of a protected community. That is, it is important in conserving the natural enemy that neither its victim nor other interacting species are endangered by the conservation efforts directed at the enemy.

Assume that the novel victim (which receives the introduced enemy species) is distributed over its geographical range as in the model presented above, and the enemy we desire to introduce is found on a different (but related) exotic victim species that has no overlap in its distribution with the focal victim. The question is, where along an exotic victim's distribution should one procure the enemy?

If, for example, the enemy is taken from peripheral, nonproductive sites, then a risk is that it will not be preadapted to invade the introduction sites successfully, especially if the productivity of the novel victim is very high through most of its distribution. If, on the other hand, there are large expanses where the novel victim is unproductive, then it is possible for the natural enemy to invade and persist. However, a potential problem may emerge if the introduced enemy has too high an impact on the nonproductive victim, especially in the most marginal sites. In a recent theoretical study on host–parasitoid interactions over geographical gradients in host productivity, it was shown that enemies can readily fragment the geographical distribution of their hosts if the host is relatively poor at defending itself against the parasitoid and the parasitoid is highly vagile (Hochberg and Ives 1999).

Now assume that the enemy is procured from highly productive sites in the exotic victim's geographical range. The problem here is very similar to the latter scenario described above: if the target sites are unproductive relative to the site of origin, then the enemy can disrupt the victim population and species interactions with the victim (e.g., Holt and Lawton 1993). Of course this problem becomes more intense (according to our model) as the discrepancy in productivity grows.

Assuming that the productivities of exotic and introduced sites are comparable and the enemy invades, then what are the evolutionary consequences? With sufficient (but not too pronounced) migration and mutation, we would expect the natural enemy to gradually spread and adapt over the victim's distribution, occupying that subset of it for which site productivities are sufficient for the enemy to persist. The race between local adaptation and population ecology is relevant here (see Holt and Gomulkiewicz 1997), and unless the migrating enemy is sufficiently preadapted and/or numerous, it will be unable to spread so as to fill its fundamental niche (i.e., that part to which it is potentially able to adapt and persist in the absence of migration). Assuming that geographical adaptation is taking place or has taken place, if the objective is to conserve maximal enemy diversity, then postintroduction measures should aim to conserve the enemy in a range of habitat types to which it is adapting or has adapted (Hochberg and Van Baalen 1998).

Introduction of natural enemies for control of pests. Like the application of conservation measures to natural enemies, the problem of pest control over the pest's geographical ranges is little explored. The problem here is that an enemy is most likely to be released in areas where the economic damage inflicted by the pest is most intense. Since our model does not consider the trophic level below the victim, it is difficult to generalize where the pest would be most damaging economically.

If we are searching for an exotic natural enemy to release against the victim (i.e., classic biological control), then where should we take it from: the productive or marginally productive sites of an exotic victim? Let us assume as a first scenario that the pest is most damaging in the nonproductive sites of its range. According to our model, if we take the enemy from productive source sites, then it will have a major impact on the pest, but only temporarily (which may be sufficient to cause local extinction of the pest). As the enemy is not adapted to the marginally productive introduction sites, it must either adapt to a form with less impact λ, or go extinct itself. If, on the other hand, the enemy comes from a nonproductive site, then it may be preadapted to the pest it is about to encounter, and some level of lasting control may be achievable with lower risks of local extinctions following population transients.

Now assume that the pest is of most concern in productive sites. Introduction of a natural enemy from a nonproductive source is unlikely to achieve any control, and unless the enemy can adapt very quickly, it may go extinct. In contrast, introduction from a productive source is likely to give substantial control (i.e., impact λ is predicted to be high). It is interesting that according to the model presented here (see also Hochberg and Ives 1999), although a successfully introduced natural enemy will have the most impact in the most productive sites, pest densities are likely to be lowest in areas where the enemy has least impact! This is simply because density-dependent limitations in this model are such that the enemy always depresses victim density to the same or lower levels as productivity decreases.

Hokkanen and Pimentel (1984) hypothesized that the introduction of natural enemies for pest control often worked so well because the natural enemy was introduced against a pest with which it had never been in evolutionary contact (see also Waage 1990). The idea is that if there had been an evolutionary or coevolutionary interaction, then the impact of the enemy on the pest would have diminished through time. Indeed, our model suggests that the introduction of an enemy from a productive site to a nonproductive one may well result in an initially impressive control, but then evolve to more moderate results. There appears to be no evidence for evolution in biological control that involves insect parasitoids as natural enemies against arthropod pests. However, scattered evidence for the evolution of resistance does exist for insect pathogens (Holt and Hochberg 1997).

Conclusion

We believe that the framework presented here for exploring enemy–victim interactions over geographical and evolutionary scales is the first step in what will prove to be an interesting and fruitful area of research. Even based upon a variety of oversimplified assumptions, the theory predicts that one must be cautious in introducing natural enemies with the goal of their global conservation or the biological control of other species.

15

Major Histocompatibility Complex: Polymorphism from Coevolution

Joost B. Beltman, José A.M. Borghans, and Rob J. de Boer

15.1 Introduction

There are many examples of pathogens adapting toward evasion of immune responses. Viruses, such as influenza, rapidly alter their genetic make-up, and each year there appear to be sufficient susceptible hosts that lack memory lymphocytes from previous influenza infections to give rise to a new epidemic (Both *et al.* 1983; Smith *et al.* 1999). During human immunodeficiency virus (HIV) infection, such alterations occur at an even faster rate, enabling the virus to escape repeatedly from the immune response within a single host (Nowak *et al.* 1991). Hosts, on the other hand, are selected for counteracting immune evasive strategies by pathogens. Since the generation time of hosts is typically much longer than that of pathogens, these host adaptations are expected to evolve much more slowly.

A well-known example commonly thought to reflect adaptation of hosts to pathogens is the polymorphism of major histocompatibility complex (MHC) molecules, which play a key role in cellular immune responses. When a pathogen infects a host cell, the proteins of the pathogen are degraded intracellularly, and a subset of the resultant peptides is loaded onto MHC molecules, which are transported to the cell surface. Once the peptides of a pathogen are presented on the surface of a cell in the groove of an MHC molecule, T lymphocytes can recognize them and mount an immune response.

The population diversity of MHC molecules is extremely large: for some MHC loci, over 100 different alleles have been identified (Parham and Ohta 1996; Vogel *et al.* 1999). Nevertheless, the mutation rate of MHC genes does not differ from that of most other genes (Parham *et al.* 1989a; Satta *et al.* 1993). Studies of nucleotide substitutions at MHC class I and II loci revealed Darwinian selection for diversity at the peptide-binding regions of MHC molecules. Within the MHC peptide-binding regions, the rate of nonsynonymous substitutions is significantly higher than the rate of synonymous substitutions; in other regions of the MHC, the reverse is true (Hughes and Nei 1988, 1989; Parham *et al.* 1989a, 1989b). Compared to the enormous population diversity of MHC molecules, their diversity within any one individual is quite limited. Humans express maximally six different MHC class I genes (HLA A, B, and C), which are codominantly expressed on all nucleated body cells. Additionally, there are maximally 12 different MHC class II molecules (HLA DP, DQ, and DR), which are expressed on specialized antigen-presenting cells (Paul 1999). The complete sequence of a human MHC has been unraveled recently (MHC Sequencing Consortium 1999). Despite the high population diversity of MHC molecules, MHC genes appear to be extremely conserved evolutionarily. Allelic MHC lineages have persisted over long evolutionary time spans, often predating the divergence of present-day species (Klein 1980; Lawlor *et al.* 1988; Mayer *et al.* 1988; Klein and Klein 1991). As a consequence, individual MHC alleles from a species tend to be more closely related to particular MHC alleles from other species than to the majority of alleles that occur within the species (Parham *et al.* 1989b).

As a result of the high population diversity of MHC molecules, different individuals typically mount an immune response against different subsets of the peptides of any particular pathogen. Pathogens that escape from presentation by the MHC molecules of a particular host may thus not be able to escape from presentation in another host with different MHC molecules. MHC polymorphism may therefore seem a good strategy by which host populations counteract escape mechanisms of pathogens. This group selection argument, however, fails to explain how such a polymorphism could have evolved (Bodmer 1972).

The mechanisms behind the selection for MHC polymorphism have been debated for over three decades. A commonly held view is that MHC polymorphism arises from selection that favors heterozygosity. Since different MHC molecules bind different peptides, MHC heterozygous hosts can present a greater variety of peptides, and hence defend themselves against a larger variety of pathogens compared to MHC homozygous individuals. This hypothesis is known as the theory of overdominance or heterozygote advantage (Doherty and Zinkernagel 1975; Hughes and Nei 1988, 1989; Takahata and Nei 1990; Hughes and Nei 1992). A recent study of patients infected with HIV-1 supports this theory. It was shown that the degree of heterozygosity of MHC class I loci correlated with a delayed onset of acquired immunodeficiency syndrome (AIDS). Individuals who are homozygous at one or more loci typically progressed more rapidly to AIDS (Carrington *et al.* 1999).

It has been argued that selection for heterozygosity alone cannot explain the large MHC diversity observed in nature (Parham *et al.* 1989b; Wills 1991). Although there is general agreement upon the significance of overdominant selection, it has been proposed that additional selection pressures must be involved in the maintenance of the MHC polymorphism (Parham *et al.* 1989b; Wills 1991). A frequently studied additional mechanism is frequency-dependent selection. The corresponding theory states that evolution favors pathogens that avoid presentation by the most common MHC molecules in the host population. Thus, there is a permanent selection force favoring hosts that carry rare (e.g., new) MHC molecules. Since hosts with rare MHC alleles have a higher fitness, the frequency of rare MHC alleles will increase, and common MHC alleles will become less frequent. The result is a dynamic equilibrium, maintaining a polymorphic population (Snell 1968; Bodmer 1972; Slade and McCallum 1992; Beck 1984).

Both selection for heterozygosity and frequency-dependent selection have been modeled extensively. Most models address either of the two hypotheses, and are so-called "top-down" models. Assuming that heterozygous individuals have a higher fitness than homozygous individuals (see, for example, Takahata and Nei 1990), or assuming that individuals carrying rare alleles have a higher fitness than individuals carrying common alleles (see, for example, Takahata and Nei 1990; Wills 1991; Wills and Green 1995), it has been shown that an existing MHC polymorphism can be maintained.

Here we take a more mechanistic approach by making no assumptions about selective advantages or disadvantages. We develop a computer simulation to study the coevolution of diploid hosts with haploid pathogens. By comparing simulations in which pathogens do coevolve with simulations in which they do not, our model allows us to study the effect of selection for heterozygosity and frequency-dependent selection on the polymorphism of MHC molecules. Starting from a population diversity of only one MHC molecule, we show that a diverse set of functionally different MHC molecules is obtained. Our analysis demonstrates that selection involving rapid evolution of pathogens can account for a much larger MHC diversity than can selection for heterozygosity alone.

15.2 Simulating the Coevolution of Hosts and Pathogens

We have developed a genetic algorithm (Holland 1975) to investigate the coevolution of pathogens and MHC molecules. Genetic algorithms are frequently applied as problem-solving tools, using the principles of evolution to find solutions in, for example, optimization problems. We instead use them here as a simulation of evolution (see also Forrest 1993; Pagie and Hogeweg 1997), and thereby take them "right back to where they started from" (Huynen and Hogeweg 1989).

In our simulations, we consider a population of N_{host} diploid hosts, each represented by a series of bit strings coding for two alleles at N_L MHC loci. Pathogens are haploid and occur in N_S independent species of maximally N_G different genotypes. For simplicity, we omit the complex process of protein degradation into peptides, and model each pathogen by N_P bit strings that represent the set of peptides that can possibly be recognized by a host. Peptide presentation by an MHC molecule can occur at different positions on the MHC molecule, and is modeled by complementary matching. Peptides are L_P bits long, and MHC molecules are L_M bits long. For each peptide of a pathogen and for each MHC molecule of a host, we seek the position at which the peptide finds the maximal complementary match. If the number of complementary bits at this position is at least a predefined threshold L_T, the peptide is considered to be presented by that particular MHC molecule. In the simulations presented here, pathogens consist of $N_P = 20$ different peptides, which are $L_P = 12$ bits long. MHC molecules are $L_M = 35$ bits long, and present a peptide if, of the 12 peptide bits, at least 11 ($= L_T$) match with the MHC. Thus, the chance that a random MHC molecule presents a randomly chosen peptide is 7.3%. [The chance that a random peptide binds at a random, predefined position of an MHC molecule is $p_b = \sum_{j=L_T}^{L_P} \binom{L_P}{j}(0.5)^{L_P}$. Thus, the chance that a random MHC molecule presents a randomly chosen peptide is $1 - (1 - p_b)^{L_M - L_P + 1} = 7.3\%$.] Also, the chance that a pathogen of $N_P = 20$ peptides escapes presentation by a randomly chosen MHC molecule is $p_e = 22\%$. Hosts that carry different MHC molecules hence typically present different peptides of the pathogens.

The quality of different MHC molecules varies. Some MHC molecules may be more stably expressed on the surfaces of host cells than others, or fold into a better peptide-binding groove. To model such MHC differences, a random quality parameter $0 < Q < 1$ (drawn from a uniform distribution) is attributed to every MHC molecule in the population. These quality differences between MHC molecules prevent extensive drift in simulations with random pathogens. The fitness contribution of a host–pathogen interaction is determined by the quality of the best MHC molecule that is able to present a peptide of the pathogen. We omit the role of lymphocytes by assuming that every presented peptide is recognized by at least one functional clonotype. The role of lymphocytes, in particular the (functional) deletion of lymphocytes during self-tolerance induction, is to be reported in a follow-up paper (Borghans *et al.*, unpublished; see also Borghans *et al.* 1999).

At each generation, every host interacts with every genotypically different pathogen. To account for the shorter generation time of pathogens, we can allow for several pathogen generations per host generation. The fitness f_h of a host is proportional to the fraction of pathogens it is able to present,

$$f_h = \sum_{j=1}^{N_{\text{path}}} Q_j / N_{\text{path}} , \qquad (15.1)$$

where N_{path} denotes the total number of different genotypes in the pathogen populations. Q_j denotes the quality of the best MHC molecule that presents at least one peptide of pathogen j; we set Q_j to zero if none of the MHC molecules of a host present pathogen j. Similarly, the fitness f_p of a pathogen is proportional to the fraction of hosts that the pathogen can infect without being presented on the host's MHC molecules,

$$f_p = 1 - \sum_{k=1}^{N_{\text{host}}} Q_k / N_{\text{host}} , \qquad (15.2)$$

where Q_k is the quality of the best MHC molecule of host k that presents at least one peptide of the pathogen. Again, Q_k is set to zero if none of the MHC molecules of host k present the pathogen.

At the end of each generation, all individuals are replaced by fitness-proportional reproduction. The sizes of the host population and all pathogen species remain constant. All fitnesses are rescaled such that the highest fitness in each host population and in each pathogen species becomes one and the lowest becomes zero. The different individuals in the host population, and the different genotypes in each pathogen species, reproduce according to a fitness-dependent reproduction function,

$$Pr(j) = \frac{e^{\overline{f}_j}}{\sum_{k=1}^{N} e^{\overline{f}_k}}, \tag{15.3}$$

where $Pr(j)$ is the reproduction probability of host j or pathogen genotype j, $\overline{f}_j$ denotes its rescaled fitness, and N is the total number of different individuals in the host population or genotypes in the particular pathogen species. Pathogen genotypes reproduce asexually; new-born pathogens come from parents of the same pathogen species. New-born hosts have two parents, each of which donates a randomly selected MHC allele. During reproduction, point mutations can occur. Both peptides and MHC molecules have a mutation chance of $p_{mut} = 0.1\%$ per bit per generation. The chance for a new-born host to receive a nonmutated MHC molecule is thus $(1 - p_{mut})^{LM} = 96.6\%$, and the chance for a new-born pathogen to receive a nonmutated peptide is $(1 - p_{mut})^{LP} = 98.8\%$. One cycle of fitness determination, reproduction, and mutation defines a generation. We study evolution over many generations.

15.3 Dynamically Maintained Polymorphism

The simulation model allows us to study the mutual influence of host and pathogen coevolution on the composition of MHC molecules in the host population, and of peptides in the pathogen species. In particular, we:

- Study whether a polymorphic set of MHC molecules can develop from an initially nondiverse host population;
- Investigate the relative roles of frequency-dependent selection and selection for heterozygosity in maintaining the polymorphism of MHC molecules.

All simulations are initialized with random pathogen genotypes, and all hosts initially carry identical MHC molecules – that is, there is neither variation between MHC molecules within the hosts, nor between the hosts. Two examples of such simulations are shown in Figure 15.1, in which the average fitnesses of the pathogens and the hosts are shown as a function of the host generation number t. To study the effect of the typically short generation time of pathogens, we consider two different cases. In one of them (Figures 15.1a and 15.1b), the pathogens evolve as fast as the hosts (i.e., one parasite generation per host generation), while in the other case (Figures 15.1c and 15.1d), the pathogens evolve 100 times faster than the hosts. Since there is no initial MHC diversity, in both simulations the pathogens immediately attain a relatively high fitness and the hosts a correspondingly low fitness. Any pathogen that is able to infect one host is able to infect all hosts, and hence it rapidly takes over the pathogen population. Under this selective pressure caused by the pathogens, the hosts develop an MHC polymorphism (as is shown in Section 15.4) and, in so doing, regain a high fitness. After about 300 host generations, a quasi-equilibrium is approached that is followed until generation $t = 1\,000$. A similar equilibrium is attained if the host population is initialized

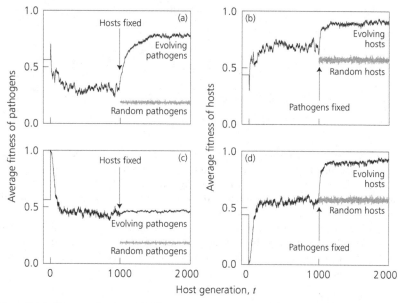

Host generation, t

Figure 15.1 Fitness of hosts and pathogens. The average fitnesses of pathogens (a, c) and hosts (b, d) in a simulation in which the generation time of the pathogens is equal to that of the hosts (a, b), and in a simulation in which the generation time of the pathogens is 0.01 times that of the hosts (c, d) are plotted against the host generation t. Note that, by Equations (15.1) and (15.2), the average host and pathogen fitnesses in a single simulation always totals one. The simulations are initialized with MHC-identical hosts and random pathogens. Coevolution is stopped at host generation $t = 1\,000$. We either stop the evolution of the hosts and let only the pathogens carry on evolving (a, c), or we stop the evolution of the pathogens and let only the hosts carry on evolving (b, d). The gray curves denote the average fitness of randomly created pathogens evaluated against the fixed host populations of generation $t = 1\,000$ (a, c), and the average fitness of random, heterozygous hosts evaluated against the fixed pathogen populations of generation $t = 1\,000$ (b, d). Other parameters: $N_{\text{host}} = 200$, $N_L = 1$, $N_S = 50$, $N_G = 10$, $N_P = 20$, $L_P = 12$, $L_M = 35$, and $L_T = 11$.

with random MHC molecules (not shown here). The average fitnesses during the quasi-equilibrium depend on the relative generation time of the pathogens. The faster the pathogens evolve, the higher their average fitness, and the lower the average fitness of the hosts (Figure 15.2). Once the pathogens evolve 100 times faster than the hosts, the average pathogen fitness saturates.

The quasi-equilibrium that is approached is a dynamic one. As in a Red Queen situation, hosts and pathogens continually counteract each other by adaptation. This follows from additional simulations in which, from $t = 1\,000$ onward, further evolution of either the hosts or the pathogens is prevented. If the pathogens and the hosts evolve equally fast, and the evolution of the hosts is subsequently halted, the pathogens markedly increase their fitness (Figure 15.1a). Such an increase of the average pathogen fitness is not observed, however, if the pathogens

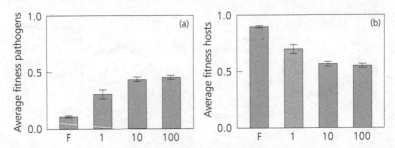

Figure 15.2 The average fitnesses of pathogens (a) and hosts (b) over the final 100 generations of the coevolution (i.e., between $t = 900$ and $t = 1\,000$). Results are shown for four different simulation types: F = fixed (nonevolving) pathogens, 1 = pathogens evolving as fast as the hosts, 10 = pathogens evolving 10 times faster than the hosts, 100 = pathogens evolving 100 times faster than the hosts. In the coevolutionary simulations, there are typically two different genotypes per pathogen species (not shown). We therefore initialized the F simulation with two randomly chosen pathogen genotypes per species. The error bars denote the standard deviations of the average host and pathogen fitnesses in time. Parameters are set as in Figure 15.1.

were evolving 100 times faster than the hosts before the evolution of the hosts was stopped (Figure 15.1c). Their short generation time apparently enables the pathogens to adapt "completely" during each host generation even before the host population is frozen. Stopping the evolution of the hosts then hardly makes a difference. Remarkably, once the evolution of the hosts is stopped, the pathogens that used to evolve as fast as the hosts attain a significantly higher average fitness (Figure 15.1a) than the pathogens that used to evolve faster than the hosts (Figure 15.1c). The reason for this difference is addressed in Section 15.4. Likewise, if the evolution of the pathogens is stopped and only the hosts carry on evolving, they evolve such that they can resist almost all pathogens – that is, they approach a fitness of one (Figures 15.1b and 15.1d). Pathogens that evolve in a nonevolving host population attain a larger average fitness than random pathogens (see the gray curves in Figures 15.1a and 15.1c). Similarly, evolving hosts in the presence of a nonevolving pathogen population attain a higher fitness than random, heterozygous hosts (see the gray curves in Figures 15.1b and 15.1d). Thus, evolving hosts and pathogens have the capacity to adapt to nonevolving populations of pathogens or hosts, respectively.

15.4 Host and Pathogen Evolution

As soon as a coevolutionary simulation is started, the number of different MHC molecules in the host population rapidly increases to reach a high quasi-equilibrium diversity (Figure 15.3). This diversification also occurs if the pathogens do not evolve at all. In that case, the high population diversity of MHC molecules results from selection that favors heterozygous hosts. The faster the pathogens evolve, however, the larger the MHC population diversity becomes (Figure 15.4a).

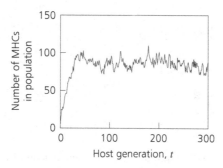

Figure 15.3 Evolution of MHC polymorphism. The number of different MHC molecules in the host population is shown from the start of the coevolution ($t = 0$) until host generation $t = 300$. The generation time of the pathogens is 100 times shorter than that of the hosts. Parameters are set as in Figure 15.1.

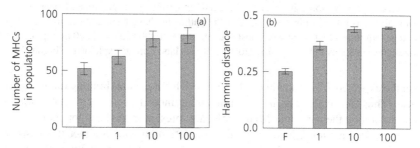

Figure 15.4 MHC molecules become functionally polymorphic. (a) The average number of different MHC molecules in the host population. (b) The average of the Hamming distances between all possible pairs of different MHC molecules in the host population. Parameters are set as in Figure 15.1. Horizontal axis labels are as explained in Figure 15.2.

To check if the MHC molecules that arise in a host population are really different from each other, and do not differ at a few mutations only, we have calculated the average genetic distance (Hamming distance) between all different MHC molecules in the host population (Figure 15.4b). Evolution of the pathogens appears to increase MHC diversity; the shorter the generation time of the pathogens, the larger the genetic distance between the MHC molecules of the hosts. Thus, rapidly coevolving pathogens trigger selection for a functionally diverse set of MHC molecules.

To measure the extent to which the pathogens evade presentation on the MHC molecules of the hosts, we calculated the average fraction of peptides from the pathogen genotypes presented by the MHC molecules in the host population. The faster the pathogens evolve, the better their evasion of presentation by the hosts' MHC molecules (see the gray bars in Figure 15.5). If the pathogens evolve, the average fraction of peptides presented by the MHC molecules of the hosts is smaller

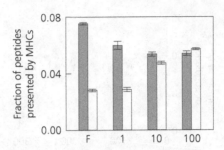

Figure 15.5 Pathogens evolve toward evasion of presentation by the particular MHC molecules present in the host population. The average presentation efficiency of the MHC molecules (i.e., the average fraction of peptides from the pathogen genotypes presented by the MHC molecules) is plotted for different pathogen generation times. The gray bars show the average presentation efficiency of the MHC molecules of coevolving hosts – that is, between host generation $t = 900$ and $t = 1\,000$ in Figure 15.1. The white bars denote the average presentation efficiency of the MHC molecules that have been frozen at host generation $t = 1\,000$ in Figure 15.1, after the pathogens have been allowed to evolve for $1\,000$ generations – that is, between host generation $t = 1\,900$ and $t = 2\,000$ in Figure 15.1. Parameters are set as in Figure 15.1. Horizontal axis labels are as explained in Figure 15.2.

than the expected 7.3% calculated above for MHC molecules binding random peptides. Thus, the pathogens in our simulations indeed evolve toward evasion of presentation by the particular MHC molecules present in the host population.

We applied a similar analysis to the simulations in which either the hosts or the pathogens are prevented from evolving. This analysis partially explains our earlier observation that pathogens evolving in a frozen host population stringently selected by rapidly coevolving pathogens (Figure 15.1c) attain a lower fitness than pathogens evolving in a host population selected only moderately (Figure 15.1a). If the pathogens do not evolve faster than the hosts, the fraction of pathogen peptides recognized by the hosts' MHC molecules decreases dramatically when the evolution of the hosts is stopped (see the white bars denoted by F and 1 in Figure 15.5). Apparently, during the coevolution the hosts specialize on the particular pathogens present in the population. This specialization enables the pathogens to escape immune recognition once the evolution of the hosts is stopped. In contrast, if the pathogens evolve faster than the hosts during the coevolution, the hosts cannot specialize on the particular pathogens present in the population. As a consequence, the pathogens fail to escape immune recognition once the evolution of the hosts is stopped (see the white bars denoted by 10 and 100 in Figure 15.5). Another reason why the evolutionary history of a frozen host population influences the escape possibilities of a pathogen lies in the polymorphism of the hosts' MHC molecules. As discussed above, the faster the evolution of the pathogens is, the more polymorphic the MHC molecules of the hosts become. Thus, pathogens evolving in a frozen host population that used to be stringently selected by rapidly coevolving pathogens have more difficulty in escaping presentation by the highly polymorphic MHC molecules of the hosts.

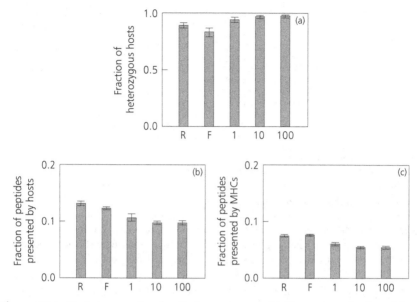

Figure 15.6 Hosts become functionally heterozygous. (a) The average fraction of heterozygous hosts. (b) The average fraction of peptides from the pathogens presented by the hosts. (c) The average fraction of peptides from the pathogens presented by the individual MHC molecules of the hosts. R denotes the simulation in which pathogens are introduced randomly at every host generation. Like the fixed pathogen population denoted by F, randomly introduced pathogen species consist of two randomly created pathogen genotypes per species. Parameters are set as in Figure 15.1. Horizontal axis labels are as explained in Figure 15.2.

15.5   Heterozygosity versus Frequency-dependent Selection

Since the evolution of pathogens can be switched off in our model, we can separately study the effect of selection for heterozygosity. In coevolutionary simulations, there is selection for heterozygosity as well as frequency-dependent selection. To exclude evolution of the pathogens, one possibility is to let the hosts evolve in response to a fixed pathogen population. As we have seen, in that case hosts adapt to the specific pathogens that are present (Figure 15.5). To exclude this specialization, we have also performed simulations in which at every host generation all pathogens are replaced by random ones (R in Figure 15.6).

The role of selection for heterozygosity appears to be strong under all conditions. During the quasi-equilibrium, the fraction of heterozygous hosts is always close to one (Figure 15.6a). To check if this heterozygosity is also functional (i.e., if the two MHC molecules of a host generally present different peptides), we compare the average fraction of peptides from the pathogens that are presented by the hosts (Figure 15.6b) with the average fraction of peptides from the pathogens presented by their individual MHC molecules (Figure 15.6c). It appears that in all simulations, the hosts (with their two MHC molecules) present nearly twice as

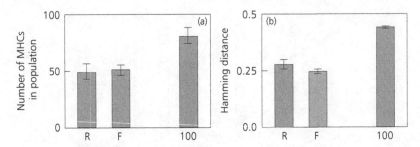

Figure 15.7 Selection for heterozygosity versus frequency-dependent selection. (a) The average number of different MHC molecules in the host population, and (b) the average Hamming distance between the different MHC molecules. We show a coevolutionary simulation in which the pathogens evolve 100 times faster than the hosts (100), and two simulations in which the pathogens do not evolve (R and F). The coevolutionary simulation represents the MHC diversity that evolves in the presence of both frequency-dependent selection and selection for heterozygosity, while the two latter simulations (R and F) represent the MHC diversity that evolves under selection for heterozygosity only.

many peptides as their individual MHC molecules. Thus, the hosts in our simulations indeed typically carry functionally different MHC molecules.

To study the relative roles of selection for heterozygosity and frequency-dependent selection, we compare the MHC polymorphism arising in the absence and presence of frequency-dependent selection. Figure 15.7a shows that heterozygosity plus frequency-dependent selection (i.e., a simulation with evolving pathogens, denoted by 100) results in a much higher degree of polymorphism than selection for heterozygosity alone (i.e., simulations with nonevolving pathogens, R and F). The average genetic differences between the MHC molecules that arise support this notion (see Figure 15.7b). Summarizing, our simulations show that a polymorphic set of MHC molecules rapidly develops in an initially nondiverse host population, and that selection by coevolving pathogens can account for a much larger population diversity of MHC molecules than mere selection for heterozygosity can.

15.6 Discussion

We have shown that both the origin and the maintenance of MHC polymorphism can be understood in a model that does not assume any *a priori* selective advantage of heterozygous hosts or hosts with rare MHC molecules. By starting our simulations with MHC-identical hosts, we have studied a "worst-case" scenario. Polymorphisms of MHC-like molecules seem to have been present since colonial or multicellular life (Buss and Green 1985). Thus, the *origin* of MHC polymorphism may not lie in immune function. For example, de Boer (1995) showed that in primitive colonial organisms the preservation of "genetic identity" is sufficient to account for highly polymorphic histocompatibility molecules.

Our simulation model demonstrates that coevolution of hosts and pathogens yields a larger MHC polymorphism than merely selection for heterozygosity. Our analysis thus supports the view that additional selection pressures on top of over-dominant selection do play a role in the evolution of the MHC polymorphism (Parham *et al.* 1989b; Wills 1991). It has been shown experimentally that many MHC alleles have persisted for significant evolutionary periods of time (Klein 1980; Lawlor *et al.* 1988; Mayer *et al.* 1988; Klein and Klein 1991). This has been used as an argument against frequency-dependent selection (Hughes and Nei 1988), but was later demonstrated to be compatible with selection for rare MHC molecules (Takahata and Nei 1990). Analysis of the persistence of particular MHC alleles in our simulations would allow us to study this in more detail.

To increase the speed of our simulations, we used a rather high mutation frequency for the hosts' MHC molecules: $p_{mut} = 0.001$ per bit per generation. Indeed, decreasing this mutation frequency resulted in a lower MHC population diversity. Increasing the host population size in our simulations, on the other hand, increased the MHC polymorphism. Using a mutation frequency for MHC molecules of $p_{mut} = 10^{-6}$ and a host population size of $N_{host} = 1\,000$ hosts, we still found a population diversity of approximately 30 different MHC molecules. Independently of the choices of p_{mut} and N_{host}, the MHC polymorphism attained in coevolutionary simulations was always considerably (e.g., fivefold) higher than the polymorphism arising under overdominant selection only (results not shown).

Regarding the enormous population diversity of MHC molecules observed in nature (Parham and Ohta 1996; Vogel *et al.* 1999), it is surprising that the number of different MHC molecules expressed per individual is quite limited (Paul 1999). In our simulations, hosts carry only one MHC gene. What would change if this number of MHC genes per individual increased? Individuals expressing more MHC genes are expected to have a selective advantage, in that more pathogens would be presented. This selective advantage vanishes, however, once the chance to present (at least one peptide from) any pathogen approaches 100%. For the parameter setting used here, the chance that a random pathogen consisting of 20 peptides evades presentation by a single MHC molecule is $p_e = 22\%$. In the absence of pathogen evolution, expression of about 10 different MHC molecules would thus be sufficient to ensure the presentation of virtually any pathogen. In co-evolutionary situations, however, the selection for expression of MHC molecules that are different from the other MHC molecules in the population would remain. This selection only disappears when the number of different MHC molecules per individual becomes so large that every host is expected to present all pathogen peptides. If individuals no longer draw different "samples" from the pool of peptides from each pathogen, pathogens may be expected to exploit this "predictability" of the hosts' immune responses (Wills and Green 1995). We have demonstrated, for instance, that increasing the number of MHC loci increases the likelihood of autoimmunity (Borghans and de Boer 2001). Extension of the current model with host self-molecules and a variable number of MHC genes sheds light on the role of such mechanisms in the maintenance of the MHC polymorphism (Borghans *et al.*, unpublished).

16

Virulence Management and Disease Resistance in Diploid Hosts

Viggo Andreasen

16.1 Introduction

Genetic variation in the host's response to infections is likely to be present in any host–parasite system. Thus, in most cases, virulence management and the associated changes in disease characteristics affect the genetic composition of the hosts over a time scale comparable to the host's life span. If virulence management is concerned with such time scales and, in particular, if the disease in question has a significant impact on host survival or fertility, host evolution cannot be ignored. In this chapter a modeling framework is developed to allow virulence managers to assess disease-induced host–evolution in a sexually reproducing diploid host, that is, to express fitness in terms of variation in epidemic quantities such as morbidity and infectivity. Thus, we take the consequences of virulence management one step further, and examine the effect of virulence management on the host population. Thereafter, we look further ahead at the consequences of changes in host genetics on pathogen strains.

As a first approximation, variation in disease-induced mortality among hosts causes variation in host survival and hence in fitness. In turn, differential fitness changes the composition of host susceptibilities and thus the prevalence of the disease. This interaction between host composition and disease prevalence is the focus of this chapter; we assume that the immediate consequences of virulence management on prevalence are known and expressed as changes in epidemic parameters.

Since the work of Haldane (1949) it has been recognized that infectious diseases may exert a major impact on host evolution. The arguments in support are quite clear: infectious diseases often cause significant mortality and simple molecular changes in the host can affect morbidity. Thus sufficient variation in the host population is likely to be present and fitness variation may be significant. In insects (see Chapter 6), mammals (Fenner and Ratcliffe 1965), and humans (Sørensen et al. 1988) heritable variation in disease-induced mortality has been observed. In plants, breeding for genetically based disease resistance is common (the methods of this chapter apply only to obligate outbreeding natural species, however).

A large body of literature deals with host–pathogen coevolution in which the host, as well as pathogen, is a haploid or an asexually reproducing species (for a review see Frank 1996c). For a sexually reproducing diploid, these haploid models reflect the genetics of the population only if the two alleles contribute additively to growth rate. In particular, polymorphism in haploids can arise only through frequency-dependent selection, and in the absence of segregation–recombination, the structure of the polymorphic equilibrium resembles that of the coexistence of competing species.

As shown later, the segregation–recombination process complicates matters considerably. Important epidemic phenomena in host selection, however, depend on both the diploid nature of host genetics and polymorphism. For example, the major histocompatibility complex (MHC) and the associated leukocyte antigens (HLAs), which play key roles in the immune system, are determined by highly polymorphic genes. This suggests that genetic recombination is essential for disease resistance, though maybe not for resistance to recently emerged pathogens (May and Anderson 1983a; Hamilton and Howard 1994; Singh *et al.* 1997). Bremermann (1980) and Hamilton (1980) stressed the point by suggesting that maintenance of sufficient variation in the immune response through recombination is a major force in the evolution of sexual reproduction.

Especially in cases of heterozygous advantage or overdominance, segregation is essential for the evolution of hosts in response to disease-induced selection. Perhaps the most well-documented example of the importance of overdominance is malaria resistance caused by a mutation in the gene coding for hemoglobin. Heterozygotes for the gene, who carry one copy of the mutant gene, the so-called S gene, suffer from a mild anemia (sickle-cell anemia) because of a slight change in the erythrocytes. The change in the red blood cells confers some protection against malaria (in particular against the malignant infections by *Plasmodium falciparum*). In areas with a high incidence of malaria, this malaria resistance increases heterozygote fitness by as much as 15% relative to that of homozygotes who carry two regular A genes. Although SS homozygotes suffer from a severe anemia that significantly reduces their fitness, the selective advantage of the heterozygote is so pronounced that up to 30% of the individuals in malaria-infested areas of central and western Africa carry the S gene. This corresponds to a gene frequency of 15%. In southeast Asia and other areas of endemic malaria, various mutations in hemoglobin with similar properties occur in appreciable frequencies (Cavalli-Sforza and Bodmer 1971; Jones 1997). Recently, it was suggested that the gene coding for cystic fibrosis confers a similar partial protection against typhoid fever (Pier *et al.* 1998). From the viewpoint of virulence management, the key problem is how modifications in malarial transmission dynamics affect the system. For example, how do changes in disease transmission affect the gene frequency of the S allele and the morbidity of the disease?

While a growing body of literature discusses genetic aspects of host–pathogen interactions (e.g., Levin and Udovic 1977; Fischer *et al.* 1998), only few workers address the interaction between host gene composition and the strength of the epidemics. The first models to include explicitly epidemic assumptions in Mendelian genetic models appear to be Gillespie (1975), Lewis (1981), Longini (1983), and Beck *et al.* (1984). These early works fall into two categories, each building on rather different population genetic approaches. In Sections 16.2 and 16.3 we discuss the modeling framework of Gillespie, Lewis, and Longini, in which epidemic information is incorporated in the fitness expressions for the discrete time model of nonoverlapping generations. Section 16.4 describes the approach of Beck *et al.* (1984), who determine how fitness depends on variation in epidemic parameters for a continuous time model with overlapping generations. Finally, in Section 16.5 we discuss ways to include selection in the pathogen and allow a description of host–pathogen coevolution. A general discussion of the relationship between discrete and continuous time genetic models can be found in most population genetic text books (e.g., Hartl and Clark 1997, Chapter 6).

16.2 Discrete-time Genetics and Epidemic Diseases

The approach of Gillespie (1975) and others determines the effect of disease on host fitness in each host generation. For simplicity, we assume that host response to a disease is determined by a single autosomal locus with two alleles, A and B. In the discrete-time model, hosts are born in distinct generations, and mating is random so that the genotypes AA, AB, and BB at the beginning of each generation occur in Hardy–Weinberg proportions $p^2, 2pq, q^2$, where p and $q = 1 - p$ denote the allele frequency of A and B. Genetic variation is represented by differential ("Darwinian") fitness, that is, contribution to the next generation, w_{AA}, w_{AB}, and w_{BB}. The total frequency p' of A in the next generation is then given by

$$p' = \frac{w_{AA}p^2 + w_{AB}pq}{\overline{w}} = \frac{p[w_{AB} + (w_{AA} - w_{AB})]p}{\overline{w}}, \qquad (16.1)$$

where $\overline{w} = w_{AA}p^2 + w_{AB}2pq + w_{BB}q^2$ denotes the average fitness.

To introduce epidemic information into the fitness, Gillespie (1975) assumes that the epidemic runs through each generation, for example resembling a population of annual insects that experiences an epidemic in each season. We assume that the disease follows an SIR-type epidemic (see Chapter 6) so that if the host population is homogeneous, the densities of susceptible S, infectious I, and recovered R hosts change according to the epidemic model

$$\frac{dS}{dt} = -\beta SI, \qquad (16.2a)$$

$$\frac{dI}{dt} = \beta SI - \mu I, \qquad (16.2b)$$

$$\frac{dR}{dt} = \mu I, \qquad (16.2c)$$

where β and μ give the transmission coefficient and the exit rate out of the infected class. The "recovered" class R includes the hosts that recover from infection as well as the hosts that die from infection. Provided that the transmission coefficient β is independent of population density, the model does not distinguish between death and recovery, as both types of hosts are lost to disease transmission; the effect of infection on viability and fitness is accounted for elsewhere in the model. In fact, R is redundant as the total density of hosts (dead or alive) is constant during the epidemic. The exit rate μ thus combines recovery and mortality caused by the infection, while we assume that death unrelated to the disease takes place after the epidemic.

Genetic variation is specified through the values of μ_{XX} and β_{XX} for each genotype XX. For example, the densities of susceptible and infected hosts of genotype AA, S_{AA} and I_{AA}, change according to the model

$$\frac{dS_{AA}}{dt} = \beta_{AA} S_{AA} I ,\tag{16.3a}$$

$$\frac{dI_{AA}}{dt} = \beta_{AA} S_{AA} I - \mu_{AA} I_{AA} ,\tag{16.3b}$$

with similar expressions for AB and BB. Here $I = I_{AA} + I_{AB} + I_{BB}$ denotes the total disease prevalence while the initial conditions are that all individuals are initially susceptible and distributed among genotypes in Hardy–Weinberg proportions. The epidemic is set off by a few infectious individuals, $0 < I(0) \ll N$ where N denotes the total population density.

To find the effect of the epidemic within a generation, one assumes that the epidemic runs through the population until it dies out. Mathematically, this corresponds to solving six coupled equations to find $S_{AA}(\infty)$, etc. The reduction in gene frequency of genotype XX caused by the epidemic can be determined from the proportion of individuals that escape the infection, that is, $S_{XX}(\infty)/S_{XX}(0)$, and the fitness-reduction for hosts that are infected, $1 - u$.

Model (16.3) cannot be solved analytically without simplifying assumptions. To obtain a model with some analogy to the sickle-cell malaria situation, assume that allele B codes for complete immunity to the infection and that B is dominant, so that $\beta_{AB} = \beta_{BB} = 0$. Following Kermack and McKendrick (1927), we find that in a population composed solely of susceptible hosts, an epidemic occurs only if the basic reproductive ratio $R_0 = \beta N/\mu$ exceeds one. For $R_0 > 1$, the intensity of the epidemic z – the fraction of the population that is affected by the epidemic – is given implicitly by

$$z = 1 - e^{-zR_0} .\tag{16.4}$$

With our assumptions, only AA homozygotes are susceptible and hence only the fraction p^2 of the contacts made by an infected host are with susceptible hosts, so that the effective reproduction ratio is $p^2 R_0$. The intensity of the epidemic now becomes

$$z(p) = \begin{cases} 0 & \text{if } p^2 R_0 < 1 \\ \text{positive solution of } \quad z = 1 - e^{-zR_0 p^2} & \text{otherwise} \end{cases} .\tag{16.5}$$

If $p^2 R_0 < 1$, the disease dies out without causing an epidemic and it must be reintroduced in the next generation. For simplicity, we assume that A is sufficiently frequent to ensure that $p^2 R_0 > 1$ in all generations. Thus, the fraction of AA homozygotes that avoids the epidemic is $1 - z(p)$. If infected individuals suffer a fitness reduction of $1 - u$, the fitness of AA in the presence of the disease becomes $w_{AA} = 1 - z(p) + (1 - u)z(p) = 1 - uz(p)$.

For the sickle-cell malaria system, carriers of the malaria resistance B gene suffer a reduced fitness in the absence of malaria. Let us assume by analogy that heterozygote fitness is reduced to $w_{AB} = 1 - \sigma$, while BB homozygote fitness is $w_{BB} = 1 - \rho$, with $0 \le \sigma \le \rho$ and $\sigma \le u$. The last inequality indicates that the fitness of infected AA homozygotes is less than the fitness of heterozygotes, ensuring B dominance at high disease prevalence. We are now able to determine the frequency of A at the birth of the next generation

$$p' = \frac{\{p[1 - z(p)u] + q(1 - \sigma)\}}{\overline{w}}, \tag{16.6}$$

where $\overline{w} = p^2[1 - z(p)u] + 2pq(1 - \sigma) + q^2(1 - \rho)$. In principle we can determine the equilibrium values of p and z by simultaneously solving Equations (16.5) and (16.6) for $p = p'$. For the sickle-cell malaria system, this determines how disease incidence and frequency of the resistance gene, $q = 1 - p$, covary. The relationship between p and z in the full model is rather complicated and we restrict our discussion to two special cases.

First assume that BB homozygotes have the same fitness as the heterozygotes (i.e., $\rho = \sigma$). In this case Equation (16.6) simplifies and we find that the equilibrium intensity z and the frequency p of A are given by

$$z(p) = \sigma/u \quad \text{and} \quad p = \sqrt{\frac{-\ln(1 - \sigma/u)}{R_0 \sigma/u}}. \tag{16.7}$$

Thus, in a situation in which the resistance gene B is dominant, the intensity of the disease is determined as the ratio between the cost of resistance and the cost of infection. For virulence management the implications are now straightforward: if disease virulence is changed in a way that reduces the cost of infection, the host population responds by increasing the frequency of the susceptible A allele, and the intensity of the infection also increases. The effect of reducing the transmissibility of the disease is more surprising: decreasing R_0 does not affect disease intensity z, but increases the frequency of the susceptible allele until A is fixed.

Next, assume that BB homozygotes are not viable, so that $\rho = 1$, and that the disease is universally fatal, $u = 1$. This corresponds to an extreme case of heterozygous advantage. The sickle-cell malaria system may be thought of as an intermediate case between these two examples. In this case, the intensity of the disease depends on the frequency of the resistance gene,

$$z = 2\sigma - 1 + (1 - \sigma)/p, \tag{16.8}$$

while the frequency of A is determined implicitly by Equation (16.9),

$$2\sigma - 1 + \frac{1 - \sigma}{p} = 1 - e^{-R_0[p^2(2\sigma-1)+p(1-\sigma)]}. \tag{16.9}$$

This model has not been analyzed in depth, but apparently the polymorphic equilibrium vanishes for small R_0, leading to fixation of A.

In our example we focused on selection in the case of a dominant resistance B gene, but a similar analysis for recessive resistance is given in Gillespie (1975) and for more general situations in May and Anderson (1983a). The basic assumption is that there is a cost to carrying the resistance B gene compared to uninfected susceptible hosts, but that this cost is less than the cost of being infected. These assumptions give rise to frequency-dependent selection, because disease prevalence increases with the frequency of A, so that A is selected for when it is rare and against when it is common. Therefore, a polymorphic equilibrium caused by frequency-dependent selection is expected. The primary effect of virulence management is to change this genetic equilibrium. In some cases this effect may be so pronounced that management may not alter disease prevalence. The benefit of virulence mangement may show up as a reduction in the resistance gene and the associated genetic diseases.

16.3 Discrete-time Genetics and Endemic Diseases

Gillespie's original model focused on epidemic diseases in which incidence varies considerably over time. Longini (1983) suggests that endemic diseases with constant incidence that follow, for example, an SIS-type dynamics can be handled in a similar framework. To be specific, we once more focus on a dominant resistance gene so that $\beta_{AB} = \beta_{BB} = 0$ and $\beta_{AA} = \beta$. In addition, we assume that hosts are susceptible immediately after recovery so that the density of susceptible and recovered individuals can be determined by

$$\frac{dS}{dt} = -\beta SI + \mu I , \tag{16.10a}$$

$$\frac{dI}{dt} = \beta SI - \mu I . \tag{16.10b}$$

Since the disease is endemic, its impact is determined by the disease incidence at equilibrium. The fractions of AA individuals that are susceptible and infected are

$$\frac{S^*}{p^2 N} = \frac{\mu}{\beta p^2 N} = \frac{1}{p^2 R_0} , \tag{16.11a}$$

$$\frac{I^*}{p^2 N} = 1 - \frac{1}{p^2 R_0} , \tag{16.11b}$$

where N is the total population density and R_0 is the basic reproduction ratio if all hosts are susceptible. Equations (16.11) hold only when $p^2 R_0 > 1$; if the frequency of the susceptibility gene is too low, the disease dies out and its possible reintroduction depends on stochastic effects. Disease-induced fitness reduction reflects how much time AA homozygotes spend in the infected class, and we obtain $w_{AA} = 1 - u/p^2 R_0$ where $1 - u$ gives the fitness of homozygotes while infected.

We can now introduce the fitness expression into the genetic model, Equation (16.1). As in the case of epidemic diseases, the conclusions depend on the genetic assumptions, and frequency-dependent selection may lead to polymorphism (see Longini 1983).

The discrete-generation approach rests on a couple of major assumptions. First, the combination of epidemic and genetic time scales is hard to follow in detail. For a population of annual insects that each year experiences an epidemic, the Gillespie model reflects the time scales well, but for epidemics in hosts with overlapping generations, such as the human population interacting with measles or malaria, epidemics are so frequent that in reality only a fraction of the hosts are available at each epidemic. The endemic approach of Longini suffers from exactly the opposite problem: disease incidence in newborns differs from that of older hosts, so that in reality the susceptible pool is larger than the Longini model suggests.

Perhaps the most critical problem with the discrete-generation approach is the lack of an explicit account of the host density effects. By keeping track only of gene frequency, the models implicitly assume that the population is subject to regulation that compensates for disease-induced deaths in such a way that the population density remains constant. This assumption is critical since transmission dynamics, in particular disease transmissibility β, are often density dependent. While variation in host density may not play a major role in most applications to human populations, the models may not describe well the genetic response to major epidemics in natural populations in which diseases can change the host density dramatically, such as myxomatosis in the Australian rabbits or the phocine distemper virus epidemic in the seals of the North Sea (Fenner and Ratcliffe 1965; Heide-Jorgensen and Harkonen 1992).

16.4   Continuous Genetic Models

As an alternative to the discrete-generation approach, Beck *et al.* (1984) consider the three genotypes as separate entities and write explicit expressions for their dynamics. Still assuming that all relevant information is carried at one autosomal locus with alleles A and B, and that the population is subject to an SI-type disease, it is straightforward to write the combined dynamics of changes in genetic composition and epidemics,

$$\frac{dS_{AA}}{dt} = B_{AA} - \beta_{AA} S_{AA} I - d_{AA} S_{AA} , \tag{16.12a}$$

$$\frac{dI_{AA}}{dt} = \beta_{AA} S_{AA} I - (\alpha_{AA} + d_{AA}) I_{AA} , \tag{16.12b}$$

with similar expressions for AB and BB. Here $I = I_{AA} + I_{AB} + I_{BB}$ denotes the total disease prevalence, while B_{AA} is the birth rate of AA individuals. We assume random mating, such that B_{AA} is given by Hardy–Weinberg proportions as $B_{AA} = bNp^2$, where N gives the total population density and b the per capita birth rate, while

$$p = [S_{AA} + I_{AA} + \tfrac{1}{2}(S_{AB} + I_{AB})]/N \tag{16.13}$$

is the frequency of the A allele. Notice that this formalism allows us to specify freely density-dependent effects in disease transmission and host mortality.

Model (16.12) suffers from two major shortcomings. Obviously, the complexity of the model prohibits any general analysis, but in addition the lack of age structure implies that new-born hosts reproduce immediately after birth. Particularly if the genetic composition changes fast over age classes, corresponding to a large variation in survival, this may give spurious effects. Continuous genetic models therefore should be used only when the genetic variation is small, so that gene frequencies change slowly compared to population processes.

To specify that genetic variation is small, we introduce a small parameter $\varepsilon \ll 1$ and replace the genotypic parameters by ones with small genotypic variation

$$d_{AA} = d + \varepsilon \hat{d}_{AA} , \tag{16.14}$$

with similar expressions for the remaining parameters.

With the assumption $\varepsilon \ll 1$, a lengthy mathematical analysis based on singular perturbation theory shows that the epidemic "quickly" settles to the endemic equilibrium values, while the genetic composition "quickly" reaches Hardy–Weinberg proportions and a uniform gene frequency in all epidemic compartments (see Beck *et al.* 1984; Andreasen and Christiansen 1993). Once the epidemic variables settle to equilibrium and a uniform gene frequency is reached in all epidemic classes, the frequency p of the A allele changes "slowly" according to

$$\frac{dp}{dt} = \varepsilon p q f(p) , \tag{16.15a}$$

with

$$f(p) = \frac{[b - \beta S^*(p)]I^*(p)\langle \hat{\beta}_{XX}|p \rangle + bI^*(p)/S^*(p)\langle \hat{\alpha}_{XX}|p \rangle}{\beta S^*(p) + bI^*(p)/S^*(p)} - \langle \hat{d}_{XX}|p \rangle , \tag{16.15b}$$

where $\langle \hat{d}_{XX}|p \rangle = p d_{AA} + (q - p)d_{AB} - q d_{BB}$ and $\langle \hat{\beta}_{XX}|p \rangle$ is defined analogously. Equations (16.15) should be compared to the classic model of slow selection in a random mating population (Norton 1928). In the slow selection model the change in p is determined as

$$\begin{aligned}
\frac{dp}{dt} &= \varepsilon p(p r_{AA} + q r_{AB} - \bar{r}) , \\
&= \varepsilon p q [p r_{AA} + (q - p)r_{AB} - q r_{BB}] , \\
&= \varepsilon p q \langle r_{XX}|p \rangle, \tag{16.16}
\end{aligned}$$

where r_{AA} denotes the Malthusian growth rate for individuals of genotype AA and $\bar{r} = p^2 r_{AA} + 2pq r_{AB} + q^2 r_{BB}$ is the average rate of growth. In this context r_{AA} is also referred to as the (Malthusian) fitness of AA, and in this sense Equation (16.15) allows us to relate genetic variation in epidemic parameters directly

to host fitness. Notice that the weights with which various types of genetic variation contribute to the fitness depend on the disease incidence at equilibrium. The
weights derive from the mathematical analysis, and they seem to have no simple
biological interpretation. When transmission dynamics lead to sustained oscillations in disease incidence, the weights can be determined by suitable time averages
(see Andreasen and Christiansen 1993).

For virulence management the models suggest how changes in disease control
practices affect the long-term genetic composition of the host population, provided
that the practices do not lead to new viral types. One simply determines the new
endemic equilibrium and, using Equations (16.15), one may predict the direction
of change in host composition. For example if disease transmission β is reduced,
Equations (16.15) suggest that the importance of genetic variation in the disease-
induced mortality α_{XX} increases relative to that of variation in transmission rate
β_{XX}. In the sickle-cell malaria situation, the model could suggest how changes
in the transmission coefficient caused by draining or insect spraying affect the
polymorphic equilibrium in the A/S susceptibility and resistance system.

To study the evolution of resistance, Gupta and Hill (1995) applied a model
similar to Beck's to the malaria system. The approach is not well-suited to the
sickle-cell problem, because genetic variation among the genotypes is large, which
generates large deviations from Hardy–Weinberg proportions and thus violates the
assumptions of the model. In a rather complicated model of scrapie transmission
dynamics in sheep, Stinger *et al.* (1998) applied an age-structured version of model
Equations (16.12). By assuming age-dependent fertility, the spurious effects of
allowing new-born individuals to give birth prior to any selection are avoided, but
as a result rather arbitrary assumptions about the age-dependent mating structure
of the population have to be made.

Several problems with the continuous approach have already been discussed.
In addition, the time scale separation suggests that the model cannot capture the
interaction between disease prevalence and genetic composition, because this separation implies that the epidemic equilibrium and host abundance do not change
significantly in response to changes in host genetics. In this sense the model provides a static picture of the host's epidemic characteristics. The strength of the
approach is that it provides a direct link between genetically based variation in
epidemic parameters and fitness.

16.5   Coevolution

We conclude this survey by discussing the inclusion of pathogen evolution into the
description. Focusing on asexual pathogens and excluding the possibility of superinfection and cross-immunity, Levin and Pimentel (1981) and Saunders (1981)
suggested that the natural unit of selection in the pathogen is the infected host.
Selection between two pathogens can now be described by an extension of the
SI model in which the number of infected hosts of each type is accounted for

separately. Denoting the two types by subscripts 1 and 2, the model becomes

$$\frac{dS}{dt} = -\beta_1 I_1 S - \beta_2 I_2 S, \tag{16.17a}$$

$$\frac{dI_j}{dt} = \beta_j I_j S - \mu_j I_j \qquad \text{for} \qquad j = 1, 2. \tag{16.17b}$$

Since we assume nonsexual reproduction in the pathogen, pathogen polymorphism is not possible, except for special types of density dependence in host–dynamics (Andreasen and Pugliese 1995).

The Gillespie and Longini models have not been extended to include coevolution, but Beck (1984) and Andreasen and Christiansen (1995) included a slowly evolving pathogen in the slow evolution model, Equations (16.15). Slow in this context means that the variation in disease parameters among strains is on the order of ε. After a lengthy mathematical analysis they concluded that the system can be reduced to two variables describing the frequency of the A-host allele, p, and the frequency of the pathogen-1-strain, $\pi = I_1/(I_1 + I_2)$. Evolution is now determined by

$$\frac{dp}{dt} = \varepsilon p q [\pi f_1(p) + (1 - \pi) f_2(p)], \tag{16.18a}$$

$$\frac{d\pi}{dt} = \varepsilon \pi (1 - \pi)(p^2 c_{AA} + 2pq c_{AB} + q^2 c_{BB}), \tag{16.18b}$$

where $f_j(p)$ is the fitness of A as introduced in Equation (16.15b) if the host is exposed solely to strain j, while c_{XX} is the (positive or negative) difference in reproduction ratio between strain 1 and 2 relative to the average reproduction ratio if all the hosts were of genotype XX.

Andreasen and Christiansen (1995) discuss in some detail the rich behavior of this coevolution model, which includes multiple stable steady states; in particular, they demonstrate that in specific situations the system may give rise to unstable oscillations with growing amplitude resembling gene-for-gene dynamics. The oscillations in Equations (16.18), however, are structurally unstable in the sense that infinitely small perturbations may alter the oscillations. This observation suggests that oscillations of the gene-for-gene type, known from many crop–pathogen systems, may be unlikely in natural systems and that oscillations in coevolution models may result from built-in assumptions about frequency or density dependence in the system.

16.6 Discussion

The two modeling approaches differ so much that a direct comparison of their dynamics makes little sense. Both models allow us to express host fitness in terms of disease characteristics that are affected directly by virulence management. Hence, both modeling approaches allow us to assess the consequences of management on host genetics, in particular changes in polymorphic equilibria such as the malaria-induced sickle-cell polymorphism.

In the discrete-time, nonoverlapping generations approach, the disease-dependent fitness expressions are derived by assuming that in each generation either the disease is at its endemic equilibrium or the disease sweeps through the population in one epidemic. In both cases, analytic solutions can be obtained only with additional assumptions about the genetics of resistance.

For the continuous time model with overlapping generations, the effects of disease on Malthusian fitness can be determined explicitly. Thus, it is straightforward to determine the rate of change of an allele with known epidemic effects. The main limitation of the continuous model is that it can be applied only to situations with small genetic effects.

Acknowledgments   This work was in part supported by grant 97-0141-2 from the Danish Natural Science Research Council.

17

Coevolution in Gene-for-gene Systems

Akira Sasaki

17.1 Introduction

Gene-for-gene (GFG) systems are genotype-specific, antagonistic interactions between hosts and parasites, widely observed in plants and their microbial parasites (Burdon 1987). Detailed studies on crop plant and fungus pathogen systems have revealed that when breeders introduce resistant hosts, a rapid evolution of parasite virulence occurs that overcomes the resistance. This process suggests a continuous coevolutionary change in both host and parasite. The spread of a resistant genotype capable of escaping a currently prevalent parasite will be challenged by a new parasite strain that harbors a virulent gene, capable of overcoming that resistance. Similarly, a host with a new resistant gene, possibly at another locus, would be able to restore resistance against the same parasite. Besides its importance in agriculture and biological control, GFG interactions play a key role in models of host–parasite coevolution. These models reveal a robust tendency toward protected polymorphisms and sustained cycles of host and parasite genotypes, which, in turn, favor higher rates of mutation, recombination, and sexual reproduction (e.g., Hamilton 1980; Hamilton *et al.* 1990; Frank 1993b; Haraguchi and Sasaki 1997).

From the perspective of virulence management, one consequence of the modeling approach described in this chapter is of potential practical importance: the results presented here reveal a wide parameter range in which polymorphism of host resistance can prevent the spread of any virulent strain of parasite and maintain a disease-free host population. This requires that the cost of virulence exceed a certain threshold, but the threshold can be lowered by increasing the number of resistant genotypes maintained in the population. Hence, for any given cost of virulence, it is theoretically possible to protect the host population from disease by keeping a sufficient number of resistant varieties. This strategy requires that none of the genotype frequencies exceeds the threshold, failing which the corresponding virulent parasites can spread. In addition, if the same variety tends to be spatially clustered, local spreading of the virulent strain might occur. Despite these potential difficulties, the principle is far more promising than the introduction of multiply resistant hosts, which has invariably failed.

The emergence of multiple drug resistance (MDR) in infectious bacteria is still a serious problem in our species (see also Chapter 23). To prevent epidemics, it is preferable to use a variety of separate antibiotics in the population as a whole rather than to use multiple drugs in the same patient. A similar principle also applies to the emergence of resistant biotypes in pest control.

Previous GFG models assumed symmetric and specific interactions between host and parasite genotypes. This assumption is challenged by empirical studies, which reveal a great asymmetry in GFG systems (Parker 1994). Some parasite genotypes have a broader host range than others. Therefore, it is often the case that a generalist parasite will locally predominate and exploit all the existing host genotypes (e.g., Espiau *et al.* 1998). Under this common type of GFG interaction, Parker (1994) argues that cycles in genotype frequencies are less likely, and that evolution of sex is unlikely to result from host–parasite interaction. In this chapter, I explore the consequence of the coevolution of host resistance and parasite virulence, taking into account the asymmetrical nature and multilocus inheritance of GFG systems. Two contrasting forms of host–parasite interactions – the matching genotype and the GFG models – are discussed in detail. It is shown that coevolution in multilocus GFG models is characterized by two processes: first, the evolutionary arms race of quantitative traits (Rosenzweig *et al.* 1987; Saloniemi 1993; Frank 1994a; Matsuda and Abrams 1994; Dieckmann *et al.* 1995; Doebeli 1996, 1997; Abrams and Matsuda 1997; Sasaki and Godfray 1999) – representing the degree of resistance and virulence – and second, the antagonistic genotype dynamics between host and parasite (Hamilton 1980; Hamilton *et al.* 1990; Frank 1993b; Haraguchi and Sasaki 1996). I also refer to Chapter 31 Section 31.2, where Jarosz introduces the well-chosen terms "matching virulence" and "aggressive virulence."

17.2 Gene-for-gene Interaction

In this section, I review some basic concepts and characteristics of GFG interactions. Neither the molecular and physiological basis of GFG interactions nor the molecular evolutionary analyses of the genes responsible for resistance and virulence (Song *et al.* 1997; Leister *et al.* 1998) are discussed in this chapter.

I first summarize the evolutionary changes that occur in pathogens associated with the introduction of resistant varieties of crop plants, and discuss a possible scenario for a resistance–virulence arms race that has come to light through field observations. I then discuss the two contrasting views of GFG systems (matching genotype versus GFG interaction), especially in relation to the Red Queen hypothesis for the evolution of sex (see Box 18.1).

Evolution in plant–pathogen systems

The introduction of resistant races of crop plants often results, over an ecological time scale, in evolutionary changes in the degree of virulence (Burdon 1987). The best known example is found in Australian wheat varieties. In the 1950s, breeders adopted those resistant varieties of wheat that contain the single-resistance gene active against the fungus pathogen *Puccinia graminis tritici*. However, as the virulence genes of the pathogen overcame resistance and spread in the pathogen population, the original wheat varieties were replaced by those with multiple resistance

Parasite: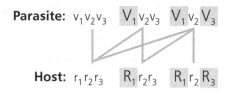

Host: $r_1 r_2 r_3$ $R_1 r_2 r_3$ $R_1 r_2 R_3$

Figure 17.1 GFG relationship between host resistance and parasite virulence. V and R denote virulence and resistance genes, and v and r avirulence and susceptible genes. Lines indicate which parasite genotype can infect which host genotype.

genes. As a result, the mean number of virulence genes in individual pathogens increased – from a mean of 1.46 in the mid-1950s to 3.18 in the mid-1960s (see Burdon 1987 and references therein).

Figure 17.1 illustrates a simple coevolutionary arms race model indicating the host resistance and parasite virulence suggested by these studies of crop plant–pathogen interactions (including GFG systems). The bottom row represents the three-locus haploid resistance genotype of the host; the upper row represents the corresponding (three-locus) virulence genotypes of the pathogen. Lines indicate where infection is possible. In each haploid genotype, R_j denotes a resistance gene of the host at the jth locus. V_j denotes a virulence gene of the parasite at the jth locus, which blocks the function of the corresponding resistance gene R_j. Small letters r_j and v_j denote susceptible and avirulence genes in the jth locus of the host and parasite. (It is conventional practice to denote virulence genes by using lower case letters and avirulence genes by using capital letters. This method serves to reflect the typical dominance relationship: indeed, avirulence is often dominant to virulence. The usage however is reversed in this chapter, which only deals with haploid inheritance and escalated phenotypes: resistance in host and virulence in parasites are denoted by capital letters, or in later sections, by the number 1.)

Let us begin with a wild-type host containing no resistance gene, and a wild-type parasite with no virulence gene (left column in Figure 17.1). If a resistance gene R_1 at locus 1 is introduced either by mutation or by breeders, it is possible that this resistant host can escape the wild-type parasite. Sooner or later, however, a new virulence gene emerges in the form of V_1. This new gene neutralizes the effect of the resistant gene R_1, and the resultant parasite can infect both the resistant and the wild-type hosts (middle column in Figure 17.1). A chain of events based on this principle is set in motion. For example, when a new resistance gene R_3 is introduced to generate a host genotype with two resistance genes, it only has a transient effect until a new parasite with two corresponding virulence genes emerges (right column in Figure 17.1). This suggests an evolutionary trend that involves increases of the numbers of resistance genes in the host and virulence genes in the parasite. One of the main objectives of this chapter is to clarify the consequence of this evolutionary arms race.

Table 17.1 Matching genotype and GFG interaction between two host genotypes and two parasite genotypes.

	P_1	P_2
(a) Matching genotype		
H_1	+	–
H_2	–	+
(b) Gene-for-gene		
H_1	+	+
H_2	–	+

Note: Plus and minus signs indicate whether infection occurs (+) or not (–) in each combination of host and pathogen genotypes.

Matching genotype versus gene-for-gene models

Although the evolutionary arms race between host and parasite discussed in the previous section is interesting, it captures only a part of the coevolutionary process of GFG interactions. The defect is obvious – evolutionary dynamics cannot simply be reduced to the number of virulence and resistance genes. In general, there are many genotypes that contain the same number of resistance and virulence genes, but at different loci. The antagonistic interactions and frequency-dependent selection associated with these genotypes are very important aspects of host–parasite coevolution – particularly in conjunction with the Red Queen hypothesis and the evolution of sex (see Box 18.1).

How then can we define the relationship between host and parasite genotypes in GFG systems? In the GFG system, one host-resistance gene corresponds to one of the parasite's virulence genes, assuming that a host with a resistance gene can evade infection by those parasites that do not possess the corresponding virulence gene (Flor 1956; Burdon 1987). There does, however, seem to be some conceptual confusion in previous models of GFG interactions. Parker (1994) pointed out that GFG models often assume matching genotype interactions, in which there is a one-to-one correspondence between *genotypes* of host and parasite. For example, a parasite genotype can infect only a perfectly matched host genotype (Table 17.1a). Alternatively, a host genotype is only resistant against a perfectly matched parasite genotype.

In typical GFG interactions studied in plant–pathogen systems, the relationship between host and pathogen genotypes is highly asymmetric (Table 17.1b). This is because the pathogen genotype capable of overcoming host resistance does not necessarily lose the ability to infect susceptible genotypes; indeed, it is capable of infecting both. Within the multilocus GFG interactions, some parasite genotypes have broader host ranges than others, while some host genotypes have the ability to resist a wider range of parasites. The ranges of parasite virulence and host resistance depend on the numbers of virulence and resistance alleles in their genomes (Table 17.2).

Table 17.2 Two-locus GFG system.

	$v_1\,v_2$	$V_1\,v_2$	$v_1\,V_2$	$V_1\,V_2$
$r_1\,r_2$	+	+	+	+
$R_1\,r_2$	–	+	–	+
$r_1\,R_2$	–	–	+	+
$R_1\,R_2$	–	–	–	+

Note: Plus and minus signs indicate whether infection occurs (+) or not (–) in each combination of host and pathogen genotypes.

Table 17.3 Host and parasite fitness in the single-locus GFG model.

	Avirulent parasite		Virulent parasite	
Susceptible host	$e^{-\eta}$	e^{ζ}	$e^{-\eta}$	$e^{\zeta-c_V}$
Resistant host	$e^{-\eta\,\sigma-c_R}$	$e^{\zeta\sigma}$	$e^{-\eta-c_R}$	$e^{\zeta-c_V}$

Note: The fitness of the host (left) and of the parasite (right) for each pair of host and parasite genotype while exposed in monocultures. Single-locus haploid inheritance is assumed.

17.3 Coevolutionary Dynamics in Gene-for-gene Systems

Two selective forces drive the coevolutionary process of host resistance and parasite virulence in the GFG system. The first is a selection that favors greater degrees of resistance and virulence as quantitative traits, and this results in an escalation in the number of both the resistance genes in the host and virulence genes in the parasite. The second selective force is a process of frequency-dependent selection, which favors new combinations of genes. This section starts with a brief review of the simplest GFG model with haploid single-locus inheritance in the host and in the parasite. By extending the model to multilocus inheritance, I explore the main topic of the chapter – namely, the consequences of coevolutionary dynamics in multilocus GFG interactions. Two aspects are embedded naturally into the model: the coevolutionary escalation of resistance and virulence as quantitative traits, and the antagonistic multilocus genotypic dynamics of host-resistance genes, together with their corresponding parasite-virulence genes.

A single-locus gene-for-gene model

The simplest model for GFG interactions assumes haploid single-locus inheritance in host resistance and parasite virulence (e.g., Jayaker 1970; Leonard 1977; see Seger and Hamilton 1988 for the matching genotype versions). I denote the resistance and susceptibility alleles of the host as 1 and 0, respectively, and the virulence and avirulence alleles of the parasite as 1 and 0. The resistance only takes effect when the resistant host is attacked by an avirulent parasite. Table 17.3 indicates the fitness of the host and parasite for each combination of their genotypes when exposed to each other in monocultures.

Assuming that the individuals in each generation take part in very many independent interaction events, that host and parasite mingle randomly, and that the

effects of these events are multiplicative, the fitness of the host and parasite genotypes are given by

$$w_{\text{susc}} = e^{-\eta} , \tag{17.1a}$$

$$w_{\text{resistant}} = e^{-\eta[\sigma(1-p)+p]-c_{\text{R}}} , \tag{17.1b}$$

$$w_{\text{avirulent}} = e^{\zeta[(1-q)+\sigma q]} , \tag{17.1c}$$

$$w_{\text{virulent}} = e^{\zeta-c_{\text{V}}} , \tag{17.1d}$$

where σ is the probability that a resistant host is infected by an avirulent parasite
($\sigma = 0$ if the resistance to an avirulent parasite is complete); η measures the fitness
loss incurred by the host, and ζ the fitness gain obtained by a parasite through
successful infection. c_{R} and c_{V} are the cost of resistance and virulence, respectively; p is the frequency of the resistance gene and q that of the virulence gene.
Each generation before the start of the interaction, the sum of the frequencies of
the host is set back to 1, and the same applies for the parasite. The frequencies of
the resistance gene and the virulence gene at the internal equilibrium are

$$q^* = c_{\text{V}}/\zeta(1-\sigma) , \tag{17.2a}$$

$$p^* = 1 - c_{\text{R}}/\eta(1-\sigma) . \tag{17.2b}$$

It can be shown that this internal equilibrium is always unstable; any trajectory
converges to the heteroclinic cycle that connects four monomorphic corners in
gene frequency space (Hofbauer and Sigmund 1988). A small amount of mutation
or migration, however, suffices to keep the trajectory away from the boundary and
to yield a stable limit cycle (Seger and Hamilton 1988).

Multilocus gene-for-gene dynamics

The multilocus system can be extended by considering how each locus contributes
in the overall resistance reaction. Resistance occurs if there is at least one pair of
resistant and avirulent alleles at corresponding loci of the host and parasite genotypes. I assume that the resistance effects at the different loci are multiplicative
– that is, the overall resistance reaction is doubled if there are two loci with a
resistance–avirulent combination of alleles at the host and parasite. On the other
hand, any other combinations of alleles at a locus – that is, susceptible–avirulent,
susceptible–virulent, and resistant–virulent – do not contribute to the resistance
reaction.

Let us consider the n resistance loci of the host, with two alleles [1 (resistant)
and 0 (susceptible)] at each locus, and the corresponding n virulence loci of the
parasite, with two alleles [1 (virulent) and 0 (avirulent)] at each locus. Host multilocus genotypes for resistance, and parasite multilocus genotypes for virulence,
can then be denoted by binary numbers x and y with n digits, with each digit

Host: x = 0 0 1 **1** 0 **1**

Parasite: y = 0 0 1 **0** 0 **0**

Figure 17.2 Host and parasite genotypes in a five-locus GFG system. Each digit in a geno-type represents the allelic state at that locus, with 1 and 0 denoting, respectively, resistant and susceptible alleles in hosts, and virulent and avirulent alleles in parasites. The shaded columns indicate pairs of host and parasite loci at which resistant reactions occur (i.e., pairs of loci with resistant alleles in the host and avirulent alleles in the parasite). In this en-counter of host and parasite, there are two pairs of effective resistance, $m(x, y) = 2$, and, hence, the probability of infection of the host by the parasite is reduced to σ^2.

describing the allelic state of resistance and virulence at the corresponding locus. Therefore, if there are three resistance and virulence loci, the host genotype $x = 101$ implies resistance alleles at the first and the third locus and a susceptible allele at the second. A parasite virulence genotype $y = 001$ represents the genotype with a virulent allele at the third locus and avirulent alleles at the first and the second. Host resistance based on the gene in the jth locus is effective only for parasites that have an avirulent allele at the corresponding locus (i.e., when $x_j = 1$ and $y_j = 0$, where x_j and y_j are the allelic states at the jth locus of host genotype x and parasite genotype y).

Let us suppose that each effective resistance gene reduces the probability of successful infection to σ ($0 < \sigma < 1$). By denoting the number of effective resis-tance genes of the host genotype x that are not masked by the parasite genotype y by

$$m(x, y) = \left\{ \begin{array}{c} \text{the number of loci with resistance allele in host} \\ \text{and avirulence allele in parasite} \end{array} \right\} = \sum_{j=1}^{n} x_j (1 - y_j) ,$$

(17.3a)

the probability of infection per contact is

$$p_{\text{infection}}(x, y) = \sigma^{x_1(1-y_1)} \ldots \sigma^{x_n(1-y_n)} = \sigma^{m(x,y)} .$$

(17.3b)

For example, if having one effective resistance locus reduces the probability of in-fection by 90% ($\sigma = 0.1$), then if the host genotype 001101 encounters the parasite genotype 001000, the probability of infection is $\sigma^2 = 0.01$, that is, 99% of hosts are protected from infection (Figure 17.2). If a host randomly encounters a parasite, the probability that the host genotype x experiences infection (the mean parasite load) is calculated as the average infection probability over all parasite genotypes,

$$\sum_y p_{\text{infection}}(x, y) p(y) = \sum_y \sigma^{m(x,y)} p(y) ,$$

(17.4a)

where $p(y)$ is the frequency of parasite genotype y. The fitness of host genotype x is assumed to decrease with the mean parasite load sum and with the number of resistance genes $|x| = \sum_j x_j$, due to the cost of resistance,

$$w_{\mathrm{H}}(x) = \exp\left[-\eta \sum_y p_{\mathrm{infection}}(x, y)p(y) - c_{\mathrm{R}}|x|\right],\tag{17.4b}$$

where η is the selection intensity for a unit increase of mean parasite load, and c_{R} is the cost incurred per resistance gene. Likewise, the fitness w_{P} of a parasite genotype y is assumed to increase with the mean host availability,

$$\sum_x p_{\mathrm{infection}}(x, y)q(x) = \sum_x \sigma^{m(x,y)}q(x),\tag{17.5a}$$

where $q(x)$ is the frequency of host genotype x, and to decrease with the number of virulence genes $|y|$,

$$w_{\mathrm{P}}(y) = \exp\left[\zeta \sum_x p_{\mathrm{infection}}(x, y)q(x) - c_{\mathrm{V}}|y|\right],\tag{17.5b}$$

where ζ is the selection intensity for a unit increase of the mean host availability, and c_{P} is the cost incurred per virulence gene in the parasite.

If the costs c_{R} and c_{V} for resistance and virulence are not introduced into the model, then the coevolutionary dynamics converge to the trivial equilibrium, with fixation, at all loci, of resistance alleles in the host and of virulence alleles in the parasite. Small costs suffice to prevent the convergence to this static equilibrium. As is observed below, the most interesting behavior in coevolutionary dynamics occurs in those cases that involve small costs for resistance and virulence. Recurrent mutations between alleles at each locus are also assumed in both species, and occur at rates of 10^{-5} per generation per locus in both host and parasite. Both populations are assumed to be infinitely large. I first examine changes in the degrees of resistance and virulence in terms of the distributions of the number of resistance alleles and virulence alleles within the population. Then I concentrate on changes in the frequencies of genotypes that have the same number of resistance and virulence alleles, but at different loci. From the perspective of the evolution of sex, this aspect is most important.

Arms races between virulence and resistance

For the case of five resistance loci in the host and five corresponding virulence loci in the parasite, a typical coevolutionary trajectory for the number of host-resistance genes and the number of parasite virulence genes is shown in Figure 17.3. The figure clearly demonstrates that the mean numbers of resistance genes and virulence genes cycle endlessly. This pattern is observed in extensive simulations for a wide range of parameters, as long as the costs of resistance and virulence are not very high (see below). In the specific case illustrated in Figure 17.3, the number of resistance genes in the host population mostly alternates between zero and one, whereas the number of virulence genes in the parasite population alternates between five and four, with excursions to three.

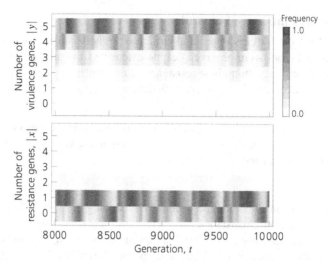

Figure 17.3 Evolutionary trajectories for the number of resistant genes in hosts and the number of virulence genes in parasites. The upper panel shows the dynamics of the frequency distribution of the number of virulence alleles in parasites. Each vertical slice in the panel represents the frequency distribution of the number of virulence genes in parasites in the generation of interest, with darker shades indicating higher frequencies. The lower panel shows the frequency distribution of the number of resistance genes in hosts. There are five resistance and virulence loci ($n = 5$) and, hence, there are six classes (from 0 to 5) for the number of virulence and resistance genes. Parameters: $\sigma = 0.2$, $\zeta = \eta = 0.3$, $c_R = c_V = 0.03$. Population sizes are assumed to be infinite. The recurrent mutation rate in each locus is 10^{-5} per generation in both host and parasite.

To describe the evolutionary cycles shown, let us start from a point at which the majority of parasites have one avirulent and four virulent alleles, and the majority of hosts have one resistance allele and four susceptible alleles. This quasi-equilibrium is broken by the spread of a super-strain of parasite, whose genotype has virulence alleles at all the loci. The predominance of the parasite super-strain then precludes the spread of a new resistance gene. The host genotype without any resistance subsequently spreads because of the cost of resistance. Once the majority of hosts become universally susceptible, a gradual decline in the number of virulence genes occurs in the parasite population because no costly virulent genes are needed to exploit the susceptible host. This lays the basis for the next phase, during which there is a spread of resistant genotypes in the host. These can avoid some of the parasite genotypes, and the coevolutionary trajectory returns to the starting point of the cycle. The sequence of events in the coevolutionary cycles is therefore characterized as follows: increased virulence in the parasite (escalation against resistance) → decreased resistance in the host (no resistance is effective against the super-strain) → decreased virulence in the parasite (virulence does not improve infectivity) → increased resistance in host (some resistance helps against avirulent parasites).

What prevents the spread of virulence genes?

Although the degrees of resistance and virulence rarely balance at an intermediate level (no actual cases were observed despite extensive simulations), a static equilibrium occurs either with no host resistance or no virulence in the parasite if the costs of resistance and virulence are sufficiently large. Indeed, it can be shown that:

1. *No resistance and avirulence* is evolutionarily stable if the ratio $\tilde{c}_R = c_R/\eta$ between the cost of resistance c_R and the selection intensity η for parasite load is larger than the efficiency of resistance,

$$\tilde{c}_R > 1 - \sigma .\qquad(17.6)$$

This simply means that the fitness gained by reducing the parasite load of a host mutant with a single-resistance gene, $\eta(1 - \sigma)$, must be smaller than the cost of resistance; otherwise, the mutant can invade the susceptible population. The evolutionary stability of avirulent parasites automatically follows suit, as there is no advantage for virulence genes against universally susceptible hosts.

2. *One-locus resistance and avirulence* is evolutionarily stable if the relative cost of resistance is relatively large,

$$\sigma(1 - \sigma) < \tilde{c}_R < 1 - c ,\qquad(17.7a)$$

and if the frequencies of single-resistance genotypes all lie below a threshold,

$$q(10\ldots0), q(010\ldots0), \ldots, q(0\ldots01) < \tilde{c}_V/(1 - \sigma) ,\qquad(17.7b)$$

with $\tilde{c}_V = c_V/\zeta$. Condition (17.7a) describes the situation in which a singly resistant host population can resist the invasion by a doubly resistant host (the first inequality) and by a universally susceptible host (the second inequality). Condition (17.7b) is necessary to protect an avirulent population of parasites from invasion by a singly virulent parasite. Indeed, if the frequency of any of the singly resistant genotypes in the host population exceeds the threshold, this may allow an invasion by the corresponding parasite genotype, which can exploit the overabundant host. This situation implies that any combination of frequencies of singly resistant genotypes is neutrally stable, as long as none exceeds the threshold. If the relative cost of virulence satisfies

$$\tilde{c}_V > \frac{1 - \sigma}{n} ,\qquad(17.7c)$$

then a combination of single-resistance genotype frequencies exists that makes the equilibrium stable. At the stability boundary, all single-resistance genotypes must be segregating with the same frequency, $1/n$.

Asymptotic states of the coevolutionary trajectories for various relative costs of resistance $\tilde{c}_R$ and of virulence $\tilde{c}_V$ are summarized in Table 17.4. This summary assumes five resistance and virulence loci, and that the probability of infection given one effective resistance locus is $\sigma = 0.2$. The boundaries for different asymptotic

Table 17.4 Phase diagram for the coevolutionary dynamics.

		Relative cost for resistance, c_R/η					
		0.05	0.1	0.3	0.5	0.7	0.9
Relative	0.05	Cycle	Cycle	Cycle	Cycle	Cycle	S/A
cost	0.1	Cycle	Cycle	Cycle	Cycle	Cycle	S/A
for	0.3	DR/A	DR/A	SR/A	SR/A	SR/A	S/A
viru-	0.5	DR/A	DR/A	SR/A	SR/A	SR/A	S/A
lence,	0.7	DR/A	DR/A	SR/A	SR/A	SR/A	S/A
c_V/ζ	0.9	DR/A	DR/A	SR/A	SR/A	SR/A	S/A

Cycle: Sustained cycles for the degrees of resistance and virulence.
DR/A: Double-resistance/avirulence equilibrium. SR/A: Single-resistance/avirulence equilibrium. S/A: No resistance/no virulence equilibrium.
Note: The number of resistance and virulence loci is $n = 5$; the efficiency of an effective resistance is 0.8 ($\sigma = 0.2$). The predicted conditions for the stability of S/A equilibrium are $c_R/\eta > 0.8$, Condition (17.6). The condition for the stability of SR/A equilibrium is $0.8 > c_R/\eta > 0.16$, Condition (17.7a), and $c_V/\zeta > 0.8/5 = 0.16$, Condition (17.7c) with $n = 5$. All these predictions agree with the actual simulation results shown in the table.

states observed in the simulations agree with the predictions provided by Conditions (17.6) and (17.7).

The main conclusion from Table 17.4 is that evolutionary cycles occur for relatively small costs of resistance and virulence. Another important conclusion can be drawn from Condition (17.7c). A static polymorphism of resistance is more likely to occur as the number of loci increases. The existence of the avirulence equilibrium raises the hope that we may be able to minimize parasitic damage by carefully mixing a large number of different single-resistance genotypes, rather than by constructing a multiple-resistance genotype. Possible limitations concerning this idea are discussed in Section 17.4.

Why does a super-strain of parasites dominate?

Within the evolutionary cycles of host resistance and parasite virulence, the mean number of the population's virulence genes is kept large, compared to the mean number of resistance genes. For example, the mean number of resistance genes in the host population is at most one in the trajectory illustrated in Figure 17.3, while the mean number of virulence genes in the parasite population varies between four and five. At first glance, the parasite virulence seems unnecessarily high, because one extra virulence gene would be sufficient to infect a host with a single-resistance gene. This apparant paradox can be explained through the polymorphism and asynchronous cycles in the frequencies of host-resistance genotypes, as explained below.

Assuming that the host population is polymorphic in the combination of resistant genes at different loci, a parasite should possess all sets of virulence genes to exploit a randomly encountered host successfully. One may expect the polymorphism of parasite genotypes to occur with a few virulence genes, each specialized to exploit one of the host's resistance genotypes. This is not the case in this model, though, and the super-strain of parasites often predominates because of its advantage (described below) – it can hedge bets in an unpredictable and changing environment (e.g., Seger and Brockmann 1987). The selection then favors a costly generalist parasite rather than the coexistence of several strains of specialist parasites.

The frequencies of host genotypes possessing the same number of resistance genes (but at different loci) fluctuate with approximately the same period. They do, however, fluctuate with different phases (Figure 17.4). This creates a fluctuating selection coefficient for each virulence gene. If the magnitude of the fluctuation is sufficiently large, the generalist parasite strategy (i.e., a super-strain) enjoys an advantage over the specialist strategies, although it has to pay for the extra cost of having many virulence genes.

Phenotypic and genotypic cycles

It is important to distinguish between the cycles shown in Figures 17.3 and 17.4. Figure 17.4 illustrates how the genotypic frequencies fluctuate wildly, while the mean numbers of resistance and virulence genes keep nearly constant (as shown in Figure 17.3).

The universally susceptible genotype is, on average, the most abundant in the host population, but its frequency does fluctuate (Figure 17.4a). The class of singly resistant genotypes constitutes the rest of the population. However, all possible single-resistance genotypes coexist (for $n = 5$, there are five such genotypes), and these show alternating periodical fluctuations with roughly equal phases of separation (Figure 17.4b). Regarding the parasite, the super-strain genotype is the most abundant, but its frequency is intermittently reduced, corresponding to the temporal predominance of universally susceptible hosts (Figure 17.4c). The second prevalent class of parasite genotypes is that composed of four virulence genes (i.e., comprising genotypes with one avirulence gene). As with single-resistance hosts, all parasite genotypes within this class coexist and fluctuate asynchronously, each being coupled to the corresponding resistance genotypes (Figure 17.4d).

17.4   Discussion

The most important contribution of the model to the theory of host–parasite coevolution and the Red Queen hypothesis for the evolution of sex (Jayaker 1970; Jaenike 1978; Bremermann 1980; Hamilton 1980, 1993; Seger and Hamilton 1988; Hamilton *et al.* 1990; Frank 1993b; see also Box 18.1) is that both genetic diversity in host and parasite genotypes and the complex cycles of their frequencies are promoted under the asymmetric GFG interactions often found in nature (Parker 1994; but see also Frank 1996a, 1996b; Parker 1996). Although I assume in this chapter that the GFG interaction is extremely asymmetric, it can still promote cycles in genotype frequencies and genetic polymorphism. The process considered here is doubly cyclic – it is the combination of evolutionary cycles in the degree of host resistance and parasite virulence, and asynchronous cycles in

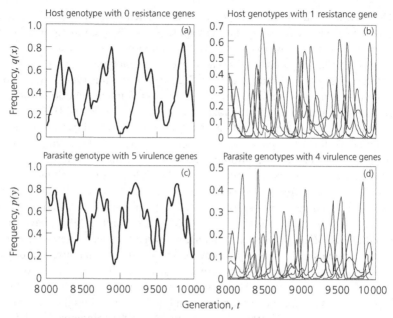

Figure 17.4 Genotype frequencies within the same class of resistance and virulence. The changes in genotype frequencies are shown for the same run as in Figure 17.3. (a) The frequency of the host genotype without any resistance gene; (b) the frequencies of host genotypes with one resistance gene; (c) the frequency of the parasite genotype with all five virulence genes (the super-strain); (d) the frequencies of parasite genotypes with four virulence genes. Parameters are the same as in Figure 17.3.

genotype frequencies under potential combinatorial diversity in multilocus inheritance. Whether the sustained asynchronous cycles and the protected multilocus polymorphism yield a sufficient short-term advantage for sex and recombination is an important question still to be explored.

Costs of resistance and virulence are necessary to ensure that the GFG interaction allows protected polymorphism of both resistance and virulence genotypes. Otherwise the best genotypes (those with all the resistance genes and those with all the virulence genes) will establish themselves in the host and the parasite populations. That the virulence of a pathogen declines after the reduction of resistance in a host (which is historically called "stabilizing selection") can be attributed to a selection process that opposes unnecessary virulence genes (i.e., to the cost of virulence). It is, however, difficult to measure the cost of virulence directly (Burdon 1987). The cost of resistance in the GFG system is even more difficult to detect, but, except in a few cases, it is considered to be small (Bergelson and Purrington 1996). At the same time, the cost of resistance may be condition-dependent, as is the case for the significant cost of encapsulation against parasitoids in *Drosophila* under starved conditions (Kraaijeveld and Godfray 1997).

The GFG and matching genotype interactions often produce cycles in host and parasite genotypic frequencies, but the population tends to converge to a heteroclinic cycle, and so is often to be found in a monomorphic corner within frequency space. A new transient occurs when a new favorable genotype emerges, which then leads the population into another corner (Seger and Hamilton 1988). Indeed, population genetic models for GFG (matching genotype) interaction do not indicate a promotion of genetic diversity (Takahata and Nei 1990; Frank 1993b). (This, incidentally, leads to a rejection of the hypothesis of parasite adaptation as the factor responsible for MHC polymorphism; see Takahata and Nei 1990.) However, a relatively small rate of migration or mutation has been found to restore diversity and to enhance the persistence of multiple alleles in the matching genotype model (Seger and Hamilton 1988). A relaxed genotypic specificity in resistance and virulence, and the evolution of more virulent strains within an infected host, also promote polymorphism (Clarke 1976; Maynard Smith 1989). I have shown that multilocus GFG dynamics can maintain a large number of genotypes. How well the dynamics of host–parasite coevolution account for the observed degrees of genetic diversity in both resistance and virulence is still an open question.

It is tempting to compare GFG coevolution at the population level with the intrahost dynamics of parasite quasi-species and the variety of immune system cells (Agur et al. 1989; Sasaki 1994; Nowak and May 1991; Haraguchi and Sasaki 1997; Sasaki and Haraguchi 2000). We are now in a position to record the evolution of viruses in a single patient. These dynamics share many characteristics of host–parasite coevolution based on genetically specific interactions, like the GFG system [e.g., the fluctuation of "genotypic" frequencies and the nonuniform speed of evolution found in the intrahost evolution of HIV (Yamaguchi and Gojobori 1997)]. There are, of course, many essential differences between the two levels of evolution. The host defense system, for example, now has a generation time and mutability comparable to that of the parasite. The question of how acquired immunity has evolved is still too difficult to answer, but the question of how the evolution of parasites is affected by the addition of acquired immunity to genetically specific defense mechanisms constitutes a tractable and important problem.

The model currently used for multilocus host–parasite GFG interactions has revealed that a condition arises in which the host resistance polymorphism can successfully prevent the spread of any virulent race in the parasite species. The condition for this success depends on the cost of virulence and on the number of resistant loci in the host species (see Table 17.4). Theoretical results show that for any cost of virulence, the evolutionary escalation of parasite virulence can be precluded by increasing the number of resistant varieties in the population [see Equation (17.7c)]. This can be compared to the use of multiple resistance against parasites, which has often failed in practice and invariably fails in the present model, which assumes additivity in the costs for multiple virulence and resistance. In short, the present model suggests an advantage of multiline resistance over multiple resistance from the perspective of preventing the evolutionary escalation of parasite virulence. However, it is possible that a strong synergetic effect between the costs of multiple virulence and resistance would change the results. In addition, in practice the spatial distribution of the resistant varieties would be an important factor in determining the success or failure of parasite control with multiline resistance. Clearly, further theoretical studies are needed to incorporate the spatial structure (e.g., the limited dispersal of fungus spores and the spatial cropping pattern of different resistant varieties) and demographic dynamics (in addition to the genetic dynamics) of the parasite.

As noted earlier, there is a clear parallelism between the problem discussed in this chapter and the emergence of multiple drug resistance in infectious diseases of humans (Anderson and May 1991). The emergence of multiple drug resistance corresponds to the emergence of a parasite super-strain in GFG coevolutionary dynamics. We have shown that the predominance of a super-strain is a robust outcome of the GFG arms race, but such a predominance can be prevented by increasing the number of resistant varieties. This suggests that a variety of antibiotics used separately in different patients, rather than multiple drugs used for each patient, is preferable to prevent epidemics. A similar principle would also apply to the emergence of resistant biotypes in pest control.

Acknowledgments   I thank David Krakauer for his detailed comments on how to improve the manuscript of this chapter. I also thank Bill Hamilton, Austin Burt, and Tetsukazu Yahara for their valuable discussions.

18

Implications of Sexual Selection for Virulence Management

Claus Wedekind

18.1   Introduction: Sex and Coevolution

In contrast to asexual reproduction, sex involves a number of quite obvious disadvantages (e.g., Williams 1975; Maynard Smith 1978; Stearns 1987; Michod and Levin 1987). The major disadvantage has been termed the "cost of meiosis" (Williams 1975): a female that reproduces sexually is only 50% related to her offspring, while an asexual female transmits 100% of her genes to each of her daughters. Hence, gene transmission is about twice as efficient in asexuals as in sexuals. The other disadvantages of sex are, for example, cellular mechanical costs, genetic damage through recombination, exposure to risks, mate choice, mate competition, etc. (see review in Lloyd 1980; Lewis 1987). Therefore, if asexuals had a survival probability comparable to sexuals, a mutation causing a female to produce only asexual daughters would, when introduced into a sexually reproducing population, rapidly increase in frequency and outcompete sexuals in numbers within a few generations (Williams 1975; Maynard Smith 1978). Why does this not happen? What are the advantages of sex, or what are the disadvantages of asexual reproduction?

One serious disadvantage of asexual clones is that they are likely to die out after some hundred or thousand generations because of a fatal mechanism called "Muller's ratchet" (Muller 1932). Roughly summarized, Muller's ratchet predicts that slightly deleterious mutations are accumulated in asexuals from generation to generation until the genome does not code for a viable organism any longer, and the population becomes extinct (e.g., Andersson and Hughes 1996). At first glance, one might therefore think that sex must be so successful because recombination and selection can result in the efficient removal of damaged genes from generation to generation. However, a danger of this benefit of sex is it may have a significant effect only when it is too late – that is, after an asexual mutation in a population has outcompeted all its sexual conspecifics (e.g., Michod and Levin 1987; Stearns 1987; Kondrashov 1993; Howard and Lively 1994; Hurst and Peck 1996). A second set of hypotheses, therefore, suggests that sex enables the spread, or even the creation, of advantageous traits. This second category of hypotheses requires that the direction of selection continuously change – that is, the main source of mortality is short-term environmental changes. This condition is especially fulfilled in coevolving systems such as parasite–host communities. The idea is that

---

Box 18.1 Sex as an evolutionary response to parasites

The many different ideas to explain why sexual reproduction is so common as compared to asexual reproduction can be grouped into two broad categories:

- Sex enables the efficient removal of deleterious genes;
- Sex enables the spread or even the creation of advantageous traits (e.g., Michod and Levin 1987; Stearns 1987; Kondrashov 1993; Howard and Lively 1994; Hurst and Peck 1996).

The second category of hypotheses typically requires that the direction of selection be continuously changing – that is, the main source of mortality arises from short-term environmental changes. This condition is fulfilled especially in coevolving systems such as parasite–host communities (i.e., the "Red Queen hypothesis," e.g., Jaenike 1978; Hamilton 1980; Hamilton *et al.* 1990). The changes could be irreversible or fluctuating. Genetic heterogeneity within a sexually produced clutch may increase the chances of that clutch containing an optimal genotype (i.e., the "lottery model," Williams 1975; Young 1981; Kondrashov 1993), and it may decrease the risk of competition between relatives (i.e., the "elbow-room model," Maynard Smith 1976; Young 1981; Kondrashov 1993). The possibility that sex reduces the risk of transmission of pathogens between relatives (because of their genetic dissimilarity; Rice 1983; Shykoff and Schmid-Hempel 1991) can be seen as a variant of Williams' lottery model.

The effect of parasite–host coevolution on sexuality has been studied almost exclusively from the host's point of view (e.g., reviews in Ladle 1992; Sorci *et al.* 1997). However, the argument can be turned around to explain that sexual reproduction in parasite populations is maintained as a diversity-generating mechanism to counteract the rapidly changing selection imposed by the hosts' immune system (Read and Viney 1996; Gemmill *et al.* 1997; Taylor *et al.* 1997b; Wedekind 1998).

---

host resistance genes that are advantageous today might become disadvantageous in the near future because parasites evolve to overcome them. Therefore, hosts must continuously change gene combinations, and sex is an efficient means to do so ("Red Queen hypothesis," e.g., Jaenike 1978; Hamilton 1980; Hamilton *et al.* 1990; Ladle 1992; Clarke *et al.* 1994). Additionally, the argument can be turned around to predict that sexual reproduction in parasite populations is maintained as a diversity-generating mechanism to counteract the rapidly changing selection imposed by the hosts' immune system. This may be so especially for multicellular parasites (e.g., Read and Viney 1996; Gemmill *et al.* 1997; Taylor *et al.* 1997b; Wedekind and Rüetschi 2000).

If this explanation is true, then the evolutionary conflict between parasites and hosts selects for diversity-generating mechanisms such as the different forms of sexual reproduction, which, in turn, prevent both parties from dying out as a direct or indirect consequence of Muller's ratchet (see also Box 18.1). It may sound absurd, but this is probably one of the reasons why our world does not look as bleak as a lunar landscape.

This idea of sex as a response to parasites is one of the reasons why increasing attention has been given in recent years to the role of genetic variation in host–parasite coevolution. However, when investigating genetic variation in the field or in the laboratory, one often faces the problem of potential biases caused by confounding variables. In parent–offspring comparisons, for example, it is often difficult to control for parental (mostly maternal) carryover effects, even when using cross-fostering experiments in which eggs or young are exchanged between clutches (see the discussion in Sorci *et al.* 1997). This problem may partly explain the discrepancy between the burst of theoretical work on the one hand, and the relatively small number of empirical studies on genetic heterogeneity in host–parasite systems on the other hand (e.g., review in Ladle 1992; Frank 1996c; Sorci *et al.* 1997).

The most important aspect of the interaction between a pathogen and its host is the virulence an infection is associated with – defined here as the loss of fitness the parasite causes the host. Although this definition sounds as if virulence is a specific trait of a parasite, it is not. Rather, it is the result of the interaction between the pathogen and its host (e.g., Bull 1994; Ewald 1994a; Lenski and May 1994; Read 1994; Ebert and Herre 1996; Frank 1996c). Sexual reproduction and the different forms of parasite-driven sexual selection have the potential to increase or decrease virulence. First, the ability of the host to reproduce sexually plays a significant, if not the most important, role (Ebert and Hamilton 1996). Sexual reproduction (outcrossing) results in a rearrangement of host genes. Pathogen populations that have adapted to one host line have to readapt to the next one, and so on. Hence, host reproduction by sex results in an existing parasite population that is suboptimally adapted to their current host population. This suggests that the pathogens are less virulent than would be expected if they were optimally adapted to their host (Ebert 1994).

This chapter stresses a number of further factors that could be important in determining the virulence in locally adapted host–pathogen systems.

18.2 Sexual Selection

Sexual reproduction is, of course, not just a harmonious venture that results in the mixing of a male's and a female's genetic material; rather, individuals must compete for, and must be chosen as, mating partners. This is so, in most cases, because males and females differ greatly in their parental investment (Trivers 1972). As a consequence, males usually have a much higher potential rate of reproduction than females (Clutton-Brock and Vincent 1991), which is the major cause of conflict both between and within the sexes (Clutton-Brock and Parker 1992). These conflicts gives rise to a new kind of selection important in the evolution of a species: if individuals are to have a chance of propagating their genes, they must survive not only all the lethal threats imposed by harsh climates, predators, pathogens, and competitors (natural selection), but they must also be able to find a mate and withstand competition from rivals of the same sex for access to mates ("sexual selection"; for a review, see Andersson 1994).

Mate choice and competition for mates are two forms of sexual selection (intersexual selection and intrasexual selection), and they both cause a number of immediately negative aspects – for example, the waste of resources associated with attractiveness (e.g., the peacock's tail) or the risk of injury or distraction from predator surveillance during intrasexual struggles or courtship behavior. Furthermore, transmission, maintenance, and growth of many pathogens are increased during the host's courtship phase – either directly by sexual behavior itself (transmittance of ectoparasites or sexually transmitted microparasites) or indirectly by a reduction of immunocompetence of the host (e.g., Grossman 1985; Folstad and Karter 1992). The reduced immunocompetence may be a consequence of an adaptive resource reallocation during the courtship phase (Wedekind and Folstad 1994; Sheldon and Verhulst 1996), mating, or reproduction (Gustafson *et al.* 1995; Richner *et al.* 1995; Oppliger *et al.* 1996). These disadvantages associated with sexual selection add to the costs of sex mentioned above. Rather than review them in more detail here, I instead concentrate on the possible impact sex and sexual selection may have on the evolutionary conflict between pathogens and their hosts. When discussing sexual selection, I therefore focus on intersexual selection – that is, mate choice and cryptic female choice mechanisms after mating.

To relate mate choice to the evolution of virulence, it is necessary to know the different criteria that determine mate choice, to understand why they are used, and to identify their relative importance. The literature usually groups the possible criteria used into three classes (e.g., Andersson 1994):

- Criteria that offer direct benefits (e.g., good parental care, nuptial gifts, etc.);
- The so-called "Fisher-traits" [i.e., criteria that are attractive to members of the other sex and do not reveal anything apart from that (Fisher 1930; Lande 1981; Kirkpatrick 1982; Pomiankowski *et al.* 1991)];
- Criteria that reveal "good genes" (Zahavi 1975; Hamilton and Zuk 1982; Grafen 1990; Iwasa *et al.* 1991; Wedekind 1994a, 1994b; Johnstone 1995).

The third class of criteria is of special interest to the discussion here: by "good genes" I mean those that are advantageous in the coevolution between pathogens and their hosts (Hamilton and Zuk 1982). Mate choice for good genes may therefore be an important factor in determining the level of virulence in a natural pathogen–host system.

This suggests that not only sex itself, but also some forms of sexual selection could be strongly influenced by the coevolutionary dynamics of parasite–host systems, and, in turn, could influence this coevolutionary relationship.

Mate choice for criteria that reveal good genes is only one possible level of parasite-driven sexual selection. Further possible levels are, for example, selection on sperm by the female reproductive tract, selective fertilization, or selective support of the embryo or the born offspring (Figure 18.1 indicates eight possible levels). All these levels could potentially be connected to parasite–host coevolution.

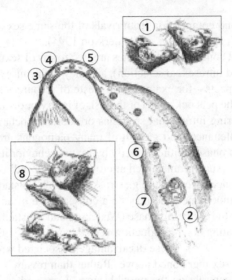

Figure 18.1 The possible selection levels at which females or their ova could select for heterozygosity, or even specific allele combinations in the offspring, on loci that are important in the parasite–host coevolution: (1) mate choice, (2) selection on sperm by the female within her reproductive tract, (3) egg choice for the fertilizing sperm, (4) second meiotic division of the egg influenced by the fertilizing sperm, (5) selection on the early embryo by the oviduct, (6) implantation, (7) nutrition supply to the embryo and spontaneous abortion, (8) selective feeding or selective killing of newborns. *Source*: Wedekind (1994b).

18.3 Hypotheses for Parasite-driven Sexual Selection

In this section three different hypotheses that offer possible explanations for parasite-driven sexual selection are discussed.

Uniform preference for health and vigor

In their original hypothesis, Hamilton and Zuk (1982) suggested that individuals in good health and vigor are preferred in mate choice because they are likely to possess heritable resistance to the predominant pathogens. By preferring healthy individuals, one may thereby produce resistant progeny. This could result in subsequent generations of hosts that are better adapted to the local pathogens – that is, they are less susceptible (Grahn *et al.* 1998).

The mechanisms Hamilton and Zuk (1982) suggested will lead to populations in which all individuals of one sex have the same mate preference (i.e., members of the opposite sex could be ranked in a universally valid order of attractiveness, and less attractive individuals would only be taken as mates if the more attractive ones are not available for some reason). However, in some species, this prediction does not appear to be fulfilled (see references below). Furthermore, an offspring's level of resistance depends on both its mother's and father's genetic contribution. At loci important for the parasite–host interaction (e.g., immunogenes) certain combinations of alleles may be more beneficial than others. If individuals choose their mates to produce such beneficial allele combinations, their preferences would depend on their own genotypes as well as their partners' genotypes. As a consequence, individuals with different resistance genes would show different preferences, and there would be no universally valid order of sexual attractiveness with respect to signals that reveal heritable disease resistance (or immunogenes).

Genetically variable preferences for complementary genes

Preferences for mates or for sperm of genetically dissimilar types have been observed in several species. Olsson *et al.* (1996) found that in a population of sand lizards (*Lacerta agilis*) in which most females mate with more than one male, the male's genetic similarity to the female correlates with the proportion of her offspring that he sires: more dissimilar males sire more offspring, both in the field and in the laboratory. They concluded that the female reproductive tissue actively selects from genetically dissimilar sperm. Another example of this can be found in the ascidian *Diplosoma listerianum* – a colonial, sessile, marine filter-feeder that disperses sperm into surrounding water. Sperm are taken up and pass via the oviduct to reach oocytes within the ovary. Autoradiography of labeled sperm revealed that sperm from the same clone were normally stopped in the oviduct, while sperm from other clones progressed to the ovary (Bishop 1996). Furthermore, a weak negative correlation was found between the mating success of pairs and their overall genetic similarity (Bishop *et al.* 1996). Gametic self-incompatibility has also been intensely studied in the hermaphroditic tunicate *Ciona intestinalis* (e.g., Rosati and DeSantis 1978; DeSantis and Pinto 1991). Self-discrimination occurs in the vitelline coat, is established there in late oogenesis, and is controlled by products of overlying follicle cells. Self-sterility in this species is not absolute, but appears to depend on still unidentified factors (Rosati and DeSantis 1978; DeSantis and Pinto 1991). Further examples that suggest nonrandom fertilization are reviewed in Eberhard (1996) and Zeh and Zeh (1997), but Simmons *et al.* (1996) and Stockley (1997) could not find it in yellow dung flies or common shrews, respectively. While the loci involved in reproductive compatibility or incompatibility are not yet known in the above examples, they are known in at least several plants and in a tunicate. Growth of the pollen tube is often affected by the stigma and depends on the combination of male and female alleles on the self-incompatibility locus (e.g., Franklin-Tong and Franklin 1993). In the tunicate *Botryllus* spp., eggs appeared to resist fertilization by sperm with the same allele on the fusibility locus for a longer time than sperm with a different allele on it (Scofield *et al.* 1982).

The best-studied example of mate preferences that depend on the chooser's own genotype occurs in the mouse. Probably the most important genes in this respect are the genes of the major histocompatibility complex (MHC). Mice base their mate choice to a large extent on odors. These odors reveal some of the allelic specificity of MHC (Yamazaki *et al.* 1979, 1983a, 1983b, 1994), and this information is used in mate choice by males and females (Yamazaki *et al.* 1976, 1988; Egid and Brown 1989; Potts *et al.* 1991). They choose according to their own

MHC types, apparently to reach certain allele combinations or to avoid certain allele combinations in the progeny. In humans, too, MHC correlates with odor production and with male and female preferences for human body odors (Wedekind *et al.* 1995; Wedekind and Füri 1997). Human noses are even able to discriminate odors of two mouse strains that are congenic with respect to their MHC – probably because there are still some similarities between murine and human MHC antigens (Gilbert *et al.* 1986). Preferences for human body odors correlated with actual mate choice in two independent test series: odors of MHC-dissimilar persons reminded the test subjects more often than expected by chance of their own current mate or former mate (Wedekind *et al.* 1995; Wedekind and Füri 1997). Recently, Ober *et al.* (1997) partly confirmed these findings in a study on American Hutterites: the MHC types of 411 couples were more often different from each other than would be expected if matings were random (in their calculation of the null expectancies, they controlled for nonrandom mating with respect to colony lineage and with respect to kinship).

MHC-correlated sexual selection need not be restricted to mate choice (Figure 18.1). Numerous studies in humans from many different regions of the world indicate a connection between the risk of experiencing a spontaneous abortion and the degree of MHC similarities between the pregnant woman and the father of her embryo (Gill 1994 cited 26 such studies in his review). The higher the degree of MHC similarity, the higher the risk that an apparently healthy embryo will abort. There is even evidence that the degree of MHC similarity between couples has an effect on earlier stages of pregnancy – that is, before an abortion would be recognized. Weckstein *et al.* (1991) found that the success rate of *in vitro* fertilization and tubal embryo transfer correlates with MHC similarity, and Ober *et al.* (1988) found in the American Hutterites who proscribe contraception that longer intervals between successive births are again associated with increased MHC similarity of the couple. All this evidence is of course correlational – that is, causes and effects are unclear since experiments on this topic are not possible in humans for obvious reasons. However, experiments on mice show that the observed associations between MHC and abortion are causal for mice (Yamazaki *et al.* 1983a, 1983b).

It is not yet clear whether MHC-correlated mate preferences optimize the offspring's immunogenetics or whether MHC merely serves as a marker of kinship to avoid inbreeding (e.g., Potts and Wakeland 1993; Apanius *et al.* 1997). If the first variant holds, this would further improve host defense against pathogen populations – that is, it would further reduce the observable level of virulence. However, even if the first variant holds, it remains unclear what kind of MHC complementarity is preferred – that is, whether individuals simply prefer other types to ensure a higher proportion of MHC-heterozygous offspring (e.g., Brown 1997) because heterozygosity of MHC appears to be beneficial on average (Doherty and Zinkernagel 1975; Hedrick and Thomson 1983), or whether mate choice aims to reach specific allele combinations that are more beneficial under given environmental conditions than others (Wedekind 1994a, 1994b; Wedekind and Füri 1997). At the moment, there are only a few indications that such beneficial allele combinations exist. Hirayama *et al.* (1987) found epistatic effects of at least two human MHC antigens to a pathogen antigen (of a schistosome). Moreover, the strong linkage disequilibrium observed between some MHC alleles could be explained by long-term epistatic fitness effects (Klein 1986; Maynard Smith 1989). There is still much need for research on the beneficial or deleterious aspects of various homo- or heterozygous combinations of MHC haplotypes under given pathogen pressures.

In general, mice and humans appear to prefer dissimilar types. If the MHC is not just used as a marker for kinship (Brown and Eklund 1994; Potts *et al.* 1994), most evidence on MHC-correlated sexual selection is most easily explained by a general preference to produce heterozygous offspring. However, this putative preference for dissimilar types could be a statistical artifact if most beneficial allele combinations under different environmental conditions are heterozygous combinations. In a recent study in which male and female students scored the odors of T-shirts worn by other students, Wedekind and Füri (1997) tested whether MHC-correlated odor preferences favored specific allele combinations in the offspring rather than simply favoring heterozygous combinations. They found evidence for the latter possibility but not for the former, confirming an earlier study on this subject (Wedekind *et al.* 1995). However, the fact that they did not find any preference for specific allele combinations does not exclude the possibility that one could find such preferences under other circumstances (e.g., in populations that are under stronger pathogen pressure than this particular group of Swiss students).

Conditional preference for complementary genes

Choice for complementary alleles would be most efficient – that is, it would result in the highest fitness return – if individuals were able to choose their mate conditionally. A well-tuned, condition-dependent mate selection could have evolved in some species because of a nontrivial fitness advantage. Conditional choice takes into account the present pathogen pressure and promotes allele combinations in the host that ensure the optimal defense against these pathogens. However, this requires physiological achievements that have not been demonstrated so far.

Two other studies on mice suggest that sexual selection takes into account not only the male and female MHC genotypes, but is also conditional since it takes into account at least one external factor that can vary over time. In an *in vitro* experiment with two congenic mouse strains, Wedekind *et al.* (1996) tested whether eggs select for sperm according to their MHC, and whether the second meiotic division of the egg is influenced by the MHC type of the fertilizing sperm (see Figure 18.2). They found that neither egg–sperm fusion nor the second meiotic division is random, but that both processes actually depend on the MHC of both the egg and the sperm. However, to the great surprise of the authors of this study, these selection levels did not simply select for heterozygous MHC combinations. Sometimes the eggs appeared to prefer homozygous combinations, and sometimes

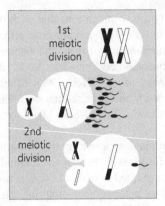

Figure 18.2 Schematic illustration of oogenesis, after cross-over has taken place and before the first meiotic division. The genetic material of the female's father and mother are illustrated as black and white chromosomes, while sperms and the male pronucleus are black ovals. In mammals and many other vertebrate species, the eggs are in a stage of meiotic arrest at the time of fertilization. The second meiotic division is completed only when the sperm has penetrated the vitelline membrane of the ovum (e.g., Bazer *et al.* 1987; Wolgemuth 1983). Therefore, the egg is often still heterozygous at some loci when the sperm enters it. This opens the possibility that the second meiotic division of the egg could be influenced by the genotype of the fertilizing sperm (selection level number 4 in Figure 18.1).

they appeared to prefer heterozygous combinations. This effect of time was statistically significant. An external factor that varied over time appeared to have had an influence on the experiment. The authors speculated that this external factor was an uncontrolled epidemic by mouse hepatitis virus (MHV) that occurred during the course of the experiment. The presence of MHV appeared to stimulate a preference for heterozygous combinations, while when absent, the mice seemed to prefer homozygous variants. To test this hypothesis, Rülicke *et al.* (1998) repeated the experiment *in vivo* with two mouse groups: some were slightly infected by MHV while the others were sham infected. When they typed the blastocysts of these mice for MHC, they found again that infected mice made more MHC-heterozygous combinations than uninfected mice. This time, however, the finding was the outcome of an experiment designed to test this *a priori* hypothesis.

Several authors have previously searched for deviations from the expected ratios of MHC heterozygosity in the progeny of controlled matings (Gorer and Mikulska 1959; Palm 1969, 1970; Hings and Billingham 1981, 1983, 1985; Potts *et al.* 1991). They reported a significant variability of MHC-heterozygote frequencies which, however, remained poorly understood because there appeared to be a general inability to replicate previous findings (see the discussion in Hings and Billingham 1985). The results of Rülicke *et al.* (1998) could lead to an explanation for the apparently controversial findings published before, since they were able to perform the experiments under defined hygienic conditions with a selective and monitored viral infection (i.e., they could control for infection, a factor that was not controlled for in the previous studies and that could have influenced their outcomes).

It is still not clear whether MHC-heterozygous offspring of the mice strains used in the above experiments have a greater resistance to MHV infection than the homozygous variants, and it is still not known whether homozygous offspring have higher survival rates in the absence of MHV. However, if so, nonrandom fusion of egg and sperm with regard to their respective MHCs and to the presence or absence of MHV could improve the health of the progeny – that is, it would further decrease the observed level of virulence in a locally adapted host–pathogen system.

18.4   The Pathogen's View

Parasites may have sex as an adaptation to combat antiparasite defense mechanisms in the host (Read and Viney 1996; Gemmill et al. 1997; Wedekind 1998). While hosts have to escape from parasite adaptations to given host genotypes, parasites have to counter the somatic evolution of the host's immune response (i.e., the acquired immunity that is most effective against the parasite genotype that originally initiated it). This may be especially so in microparasites. In helminth infections, there is less evidence for genotype-specific immunity, probably because this has been less intensely studied (review in Read and Viney 1996). However, Gemmill et al. (1997) found an interesting connection between the degree of sexuality of a parasitic nematode and its hosts' immune system: the nematode *Strongyloides ratti*, a parasite of rats, can have a direct lifecycle with clonal reproduction or an indirect lifecycle with free living sexual adults. In a series of experiments, Gemmill et al. (1997) manipulated the rats' immune status using hypothymis mutants, corticosteroids, whole-body γ-irradiation, and previous exposure to *S. ratti*. They found that parasite larvae from hosts that acquired immune protection are more likely to develop into sexual adults than larvae from hosts that have an experimentally compromised immune status. This suggests that the hosts' immune response selects for sex in this nematode.

Wedekind and Rüetschi (2000) used the cestode *Schistocephalus solidus* and its first intermediate host, the copepod *Macrocyclops albidus*, to test whether parasite heterogeneity affects infection success and the occurrence of within-host competition. Genetic heterogeneity in parasite larvae would be the result of sexual reproduction, especially so if the worms were allowed to outbreed. However, outbred offspring further differ in properties from inbred (or asexually) produced ones, as demonstrated by Wedekind et al. (1998) and Schärer and Wedekind (1999). Worms that were allowed to reproduce in pairs produced on average larger eggs that contained larger embryos (even larger if measured as percentage of egg volume) than worms that were forced to reproduce alone. This suggests a strategic egg production in this parasite that is dependent on the social situation a worm finds itself in when mating. Therefore, the lower hatching rates of inbred eggs observed under laboratory conditions (Wedekind *et al.* 1998) could result from in-breeding depression or lower maternal investment into individual eggs when the mother could not outbreed, or both.

To decouple such potentially confounding effects from the degree of hetero-geneity, Wedekind and Rüetschi (2000) used 10 parasite sibships to infect cope-pods each with six larvae that stem either from a single clutch (pure exposure, i.e., low heterogeneity) or a mix of two clutches (mixed exposure, i.e., high hetero-geneity). They found that infection was more likely in mixed exposure (increase of prevalence: >50%), but infections are more often multiple – that is, individu-als had more often to compete within a single host, in the case of pure exposure. Parasite transmission rates were therefore only slightly increased in mixed expo-sure (22%). However, since parasite growth was reduced in multiple infections, parasites from mixed exposure were on average more than 50% larger than those from pure exposure at a time when they were infectious for the next intermediate host. This demonstrates two important benefits from sexual reproduction in this parasite:

■ The offspring can infect a broader range of hosts and even achieve slightly higher transmission rates than with asexual reproduction (supporting the "lot-tery model," Williams 1975; Young 1981);
■ Infections result in better growth because multiple infections and, hence, within-host competition were less likely (supporting the "elbow-room model," Maynard Smith 1976; Young 1981).

The observed effects may be in a range that could compensate for the costs of sex in this parasite.

Sex in the pathogen population, followed by selection, may increase its viru-lence while sex in the host may decrease it. Furthermore, any nonrandom mating that improves the virulence genetics of the next generation of a pathogen may be more beneficial to the parasite than random mating or a simple preference for out-crossing. However, in many parasite species, mate choice may be so costly that its benefits cannot outweigh its costs in evolutionary terms. In species for which this does not hold, it may, in principle, be possible that mate choice has evolved to its most sophisticated form (i.e., to the conditional choice for complementary virulence genes), which is analogous to the conditional choice for complemen-tary resistance genes possible in some host species. Whether this is true, or even whether there is any kind of mate preference in a parasite species that improves the virulence of the next generations, remains to be shown. At the moment, few studies address the interesting problem of mate choice in parasite species (see, e.g., Lawlor *et al.* 1990; Tchuem Tchuenté *et al.* 1995, 1996).

Box 18.2 summarizes the different forms of reproduction and their expected implication for virulence from the host's point of view and from the parasite's point of view.

Box 18.2 The dependence of virulence on the mode of reproduction and sexual selection

Increasingly sophisticated forms of reproduction of hosts and parasites and their expected implications for virulence from the host's view and from the parasites' view:

Different forms of reproduction/sexual selection	Expected implication for virulence: host's view parasite's view
Parthenogenesis	High Low
Selfing/inbreeding	
Random mating	
Inbreeding avoidance	
Preference for health and vigor	
Preference for complementary resistance genes/virulence genes	
Conditional preference for complementary resistance genes/virulence genes	Low High

(Host's view: Virulence decreasing from High to Low; parasite's view: Virulence increasing from Low to High.)

18.5 Implications for Virulence Management

There are many examples of human interference with animal and plant reproduction. Natural mate choice is usually circumvented in many farm animals and plants, and it is often rather restricted in zoo animals. Even in our own species, there are cases in which free mate choice is prevented for cultural reasons. From a genetic point of view, this can also occur as a result of infertility treatment with some forms of assisted reproductive technology (ART) – especially so in donor insemination and in egg or embryo donation (donors are usually anonymous), but possibly also in intracytoplasmic sperm injection, in which potential egg choice is not allowed for.

It may be too early to speculate about the evolutionary consequences of such interference in our own species. As this chapter may reveal, the implication of sexual selection on parasite–host coevolution is not well understood in natural systems, and it is even less understood in a culturally shaped species like our own one. Moreover, while the evidence for cryptic female choice (points 2–8 in Figure 18.1) is increasingly convincing in some plants and animals, the evidence for it in humans is only correlational (i.e., causes and effects are unclear). The existing data can therefore be interpreted in a number of ways (Hedrick 1988; Verrell and McCabe 1990; Wedekind 1994b).

ART is now responsible for tens of thousands of new births annually. Gosgen *et al.* (1998) discussed the possibility of genetic costs of ART in humans. The authors concluded that such costs are not obvious in the context discussed here. (Moreover, the incidence of birth defects in children is not higher than in those conceived naturally.) However, the success of ART may depend on the respective MHC types of the genetic parents of an embryo. Gosgen *et al.* (1998) finished their article with a call for more research on the impact of new reproductive technologies on individuals and the population, and whether or not donor insemination programs should reflect female choice.

The implication of sexual selection on virulence evolution should also be studied in the context of wildlife conservation. While the most important strategy for the conservation of endangered species is certainly the protection of natural populations (Gibbons *et al.* 1995), for some species, captive proliferation followed by reintroduction into the natural environment may be the only way to prevent extinction (e.g., in the river dolphins, see Ridgway 1995). A number of endangered species are now bred in captivity according to techniques published in textbooks, and often through the use of ART. Moreover, several authors predict that biotechnological procedures like embryo splitting or recombinant DNA technology is already finding applications in the promotion of endangered species (e.g., Gee 1995; Durrant 1995; Kholkute and Dukelow 1995). These authors and most other conservation biologists seem to agree that attempts should be made to maintain as much genetic diversity as possible.

Many studies on sexual selection suggest that genetically dissimilar mates are sexually preferred (in a number of vertebrates such as mice, lizards, and humans, as well as in several invertebrates and plant species; see references cited in Section 18.3), probably because high genetic diversity in the next generation is beneficial. However, a preference for genetic dissimilarity has not always been observed (e.g., Yamazaki *et al.* 1976; Wedekind *et al.* 1996; Rülicke *et al.* 1998). It is not yet clear whether these exceptions would have lead to higher viability in the offspring under given environmental conditions. Nevertheless, the possibility exists that in some cases of captive proliferation, one should not simply try to reach as much genetic diversity as possible, but should follow the natural mate preferences of individuals.

This implies that conservation biologists give the animals the chance of choosing their mate from a larger sample. Although it is often not possible to transport animals from one zoo to another, odor samples could in many cases suffice for a preference test. A higher success rate for the more recently developed ART methods could be a by-product of such efforts.

18.6 Discussion

Sex is important in the coevolution of parasites and hosts for several reasons. First, it creates genetic diversity. This hinders parasites from evolving toward higher levels of virulence, because parasite adaptation to one host genotype is often of not much help against another one. Analogously, parasite diversity hinders the host from evolving toward very low levels of virulence. Both parties can therefore benefit from outbreeding as compared to random mating. A preference for outbreeding is already a very simple form of sexual selection. Several more sophisticated kinds of sexual selection that can be relevant for parasite–host coevolution have been proposed, some of them even after mate choice has occurred (i.e., before, during, and after fertilization). These possibilities have been investigated in some model species, especially with respect to MHC (i.e., a set of loci that is crucial in the parasite–host interaction). However, sexual selection not only concentrates on genetic aspects, but is also often connected to life history decisions about the use of resources, as, for example, in peahens that lay more eggs if mated to a more attractive peacock (Petrie and Williams 1993). Such plasticity in life history decisions is also expected to interact with the progress of a parasite–host interaction. Moreover, sexual selection can interact with sexually transmitted diseases. All this is expected to have implications for management decisions in the breeding programs for animals and plants.

It may be too early to reach strong conclusions that could directly be adapted to management decisions. However, the available evidence suggests that free natural mate choice could be important for the health of host populations. It may enable host populations to react to coevolving pathogens in evolutionary time. A consequence of free mate choice in a host may be that the next generations suffer less from the virulence caused by pathogens. I therefore suggest that this should be taken into consideration in the breeding programs of endangered species (e.g., Gibbons *et al.* 1995) and of farm animals. It could even be important in wild animals whose reproduction is artificially "supported," as, for example, in many fish species where ripe females are stripped and the eggs fertilized by sperm of any available male (and not necessarily one of the males the female would have chosen). On the other hand, the possibility of mate choice in some parasite species may deserve more attention since one of the criteria of such a mate choice may be the individuals' virulence genetics (analogous to the resistance genetics of the hosts). If so, the opportunity of free mate choice in a parasite population may increase its average degree of virulence.

19

Molecular Phylogenies and Virulence Evolution

Bruce Rannala

19.1   Introduction

The effective management and prevention of outbreaks of virulent strains of microbes depends on information about when, where, and how such strains arise. In the case of newly emerging human pathogens, this might involve tracing the source of a zoonotic infection to an animal population – such was the case with a hantavirus outbreak in the US state of New Mexico (Nichol *et al.* 1993). In other cases, an existing, possibly benign, microbe infecting humans or livestock may suddenly give rise to a highly pathogenic (virulent) strain – such was the case with the 1918 "Spanish" flu pandemic. In this second case, information about the mechanism by which virulence arose can provide practical guidance to epidemiologists developing strategies to prevent, or forestall, future epidemics. As well, information about the time and location of origination of virulent strains can inform us about how quarantine measures ought to be applied in the future, or how effective such measures have been in particular instances in the past.

In recent years, molecular phylogenies have come to play an increasingly important role in epidemiological studies of microbial pathogens, as they provide information about the location, timing, and mechanisms by which virulent strains arise. In particular, sequences from disease-causing viruses and bacteria that infect humans and livestock have been studied extensively, with hundreds of phylogenies published in medical and veterinary journals in 2000 alone. (A search of PubMed for articles published in 2000 that contained both the words "virus" and "phylogeny" produced 699 articles.) There are at least two major reasons for this rapid growth in the use of phylogenies by epidemiologists. The first is that many viruses [especially ribonucleic acid (RNA) viruses] and bacteria experience mutations at a much higher rate than eukaryotes. This difference is compounded by generation times that are typically orders of magnitude shorter. The expected substitution rate per-site per-year, which for neutral genes is roughly the per-site mutation rate divided by the generation time (in years), is therefore much higher for viruses and bacteria than for eukaryotes, even if, as is the case for certain bacteria, their mutation rates are roughly equal. Synonymous substitution rates per-site per-year for nuclear genes in mammals, for example, are about 10^{-9}, whereas rates for RNA viruses such as influenza A and human immunodeficiency virus (HIV) are about 10^{-2} (Li 1997). The high rates of substitution found in viruses and bacteria allow phylogenies to be reconstructed for sequences that have diverged only recently. Phylogeny has therefore become relevant to the questions typically addressed by epidemiologists such as the source of origin, and the rate of spread, of pathogenic strains of microbes. A second reason for the recent growth in the application of phylogenetics to microbial epidemiology is the development of new methods for isolating, amplifying, and sequencing nanogram quantities of deoxyribonucleic acid (DNA; and also RNA) obtained from blood or tissue samples, and, in particular, the polymerase chain reaction (PCR) method of amplifying DNA (Mullis 1986).

Advances in molecular genetic studies of viruses and bacteria have been paralleled by advances in the computational methods used to analyze nucleotide sequences and reconstruct phylogenetic trees of divergent strains or species. Explicit probabilistic models of nucleotide substitution have been developed (Jukes and Cantor 1969; Kimura 1980; Swofford *et al.* 1996) and used to derive quantitative statistical methods that allow phylogenies to be inferred for sequences under well-established criteria using likelihood (Felsenstein 1981) or Bayesian approaches (Rannala and Yang 1996). With these advances have come new opportunities to apply existing principles from the fields of evolutionary biology and population genetics to the effective management of virulent strains of microbes. This chapter focuses on two aspects of virulence management that have benefited from a phylogenetic perspective:

- Tracing when and where virulent strains arose;
- Identifying the genetic mechanisms by which they arose.

This information may often suggest practical forms of intervention to reduce the likelihood that virulent strains will emerge in the future. Examples are given that analyze sequences of virulent strains of influenza A from chickens and humans.

19.2 Phylogenetic Tools

To apply parametric statistical methods to estimate phylogeny using sequence data, a mathematical model is needed. The basic components of the model employed in a typical analysis are as follows:

- A set of potential phylogenetic trees, with branch lengths measured in units of the expected number of substitutions;
- A model of the process of nucleotide substitution that assigns a probability to any observed set of sequences given a phylogenetic tree (see Box 19.1).

In this chapter, the substitution model proposed by Hasegawa *et al.* (1985) is used, which allows for biases in transitional substitutions (e.g., A to T) versus transversional substitutions (e.g., A to C). The method of Yang (1994) is implemented with this model to allow for gamma distributed rate variation among sites (HKY+Γ). Likelihood ratio tests (LRTs; see below) can be used to choose a substitution model that best fits the observed sequences without introducing superfluous parameters (Goldman 1993), thus reducing the arbitrary aspects of model choice in a phylogenetic analysis. In this chapter, maximum likelihood (ML) is used to estimate the phylogenetic tree and branch lengths. The researcher chooses as the best estimates

Box 19.1 Likelihood methods for phylogenetic inference

The starting point for estimating a phylogenetic tree from DNA sequence data by ML is a sample of a aligned sequences, each n nucleotides in length. The sequence data may be summarized as an $a \times n$ matrix $X = \{x_{jk}\}$, where x_{jk} is the nucleotide at the kth site of the jth sequence. A data matrix of only three sequences might have the form

$$X = \begin{pmatrix} G & T & T & \ldots & C \\ C & T & T & \ldots & C \\ C & T & T & \ldots & A \end{pmatrix} ,$$

where $x_{11} = G$, $x_{21} = C$, $x_{22} = T$, etc. One of the three possible distinct rooted phylogenetic trees, to be denoted by τ_j, $j = 1, \ldots, 3$, for three sequences with branch lengths $V = \{v_1, v_2, v_3, v_4\}$ is

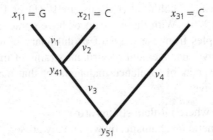

where the nucleotide observed at the first position of each sequence is shown at each tip of the tree (x_{11}, x_{21}, and x_{31}) as well as the ancestral nucleotides (y_{41} and y_{51}). The states of the ancestral nucleotides y_{41} and y_{51} are unobserved, and so the probability of the nucleotides at tips x_{11}, x_{21}, and x_{31} is calculated by summing over the probability obtained for each possible assignment of the ancestral nucleotides. The probability that the nucleotide observed at the root y_{51} is denoted by $\pi_{y_{51}}$, and is assumed to be that of the equilibrium distribution for the substitution model (this probability is usually estimated using the empirical frequencies of the nucleotides averaged over all sites in all sequences). The overall probability of the nucleotides observed at the first site in the example tree, which we denote as τ_1, is

$$\Pr(X_1 | \tau_1, V, \Theta) =$$

$$\sum_{y_{51}, y_{41}} \pi_{y_{51}} p_{y_{51}y_{41}}(v_3|\Theta) p_{y_{51}C}(v_4|\Theta) p_{y_{41}G}(v_1|\Theta) p_{y_{41}C}(v_2|\Theta) ,$$

where $X_1 = \{G, C, C\}$, $p_{AC}(v_j|\Theta)$ is the probability that nucleotide C is substituted for A over a branch of length v_j, and Θ is a vector of the parameters of the substitution model.

An example of a simple substitution model with a single parameter $\Theta = \mu$ is that of Jukes and Cantor (1969), which assumes that all possible nucleotide substitutions occur with an equal rate. A substitution model that often provides a

continued

Box 19.1 *continued*

good fit to real sequences, proposed by Hasegawa *et al.* (1985), has two parameters $\Theta = (\mu, \kappa)$, an overall substitution rate μ (proportional to the branch lengths), and a parameter κ, the bias in rates of transition versus transversion. Most methods of likelihood analysis assume that substitutions at different sites in a sequence are independent and identically distributed. The probability of observing the complete sample of sequences, X for a tree τ with branch length vector V, is then a product over the probabilities of the nucleotides observed at each successive site

$$\Pr(X|\tau, V, \Theta) = \prod_{j=1}^{n} \Pr(X_j|\tau, V, \Theta) .$$

The likelihood is defined as the probability of the observed sequences treated as a function of the model parameters τ, V, and Θ,

$$L(\tau, V, \Theta|X) = \Pr(X|\tau, V, \Theta) .$$

The likelihood function is maximized as a function of the parameters to obtain maximum likelihood estimates (MLEs) of the substitution model parameters, the branch lengths, and the phylogeny. The likelihood method proposed by Felsenstein (1981) estimates the branch lengths and parameters of the substitution model separately for each phylogenetic tree, and the MLE of phylogeny is chosen to be the tree with the highest relative likelihood. The logarithm of the likelihood is often used when probabilities are small.

---

of the phylogeny, branch lengths, and parameters of the substitution model those values that maximize the probability of the observed data (see Box 19.1).

Hypothesis tests using sequences

In a likelihood framework, one can also examine the support of the sequence data for different evolutionary hypotheses that may depend on the phylogeny, or the substitution model, by use of an LRT (reviewed by Huelsenbeck and Rannala 1997). The basic procedure is to calculate the relative probability of the observed sequence data under the null hypothesis versus the alternative hypothesis. Often the hypotheses in question are "nested," so that the null hypothesis is a special case of the alternative hypothesis. By considering the probability distribution of the LRT statistic under the null hypothesis, the significance of the value obtained for the sampled sequences can be determined. For nested hypotheses, the null distribution of the test statistic $-2 \ln \Lambda$, where Λ is the ratio of the likelihood under the null hypothesis to that under the alternative hypothesis, is approximately χ^2 with k degrees of freedom, where k is the difference in the number of free parameters under the null and alternative hypotheses. For non-nested hypotheses, the parametric bootstrap can be used to generate the null distribution of the likelihood ratio (see Goldman 1993).

In this chapter, LRTs are applied in some familiar ways, such as to test the fit of a molecular clock to sequence data, or of different models of substitution that incorporate effects such as transitional bias, or rate variation among sites (Goldman 1993; Huelsenbeck and Rannala 1997). The LRTs are also applied in some less familiar ways, such as to test whether virulent strains of a virus from a particular epidemic had a single recent (and perhaps local) origin, or instead were introduced multiple times, and to examine the agreement between phylogenies of viral sequences obtained using different genes to test whether recombination (exchanges of segments in individuals infected with multiple strains) is an important source of new virulent strains.

Molecular clock for virus sequences

The molecular clock hypothesis assumes that rates of substitution do not vary among phylogenetic lineages. If a molecular clock is imposed in a likelihood analysis, the branch lengths are constrained and the root of the tree chosen so that the sum of the branch lengths along any ancestor–descendent path from the root of the tree to any tip is the same. Many samples of viruses are temporally stratified – that is, they are made up of sequences isolated at different times (typically strains of viruses isolated in different years). As substitution rates for viruses and bacteria are so high, differences in sampling times can have an important effect on branch lengths. The likelihood-based molecular clock proposed by Felsenstein (1981), which assumes that the sequences are sampled simultaneously, is not expected to fit such data, even if substitution rates are constant among lineages.

Rate variation among lineages can be accommodated by performing an "unconstrained" likelihood analysis, in which the length of each branch in a tree (the product of the branch duration and the substitution rate) is treated as a separate parameter to be estimated jointly from the sequence data (Felsenstein 1981). Rambaut (1996) refers to this as the different rate (DR) model. Alternatively, if the times are specified at which the sequences are sampled, the ages of the tips of the tree can be set equal to the sampling times; this allows a joint estimation of the branch lengths under the constraint that the length of any path from the root of the phylogeny to any tip is proportional to the time at which the sequence at that tip was sampled minus the time at which the root ancestor existed. This can potentially distinguish a deviation from the molecular clock caused by differences in age among sequences from one caused by rate variation among lineages, and can allow the ages of common ancestors in the phylogeny, and of the root, to be estimated (Rambaut 1996). This model is referred to as the single rate, fossil sequence (SRFS) model (Rambaut 1996).

Sources of phylogenetic uncertainty

The most important factors that affect the accuracy of hypothesis tests involving phylogeny can be grouped into four broad categories:

■ Errors through the finite length of sequence sampled;

▪ Errors resulting from an inaccurate model of either the substitution process or the evolutionary process;
▪ Errors in sequence alignment;
▪ Errors from population sampling.

Sampling errors of the first type are adequately accounted for when using likelihood methods. It is more difficult to guard against errors of the second and third types. Improved models of nucleotide substitution can be used to account more adequately for many of the nonuniform substitution patterns commonly observed among sampled sequences, including transition versus transversion biases among nucleotides and rate variation among sites or genomic regions; LRTs can be used to decide when a more complicated substitution model provides a significant improvement in the fit of the sequence data. However, certain complications of the substitution process, such as intragenic recombination and nonindependence of substitutions among sites, cannot be easily accommodated using presently available methods; if such factors are important, overly simple models could potentially lead to incorrect conclusions (see the review by Huelsenbeck and Rannala 1997). Alignment errors are neglected in most studies, although, in some cases, they may be an important source of phylogenetic uncertainty (Goldman 1998).

The fourth source of uncertainty arises because phylogenies are typically constructed for a very small sample of sequences that represents only a fraction of the total population. This is particularly important for viruses and bacteria, which may have very large (and subdivided) populations even over a limited geographical scale. Each population sample has a phylogeny and branch lengths associated with it the form of which may vary substantially among samples. As a result, the outcomes of hypothesis tests involving phylogeny also typically vary from sample to sample, which introduces additional uncertainty into the analysis. In the strictest sense, the statistical tests discussed in this chapter, only apply to the samples of sequences and the models being considered and should not be too readily extended to the population of sequences in a geographical region or an epidemic as a whole.

19.3 Case Studies

Below we consider three case studies that show how the framework outlined above can shed light on the evolution of three related viral diseases.

Influenza A

The influenza A virus is of great medical and economic importance. In the 20th century alone, four successive pandemics of human influenza resulted in the deaths of between 20 million and 40 million persons. In addition, virulent strains of avian influenza A frequently arise that, in extreme cases, may kill millions of birds; the global costs of losses for poultry producers related to influenza A can reach millions of dollars annually. Influenza A is a negative-strand RNA virus with eight segments carrying a total of 10 protein-coding genes that make up the influenza A genome (see Voyles 1993). Two of these genes code for proteins that are expressed in the viral envelope, which is derived primarily from the host cell membrane.

These two proteins, hemagglutinin (HA) and neuraminidase (NA), are of critical importance as antigenic determinants of the host immune response. HA is a trimer that appears to be involved in host cell recognition, and NA is a tetramer that is possibly involved in mediating the release of newly formed viruses.

Influenza A was first isolated from humans in the 1930s, and it was recognized early on that different genetic variants, or epitopes, of the virus at the HA and NA loci elicited different immune responses. The variants were originally classified according to whether exposure to one produced antibodies that were cross-reactive to the other. Numerous epitopes were identified that were not cross-reactive, and such serologically novel strains were sequentially numbered according to their HA and NA types. Examples are H1N1 (1918 "Spanish" flu) and H2N2 (1957 "Asian" flu; see Levine 1992). It was soon recognized that antigenic shifts could occur through mutational changes to new subtypes (N1 to N2, H1 to H2, etc.), a process known as antigenic drift (see e.g., Both *et al.* 1983), as well as by the exchange of viral segments between strains in individuals infected with multiple strains (H1N1/H2N2 giving rise to H1N2, H2N1, etc.; see e.g., Li *et al.* 1992). Additionally, virulent strains of avian influenza are known to arise by mutations in the HA gene that increase its cleavability. This appears to be a critical step in facilitating the spread of the viral infection from the respiratory tract in progression to a more severe systemic infection (Bosch *et al.* 1981; Kawaoka *et al.* 1987).

Phylogenies have been used to study the evolution of virulent strains of influenza A in several different ways. The most common applications are in studies of the geographical or zoonotic origins of certain virulent strains (see e.g., Rohm *et al.* 1995), and to the study of the mechanisms by which virulent forms have arisen in particular cases – whether by reassortment of segments among strains in swine that were multiply infected with viruses from ducks and humans (see e.g., Yasuda *et al.* 1991), for example, or instead by point mutation (see e.g., Horimoto *et al.* 1995). Phylogenies have also been used in attempts to study the propensity of different lineages to give rise to new genetic forms that are novel antigens and potentially capable of producing a pandemic (Fitch *et al.* 1997). In this chapter, I consider some simple ways that phylogenetic trees can be used to study the question of where and when virulent strains arose, as well as the question of how (i.e., whether by recombination or point mutation).

Mexican chicken flu

The most recent major North American epidemic of highly virulent H5N2 avian influenza occurred in 1983–1984 among turkeys and chickens in Pennsylvania (Bean *et al.* 1985). The indirect costs of this epidemic to the poultry industry have been estimated at over a quarter of a billion dollars (Horimoto *et al.* 1995). In 1993, an outbreak of type H5N2 avian influenza occurred among Mexican chickens. Most isolates of the virus produced only mild respiratory symptoms and, for economic reasons, infected chickens were not eliminated, nor were infected poultry farms quarantined. As a result, the virus was able to spread unchecked and several highly pathogenic isolates ultimately appeared (Horimoto *et al.* 1995). At least two pathogenic strains of H5N2 from Mexican chickens isolated in 1994 and 1995 (labeled CP607 and CQ19, respectively, see below) appear to have arisen by an insertion in the HA connecting peptide, which rendered it highly cleavable (Horimoto *et al.* 1995).

Horimoto *et al.* (1995) examined HA gene segments for three H5N2 isolates from Mexican chickens. One of the isolates, CQ19, was highly pathogenic and contained an insertion coding for two additional amino acids at the HA cleavage site. A second, CP607, was mildly pathogenic and also contained the insertion. A third, CM1374, was nonpathogenic and did not contain the insertion. Horimoto *et al.* (1995) compared sequences encoding the HA1 subunit for these three Mexican strains as well as 10 additional strains from other regions of North America (the USA and Canada), Europe, and Africa, using a maximum parsimony method of phylogenetic analysis. An important epidemiological question these authors attempted to address is whether the virulent Mexican flu strains arose locally; in that case, they would share a most recent common ancestor (MRCA) unless some of their ancestral strains were reintroduced into Canada or the USA. If the strains arose in the USA or Canada, with a subsequent introduction into Mexico, they would not share an MRCA unless none of the intervening ancestral strains from the USA and Canada were sampled. In this section, I reanalyze a subset of the sequences originally examined by Horimoto *et al.* (1995) using an LRT to examine support for the hypothesis that the strains of H5N2 isolated during the recent Mexican chicken flu epidemic did not arise from strains in the USA and Canada. Additionally, I estimate the times at which the virulent strains arose.

Sequences for 11 of the isolates of influenza A examined by Horimoto *et al.* (1995) were obtained from Genbank and aligned using ClustalW (Higgins *et al.* 1991). The isolates are as follows: chicken/Mexico/26654-1374/94 (CM1374); chicken/Puebla/8623-607/94 (CP607); chicken/Queretaro/14588-19/95 (CQ19); A/chicken/Pennsylvania/13609/93 (CP13609); chicken/Florida/25717/93 (CFLA 93); A/ruddy turnstone/Delaware/244/91 (RD244); chicken/Pennsylvania/1/83 (CP1); chicken/Pennsylvania/1370/83 (CP1370); turkey/Ontario/7732/66 (TO66); tern/South Africa/61 (TS61); chicken/Scotland/59 (CS59).

An ML tree was constructed using the program PAUP* (Swofford 1998) and applying the HKY+Γ substitution model with no constraints on the branch lengths (molecular clock not enforced). The likelihoods obtained for these sequences using several different substitution models are shown in Table 19.1. The HKY+Γ model provided a significantly better fit to the sequences than the other models considered. Parameters of the model were estimated from the data. The shape parameter of the gamma distribution describing the among-site rate variation was $\alpha = 0.501$, indicating considerable rate variation. This may arise because, for influenza A viruses, antigenic sites may experience positive selection (and consequently increased substitution rates) by comparison with nonantigenic sites (Ina and Gojobori 1994). The transition–transversion bias was $\kappa = 8.085$. The ML tree has a log-likelihood of -5739.47, and places the Mexican isolates as forming a monophyletic group and sharing a MRCA with a subclade of isolates from birds

Table 19.1 Likelihood ratio tests of the fit of several models of substitution and the molecular clock to HA sequences of avian influenza A strain H5N2. The log-likelihoods under the null and alternative models are denoted by $\ln L_0$ and $\ln L_1$, respectively, and Λ is the ratio of the likelihoods under the null versus the alternative model. HKY denotes the Hasegawa *et al.* (1985) model of nucleotide substitution with $\kappa = 1$ (model 0: no bias in rates of transition versus transversion) and $\hat{\kappa} = 3.45$ (model 1: the value of κ estimated by ML). HKY+Γ denotes the HKY model with among-site rate variation following a gamma distribution with shape parameter α (Yang 1994). Note that $\alpha \rightarrow \infty$ (model 0) implies no rate variation among sites and $\hat{\alpha}$ is the MLE of the rate variation (shape) parameter (model 1). SRFS denotes the model of Rambaut (1996) (model 0), and DR is the Felsenstein model (1981), which allows different rates among lineages (model 1). ** denotes significance at the 0.001 level.

Model of DNA substitution	$\ln L_0$	$\ln L_1$	$-2\ln\Lambda$
Test of equal transition–transversion rate			
HKY($\kappa = 1$) vs HKY($\hat{\kappa} = 3.45$)	-6079.78	-5801.36	556.84**
Test of equal rates among sites			
HKY+Γ($\alpha = \infty$) vs HKY+Γ($\hat{\alpha} = 0.576$)	-5801.36	-5739.63	123.46**
Test of molecular clock			
SRFS vs DR	-4290.37	-4278.45	23.84**

in the eastern USA. Three other strains CP1, CP1370, and TO66, from chickens and turkeys in the USA and Canada, form a separate (monophyletic) group. The tree was rooted using two sequences from South Africa (TS61) and Scotland (CS59). The topology of the ML tree for the nine North American strains is identical to that of the tree shown in Figure 19.1, although that tree is a partially constrained (SRFS) tree with branch lengths shown proportional to time.

To test the hypothesis that the Mexican strains did not arise from a recent source in the USA or Canada, an LRT was performed. Under the null (constrained) hypothesis, the Mexican isolates do not share an MRCA (other than the root ancestor) with any subset of the strains from the USA and Canada, while under the alternative hypothesis, they may have any ancestry. The best tree under the constrained (null) hypothesis has a log-likelihood of -5746.06. The test statistic is $T = -2\ln\Lambda$, where Λ is the ratio of the likelihood under the null (numerator) versus alternative (denominator) hypotheses. For the HA1 sequences examined, $T = 13.18$. As the hypotheses are not nested, a parametric bootstrap method was used to evaluate the significance of T. A total of 100 simulated data sets were generated using the PAML program (Yang 1997), with the same substitution model as was used to analyze the original data (where MLEs of the parameters of the substitution model were substituted for the true parameter values). The model tree for the simulations was the phylogeny obtained under the constrained (null) hypothesis. The original aligned sequences were of variable length with several insertions and deletions inferred from a ClustalW alignment, and a program was written to remove missing data and insertions and deletions from the simulated sequences to make them identical to the original sequences. T was calculated for each simulated

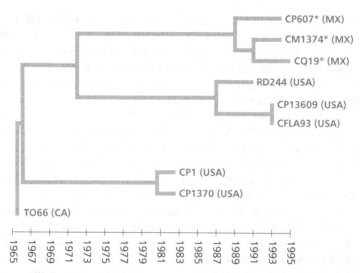

Figure 19.1 Maximum likelihood estimate of phylogeny of nine North American strains of H5N2 avian influenza. The strain abbreviations are given in the text. An asterisk indicates that a strain is pathogenic. The geographical sources for the strains are indicated by the abbreviations: MX = Mexico, USA = United States of America, CA = Canada. The divergence years were estimated using a partially constrained molecular clock (i.e., the SRFS model) and the branch lengths are calibrated in units of years.

dataset, and the proportion of times that the value of T obtained for the original dataset was exceeded by a value of T obtained for a simulated dataset was taken to be the significance of the test (i.e., a value of T at least as large as that observed for the original data would be observed under the null hypothesis with probability p, the significance). Since none of the simulated values of T exceeded the observed value, the null hypothesis can be rejected with $p \leq 0.01$. The Mexican strains do not appear to form a separate monophyletic group from the remaining North American strains. This agrees with the suggestion of Horimoto *et al.* (1995) that H5N2 influenza might have been introduced into Mexican chickens by their contact with migratory waterfowl from the USA; the RD244 strain shares an ancestor with the Mexican strains and is from a US shorebird.

One can also attempt to estimate the most recent time at which the Mexican strains of H5N2 might have been introduced from US birds. The program SPATULA (Rambaut and Grassly 1996) was used to calculate the likelihood of the tree in Figure 19.1, using only the nine North American strains and constraining the tips of the tree to be equal to the times at which the viral strains were sampled. This allowed the ages of the ancestors in the phylogenetic tree to be estimated. The likelihood under a clock hypothesis using the HKY+Γ substitution model, and allowing for the fact that the sequences have been sampled at different times using the SRFS model of Rambaut (1996), is $L_0 = -4290.37$. Relaxing the clock assumption by allowing each branch in the phylogeny to have a different substitution rate (Felsenstein 1981, the DR model), again using the HKY+Γ substitution model, gives $L_1 = -4278.45$. There are nine degrees of freedom (df) under the (null) SRFS model ($s - 1$ internal node times, where s is the number of sequences, and one overall rate parameter must be estimated; Rambaut 1996) and 15 df under the (alternative) DR model ($2s - 3$ branch-specific rates must be estimated; Felsenstein 1981). The LRT statistic for a test of the molecular clock is $T = 23.84$. Because we assume the phylogeny is the same under both hypotheses, the models are nested and the distribution of the test statistic is approximately χ^2 with 6 df (the difference in df between the null and alternative hypotheses). The SRFS molecular clock can be rejected in this case with $p \leq 0.01$. The SRFS tree with branch lengths scaled in units of years is shown in Figure 19.1. The estimated substitution rate per year is 5.18×10^{-4}. Ignoring possible rate variation among lineages indicated by a LRT, a rough estimate of the time at which the H5N2 flu strain might have been introduced into Mexico is about 1972. The results of this analysis suggest that, although nonvirulent H5N2 influenza was probably introduced to Mexico from the US, the virulent strains of H5N2 that subsequently appeared during the Mexican chicken flu epidemic likely arose locally. The practical implication of this result is that, to reduce the threat of a major outbreak of highly virulent H5N2 influenza, poultry producers should attempt to contain local outbreaks of even mildly pathogenic strains.

1918 Spanish flu and 1997 Hong Kong flu

The so-called "Spanish" influenza pandemic of 1918 resulted in the deaths of over 20 million people, with mortality rates over 25 times higher (about 2.5%) than for a typical influenza strain (about 0.1%). The reason for this virulence is not well understood. One suggestion is that the strain was not an unusual one, but global malnourishment and urban overcrowding following World War I created an immunologically suppressed population and conditions suitable for influenza transmission. The population of the USA was largely unaffected by the war in Europe, however, and yet still suffered high mortality during the 1918 influenza pandemic. Another suggestion is that the 1918 Spanish influenza pandemic was caused by a new highly virulent strain that arose by recombination between human (or swine) and avian strains. The earliest samples of human influenza date from the 1930s, and so genetic analysis has not, until recently, been available to study the origin of the 1918 influenza. Early analyses of antibody titers from survivors of the 1918 influenza did suggest, however, that the strain was probably an H1N1 subtype (see Taubenberger et al. 1997). Recently, Taubenberger et al. (1997) isolated viral RNA from paraffin-embedded tissue samples from a patient who died of 1918 influenza. They successfully amplified and sequenced fragments of several genes including HA and NA. The strain was designated A/South Carolina/1/18 (SP18).

In 1997, a highly pathogenic strain of chicken influenza, H5N1, emerged as a source of virulent influenza infections in humans exposed to infected chickens. At least 12 confirmed cases of human infection with the strain have since been documented, six of which were fatal. Subbarao et al. (1998) first isolated this virus from a 3-year-old boy, who subsequently died. The isolate, designated A/Hong Kong/156/97 (HK97), was sequenced for segments of several genes, including HA and NA, to investigate the genetic properties of the strain and, in particular, whether it arose by recombination between human (or swine) and avian strains. We analyze the HA and NA sequences for this strain, for the SP18 strain, and for several additional reference strains from humans, swine, and birds, to examine the evidence that either strain HK97 or SP18 arose by recombination between animal or human and avian strains. We also examine whether the SP18 strain shares a recent ancestry with the classic H1N1 strains isolated in the 1930s, as has been suggested.

Reference isolates that had been sequenced for both the HA and NA genes were chosen for the analysis. In the absence of recombination, the phylogeny of the isolates obtained by an analysis of each gene should agree; recombination can generate disagreements between the two gene trees. An LRT was used to quantitatively assess the evidence for recombination (different underlying gene trees) taking into account phylogenetic uncertainty (Huelsenbeck and Bull 1996). Eight strains were analyzed in total, including the HK97 and SP18 strains. The additional six strains are A/swine/Ehime/1/80 (SwEhm80), a swine influenza isolated in 1980 (H1N1); A/duck/Alberta/60/76 (DkAlb76), a North American duck influenza isolated in 1976; A/WSN/33 (WSN33), a mouse-adapted human influenza isolated in 1933 (H1N1); A/Puerto Rico/8/34 (PR34), a human influenza isolated in 1934 (H1N1); A/Yamagata/32/89 (FLA89), a Japanese swine influenza isolated in 1989 (H1N1); and A/WI/4754/94 (AWI94), a swine influenza with documented transmission to humans isolated in 1994 (H1N1).

Sequences were aligned using ClustalW (Higgins *et al.* 1991). The HA gene for SP18 was partially sequenced as three nonoverlapping fragments of variable length (Taubenberger *et al.* 1997), and these were combined to construct a composite HA sequence, with the unsequenced regions between fragments represented as missing data. An ML phylogenetic analysis was performed using PAUP* (Swofford 1998) for each gene separately and for a combined dataset of both genes. The ML tree for HA is shown in Figure 19.2.

The log-likelihood obtained in an unconstrained analysis (no molecular clock imposed; DR model) was −8315.52, with the transition–transversion bias estimated to be $\hat{\kappa} = 4.501$ and the shape parameter of the gamma distribution estimated to be $\hat{\alpha} = 0.573$.

The ML gene tree for NA is shown in Figure 19.3. The log-likelihood obtained for an unconstrained analysis was −6293.18, with the transition–transversion bias estimated to be $\hat{\kappa} = 6.208$, and the shape parameter of the gamma distribution estimated to be $\hat{\alpha} = 0.384$.

Both gene trees group together the human influenza sequences PR34, WSN33, and SP18. However, the HA gene groups the Japanese swine sequence FLA89 with the human sequences, whereas the NA gene does not. This suggests that the HA gene sequenced for FLA98 might have been introduced into this strain by recombination with a human strain. The HK97 strain diverges before the human

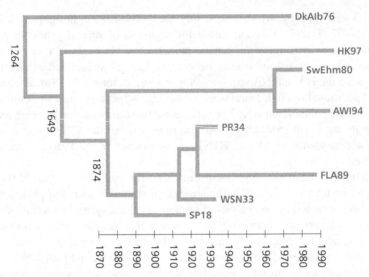

Figure 19.2 Maximum likelihood estimate of phylogeny of eight strains of influenza A isolated from humans, swine, and birds based on an analysis of the HA gene. The strain abbreviations are given in the text. The divergence years prior to 1870, estimated using a partially constrained molecular clock, are shown at the left of the branch. The branch lengths (after 1870) are calibrated in units of years (scale at bottom).

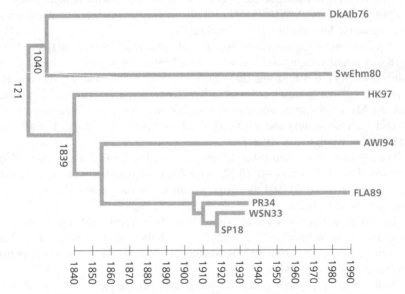

Figure 19.3 Maximum likelihood estimate of phylogeny of eight strains of influenza A isolated from humans, swine, and birds based on an analysis of the NA gene. The strain abbreviations are given in the text. The divergence years prior to 1840, estimated using a partially constrained molecular clock, are shown at the left of the branch. The branch lengths (after 1840) are calibrated in units of years (scale at bottom).

strains in both gene trees and before the swine strains as well, except in the NA gene tree, which places the SwEhm80 strain with the DkAlb76 strain; this could be either evidence for recombination of SwEhm80 with a duck strain or could be an error in the phylogeny, perhaps because of long branch attraction, as these sequences are very divergent. The HA gene tree of Figure 19.2 suggests that the human influenza strains PR34 and WSN33 share a recent ancestry with SP18, which could have arisen by recombination with an avian strain. The NA gene tree of Figure 19.3, on the other hand, suggests that the SP18 NA ancestor arose from an ancestral strain that is descended from the ancestor of PR34, and therefore is not of direct avian origin.

An LRT was used to test the hypothesis that recombination (exchange of segments) among strains, involving either the HA or the NA genes, has occurred at some point in their shared ancestry (Huelsenbeck and Bull 1996). Under the null hypothesis, the two gene trees are identical and the log-likelihood is −15281.29. Under the alternative hypothesis, each gene may have a different tree and the log-likelihood is the sum of the log-likelihoods obtained in the unconstrained analyses of the two genes, which is −14608.70. The LRT test statistic is then $T = 1345.18$, which is significant at the $p \leq 0.01$ level (based on 100 simulated datasets). The LRT, which takes into account phylogenetic uncertainty, therefore provides strong evidence for past recombination between strains.

The program SPATULA (Rambaut and Grassly 1996) was used to estimate the times at which different strains diverged under the SRFS model. For the HA gene, the log-likelihood under this model was −8330.06, with the rate of substitution estimated to be 1.48×10^{-3}. An LRT of the DR model versus the SRFS model gives $T = 56.64$, which is significant at the $p \leq 0.01$ level. The ML HA gene tree of Figure 19.2 has branch lengths scaled in units of years, and the estimated years at which different lineages diverged are indicated. This tree suggests that if SP18 arose by recombination with an avian lineage, this occurred quite recently (about 1890). The HK97 strain, on the other hand, appears to have diverged from the human and swine influenza strains roughly 200 years earlier. The ML NA gene tree of Figure 19.3 has a log-likelihood under the SRFS model of −6305.32, with the rate of substitution estimated to be 1.08×10^{-3}. An LRT of the DR versus SRFS model for this gene gives $T = 24.28$, which is significant at the $p \leq 0.01$ level. The human influenza strains appear to have diverged from the swine strains (apart from SwEhm80) in about 1910, and the HK97 strain appears to have diverged roughly 100 years earlier.

19.4 Discussion

In this chapter, several examples are given to illustrate how phylogenetic methods may be used to study the evolution of virulence. In the first example, a chicken influenza outbreak in Mexico, it is shown that a phylogenetic analysis strongly suggests that the virulent strains appearing during that epidemic originated locally; this is probably because a mildly pathogenic strain of H5N2 avian influenza was allowed to spread unchecked through the chicken population. This result suggests that measures should be taken to contain even mildly pathogenic outbreaks of chicken influenza when they arise to prevent the eventual evolution of more virulent forms.

In the example of two highly virulent influenza strains affecting humans, the 1918 Spanish influenza SP18 and the recent Hong Kong chicken influenza HK97, capable of infecting humans who came into direct contact with infected chickens, it is shown that, although recombination appears to be involved in creating new pandemics, it is not necessarily the cause of virulence, or infectivity, in these two strains. The SP18 strain appears closely related to other less-virulent human influenza strains at both the HA and NA loci, and a novel recombination event with an avian strain does not appear to be an explanation for its virulence (Taubenberger *et al.* 1997). It appears possible that the HA gene in all the human influenza strains arose by recombination with an avian strain, but this does not explain why SP18 is so much more virulent than the others. The HK97 strain, on the other hand, appears to be a typical avian influenza with genes at both HA and NA very distantly related to those of both human and swine strains. This suggests that HK97 is unlikely to become a pandemic strain in humans without first undergoing further genetic changes. There is still a significant risk that a recombination event between HK97 and a human influenza strain in an individual who is multiply infected could produce a highly virulent pandemic strain, however, and that risk alone makes a rapid response aimed at eliminating the HK97 strain from both chickens and humans of critical importance (Subbarao *et al.* 1998).

Part E

Multilevel Selection

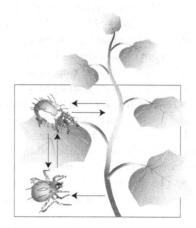

Introduction to Part E

The implications of multilevel selection for virulence evolution deserve closer attention, especially where selection leads to conflicts of interest between organisms at different organizational levels. In pathogen–host interactions this conflict is self-evident, but in numerous cases parasites have evolved to act as commensalists or even as mutualists. The latter case, however, does not imply that interests exactly match.

In Chapter 20, Hoekstra and Debets consider mutants of mitochondria that slow down the growth of their host, a bread mold fungus. In spite of this apparent disadvantage, these mutants outcompete normal mitochondria in crosses between fungi that contain wild-type and mutant mitochondria. If such crosses occur frequently enough in nature, the resultant intragenomic conflict may lead to the interesting phenomenon that, through the lower-level selection process, fungal hosts with relatively slow growth can increase in frequency in the population. Similar processes can occur via mitochondrial plasmids that cause senescence, a phenomenon normally absent in fungi. The persistence of these obviously harmful plasmids is striking in some genera of fungi; the key to understanding this observation is probably the existence of horizontal transmission by anastomosis between different fungal units. Hoekstra and Debets suggest that, once horizontal transfer is open to manipulation, the performance of fungal diseases could be changed through intragenomic conflict.

Chapter 21 describes the evolutionary dynamics of the tritrophic interaction between chestnut trees, a blight fungus, and a double-stranded RNA virus. Whereas the fungus infects the chestnut trees and greatly reduces their growth, the virus infects the fungi and greatly reduces their virulence. The virus is transmitted vertically, but also horizontally by anastomosis between fungi infesting the same tree. So, even though the virus more or less debilitates its host, the infected fungi do not necessarily lose the competition with other fungi within a tree through horizontal transmission. The virus therefore appears to be a potential candidate for the natural biological control of chestnut blight. However, Taylor shows in this chapter that only a careful analysis of the various trade-offs, of the rates of horizontal and vertical transmission, and of feedbacks involved in this system can explain why, despite the presence of the virus, chestnut blight continues to be such a devastating disease in the USA, whereas it is controlled by the virus in Southern Europe.

Multilevel selection not only has the potential to generate conflict of interest, but also forms of "conspiracy" may arise. This can happen when organisms are hierarchically organized in more than two trophic levels in a linear food chain: species at different trophic levels can then join forces against those sandwiched between them.

In Chapter 22, Sabelis, Van Baalen, Pels, Egas, and Janssen analyze how plants and predators evolved to conspire against herbivores. Plants invest in attracting, retaining, feeding, and protecting the herbivore's enemies; as this occurs in so many plant species it may help explain why herbivores are predator-controlled, and why, therefore, "the world is green." The authors ask why plant–predator mutualisms are ubiquitous and model the tritrophic interaction as a series of games: defensive allocations among neighboring plants, avoidance of plant defense and predation risk among herbivores, and resource exploitation among herbivores and among predators. They predict low-cost plant defenses when herbivores and predators are sufficiently mobile, and prudent (imprudent) exploitation strategies when single (multiple) strains exploit the same resource: what matters is the degree to which a strain monopolizes exploitation of a resource. The model needs extension to include population dynamics, and the outcome of such a model is not at all self-evident – plant–predator mutualisms therefore do not simply evolve because "it is both in the interest of plants to be rid of the herbivores and in the interest of the predators to find herbivores as prey."

Multilevel selection inevitably plays an important role in molding organismal traits in a way that we would not be able to understand by considering one-level selection only. In this sense multilevel selection poses a challenge to the experimental biologist to identify those biological levels at which relevant selection pressures may operate. Indeed, most organisms may harbor influential "passengers," and one may wonder which organisms are actually passenger-free. Moreover, organisms are part of a food web, so selection within the population of one organism influences that of others in the food web and vice versa. Theoreticians can help to predict phenomena from first principles (i.e., natural selection); when these predictions are not compatible with biological observations, possible causes and alternative mechanisms have to be identified. Multilevel selection is a likely candidate to help achieve this.

20

Weakened from Within: Intragenomic Conflict and Virulence

Rolf F. Hoekstra and Alfons J.M. Debets

20.1 Introduction

Pathogen virulence is a product of selection that operates not only at the level of the pathogen, but also at those of the host and other interacting populations. For example, if a pathogen population within a host contains genetic variation for virulence, selection may favor the most virulent type. However, selection among infected hosts favors host individuals that resist pathogens virulent against other hosts. Once these resistant hosts have become more numerous, they effectively decrease the pathogen's virulence. The combined outcome in such a two-level selection system is likely to be some intermediate level of virulence (see Chapter 17). A more complex situation of multilevel selection involving three levels is discussed in Chapter 22.

In this chapter we focus on a special type of multilevel selection, in which natural selection operates simultaneously on several levels *within* an organism. Organisms can be viewed as nested hierarchies of replication levels. A multicellular organism contains cells; cells contain nuclei and mitochondria; nuclei contain chromosomes; chromosomes contain genes and noncoding sequences. Mitochondria also contain chromosomes, which contain genes. The important point is that replication takes place at all these levels. Moreover, all these structures possess heredity and may vary within the higher-level unit that contains them: an organism contains different cell types, and a cell may contain genetically different mitochondria. Thus natural selection may operate at many of these levels. A familiar example of multilevel selection *within* an organism is cancer. When a genetic variant arises by somatic mutation in the population of, say, intestinal epithelial cells and shows aberrant continued cell division, it may outcompete the normal epithelial cells and form a tumor. In this case natural selection at the level of the epithelial cells favors the cancerous type, while selection at the individual level works against this cancer cell type. Thus there is *conflict* between selection at the cellular level and selection at the individual level. In this example the outcome is clear: although the cancer cell type may enjoy a temporary advantage (at the cell level), it ultimately loses the competition with normal cells, because individuals who contain tumors have a lower fitness than those without tumors. In Section 20.2 we discuss more fully an example of two-level selection that generates intragenomic conflict within individuals and we point out some essential aspects that may be relevant for virulence management.

20.2 "Poky" Mutations in *Neurospora crassa*

In 1952 Mitchell and Mitchell discovered a slow-growing mutant strain of the bread mold *Neurospora crassa* and called it *poky*. The slow growth appeared to be associated with impaired mitochondrial function. Crosses between wild-type and poky strains suggest that the trait is maternally inherited: if the female parent is poky and the male parent is wild-type, all progeny colonies are poky. The reciprocal cross produces normal wild-type colonies. Heterokaryons that result from somatic hyphal fusions between a poky and a wild-type strain initially appear to grow normally, but as growth proceeds the heterokaryon and its asexual descendants become progressively more abnormal until they finally exhibit the poky phenotype. Somehow the defective poky mitochondria seem to outcompete or suppress the normal mitochondria in the heterokaryon. This situation can be viewed as a clear example of intragenomic conflict. At the level of mitochondria, poky is selectively favored over the wild type, but at the level of the individual fungal colonies, poky is selected against because of its adverse effect on the growth rate and reproduction.

N. crassa is not a pathogen, but if it were, poky variants would likely represent strains of reduced virulence because of their poor growth. Moreover – and this is a very important aspect – if interindividual hyphal fusions (a process called anastomosis) leading to the formation of heterokaryons occurred at a sufficient rate, poky could reach non-negligible frequencies in the fungal population because of its suppressiveness with respect to wild-type mitochondria. One of the most interesting aspects of within-individual genomic conflict is that phenotypes with relatively low fitness at the individual level may nevertheless increase in frequency, driven by the lower-level selection process.

20.3 Senescence Plasmids in Fungi

Many plant pathogens (and few animal pathogens) are fungi. Fungi normally do not senesce and are capable of unlimited somatic propagation. However, in a few genera (*Neurospora*, *Podospora*, and *Aspergillus*) senescence has been observed (for a review see Griffiths 1992). These genera are among the best-studied fungi in the laboratory, so the actual occurrence of senescent fungal strains may well be more widespread. In *P. anserina* all natural strains appear to senesce. The senescent phenotype shows up after sporulation by a slowing down of mycelial growth, pigmentation changes, and finally death of the mycelium. In *N. intermedia* and in *N. crassa* polymorphism for senescence occurs: some strains senesce while others do not. Although the molecular details are different, in all fungal species investigated senescence is associated with the presence of mitochondrial plasmids. It has been shown that senescence is caused by progressive degeneration of mitochondrial function as a consequence of mutations in the mitochondrial DNA induced by these plasmids (Griffiths 1992). The concentration of plasmids increases during mycelial growth. The precise mechanisms involved in this within-mycelial spread of the senescence-causing plasmids are not yet fully understood. Thus, although the details differ, fungal senescence is superficially analogous to the above-described poky phenotype in *N. crassa*: progressive impairment of mitochondrial function caused by the within-organism spread of genetic elements that have a harmful effect on the individual organism. Mitochondrial inheritance is uniparental (through the maternal parent), which implies that sexual reproduction cannot contribute to the spread of senescence plasmids in fungal populations. In crosses between a senescent and a nonsenescent strain the offspring inherits the senescence plasmids on average half of the time, which implies that the transmission from parents to offspring has no systematic upward or downward effect on the frequency of senescent strains in a population. The average lifetime of a senescent colony is shorter than that of a nonsenescent fungus. No other aspects of the life history of the fungus seem to be affected by the senescence plasmids. A senescent individual, therefore, has a lower reproductive output than a nonsenescent individual, so some force(s) must be operating to counter the selective elimination of these plasmids. A likely candidate is horizontal transmission of genetic material between fungal individuals, which is possible if they are vegetatively compatible. Then the hyphae of both colonies can readily fuse, creating cytoplasmic continuity between the colonies. Vegetative compatibility between different individuals is rare, but a low rate of horizontal transmission of senescence plasmids has been demonstrated also in vegetatively incompatible combinations (Debets *et al.* 1994).

20.4 Population Genetics of Senescence Plasmids: A Model

For a better insight into the dynamics of senescence plasmids in a fungal population, we analyze a simple population genetic model. In this model we can incorporate the fitness effects resulting from senescence, and study the effect of the rates of novel plasmid infection and of production of plasmid-free spores by infected strains.

We assume a population consists of two types of colonies, either infected with senescence plasmids (relative frequency i) or nonsenescent (relative frequency $1-i$). Furthermore we suppose that all fungal colonies are subject to the following life cycle (Figure 20.1): upon germination they may encounter a conspecific during their vegetative growth; following this they sporulate, giving rise to the next generation. Pairwise contacts occur randomly.

Therefore two senescent strains meet with a probability proportional to i^2, two nonsenescent ones with a probability proportional to $(1-i)^2$, and a senescent strain grows together with a nonsenescent one with a probability proportional to $2i(1-i)$. We assume that in the third category close contact may result in infection of the nonsenescent colony with probability β, either as a consequence of anastomosis (in the case of vegetative compatibility), or otherwise. Then all colonies sporulate; a senescent colony produces $1 - s_{sel}$ times the number of spores from an uninfected colony. Finally, we assume that a fraction θ of the spores produced by a senescent colony fail to include senescence plasmids (such spontaneous loss has been observed to occur at a low rate in *Neurospora*).

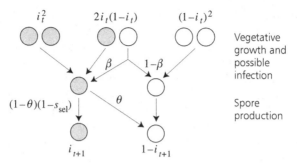

Figure 20.1 Schematic model of the dynamics of senescence plasmids in a fungal population. Gray circle = senescent mycelium, white circle = nonsenescent mycelium, i_t = fraction of senescent strains at generation t, β = rate of horizontal transmission of senescence plasmids to nonsenescent strains, s_{sel} = selective disadvantage caused by senescence, θ = spontaneous loss of senescence plasmids.

From these assumptions we deduce the following equation for the change of the relative frequency of senescent strains over one generation,

$$i_{t+1} = \frac{(1 - \theta)(1 - s_{sel})i_t[1 + \beta(1 - i_t)]}{1 - s_{sel}(1 + \beta)(1 - \theta)i_t(1 - i_t)} \, . \tag{20.1}$$

Solving Equation (20.1) analytically is possible but yields no intuitive insight. Instead we summarize the following conclusions based on linearization at sufficiently small values of i_t and standard stability analysis.

1. If $\theta = 0$ (no spontaneous loss and plasmids are included in all spores produced by a senescent colony), then a stable coexistence of senescent and nonsenescent strains is not possible. Eventually there will only be nonsenescent strains [if $\beta < s_{sel}/(1 - s_{sel})$] or only infected strains [if $\beta > s_{sel}/(1 - s_{sel})$].
2. If $\theta > 0$ (at least some plasmid-free progeny from senescent colonies), then two outcomes are possible:

   2a. If $(1 - \theta)(1 - s_{sel})(1 + \beta) > 1$ (i.e, if the rate of horizontal transfer β is sufficiently high to compensate the virus loss caused by spontaneous loss and by an impaired fitness of senescent colonies), then a stable coexistence of senescent and nonsenescent strains is possible.
   2b. If $(1 - \theta)(1 - s_{sel})(1 + \beta) < 1$ (i.e., if the rate of horizontal transfer β is too low), then the plasmids are expected to disappear from the population.

20.5 Intragenomic Conflict and Virulence Management

This chapter considers the potential exploitation of intragenomic conflicts for the management of pathogen virulence. It is necessarily rather speculative because no cases are known in which intragenomic conflict affects pathogen virulence. However, we believe that the key idea is well supported by empirical evidence. This shows that some forms of intragenomic conflict can have two consequences that form an interesting combination in terms of virulence management: a reduction of individual fitness and (despite this reduction) a possible spread of this trait in the population. The reduction in individual fitness in a pathogen probably implies a reduction in virulence, while the spread of reduced virulence is exactly the goal of virulence management. The snag, of course, is that well-studied cases of genomic conflict, such as those discussed in this chapter, are all in nonpathogenic organisms. However, this is probably largely because the model laboratory organisms used in fungal genetics (yeast, *Neurospora*, *Aspergillus*) are nonpathogenic. Similar phenomena may also occur in pathogenic fungi. Moreover, it may be worthwhile to try to transfer genetic elements that are effective in creating intragenomic conflict into pathogens. In this respect it is interesting that the mitochondrial plasmid that causes senescence in *N. intermedia* has been transferred into the related species *N. crassa* (Griffiths *et al.* 1990). This proves that interspecies transfer of such elements is possible, at least between related species, and that these elements are stably maintained in the new host, producing the same phenotype as in the original host. Interspecies transfer of senescence plasmids is currently being studied in our laboratory.

20.6   Discussion: Host Senescence and Pathogen Virulence

It is not accidental that our discussion so far has centered on fungi. We believe fungi are preadapted for intragenomic conflict, because of their inherent capacity for somatic fusion (anastomosis). Anastomosis allows horizontal transfer of genetic elements, which can provide an essential aspect in the genomic conflict because it may be the route for interindividual spread of the "selfish" lower level "player" in the conflict. It is true that anastomosis is frequently prevented by vegetative (somatic) incompatibility between strains, but this seems to fully block only the transfer of nuclei. Where two vegetatively incompatible strains meet, the cells are locally killed, but in most cases this is a postfusion response, leaving a short time window for effective cytoplasmic contact. Some "leakage" of cytoplasmic elements, like viruses and mitochondrial plasmids, has been demonstrated (Griffiths *et al.* 1990; Debets *et al.* 1994).

Can the connection between senescence and virulence, as hypothesized here for fungal systems, be generalized to other organisms? Senescence in animals is widespread and – in contrast to the situation in fungi – not caused by a genetic element that is potentially infectious. Instead, an unknown but probably large number of genes is involved in producing the senescent phenotype in animals. General evolutionary considerations (see, e.g., Williams 1957; Kirkwood and Holliday 1979) predict a trade-off between the level of somatic maintenance and reproductive output: high early fecundity is associated with a relatively short life span and decreased early fecundity implies an increased life span. These theoretical ideas have been supported by empirical evidence (e.g., Zwaan *et al.* 1995). Thus it is likely that genotypes characterized by a fast rate of senescence are relatively fertile early in life, while long-lived genotypes are less fertile. In either case, the senescent state is characterized by the effects of (probably many) mutations, with deleterious effects at late age. It is thus likely that older hosts have an increased susceptibility to pathogens. From these general considerations we can infer potential evolutionary connections between senescence and virulence at two levels. First, selection pressures on pathogen virulence may change depending on the host's age. The relative frequency of senescent individuals in a host population may thus influence the evolution of pathogen virulence. Quantitative models that address this problem specifically are required to explore in which direction pathogen virulence might change because of variation in relative abundance of senescent hosts. Second, parasite species that show senescence themselves may be subject to the above-mentioned trade-off between fecundity and longevity. In many instances high parasite fertility implies high virulence, which suggests that highly virulent strains have a shorter life span. If this is true, virulence management should try to implement measures that select for a longer parasite life span, which is certainly not easy to do. It requires creating conditions in which long-lived pathogen variants outcompete short-lived variants.

To summarize, although as yet there are no recorded examples of reduced virulence that spreads through a pathogen population by intragenomic conflict, the potential is there. It follows that a profitable strategy may be to search for intragenomic conflict that occurs in pathogens or to induce such conflict by introducing genetic elements that may be suitable to generate genomic conflict, like the fungal senescence plasmids discussed above. Another approach is to study ways to enhance horizontal transfer rates, as this would greatly increase the scope for intragenomic conflicts of the type discussed here. However, such manipulation is still a remote possibility since it requires a much deeper understanding of the mechanisms that normally prevent such somatic genetic transfer, or of the controlled use of (viral) vectors that could mediate horizontal transfer, than is currently available.

21

Ecology and Evolution of Chestnut Blight Fungus

Douglas R. Taylor

21.1 Introduction

There is a vast body of literature on the ecology and evolution of infectious diseases. Most of this literature, however, focuses on single species of hosts and pathogens, and, more specifically, on the dynamics of pathogen transmission among hosts, within-host dynamics, the evolution of host resistance, and the evolution of pathogen virulence (Levin and Bull 1996). It is often overlooked that in many natural populations, diseases exist within a complicated ecological community involving various forms of competing pathogens, pathogens of pathogens, etc. Hyperparasitism, for example, is an interaction that involves a pathogen that is in turn infected with a parasite of its own – that is, a hyperparasite. Hyperparasites may have deleterious effects on pathogens and, thereby, affect other species down the trophic chain in a manner similar to the top-down effects that predators can have on communities (Hairston et al. 1960; Fretwell 1977; Oksanen et al. 1981; Powers 1992; see Holt and Hochberg 1998).

Hyperparasitism is widespread in natural populations, and it is important in management situations. Hyperparasitoids, for example, frequently occur in arthropod food chains, often undermining efforts to employ parasitoids as biological control agents (Beddington and Hammond 1977; May and Hassell 1981). Many of the numerous examples that involve hyperparasitism of fungal pathogens of plants (Hollings 1982; Buck 1986) have attracted attention as potential agents of biological control (Nuss 1992). Many bacterial species are also infected by various plasmids and viruses (Levin and Lenski 1983), some of which might be useful for controlling infections (Levin and Bull 1996). The entire discipline of biological control involves the application of predation (e.g., ladybird beetles to control aphids) and hyperparasitism (e.g., *Bacillus* to control the gypsy moth) to the management of diseases and pests. In fact, given that the distinction between hyperparasitism and predation is often rather arbitrary, especially in the mathematical sense (Holt and Hochberg 1998), it is surprising that interactions among trophic levels have not received greater attention in studies of host–pathogen systems.

The theoretical work carried out on hyperparasitism emphasizes the conditions under which hyperparasites can invade, limit the density of their pathogen hosts, and influence the dynamical stability of the system (Beddington and Hammond 1977; May and Hassell 1981; Hochberg and Holt 1990; Holt and Hochberg 1998), rather than the effect that hyperparasites may have on pathogen virulence (Taylor *et al.* 1998a). In this chapter, I examine the implications of trophic interactions for the evolution and management of disease virulence. Additionally, I discuss how a general model can be applied to a specific experimental system – in this case, the chestnut blight host–pathogen system – and the implications this has for virulence management.

21.2 Ecology and Evolution of Virulence with Hyperparasites

For convenience, I distinguish between the ecological versus the evolutionary effects hyperparasitism can have on pathogen virulence. The ecological effects refer to the debilitating effect hyperparasites have on their pathogen hosts, which can reduce the severity of infection by the pathogen. In a formal sense, this is no different from models that consider the numerical dynamics of hyperparasites. This is relevant to virulence evolution, however, because:

- A reduced pathogen load resulting from hyperparasitism can be mistaken for an evolutionary reduction in pathogen virulence (especially when the hyperparasite is intracellular and, hence, difficult to detect);
- The fate of a hyperparasite that reduces pathogen transmission and virulence will be influenced by how those features affect the fitness of its pathogen host (Box 21.1).

One of the most interesting aspects of the ecology of hyperparasites is that the conditions that favor hyperparasitism, and therefore reduced pathogen virulence, are those that favor higher virulence in conventional evolutionary models (Taylor *et al.* 1998a). Models for the evolution of virulence generally conclude that higher virulence is favored when the host population density is high and when multiple infections per host are common. Higher host density increases the opportunity for horizontal transmission, which shifts the selective balance to favor rapid infection of new hosts rather than the continued survival of infected hosts (Lenski 1988; Lenski and May 1994). Multiple infections within a host promote competition between pathogen genotypes, which tends to favor faster growing, more virulent pathogens (Bremermann and Pickering 1983; Nowak and May 1994; see also Frank 1992a, 1996c; Herre 1993; Van Baalen and Sabelis 1995a). However, high host density and multiple infections within a host also promote the transmission of a hyperparasite, which reduces pathogen virulence. From the standpoint of a hyperparasite, a high density of virulent pathogens is a resource to be exploited, and any factor that increases the density of this resource will increase hyperparasite transmission. Multiple infections within a single host, on the other hand, place the virulent and hypovirulent (hyperparasitized) pathogens in close contact within a single host and, therefore, provide a greater opportunity for hyperparasites to spread via horizontal transmission (Taylor *et al.* 1998a).

The evolution of virulence is more complicated when hyperparasitism occurs, because the pathogen and the hyperparasite may each have some optimum balance of transmission and virulence that maximizes their own fitness. Additionally, the

Box 21.1 Modeling hyperparasitism and disease virulence

To illustrate models of hyperparasitism, we consider the canonical model by Anderson and May (1981) of hosts and pathogens, with the addition of a parasite of the pathogen – that is, a hyperparasite. The interactions between parasite and hyperparasite can be described by modifying a simple model of competing pathogens (Levin and Pimentel 1981; see also Box 9.2, and for applications see Hochberg and Holt 1990; Taylor *et al.* 1998a).

Suppose there are three classes within the host population: uninfected hosts S, hosts infected by a virulent pathogen I, and hosts infected with a pathogen that is itself infected by a hyperparasite H. The rates of change in the densities of these three classes can be described by the following set of equations,

$$\frac{dS}{dt} = bN - \beta_V SI - \beta_H SH - dS , \tag{a}$$

$$\frac{dI}{dt} = \beta_V SI - \sigma IH - dI - \alpha_V I , \tag{b}$$

$$\frac{dH}{dt} = \beta_H SH + \sigma IH - dH - \alpha_H H . \tag{c}$$

Hosts are born uninfected at rate bN, where N is the total density of hosts, $N = S + I + H$. The rate of infectious transmission of a pathogen that is hyperparasite free is denoted by β_V and the rate of infectious transmission of a hyperparasitized pathogen by β_H. All hosts die at rate d through causes other than infection, while infected hosts are subject to an extra mortality at rate α_V for pathogen type I and α_H for pathogen type H.

The model is most appropriate when the hyperparasite is an internal microparasite, and it incorporates the possibility that the parasite is both horizontally and vertically transmitted. Vertical transmission occurs at rate β_H when a pathogen that carries a hyperparasite H establishes a new infection. Horizontal transmission occurs at rate σ when a hyperparasite establishes itself within an existing infection. This general formulation has been used to study the conditions under which hyperparasites invade and persist, the dynamical stability of the system, and the effects of hyperparasitism on pathogen virulence (Hochberg and Holt 1990; Holt and Hochberg 1998; Taylor *et al.* 1998a).

To examine the effects of hyperparasitism of pathogen virulence, I introduce the assumption, conventionally made in models of virulence evolution, that there is a relationship between pathogen transmission β_j and virulence α_j, with $j = V, H$. Specifically, I assume that the extra death rate α_j caused by infection is a quadratic function of the pathogen's rate β_j of transmission (Lenski and May 1994),

$$\alpha_j = c_0 + c_1 \beta_j + c_2 \beta_j^2 . \tag{d}$$

Notice that the parameters c_0, c_1, and c_2 above are constants, reflecting the idea that there is a single functional relationship between transmission and virulence for all pathogen types. The hyperparasite, therefore, debilitates the pathogen and reduces

continued

Box 21.1 *continued*

pathogen transmission and virulence, thereby converting virulent infections I to hypovirulent infections H.

Given these assumptions and conditions, it follows that a rare hypovirulent (hyperparasitized) pathogen can invade an equilibrium population of virulent pathogens, $\frac{1}{H}\frac{dH}{dt} > 0$ at $I = I^*$, if

$$\sigma > \frac{(\beta_V - \beta_H)(d + c_0 - c_2\beta_H\beta_V)}{\beta_V I^*}. \tag{e}$$

Similarly, a rare virulent pathogen can invade an equilibrium population of hypovirulent pathogens, $\frac{1}{I}\frac{dI}{dt} > 0$ at $H = H^*$, if

$$\sigma < \frac{(\beta_V - \beta_H)(d + c_0 - c_2\beta_H\beta_V)}{\beta_H H^*}. \tag{f}$$

Coexistence of the two pathogen types occurs when either pathogen invades when rare, that is, when the two inequalities above are both satisfied. Notice that the coefficient c_1 does not enter the invasion conditions; the corresponding linear terms of the two pathogens cancel in the course of the calculation.

These two invasion conditions illustrate several of the general features that influence systems involving hyperparasites. First, hyperparasites are obviously more likely to spread when both transmission rates are high (high β_H, high σ), whereas coexistence is most likely when the hyperparasites specialize to rely on horizontal transmission (low β_H, high σ). Second, hyperparasites are more likely to spread when the density of the resident pathogen is high (virulent pathogens are essentially resources that hyperparasites exploit via horizontal transmission), but a virulent pathogen is more likely to spread when the density of hyperparasites is low. Finally, the spread of the hyperparasite is influenced by the impact the reduction in pathogen virulence has on pathogen fitness (Anderson and May 1981). When selection on the pathogen favors maximal transmission ($c_2 = 0$), a hyperparasite invariably lowers pathogen fitness, and the hyperparasite requires horizontal transmission to persist. However, when selection favors intermediate pathogen transmission ($c_2 > 0$), the hyperparasite may push the pathogen closer to its evolutionarily stable strategy. In that case, the hyperparasite is essentially a mutalist of the pathogen and does not require horizontal transmission to spread.

fitness of each of the two organisms may be influenced by the transmission properties of the other. There may also be resistance of the host (to pathogens) or the pathogen (to hyperparasites), as well as host or pathogen recovery.

To simplify the situation somewhat, consider the scenario in which there is a trade-off between pathogen transmission and virulence that defines an evolutionarily stable strategy from the perspective of the pathogen (Anderson and May 1981). We can then ask under what circumstances the evolutionary interests of the hyperparasite are the same as or are in conflict with those of the pathogen. Consider an equilibrium population with uninfected hosts S and hosts infected with pathogens that carry hyperparasite j, H_j. β_j is the rate of infectious transmission of a pathogen that carries hyperparasite j. All hosts die at the rate d through causes unrelated to infection. Infected hosts are subject to extra deaths, which are a quadratic function of pathogen transmission (i.e., $c_0 + c_1\beta_j + c_2\beta_j^2$; see Box 21.1 for further details). The per capita rate of change of a rare hyperparasite, k, is

$$\frac{1}{H_j}\frac{dH_j}{dt} = \beta_k S^* - d - (c_0 + c_1\beta_k + c_2\beta_k^2) + \sigma_{jk}H_j - \sigma_{kj}H_j \,, \qquad (21.1)$$

where σ_{jk} is the rate that hyperparasite k displaces hyperparasite j via horizontal transmission, and vice versa for σ_{kj}. Obtaining the equilibrium density of unin-fected hosts, $S^* = (d + c_0 + c_1\beta_j + c_2\beta_j^2)/\beta_j$, by setting Equations (a) to (c) in Box 21.1 equal to zero, and rearranging Equation (21.1), we obtain

$$\frac{1}{H_k}\frac{dH_k}{dt} = \frac{(\beta_k - \beta_j)(d + c_0 - \beta_j\beta_k c_2)}{\beta_j} + H_j(\sigma_{jk} - \sigma_{kj}) \,. \qquad (21.2)$$

Whether or not a new hyperparasite can invade – that is, whether the right-hand side of Equation (21.2) is positive – depends on some combination of vertical transmission (and how it affects pathogen fitness) and horizontal transmission. To illustrate this, consider two contrasting situations. In the first, assume that the two hyperparasites are equal in their ability to displace each other via horizontal trans-mission [$\sigma_{jk} = \sigma_{kj}$, and the second term in Equation (21.2) is zero]. If selection favors higher transmission in the pathogen ($c_2 = 0$), then a new hyperparasite can only invade if it allows the pathogen to have a higher rate of transmission ($\beta_k > \beta_j$). If selection favors some intermediate transmission in the pathogen ($c_2 > 0$), then a hyperparasite that further reduces pathogen transmission can in-vade, but only if the transmission rate conferred by the new hyperparasite is closer to the pathogen's evolutionarily stable strategy (ESS; which is $\sqrt{(d + c_0)/c_2}$, see Taylor *et al.* 1998a). Thus, if the hyperparasites do not compete with each other via horizontal transmission, selection favors hyperparasites that are less detrimen-tal to their pathogen hosts. In the second case, assume that the two hyperparasites have the same rate of vertical transmission [$\beta_k = \beta_j$, and the first term in Equa-tion (21.2) is zero]. In this case, selection favors a mutant hyperparasite if it has a higher rate of horizontal transmission (i.e., $\sigma_{jk} > \sigma_{kj}$). In the absence of any constraints, therefore, hyperparasite evolution should tend to maximize both hor-izontal and vertical transmission – the latter by minimizing its deleterious effects on pathogen fitness. It is straightforward to imagine that if hyperparasites have any element of vertical transmission at all, debilitating the pathogen is not evolu-tionarily stable unless there is some negative trade-off between the two modes of transmission.

From the perspective of virulence management, biological control efforts seek hyperparasites that can persist in populations while maximizing deleterious effects on pathogen fitness. Therefore, one favorable situation for evolutionarily stable biological control is when the reduction in pathogen fitness is an unavoidable con-sequence of horizontal transmission by the hyperparasite. Although horizontal transmission of the hyperparasite then reduces its own fitness via vertical trans-mission, one can imagine that a hyperparasite could still persist if the fitness gain via horizontal transmission more than offsets its loss in fitness from reduced vertical transmission. It logically follows that, from a practical standpoint, the least favorable situation occurs when both horizontal and vertical transmissions rely on pathogen fitness, because natural selection inevitably favors hyperparasites that minimize the deleterious effects on the pathogen.

In the sections that follow, I discuss how this general theoretical framework can be used to evaluate previous virulence management efforts. I then evaluate the likelihood of long-term success with biological control in a model system that involves hyperparasitism: the chestnut blight host–pathogen system.

21.3   Chestnut Blight as a Pandemic in the USA

The American chestnut (*Castanea dentata*) was once the dominant canopy hardwood in the deciduous forests of the eastern USA, but it was heavily decimated during the first half of the 20th century by the chestnut blight fungus, *Cryphonectria parasitica*.

The chestnut blight fungus was first reported on *C. dentata* near New York City in 1904. It was imported on nursery stock from Asia, where it infects the Asian chestnut (*C. mollissima*), but is only occasionally destructive. Within 5 years, the fungus spread to six nearby states, with long-distance dispersal apparently facilitated by the movement of nursery stock. During the peak of the epidemic, the fungus population increased at a rate of approximately 600% per annum. By the 1950s, the chestnut blight fungus occurred throughout the entire natural range of the American chestnut, and had destroyed approximately 3.5 billion trees (see Anagnostakis 1982; Roane *et al.* 1986).

Cr. parasitica is an ascomycete fungus. Infection occurs when haploid spores (sexual ascospores or asexual conidia) germinate and the fungal hyphae gain entry by growing through a fissure in the bark. The fungus reproduces by producing either sexual or asexual spores within fruiting bodies that erupt from the bark, producing characteristic orange-mottled cankers. The infections (cankers) are localized, but when the fungus enters the vascular cambium and encircles the tree, the tissue above the site of the infection is girdled and dies. Cankers often occur at the base of the trunk, and the entire above-ground tissue is destroyed. Nevertheless, the rootstock generally survives and sprouts new shoots. As a result, *C. dentata* is now an abundant understory shrub in the eastern forests, with the remaining rootstocks of the original chestnuts still producing new shoots that are continuously pruned by the fungus. The trees are rarely large enough to set seed, however, so the number of surviving rootstocks is steadily declining (Parker *et al.* 1993).

21.4   Hyperparasitism in the Chestnut Blight System

Beginning in the 1930s, the chestnut blight epidemic spread through European chestnut (*C. sativa*) populations, albeit more slowly than in the USA (Roane *et al.* 1986). The European chestnut, though heavily infected with *Cr. parasitica* throughout Europe, has recovered in many regions.

The recovery of the European chestnut resulted from the naturally occurring hypovirulent strains of the fungus (Bissegger *et al.* 1996). Hypovirulent strains of *Cr. parasitica* are infected with a naked double-stranded ribonucleic acid (dsRNA) molecule, or *hypovirus* (so-called because it causes *hypo*virulence). The hypovirus acts as a hyperparasite. It resides in the cytoplasm of the fungus and can reduce vegetative growth and spore production. Most importantly, cankers of fungi that carry the hypovirus are often less likely to penetrate the vascular cambium and kill the host.

The hypovirus of *Cr. parasitica* has no independent existence outside its host, but it has two modes of transmission. First, it is vertically transmitted within the asexual propagules (conidia) of the fungus, so that new infections from hypovirulent strains often carry the hyperparasite. Second, the hypovirus has an element of infectious transmission between fungal infections via hyphal anastomosis (Anagnostakis and Day 1979). When hyphae from two infections come into physical contact, they can fuse and exchange cytoplasmic material, including the hypovirus, which spreads from one infection to the other.

Hyphal fusion between fungi, and hence horizontal transmission of the hypovirus, is influenced by a self-recognition system in the fungus – so-called vegetative compatibility (see Anagnostakis and Day 1979; Milgroom 1994). Seven known loci cause vegetative incompatibility between fungal strains (with two alleles per locus) in *Cr. parasitica*. Fungal genotypes that are identical at all seven loci grow together upon physical contact, and their hyphae fuse. For fungal genotypes that differ at some combination of these loci, hyphae might not fuse, and a necrotic zone may form where the two infections meet. The system is leaky and complex, with circuitous networks of vegetative compatibility among fungal genotypes. *Cr. parasitica* strains, therefore, belong to vegetative compatibility groups; two fungi are in the same vegetative compatibility group if their hyphae fuse and if they have the same spectrum of incompatibility reactions with an assortment of other fungal genotypes. The importance of vegetative incompatibility in the current context is that the rate of horizontal transmission of the hypovirus is lowest between fungal strains that differ the most at vegetative compatibility loci (Liu and Milgroom 1996). Infectious (i.e., horizontal) transmission of the hyperparasite may, therefore, be a function of the diversity of fungal vegetative compatibility groups in natural populations (Milgroom 1994).

21.5 Previous Efforts at Virulence Management

Artificial strategies to reduce the negative impact of the chestnut blight epidemic have met with little or no success. During the peak of the epidemic, infected trees were rapidly harvested and stripped of their bark to slow progress of the infection (Gravatt 1914), but these efforts failed. Topical and systemic fungicides have not proved useful for the long-term treatment of the disease (Anagnostakis 1982). Chestnut breeding programs began in the 1930s, and there were numerous public and private efforts to introduce alleles that confer blight resistance to the Asian chestnut into the American chestnut. Although some progress has been made, it is obvious that successful control using this approach will require a sustained long-term effort. The use of the dsRNA hypovirus to control the chestnut blight was heralded as a success story for biological control (Tartaglia *et al.* 1986), but in reality the spread of the hypovirus was primarily a natural occurrence (Bissegger *et al.* 1996).

The hypovirus is now attracting much attention as a potentially powerful agent of biological control in those areas where it has not spread naturally. Recent work on the *Cryphonectria* hypovirus include studies of the effect of the hypovirus on its fungal host (Elliston 1985; Chen *et al.* 1996), studies of hypovirus transmission and epidemiological studies (Bissegger *et al.* 1996; Liu and Milgroom 1996), molecular analyses of hypovirus diversity (Tartaglia *et al.* 1986; Paul and Fulbright 1988; Chung *et al.* 1994; Chen *et al.* 1996), and explicit attempts to genetically engineer biological control agents (Choi and Nuss 1992; Chen *et al.* 1994a).

One of the most important questions that pervades this literature is why the hypovirus has effected so much recovery in European populations, but has failed to do the same in North America. The conventional answer is that the higher diversity of vegetative compatibility groups in US populations of the fungus reduces horizontal transmission of the hypovirus, thereby preventing it from invading (Anagnostakis *et al.* 1986; Milgroom 1994). There is generally a higher diversity of vegetative compatibility groups within US populations of the fungus (Roane *et al.* 1986), and it is known that transmission of the hypovirus between fungal genotypes is reduced in proportion to the difference between the fungi at the compatibility loci (Liu and Milgroom 1996). Moreover, a few *Cryphonectria* populations in the US state of Michigan have been invaded by debilitating hypoviruses (Fulbright *et al.* 1983; Elliston 1985), and, perhaps not coincidentally, these have a low diversity vegetative compatibility groups (Milgroom 1994).

There is, however, some uncertainty about whether the diversity of vegetative compatibility groups is responsible for the failure of the hypovirus to invade in the USA. First, vegetative compatibility groups are an imperfect barrier to hypovirus transmission; they are leaky with hypovirus transmission between fungi of different vegetative compatibility groups, and they form complex intransitive networks of compatibility that allow hypovirus transmission throughout the fungal population (Anagnostakis and Day 1979). Second, hypoviruses suppress sexual reproduction in the fungus, so low fungal diversity may be a consequence of, rather than the cause of, invasion by the hypovirus (Anagnostakis and Kranz 1987; Milgroom 1994). Finally, hypoviruses are already common in US populations, but they appear to have a less debilitating effect on their fungal hosts (Enebak *et al.* 1994a; Chung *et al.* 1994). Thus the reasons why the hypovirus has not effected a recovery of chestnut populations in the USA may not only involve an ecological explanation – that is, different epidemiological parameters (such as lower horizontal transmission) that oppose the spread of any hypovirus – they may also involve an evolutionary explanation of why certain hypoviruses predominate over others in natural populations.

21.6   Virulence Management: Suggestions from Theory

Several lessons may be applied from the general theoretical studies of hyperparasitism (see Box 21.1) to the chestnut blight host–pathogen system. The general message is that biological control efforts would benefit from consideration of the conflicting or overlapping "interests" of all the parties involved. In the chestnut blight system, this involves understanding the trade-off between transmission and virulence in the fungus, and how these affect fungal fitness. The relationship between fungal transmission and virulence is an important determinant of fitness for the hypovirus (the organism that humans would like to employ to reduce fungal virulence, but which depends on fungal transmission for its existence). We also need to understand the degree to which the hypovirus depends on vertical versus horizontal transmissions (both of which rely on fungal transmission in this system), and what factors influence these parameters. Clearly, the management of virulence in the chestnut blight system would benefit from a theoretical understanding of how different transmission modes affect fitness in the hypovirus and its fungal host. Some progress can be made in this regard simply by recasting a general model of hyperparasitism (Box 21.1) with specific assumptions and terminology to fit the chestnut blight system.

Consider the ecological conditions that favor the spread of hypoviruses in *Cryphonectria* populations. Some conditions that favor the transmission of the hypovirus are more prevalent in the European populations of the chestnut blight fungus than in the US populations. First, it is well known that *Cryphonectria* populations in Europe have a lower diversity of vegetative compatibility groups (Milgroom 1994), which would enhance horizontal transmission σ [Equation (e) in Box 21.1]. Second, chestnut populations in Italy (and orchards in France) are often nearly monocultures (Anagnostakis 1982); such localized high host densities favor invasion of the hypovirus [Equation (e) in Box 21.1]. Finally, despite the higher densities, the original chestnut blight epidemic was much slower in Europe (Roane *et al.* 1986), which implicates a lower rate of transmission of the virulent fungus β_V, which favors the spread of the hypovirus [Equation (e) in Box 21.1]. These conditions may have permitted hypoviruses to spread in Europe even though the hypoviruses there have a rather debilitating effect on their fungal host (i.e., low β_H). By contrast, *Cryphonectria* populations in the USA are less favorable to the spread of a debilitating hypovirus. This may be one reason why hypoviruses that have become established in the USA are generally only those with less severe effects on the fungus, and so have not caused recovery of the chestnut populations (Enebak *et al.* 1994a). Many of the patterns of infection and recovery, therefore, are consistent with the general expectations from theory. The modeling effort would make an even more substantial contribution if it could direct empirical studies toward examining the specific conditions that promote horizontal versus vertical transmissions of the different strains of the hypovirus.

For virulence management, it is not enough for a hypovirus to invade; it must also debilitate the fungus and allow the host tree to recover. This presents a practical problem in the chestnut blight system because the mechanism of hypovirus transmission dictates that debilitating the fungus generally reduces hypoviral fitness. Recall that both horizontal and vertical transmissions of the hypovirus involve the successful transmission of fungal spores. Vertical transmission of the hypovirus occurs when a fungal infection that harbors a hypovirus produces offspring that carry the hypovirus with them when they establish a new infection. Horizontal transmission of the hypovirus occurs by the same physical mechanism, except the fungal offspring carry the hypovirus to the site of an existing infection, and the virus is then transferred to the existing infection via fungal anastomosis. That both elements of hypovirus transmission are so closely connected to fungal reproduction means the hypovirus should evolve to minimize any deleterious effects on the fungus.

The general theoretical model and what is known about the transmission properties of the hypovirus suggest that hypoviruses useful for biological control may be evolutionarily unstable. One implication of this result is that the failure of the chestnut to recover in the USA may not be because hypoviruses have failed to invade. Rather, hypoviruses in the USA, which debilitate their host less, might be closer to the hypoviral ESS, and generally outcompete more debilitating varieties. Support for this is that it has been possible to artificially establish debilitating hypoviruses very locally in the USA (Anagnostakis 1982), and some hypoviruses have been found in natural populations (Chung *et al.* 1994; Enebak *et al.* 1994a); but they have not spread or caused long-term recovery. This all leads to the testable prediction that debilitating hypoviruses (such as those in Europe) are susceptible to invasion by less debilitating ones.

Although the prediction that hypoviruses should evolve away from debilitating the chestnut blight is discouraging, in reality, the hypoviruses have been a major factor in the recovery of the European chestnut. There are several ways to reconcile these two observations. First, the model may be correct, but the system may not be at equilibrium. If this is the case, the model predicts that the recovery of the European chestnut is only a transient phenomenon that could be reversed by evolution of the hypovirus. Certainly, extreme caution should be taken to avoid the introduction of less debilitating North American hypoviruses to European *Cryphonectria* populations. Secondly, the model predicts that the evolution of the hypovirus will minimize its deleterious effects on the fitness of the fungus, and this could involve a reduction in fungal transmission and virulence in circumstances in which selection favors an intermediate optimum transmission and virulence for the pathogen (Taylor *et al.* 1998a). The problem with this scenario is that the optimum pathogen transmission and virulence is that which harms the host population most (Lenski and May 1994; Box 21.1). So for advocates of biological control, little solace is found in the fact that hypovirus evolution can drive the system toward that optimum.

There is also the possibility that the model is incorrect in one or more important assumptions. The assumption that has the most serious ramifications for biological control of the chestnut blight is that horizontal and vertical transmissions of the hypovirus are influenced by the same processes during fungal reproduction, and, therefore, are positively correlated. Based on the biology of the system, this assumption is likely to be true when only a single hypovirus is present, but factors that influence the dynamics of competing hypoviruses have never been explicitly investigated. Single fungal infections can harbor more than one hypovirus (Enebak *et al.* 1994b; Smart and Fulbright 1995), but it is not clear whether the displacement of one strain by another within an infection is related to other aspects of transmission. The best hope is that there is a negative relationship between horizontal and vertical transmissions of the hypovirus, so that strains that debilitate the fungus can persist in populations and predominate against less-debilitating strains via horizontal transmission.

21.7 Discussion

It seems clear that hyperparasitism can be either a useful tool or a complicating factor in virulence management efforts. For those of us who study host–pathogen systems, how often can we be certain that the pathogen is completely parasite free? Or, if such complexities do exist, how can we be certain that they do not alter the dynamics of the system in some fundamental way? When dealing with a specific host–pathogen system, an important consideration is that the spread of hyperparasites can often mimic an evolutionary reduction in virulence; this is most likely to occur in situations for which evolutionary models predict that selection would favor higher virulence (Taylor *et al.* 1998a). For example, efforts to establish conditions that minimize selection for higher virulence could establish conditions that retard the spread of beneficial hyperparasites. As a corollary, establishing conditions of high density and frequent superinfection that promote the spread of hyperparasites may create selection for more virulent pathogens.

Hyperparasites have been viewed as either a nuisance, because they parasitize biological control agents (Beddington and Hammond 1977), or as potentially powerful agents of biological control themselves (Nuss 1992). For those investigators who view hyperparasites as potentially powerful agents of biological control, the most important consideration is how the fitness of the hyperparasite is influenced by the debilitating effect it has on the pathogen host. It is difficult to imagine how an obligately internal hyperparasite can persist when it seriously debilitates the pathogen in which it resides. Efforts to utilize such organisms as biological control agents seem to force biological systems to run in opposition to the evolutionary process. Nevertheless, the recovery of the European chestnut stands out as one of the most successful, albeit naturally occurring, examples of biological control. The theory described here suggests the recovery of the chestnut is unlikely to be evolutionarily stable unless there is some trade-off between the horizontal and vertical transmissions of the hyperparasite. With such a trade-off, it may still be possible for hyperparasites that debilitate their pathogen hosts the most (reducing their own vertical transmission) to spread via higher rates of horizontal transmission. More theoretical and experimental studies on the epidemiology of this system are required to understand how this recovery occurred in Europe, why it failed elsewhere, and whether or not it can be expected to last.

22

Evolution of Exploitation and Defense in Tritrophic Interactions

Maurice W. Sabelis, Minus van Baalen, Bas Pels, Martijn Egas,
and Arne Janssen

22.1 Introduction

Why do plants cover the earth and give the world a green appearance? This question is not as trivial as it might seem at first sight. Hairston *et al.* (1960) hypothesized that herbivores cannot ransack the earth of its green blanket because they are kept low in number by predators. They tacitly ignored the possibility that plants defend themselves directly against a suite of herbivores and together exhibit such great diversity in defense mechanisms that "super" herbivores able to master all plant defenses did not evolve and those that overcome the defenses of some plants are limited by the availability of these plants. Strong *et al.* (1984) recognized both possibilities in their review on the impact of herbivorous arthropods on plants, but they also favored the view that predators suppress the densities of herbivores, and thereby reduce the threat of plants being eaten.

The two explanatory mechanisms (plant defense versus predator impact), however, may well act in concert. Ever since the seminal review paper by Price *et al.* (1980) ecologists have become increasingly aware that plant defenses include more than just trickery to reduce the herbivore's capacity for (population) growth. For example, the plant may provide facilities to promote the foraging success of the herbivore's enemies. This form of defense is termed *indirect* as opposed to direct defense against herbivores. Examples of direct defenses are:

- Plant structures that hinder (feeding by) the herbivore (e.g., cuticle thickness, "smooth" cuticle surfaces that do not provide a holdfast, impenetrable masses of leaf trichomes, glandular trichomes acting as sticky traps);
- Secondary plant compounds that modify the quality of ingested plant food (digestion inhibitors), intoxicate the herbivore, or signal the plant's well-defended state to "discourage" the herbivore (feeding deterrents).

Indirect plant defenses bypass the direct defense route against the second trophic level by promoting the effectiveness of the third. For example, plants may retain the herbivore's enemies by providing protection and/or food and they may attract these enemies by betraying the presence of prey via herbivory-induced chemical plant signals. When Price *et al.* (1980) published their review paper, certain ant–plant interactions provided the best-known examples of indirect defenses [Janzen

1966; Bentley 1977; see also reviews by Beattie (1985) and Jolivet (1996)]. Now, more than 20 years later, it has become increasingly clear that the conspiracy between plants and predators against herbivores is a widespread phenomenon; a wide range of plants from many different families invest in promoting the effectiveness of a suite of predatory arthropods (Beattie 1985; Buckley 1986, 1987; Dicke and Sabelis 1988, 1989, 1992; Dicke *et al.* 1990; Koptur 1992; Drukker *et al.* 1995; Takabayashi and Dicke 1996; Turlings *et al.* 1995; Jolivet 1996; Walter 1996; Scutareanu *et al.* 1997; Sabelis *et al.* 1999a, 1999b, 1999c, 1999d).

In this chapter, we outline how analysis of the way plants defend themselves against herbivores can shed light on certain issues in host–parasite interactions. Many analogies exist between plant–herbivore interactions and host–parasite interactions. In fact, arthropod herbivores can easily be considered as "parasites" of the plants: compared to their host they are small, and the detrimental effects incurred result not so much from the effect of a single herbivore, but rather from the combined effects of the population that develops. Thus, it is in the interest of the plant to slow local herbivore dynamics as much as it is in the interest of an animal host to block within-host parasite dynamics. As argued above, plants may do so by direct means of defense, but they may also solicit the help of the predators of the herbivores. Thus, predators may function effectively as a kind of "immune system" for the plants. Insights into plant–herbivore–predator interactions may therefore provide clues to understand evolutionary aspects of host–parasite interactions. Our approach, as it is explicitly based on models for local population interactions between herbivore and predator, can be used to assess how, by changing certain parameters, the plant can manipulate the interaction to its own benefit. We use this approach not only to expose the game-theoretic aspects of the interaction between trophic levels, but also the interactions *within* these levels. Local competition among predators or herbivores has a clear link with the evolution of virulence, as this can also be strongly affected by within-host competition (Nowak and May 1994; Van Baalen and Sabelis 1995a).

We treat the system as a simple linear food chain of plants, herbivores, and predators, and hence ignore the many and varied ways of "cheating and misusing" that exist in complex food webs of arthropods on plants. We prefer to concentrate here on the evolution of food exploitation strategies of the herbivores and predators in response to investments in direct and/or indirect defenses on the part of the plant. Typically, herbivorous arthropods may (evolve to) be mild or malignant parasites of the plant and predatory arthropods may (evolve to) be prudent or wasteful exploiters of the population of herbivorous arthropods on a plant. Clearly, for a plant to invest in direct and/or indirect defense, it matters how virulent the herbivorous arthropods are to the plant and how virulent the predatory arthropods are to the population of herbivores on a plant. We argue that, to understand the evolution of mutualistic interactions between plants and the natural enemies of herbivorous arthropods, we should identify the advantages to the individual plant and the individual predator, predict the consequences for the population dynamics of herbivorous and predatory arthropods, and elucidate how dynamics in turn affects the evolution of plant–predator mutualism and the herbivore's response to this conspiracy. Whereas a definitive solution is not within reach, we hope to convince the reader that there are many possible outcomes for the evolution of defense and virulence in this tritrophic system, and we discuss the consequences of these insights for the "world is green" hypothesis and the common notion that plant–predator mutualisms readily evolve because "it is both in the interest of plants to get rid of the herbivores and in the interest of the predators to find herbivores as prey" (Price *et al.* 1980).

First, we briefly discuss that the players in the tritrophic game operate on very different spatial and temporal scales. Second, we introduce a simple model of local predator–prey dynamics on an individual plant based on specific scale assumptions and use this model to identify the main categories of defensive strategies of a plant, as well as the main strategies of food exploitation by the herbivores and the predators. Third, we identify evolutionarily stable strategies (ESSs) of exploitation (predator–herbivore, herbivore–plant), migration (predator, herbivore), and defense (plant, herbivore) in tritrophic systems. Finally, we speculate on the consequences for virulence management.

22.2 Spatial and Temporal Scales of Interaction

Plants, herbivorous, and predatory arthropods are engaged in interactions with widely different temporal scales. Plants usually have much longer generation times than arthropods. Hence, plant population change tends to be slow relative to that of the arthropods. Hence models of arthropod predator–prey dynamics are usually decoupled from plant population dynamics by assuming a pseudo-steady state. The generation times of herbivorous and predatory arthropods may also differ, but are usually close enough to justify modeling as a ditrophic system (Hassell 1978; Sabelis 1992).

The spatial scale of predator–prey interaction is set by the distribution of herbivorous arthropods. Many herbivorous arthropods have strongly clumped distributions over their host plants. This may be the result of:

- Aggregation toward weakened host plants or hosts whose defenses are overwhelmed by pioneer attacks, as in bark beetles (Berryman *et al.* 1985; de Jong and Sabelis 1988, 1989);
- Large egg clutches deposited by a female, as in various species of moths (larch bud moths, gypsy moths, ermine moths, tent caterpillars, and brown tail moths);
- Multigeneration congregations that result from one or a few founders with a high intrinsic capacity of population increase (relative to the rate of emigration) and low *per capita* food demands, as in many small herbivorous arthropods with short generation times (scale insects, mealybugs, aphids, leafhoppers, whiteflies, thrips, spider mites, and rust mites).

These groups of herbivorous arthropods may occupy a leaf area less than that of a plant, or cover several neighboring plants. Moreover, group size and the leaf area occupied increase with the number of generations spent in the group and with the extent to which the herbivores move to neighboring host plants instead of dispersing far away. These traits are of crucial importance to understand plant defenses, because individual plants (or kin groups of plants) are the units of selection and the selective advantage of defense depends on the extent to which a plant can influence the local dynamics of herbivores and predators by direct and/or indirect defenses. Indeed, this influence is limited because predators and herbivores are independent players in the game: they may decide to stay, to move to neighboring plants, or to disperse far away. Hence the question is: does the defense of an individual plant initially affect the herbivore population it harbors (and much later the herbivore population as a whole) or does its impact on the herbivores permeate population-wide and without delay (as a consequence of high herbivore mobility)? Much the same questions can be formulated with respect to the degree in which individual plants can monopolize the advantages from attracting natural enemies of the herbivores. To analyze the complexities that arise from spatial and temporal scales we start by assuming that plants investing in direct and indirect defense acquire all the benefits, and later we consider the case in which neighboring plants may profit too.

22.3 Predator–Herbivore Dynamics on Individual Plants

To understand the range of possible plant defense strategies it is instructive to model the dynamics of small arthropod herbivores at the scale of an individual plant (or a coherent group of clonal plants). For reasons of simplicity we assume that the spatial scale covered by a local predator and prey population is smaller than that of an individual plant (or group of ramets) and that the individual plant does not impose a carrying capacity on the arthropod population that inhabits the plant. Moreover, we limit our discussion to the case of local populations with multiple and overlapping generations, as is realistic for small arthropods. Our aim is to find the simplest possible model for the dynamics of herbivorous and predatory arthropods and then ask the question how an individual plant can reduce damage by influencing the herbivores and their predators.

We assume that prey form patchy aggregations, which consist of clusters of herbivore-colonized leaves, and that predators entering such patches can freely move around, and spend little time in moving between herbivore-colonized leaves relative to the time spent within the herbivore colonies. Moreover, predators tend to avoid each other (Janssen *et al.* 1997a) and thereby interference (Nagelkerke and Sabelis 1998). Thus, the cluster of herbivore-colonized leaves can be considered as a coherent and homogeneous arena for strongly coupled predator–prey interactions (Figure 22.1). The population of herbivores in such a cluster presents to a predator in the same way as a host does to a pathogen. In fact, local predator–prey dynamics are much more transparent, and have been much more thoroughly studied, than within-host parasite dynamics. Usually, the latter are treated as a black box [but see Box 12.1 and, in addition, Nowak *et al.* (1991), Sasaki and Iwasa (1991), Antia and Koella (1994), for models that take within-host dynamics explicitly into

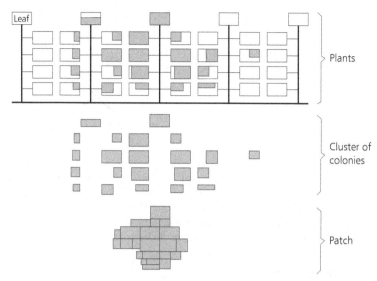

Figure 22.1 A patchy infestation of small herbivorous arthropods in a row of plants (side-view). It is inspired by observations on spider mites, but in essence applies to many other herbivorous arthropods. The infested leaf parts are excised (middle) and then put together as a jigsaw puzzle (bottom). The latter forms the patch or arena within which the interaction between predator and prey takes place (assuming negligible time spent in moving between infested leaves).

account], whereas various simple models capture the essence of local predator–prey dynamics.

To derive one example of such a model, we assume that the predators search for the highest prey densities and minimize interference with (intraspecific) competitors. In addition, we assume that within newly (and expanding) infested leaf areas herbivore density is typically constant, a characteristic determined by the herbivore or the combination of herbivore and plant (Sabelis 1990; Sabelis and Janssen 1994). Thus, per unit of plant area, herbivores raise a fixed amount of offspring and predators reaching a freshly colonized leaf site continue to eat prey until they do better by moving to a site nearby on the same plant. These assumptions lead to a constant rate of predation, which is much like "eating a pancake": a constant amount of food at each bite until there is nothing left.

As long as the pancake is not completely eaten, predators maximize the per capita rate of predation, development, and reproduction. Hence, under conditions of a stable age distribution they achieve their intrinsic rate of population increase. Similar assumptions are made with respect to herbivore growth capacity in the absence of the predators. For the case in which predators stay until all the prey are eaten, the dynamics of predator and herbivore numbers can be described by the two linear differential Equations (22.1a) and (22.1b). These differ from the classic Lotka–Volterra models in that the predation term now only depends on the number of predators (Metz and Diekmann 1986, example III.1.10; Janssen and

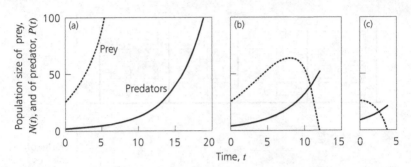

Figure 22.2 Three types of local predator–prey dynamics (continuous curve for predators and dashed curve for prey) according to the pancake predation model [with parameters $r_N = 0.3$, $k = 1$, $r_P = 0.25$, $m_P = 0$, and $N(0) = 25$]. (a) Prey increase, $P(0) = 1$, (b) delayed prey decline, $P(0) = 3$, and (c) immediate prey decline, $P(0) = 8$. The general conditions for each of these types of dynamics are discussed in the text.

Sabelis 1992; Sabelis 1992),

$$\frac{dN}{dt} = r_N N - kP \,, \tag{22.1a}$$

$$\frac{dP}{dt} = r_P P \,. \tag{22.1b}$$

This is the so-called "pancake predation" model with t = the time since start of the predator–prey interaction, $N(t)$ = number of prey at time t, $P(t)$ = number of predators at time t, r_N = rate of prey population growth, k = maximum predation rate, and r_P = rate of predator population growth. Analytical solutions for the number of predators and prey since the start of the interaction are readily obtained,

$$N(t) = N(0)e^{r_N t} - P(0)\frac{k}{r_P - r_N}(e^{r_P t} - e^{r_N t}) \,, \tag{22.2a}$$

$$P(t) = P(0)e^{r_P t} \,. \tag{22.2b}$$

Three types of dynamical behavior of the prey population may occur:

- Continuous increase (but at a pace slower than the intrinsic rate of prey population growth; Figure 22.2a);
- Initial increase, followed by decrease until extinction (Figure 22.2b);
- Continuous decay until extinction (Figure 22.2c).

In the first case only, predatory arthropods cannot suppress local prey population outbreaks, but in the two other cases they eliminate the prey population and grow exponentially until a time τ when all the prey are eaten and all the predators emigrate. This time τ from predator invasion to prey extinction can be expressed as a function of r_N, k and r_P and the initial numbers of predator and prey, $N(0)$ and $P(0)$,

$$\tau = \frac{1}{r_P - r_N} \ln \left[1 + \frac{r_P - r_N}{k} \frac{N(0)}{P(0)} \right]. \tag{22.3}$$

Immediate decline of the herbivores occurs when the net growth rate of the herbivore population is negative, $r_N N(0) < k P(0)$, or

$$\frac{P(0)}{N(0)} > \frac{r_N}{k}. \tag{22.4a}$$

Thus, for the herbivore population to decline immediately, the ratio of predators to herbivores should exceed the ratio of the per capita population growth rate of the herbivore and the maximum per capita predation rate.

The conditions for continued herbivore increase are found by calculating the condition for which the time to prey extinction has a finite value. Provided that the plant is not overexploited during predator–prey interaction, this condition is

$$\frac{P(0)}{N(0)} \leq \frac{r_N - r_P}{k}. \tag{22.4b}$$

This condition cannot hold when the growth rate of the prey does not exceed that of the predator, $r_N \leq r_P$. Whenever the condition is met, however, herbivores continue to increase and predators "surf" on the "population growth wave" of the herbivore. Inevitably, this increase stops when the plant becomes overexploited.

The total damage incurred by the plant over the whole interaction period can be expressed in the number of herbivore-days D, that is, the area under the curve that expresses the temporal changes in the size of the herbivore population,

$$D(\tau, r_N, k, r_P, N(0), P(0)) = \frac{1}{r_N} \left[P(0) \frac{k}{r_P} (e^{r_P \tau} - 1) - N(0) \right]. \tag{22.5}$$

Note that this measure of the damage strongly depends on the exponential term and thus on the time to prey extinction τ and the per capita growth rate of the predator population r_P.

Thus, given the initial population sizes $N(0)$ and $P(0)$ and estimates of the parameters r_N, k, r_P, it is possible to assess the overall damage by the herbivore and the predator's potential to suppress the prey population immediately, with a delay, or not at all. Now, we may ask how a plant can influence the local dynamics so as to minimize herbivore damage. Under the assumption that the parameters can be modified independently, the answer is straightforward. It should:

■ Make the predator-to-herbivore ratio as high as possible;
■ Increase the predation rate or the predator growth rate;
■ Decrease the growth rate of the herbivore.

To illustrate this, the plant may attract and retain the predators by providing SOS signals upon herbivore attack, it may provide food and shelter for the predators, and produce toxins or digestion inhibitors.

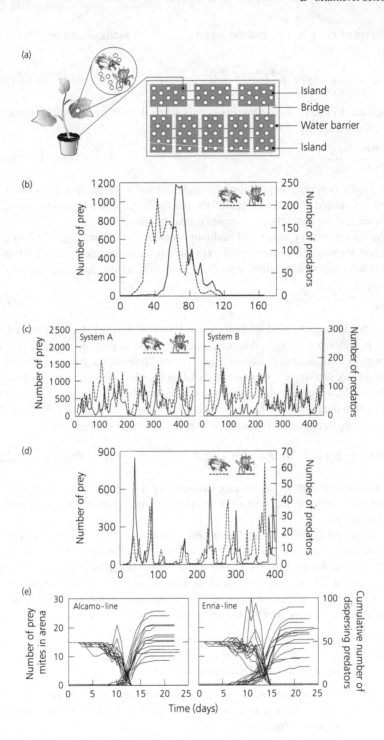

Figure 22.3 Dynamics of predatory mites (*Phytoseiulus persimilis* Athias-Henriot; continuous curves) and herbivorous mites (*Tetranychus urticae* Koch; dashed curves) at various spatial scales. (a) Circular system of eight interconnected islands (trays), each with 10 Lima-bean plants maintained in the two-leaf stage (by frequent removal of the apex and replacement of plants exhausted as a food source). *Source*: Janssen *et al.* (1997b). (b) Extinction-prone predator–prey dynamics on one super-island (the size of eight trays together). (c) Two replicate experiments showing persistent predator–prey metapopulation dynamics on the eight-island system in (a). (d) Extinction-prone predator–prey dynamics on one of the trays shown in (a). *Source*: Janssen *et al.* (1997b); see also Van de Klashorst *et al.* (1992). (e, left) Extinction-prone predator–prey dynamics on a single leaf in a wind tunnel, using a field-collected predator line selected for nondispersal before prey extermination. *Source*: Pels and Sabelis (1999); see also Sabelis and Van der Meer (1986). (e, right) As (e, left), but now using a field-collected predator line selected for dispersal before prey extermination. *Source*: Pels and Sabelis (1999).

Trade-off relations between parameters may complicate matters. For example, decreasing the growth rate of the herbivore r_N by toxins may also intoxicate the predator, thereby decreasing the predation rate k and/or its growth rate r_P. In the extreme, the herbivore may even use the plant-provided toxins to defend itself against predators. Thus, the plant does not always profit from decreasing the growth rate of the herbivore. It only profits if it decreases the herbivore's growth rate proportionally stronger than the predation rate and the growth rate of the predator.

Another message gleaned from the equations is that plants may benefit from promoting the presence of predators with high predation rates and high growth rates. Often, these demands are in conflict with each other, because the predation rate tends to increase with body size, whereas the intrinsic rate of population increase tends to decrease with body size (Sabelis 1992). Such relationships with body size are clear from analyzing published data on predators of phytophagous thrips, such as mirids, anthocorids, predatory thrips, and predatory mites (Sabelis and Van Rijn 1997). At lower taxonomic levels (within family, within genus) the picture may be different. For example, within the *Phytoseiidae* – a family of plant-inhabiting predatory mites – positive high correlations between k and r_P exist (Janssen and Sabelis 1992; Sabelis and Janssen 1994). It may be possible that the plant could selectively attract one species of predator over the other and thereby profit from selecting the more effective predators. However, how a plant could do so, given that predators will seek the most profitable prey, remains to be shown.

Predators are independent players in the tritrophic game and they decide whether it is profitable to stay on a plant or not. The pancake predation model is based on the assumption that predators are strongly retained and stay until all the prey are eaten. This scenario is not implausible, because it may be risky to disperse and search for new herbivore patches. Indeed, it is frequently observed, as in interactions among predatory mites and spider mites (Figure 22.3a; Sabelis and Van der Meer 1986). One may, of course, expect predators to leave somewhat earlier than the exact moment of prey extinction. This would relieve the herbivores from

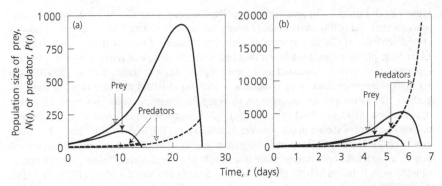

Figure 22.4 Influence of emigration on local predator–prey dynamics according to the pancake model. (a) Predator emigration rate m_P equals zero for predator and prey curves indicated by filled arrowheads, whereas it is 0.04 for the predator and prey curves indicated by open arrowheads. Parameters: $m_P = 0$ or 0.04, $r_N = 0.3$, $k = 3$, $r_P = 0.25$, $N(0) = 30$, $P(0) = 1$. (b) Prey emigration rate m_N equals zero for dashed predator and prey curves (open arrowheads) and 0.1 for the continuous predator and prey curves (filled arrowheads). Parameters: $m_N = 0$ or 0.1, $r_N = 1$, $k = 1$, $r_P = 1.5$, $N(0) = 50$, $P(0) = 1$.

predation pressure and predator-to-prey ratios may become so low that the herbivore population increases again, thereby giving rise to cyclic dynamics. The plant would then accumulate damage over the predator–prey cycles whereas it would be better off when predators exterminate the herbivores or maintain them at a very low level. Examples of local predator–prey dynamics are shown in Figure 22.3 for interactions between phytoseiid predators and spider mites. Similar examples are known from interactions between phytoseiid mites and thrips, and anthocorid predators and thrips (Sabelis and Van Rijn 1997). All these examples convincingly show that local herbivore populations are strongly suppressed by predators. Thus, the dynamics of the "pancake predator" model seems to be a good caricature of the initial predator–prey population cycle. Thus, for all cases in which one cycle of predator and prey colonization occurs before extermination, this model is useful to understand the role of indirect and direct defenses of a plant.

Predator retention during the interaction with the herbivores is decisive in the success of indirect defense strategies of the plant. For example, if we extend the pancake predator model with a constant predator emigration rate (i.e., independent of prey availability), then emigration acts to decrease the effective population growth rate of the predators. As shown in Figure 22.4a small decreases of the population growth rate have dramatic consequences for the duration of the interaction and even more for the number of herbivores attacking the plant. The overall damage to the plant increases nonlinearly with a reduction of the predator's population growth rate. Hence, it is important to observe that several species of predators are strongly retained in herbivore-colonized patches and tend not to leave until the prey population is near extinction (Sabelis and Van der Meer 1986; Sabelis and Van Rijn 1997; Pels and Sabelis 1999; Figure 22.3e).

The herbivore can always make the last move in the tritrophic game. Not only may they develop resistance against the predators and overcome barriers raised by the plant, but ultimately, they may also decide to leave the plant. If the plant drives herbivores away by attracting predators, and also by stimulating herbivore emigration, then it benefits disproportionately, as illustrated in Figure 22.4b.

In conclusion, there are various ways in which a plant can benefit by influencing behavior and dynamics of predators and herbivores. Direct plant defenses do not merely slow down the herbivore's growth rate, they may also affect predator impact, either positively (higher predator-to-herbivore ratio) or negatively (plant toxins protecting herbivores against predators). Indirect defenses do not merely affect predator performance, they may also affect selection on herbivores, be it positive (enemy avoidance) or negative (resistance to predators) for the plant. Hence, to understand the plant's allocation to direct and indirect defenses we should not only assess the costs, but also elucidate how these two types of defenses interact in their impact on overall herbivore damage.

22.4 Tritrophic Game Theory and Metapopulation Dynamics

The evolution of direct and indirect plant defenses against arthropod herbivores is by no means a simple process. For one thing, we have to take the defense and exploitation strategies of all three trophic levels into account. For another, time scale and spatial scale arguments force us to consider metapopulation models with the full tritrophic structure, and – as we emphasize later – to incorporate plant dynamics adds new dynamical behavior to the repertoire of the otherwise ditrophic models.

Indeed, evolution in structured populations can only be properly understood by taking metapopulation structure into account. The strategies play against each other at the patch level and their relative success determines metapopulation dynamics. In turn, metapopulation processes determine which strategies will meet and compete again at the patch level. This chain of processes is referred to herein as the ecological feedback.

To gain insight into this complex problem, carefully planned simplifications are required. Our strategy is first to consider the evolution of relevant traits at one trophic level in pairwise interactions with one other level (predator–herbivore; herbivore–predator; plant–herbivore; herbivore–plant; plant–predator; predator–plant), thereby assuming a steady state at the metapopulation level. We conclude with a tentative, verbal discussion of what may happen evolutionarily in the full tritrophic system.

The predator's dilemma: to milk or to kill?

It is one thing for a plant to attract and retain predators, but it is another to lure predators that also effectively suppress the herbivore population on the plant. Clearly, it is the predator who reacts to the lure, and decides how fast it will consume herbivores, how fast it will multiply, and how long it will stay. In other words, the effectiveness of indirect plant defenses depends on the herbivore-exploitation strategies present in the predator population. In principle, these exploitation strategies can be many and varied. To illustrate this point it is instructive to use the "pancake" model to examine local predator–herbivore dynamics. The strategy set is determined by combinations of the per capita growth rate of the predator r_P and the per capita predation rate k. However, a spatially structured environment has one more parameter: the per capita emigration rate of the predator. Of course, all predators disperse away when herbivores are exterminated and there is no other food, but they can also decide to move away during the interaction. Such increased emigration of the predator relieves the prey of predation pressure and thereby the prey population represents a larger future food source for the predator (Van Baalen and Sabelis 1995c). This effect can also be achieved by decreased predation (Gilpin 1975). However, increased emigration during the interaction seems a more sensible strategy, because it does not affect the intrinsic population growth rate, whereas decreased predation implies lower food intake and thereby a lower population growth rate. Moreover, the extra dispersing propagules generated under the former strategy will promote the founding of new populations. In what follows we therefore focus on the predator's migration trait alone.

Let us assume for simplicity that the per capita emigration rate is a constant m_P. The effective per capita population growth rate of the predators is reduced by m_P so that the effective predator growth rate now equals

$$\frac{dP}{dt} = (r_P - m_P)P \ . \tag{22.6}$$

As decreased values of r_P have been shown to drastically alter local predator–prey dynamics, so do increased values of m_P. As illustrated in Figure 22.4a, a small increase of m_P from 0 to 0.04 day^{-1} greatly alters the area under the prey curve and thereby also the damage to the plant. Hence, a plant benefits from stimulating predators to stay until all the prey are eliminated, but whether it succeeds depends on what is best for the predators (as well as on the ecological feedback). The local success of a predator's exploitation strategy may be expressed as the number of dispersers produced during the interaction with prey, plus those that disperse after prey extermination. As shown in Figure 22.5, the production of dispersers increases disproportionally with m_P and reaches an asymptote when the per capita emigration rate is so high that the predators cannot suppress the growth of the prey population any more, that is, when $m_P = r_P - r_N + kP(0)/N(0)$. Thus, predators that suppress emigration during the interaction reach their full capacity to suppress the local prey population, but they produce the lowest number of dispersers per prey patch. In terms of production of dispersers this so-called *killer* strategy does less well than the strategy of a *milker*, which typically has a nonvanishing emigration rate during the predator–prey interaction period. However, if killers enter a prey patch with milkers, then they would steal much of the prey the milkers had set aside for future use. Therefore, if there is a risk of invasions by killers, it pays to anticipate such events and selection will favor exploitation strategies that are less milker-type and more killer-type.

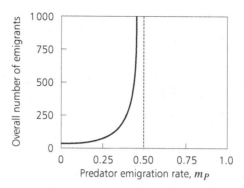

Figure 22.5 The relation between the overall production of predator dispersers and the per capita rate of emigration m_P during the predator–prey interaction period (from predator invasion to prey elimination. *Source*: Van Baalen and Sabelis (1995c).

   The outcome of the milker–killer dilemma is determined by a complex interplay of local competition between the exploitation strategies and global (= metapopulation) dynamics. It depends on the probability of multiple predator invasions in the same prey patches (or, alternatively, on the probability of exploiting a prey patch alone), on the resultant production of dispersers per prey patch, and on metapopulation dynamics, as this in turn determines the probability of multiple invasions. The complexity of this ecological feedback is staggering; to keep track of the numbers of each strategy type when competing in local populations, dispersing into the global population, and invading into local populations requires a massive bookkeeping procedure. Hence, we are bound to simplify to obtain some insight. For example, Van Baalen and Sabelis (1995c) assumed that all patches start with exactly the same number of predators and prey (which assumes metapopulation-wide equilibrium and ignores stochastic variation in the number of colonizers) and that the predators have enough time to reach their full production potential per prey patch (the assumption of sequential interaction rounds). In this setting they considered the reproduction success of one mutant predator clone with a per capita emigration rate, $m_{P,\text{mut}}$, relative to the mean success of predators in the population of the resident clone, which possesses another per capita rate of emigration, $m_{P,\text{res}}$. The question is whether there exists an ESS value of $m_{P,\text{res}}$ for which it does not pay any mutant to deviate (Maynard Smith 1982). In particular, Van Baalen and Sabelis (1995c) calculated the combinations of parameters $P(0)$, $N(0)$, r_N, k, and r_P for which it does not pay to increase m_P away from zero, that is, the conditions that favor selection for killers. As illustrated in Figure 22.6, the general outcome is that killers are usually favored by selection, except when the number of predator foundresses is low and the number of prey foundresses is high. In other words, the milkers are favored as long as they have a sufficiently large share in the local populations to maintain control over the time to prey elimination τ.
   This analysis whets the appetite for more elaborate considerations that account for:

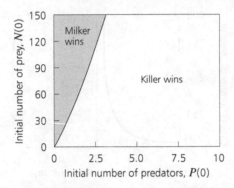

Figure 22.6 When does it pay to increase the per capita emigration rate of the predator m_p away from zero? This diagram shows that milker strategies are only advantageous when $P(0)$ is low and $N(0)$ sufficiently high. *Source*: Van Baalen and Sabelis (1995c).

■ Asynchrony in local dynamics;
■ Stochastic variation in predator and prey colonization rates (since these are probably low);
■ An upper boundary to prey population size set by the local amount of food;
■ Metapopulation dynamics.

Such extensions are likely to show that milkers which achieve a longer interaction period are also exposed for a longer period to subsequent predator invasions (and thus face competition with killers sooner or later), that stochastic rather than uniform invasions help to isolate milkers, which thereby gain full advantage of their exploitation strategy, and that limits to the amount of food available for the prey decrease opportunities for a milker, as it loses full control over the interaction period τ. As these factors have opposite effects it is not immediately clear whether killers or milkers will win the battle or whether they may even coexist. Computer simulations of the metapopulation dynamics of milkers and killers using a model parameterized for phytoseiid mites (predators) and spider mites (prey; Pels *et al.*, in press) showed that the full ecological feedback gives rise to prey and predator densities in which multiple predator invasions are sufficiently rare to make "milkers" the more successful strategy. The results of a large number of computer experiments to determine the average number of dispersers born from mutants with different emigration strategies released randomly in the metapopulation of the residents (but after 2 000 days to avoid the initial phase of transient metapopulation dynamics) is summarized in the pairwise invasibility plot shown in Figure 22.7. This shows that metapopulation dynamics can force the predator–prey system into a state [i.e., number of predator and prey colonizing a patch, $N(0)$ and $P(0)$, in Figure 22.6] in which "milkers" are the winners of the competition.

Apart from the need for more theoretical work, experimental analysis of variation in the exploitation strategies of predators seems a promising avenue for future research. Such an analysis, carried out for the predatory mite *Phytoseiulus persimilis* Athias-Henriot, revealed that laboratory cultures harbor exclusively predators

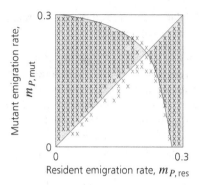

Figure 22.7 A pairwise invasibility plot for mutants that differ from the resident predators with respect to the emigration rate m_P. The resident's emigration rate is given on the horizontal axis and the mutant's emigration rate on the vertical axis. Gray areas indicate combinations in which the mutant invades the resident population (the shape of the gray areas is based on the simulation results indicated by crosses). The results shown are obtained for a value of the predator's survival rate that allows the predator–prey system to persist.

of the killer type (Sabelis and Van der Meer 1986), whereas field-collected populations in the Mediterranean (Sicily) exhibit some variation in the onset of emigration before or after prey elimination (Pels and Sabelis 1999). Interestingly, most populations collected along the coast, where predators are more abundant, initiated emigration only after elimination of the prey, whereas those collected inland, where local predator populations are scarce and hence more isolated, showed some emigration before prey elimination! These results are in qualitative agreement with the analytic ESS analysis of Van Baalen and Sabelis (1995c; Figure 22.6), but they seem to contradict the results of the more "realistic" computer simulations that are not only parameterized for this particular mite system but also take into account stochastic colonization and the full ecological feedback (Pels *et al.*, in press; Figure 22.7). This discrepancy probably results from a variety of factors that cause the predators to loose control over the exploitation of the local prey population. Examples are environmental disasters (heavy rain, wind, or fire), overexploitation of plants by large herbivores, and also exploitation competition with other predator species or herbivore diseases. The discrepancy between the simulations (Pels *et al.*, in press; Figure 22.7) and the analytic treatment (Van Baalen and Sabelis 1995c; Figure 22.6) may emerge because:

- The simulations were obviously only carried out for persisting resident populations, whereas the analytic treatment implicitly assumed equilibrium (and thus persistence);
- The simulated predator–prey feedback causes patches to be invaded by very low numbers of predators, whereas the analytic treatment presupposed a certain invasion scenario (equal for all patches);
- The stochastic colonization process of the predators allows some patches to be invaded singly and gives the single invader full control over the exploitation

of the local prey population, whereas the analytic treatment ignored stochastic variation in the number of colonizers.

Obviously, the prevalence of killers is of great importance for the evolution of indirect plant defenses. By providing protection and food to predatory arthropods and by signaling herbivore attack to predators, plants increase the predator invasion rate into young colonies of the herbivorous arthropods. This promotes the probability of coinvasions of milkers and killers, which – other things being equal – ultimately favors the latter. Yet, there may be a pitfall in that, so far, neither theoretical nor experimental analyses addressed the possibility of more flexible strategies, such as: "milk when exploiting the prey patch alone, and kill when other (e.g., nonkin) predators have entered the same patch."

In summary, how "virulent" predators should be (as "parasites" of local herbivore populations) depends on whether they are able to monopolize this resource. Sharing of the resource with other clones ("multiple infection") favors increased virulence. How often such sharing occurs depends on the ecological feedback loop. To a certain extent, this conclusion is more robust than those based on other models published in the epidemiological framework (e.g., Nowak and May 1994; Van Baalen and Sabelis 1995a), because it is based on an explicit consideration of how the predator's exploitation strategies affect the interaction time in the patch (and not on some *a priori* assumption about the relation between parasite transmission and host mortality).

The herbivore's dilemma: to stay or to leave?

Just like the predators, the herbivores are independent players in the tritrophic game. When their local populations are discovered and invaded by predators of the milker type, possibilities to achieve reproduction success remain, especially if the milker has such a high emigration rate that it cannot suppress the herbivore population. However, when killers enter the herbivore population, it may pay the herbivores to invest in defense against the killer-like predators or to leave the prey patch in search for enemy-free space. For simplicity, we consider the last type of response only. Consider the pancake predation model again, but now extended with a per capita emigration rate m_N of the herbivore:

$$\frac{dN}{dt} = (r_N - m_N)N - kP , \qquad (22.7a)$$

$$\frac{dP}{dt} = r_P P . \qquad (22.7b)$$

As shown in Figure 22.4b, an increase in m_N causes the time to prey elimination to decrease, as will the overall, local herbivore population (i.e., the area under the herbivore population curve) and the number of predators that will disperse. We may now ask whether there is an evolutionarily stable (ES) emigration rate. To obtain an answer we should first define reproduction success as the per capita emigration rate m_N multiplied by the area A under the herbivore curve (which itself depends on m_N), divided by the initial number of herbivores $N(0)$. This fitness measure always shows a maximum for intermediate values of m_N, because A decreases rapidly with m_N. Suppose, for simplicity, that all patches start synchronously with the same initial number of predators and herbivores and that each patch is colonized by $N(0)$ herbivore clones with $m_{N,\mathrm{res}}$ and just one herbivore mutant clone with $m_{N,\mathrm{mut}}$. Further, assume that the two types of herbivore clones are attacked in proportion to their relative abundance, but that the mutant is so rare that $N_{\mathrm{res}} + N_{\mathrm{mut}} \approx N_{\mathrm{res}}$. This makes herbivore dynamics in the patch and time to prey elimination τ entirely dependent on the resident population. The mutant's presence does not influence the growth of the resident herbivore population and neither does it affect the growth of the predators. Herbivore dynamics in the patch and time to prey elimination τ are thus entirely dependent on the traits of the resident prey population. Now, we ask whether there is a resident herbivore population with $m_{N,\mathrm{res}}$ that cannot be invaded by a mutant with another value of $m_{N,\mathrm{mut}}$. The results presented in Figure 22.8 (Egas *et al.*, unpublished; for model details see Appendix 22.A) show that the ES emigration rate m_N^* increases with:

- Decreasing per capita population growth rate of the herbivores r_N;
- Increasing per capita predation rate k;
- Increasing per capita population growth rate of the predators r_P, and, thus, with a decrease in time to prey elimination τ.

Moreover, the larger the initial number of herbivores, the longer the time to prey elimination and the smaller the ES emigration rate m_N^*.

These results provide some important clues as to how the ES emigration rate m_N^* will change with the exploitation strategy of the predators. This is because predator emigration affects the effective predator growth rate as experienced by the prey. Milkers are predators with an effectively lower per capita rate of population growth, $r_P - m_P$, because of nonvanishing emigration, and the lower the predator's population growth rate, the lower the ES emigration rate m_N^* of the herbivore will be. Thus, a prevalence of milkers in the predator population causes selection for lower emigration rates of the herbivores (i.e., an increased tendency to stay in the herbivore aggregation), whereas a prevalence of killers causes selection for higher herbivore emigration rates. The ES emigration rate m_N^* appears to be always intermediate between 0 and r_N. Thus, herbivores may still aggregate in the face of killer-like predators.

Much remains to be learned as to how the evolution of plant defense strategies interferes with that of herbivore emigration. Increased efforts in direct plant defense probably decrease the per capita rate of herbivore population growth r_N, and as a by-product this triggers selection for a higher ES emigration rate of the herbivores m_N^*. Thus, ultimately the effective per capita rate of herbivore population growth, $r_N - m_N^*$, decreases even more. This paves the way for the evolution of feeding deterrents. The same applies to increased investments in indirect plant defenses. When plants promote the per capita predation rate k or the per capita

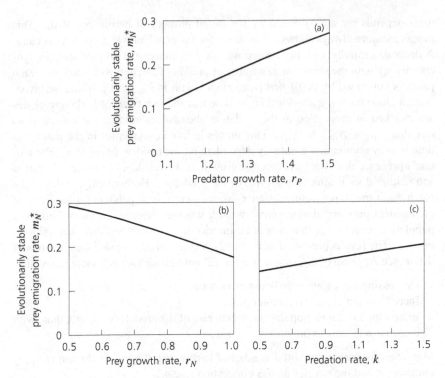

Figure 22.8 ES patch-emigration rates of prey in a predator–prey metapopulation. (a) The ES prey emigration rate m_N^* as a function of predator growth rate r_P. (b) Relation between ES emigration rate m_N^* and the prey growth rate r_N. (c) Relation between ES emigration rate m_N^* and the prey consumption rate k. Default parameter values: $N_{res}(0) = 49$ (a) or 99 (b, c), $N_{mut}(0) = 1$, $P(0) = 1$, $r_N = 1$, $r_P = 1.5$, $k = 1$. *Source*: Egas *et al.* (unpublished).

rate of predator population growth r_P, the by-product is that selection favors increased ES emigration rates of the herbivores, thereby lowering the effective rate of herbivore population growth, $r_N - m_N^*$. Thus, plants may also invest in releasing herbivore deterrents, signaling a high risk of being eaten by predators, and the herbivores are selected for vigilance in detecting the actual presence of predators. There are several examples of herbivorous arthropods that prefer plants with a lower risk of falling victim to natural enemies, despite their lower food quality (Fox and Eisenbach 1992; Ohsaki and Sato 1994). However, it is still unclear why the low quality plants are visited less frequently by the natural enemies of the herbivores. Another speculative, but potentially nice, illustration of plants signaling their predator-defended state to herbivores is found in the work of Bernasconi *et al.* (1998). When corn leaf aphids feed upon them, maize plants respond by releasing a blend of volatile compounds that repels other corn leaf aphids in search for hosts and also attracts parasitoids and lacewings. Interestingly, the blend of plant volatiles contains a monoterpene, (E)-β-farnesene, that corresponds to the alarm signal released by the corn leaf aphid upon predator attack. Another potential example of plants signaling predation risk is given by Pallini *et al.* (1997), who found that spider mites prefer odor from spider-mite infested cucumber plants over odor from thrip-infested cucumber plants, whereas the thrips show no preference. Both spider mites and thrips are herbivores, but the thrips can also act as a predator of spider-mite eggs. Thus, the olfactory avoidance-response of the spider mites may be to avoid competition as well as predation risk. Recently, Pallini *et al.* (1999) found more evidence for the avoidance of predation risk in spider mites. They demonstrated that spider mites prefer odor from plants with spider mites alone over odor from plants with spider mites and the predatory mite *P. persimilis*. Possibly, the odor signal comes from conspecific spider mites that had direct contact with the predators, but this remains to be shown.

The plant's dilemma: direct, indirect, or no defense?

By investing in direct and indirect defenses, a plant gains protection against herbivory, but in doing so it also benefits its neighbors. If these are close kin, an individual plant also increases its inclusive fitness by investment in defense; but if not, it may well promote the fitness of its competitors for the same space and nutrient sources. Thus, the neighbor gains associational protection (Atsatt and O'Dowd 1976; Hay 1986; Pfister and Hay 1988; Fritz and Nobel 1990; Fritz 1995; Hjältén and Price 1997). This leads to the plant's dilemma: should it defend itself, thereby benefiting its neighbors as well, or decrease its defensive efforts? The solution is simple: the defenses should protect the plant without benefiting palatable neighbors too much. When its neighbors are well defended, a plant can afford to invest less itself.

When plants are (constrained to be) either undefended or well defended and herbivores do not discriminate between them, three outcomes are possible:

- All plants are palatable;
- All plants are well defended;
- There is a stable mixture of palatable and well-defended plants.

Coexistence of the two types is possible when either of them increases when rare; in a population of well-defended plants a rare palatable plant can easily gain cheap protection by associating closely with a well-defended plant, whereas in a population of palatable plants a rare, well-defended plant does better as it benefits only few of the palatable individuals and, hence, increases the average fitness of the palatable plants only very little. When either of the two types gradually increases its share in the total population, the benefits wane and ultimately balance the costs, thereby giving rise to a polymorphic plant population (Sabelis and de Jong 1988; Augner *et al.* 1991; Tuomi and Augner 1993; Augner 1994).

The conditions for polymorphism are quite broad, but some of the critical assumptions are not generally valid. For example, herbivores are likely to distinguish between palatable and well-defended plants. In that case, an individual plant is likely to benefit from direct defenses and may even drive the selective herbivores

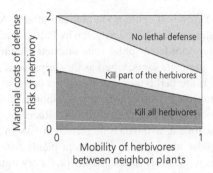

Figure 22.9 The effect of herbivore mobility (horizontal axis) on the parameter areas in which the ESS is to kill the herbivore (dark gray region), to kill part of the herbivores (intermediate, white region), and not to invest in killing herbivores at all (light gray region). The extent of these areas depends on the ratio of marginal costs of plant defense and the initial risk of herbivory. The cost of herbivory equals one if the herbivore survives, and 0.5 if it dies. *Source*: Tuomi *et al.* (1994).

toward the palatable plants. It should be noted that this mechanism does not apply to indirect defenses. Predatory arthropods are usually more mobile than the prey (stages) they attack and they readily move from the plant that employs them as bodyguards to a neighbor plant when the latter is under herbivore attack. Thus, even when the herbivore is a selective feeder, palatable plants profit more easily by settling close to a plant defended by predatory arthropods as bodyguards. Clearly, this mechanism promotes polymorphism (Sabelis and de Jong 1988). However, when defensive plant strategies are not discrete, but continuous (i.e., they cover the full range of possible investment levels), then there may be no polymorphism because ultimately all the plants will exhibit the best average defensive response.

The latter case was analyzed by Tuomi *et al.* (1994). They assumed that the cost of defense increases linearly with the probability of killing the herbivore, the slope being referred to as the marginal cost of defense (i.e., how fast costs increase with the impact of defense on the herbivore). The ES lethality level depends on the risk of herbivory, the marginal cost of defense, and the mobility of the herbivores between neighbor plants, as shown in Figure 22.9. When herbivory risks are low and marginal costs of defense high, then it does not pay to kill the herbivore. However, when the risk of herbivore damage is high and marginal defense costs are sufficiently low, then it pays to kill the herbivore. For intermediate ratios of marginal defense costs and risk of herbivory, ES lethality depends on the mobility of the herbivore between neighboring plants (Tuomi *et al.* 1994). Obviously, high mobility causes neighboring plants to share the same herbivores and selects for lower lethality levels, whereas low mobility selects for increased lethality. It is becoming increasingly clear that neighboring plants may communicate via damage-related signals (Farmer and Ryan 1990; Bruin *et al.* 1992, 1995; Shonle and Bergelson 1995; Adviushko *et al.* 1997; Shulaev *et al.* 1997), so it may well be that plant defenses include strategies conditional upon the neighbor's state, as defined by:

- Whether it is actually under attack;
- Its defensive response.

This is a largely open problem in need of further theoretical and experimental work.

Whether the effects of defensive efforts occur in discrete jumps or are more gradual is an important determinant of the existence of polymorphism, but the most relevant message is that in both cases associational protection may lead to a lower average investment in defenses. This applies to direct defenses (Tuomi *et al.* 1994), as well as to indirect defenses (Sabelis and de Jong 1988).

But coevolution may act as a boomerang . . .

The most elusive unknown of all is the interplay between metapopulation dynamics and evolution at all three trophic levels. To assume a steady state metapopulation is clearly an oversimplification in view of the complex dynamics (e.g., bistability, chaos) that may arise by adding an extra trophic level to ditrophic models (Sabelis *et al.* 1991; Jansen and Sabelis 1992, 1995; Klebanoff and Hastings 1994; Jansen 1995; Kuznetsov and Rinaldi 1996) or the complexities that may arise from the interactions within food webs of arthropods on plants (intraguild predation; "prey-eats-predator"; apparent competition; Holt 1977; Polis *et al.* 1989; Polis and Holt 1992; Holt and Polis 1997). Thus, where the tritrophic system will settle evolutionarily is very hard to predict.

To illustrate this, it is worthwhile to carry out an – admittedly speculative – thought experiment. Consider what will happen when plants evolve to invest more in direct and indirect defenses, and predators are initially of the killer type. First and foremost, increased defensive efforts by the plant decrease the size of the herbivore's metapopulation. Subsequently, the size of the predator's metapopulation decreases which in turn causes a drop in the rate of predator invasion into herbivore patches. As a consequence, the probability of coinvasion of predators with different prey-exploitation strategies into the same herbivore patch decreases, thereby providing a selective advantage to predators that are more milker-like. In addition, increased plant defense promotes herbivore emigration and decreases the size of local herbivore populations. This also decreases the probability of predator coinvasion and thus selects for milkers. Thus, the plant's investment in defense may ultimately result in ineffective predators; this we call boomerang coevolution. It is typically the consequence of adding one more trophic level to an exploiter–victim system and allowing strategies of exploitation and defense to vary at each trophic level.

A similar thought experiment can be carried out for the case that not only does the plant benefit from its own investment in direct and indirect defenses, but so do its neighbors, who may well be competitors (Sabelis and de Jong 1988; Augner *et al.* 1991; Tuomi *et al.* 1994). Again, increased investment in plant defense causes a boomerang effect because neighboring plants profit and allocate the energy saved directly to increase their seed output or indirectly by increasing their competitive ability.

Boomerang coevolution arises through the impact of plant defenses on alternative allocation strategies of neighboring plants and via the positive effect on the milker-like prey-exploitation strategies. This may well be the evolutionary reason why many plant species channel so little of their energy resources into defense against herbivores (e.g., "cheap" carbon-demanding defenses rather than "expensive" nitrogen-demanding defenses), whether this be direct defense (Simms and Rausher 1987, 1989; Herms and Mattson 1992; Simms 1992) or indirect defense (Beattie 1985, p. 52; Dicke and Sabelis 1989). Hence, we hypothesize that boomerang coevolution constrains the plant's investment in direct and indirect defenses. Low investment, however, does not necessarily imply that plant defenses have a low impact. This entirely depends on the quantitative details of how the offensive and defensive traits of the interacting organisms at all three trophic levels settle evolutionarily. In other words, the impact of the plant's defenses increases if herbivores become more milker-like and predators more killer-like, and the impact decreases when herbivores become more killer-like and predators more milker-like.

22.5 Discussion

In this chapter we give a game theoretical view of the evolution of exploitation and defense strategies in tritrophic systems. Do we now understand why plants invest in promoting the effectiveness of the herbivore's enemies and can we learn from these insights to manipulate the virulence of the herbivore to the plant and that of the predator to the herbivore?

Are tritrophic systems prone to evolve conspiracy?

It is commonly believed that plant–predator mutualisms readily evolve because it is in the interests of both the plant and the predator to act against the herbivores (Price *et al.* 1980). Indeed, much evidence shows that plants can provide alternative food, shelter, and SOS signals utilized by the natural enemies of the herbivores, but what is still lacking is a critical assessment of the overall benefits to the plant. Much work is required to detect the role of cheating. Clearly, the plant cannot control who is benefiting from the facilities offered by the plant. Alternative food, shelter, and SOS signals are all open to "misuse" by the plant's enemies or inefficient natural enemies of the herbivores (Sabelis *et al.* 1999a, 1999b). In addition, there is a need to analyze how investments in plant defense influence competition among neighboring plants, since one plant may profit from the bodyguards retained and attracted by the other. Again, the investor cannot monopolize the benefits that accrue from bodyguards, since they move to wherever their victims are. Finally, the benefits to the plant depend on the number of predators in the surrounding environment and these numbers fluctuate. Costs and benefits of indirect plant defense are therefore expected to vary greatly in time and space (Bronstein 1994a, 1994b).

Thus, even though plenty of evidence shows that plants invest in attracting, retaining, feeding, and protecting bodyguards and that the bodyguards can make good use of the facilities offered by the plant, it is not an easy task to demonstrate in the field that predators assume the role of the plant's immune system.

Whereas a net benefit of indirect plant defenses is still to be shown experimentally, the rationale that underlies the evolution of indirect defenses is not fully established either. In this chapter we analyze the interaction between a plant, its neighbors, its herbivores, and the herbivore's enemies as a game of defense (among neighboring plants), escape (among herbivores), and resource exploitation (among herbivores and among predators). We explain that the mobilities of the herbivore and its predator play a crucial role in determining the extent to which a plant can reap the benefits from investing in direct or indirect defense (Tuomi *et al.* 1994). In addition, we discuss how prevailing resource exploitation strategies may change through metapopulation structure dynamics and migration via their impact on the probability of coinvasion of exploiters with different strategies of exploiting the resource. We argue that there is an interplay between plant defense, plant competition, exploitation of host plant, and prey in tritrophic systems. We also speculate that there is room for unstable evolutionary dynamics (e.g., boomerang coevolution), which may give way to selection for low investment in plant defense and milker-like predators. Indeed, empirical observations indicate that indirect defenses are not very costly, but killer-like, not milker-like predators seem to prevail in the field. This contrast between prediction and observation may indicate that predators have no control over the exploitation of their resources. This may arise from external causes, such as mortality through abiotic factors (wind, rain, fire) and competition with other natural enemies (pathogens, parasites, predators). To specify the conditions under which low-cost indirect defenses and killer-like predators evolve, our game-theoretical analysis needs extension to include evolution in the full tritrophic system and its interaction with ecological dynamics. This approach may lead to a more sound rationale for the "world is green" hypothesis. The commonly accepted hypothesis that mutualism readily evolves in plant–herbivore–predator systems because "it is both in the interest of plants to get rid of the herbivores and in the interest of the predators to find herbivores as prey" is as yet unfounded.

Perspectives for virulence management

Given that the theory on evolution and defense in tritrophic systems is rather immature, it is too early to consider direct applications, but there are several potentially important implications for virulence management. First, we may ask how two strategies, biological control and breeding for plant resistance, to combat plant pests influence ultimate success in crop protection. One-sided measures to increase direct plant defense may ultimately select for mild predators and, therefore, increased herbivory may be the end result. It seems wise to also breed for increased indirect defenses, because this promotes multiple colonization by predators, which will in turn select for increased predator virulence. Second, an old debate among biocontrol workers is whether to release single or multiple species of predators. The latter seems best on the condition that it promotes local competition between predators, because this will increase their virulence. Third, we may wonder how mass rearing influences the virulence of predators before their release in the field. We suspect that mass rearings are like an undepletable prey patch and that leaving that patch is unlikely to promote within-rearing success. Hence, there will be selection for predators that suppress their tendency to disperse as long as there is food. This inadvertent selection for killer-like predators is good news for biological control workers aiming at fast suppression of the plant pest near the site of predator release, but are the aims of biological control over large areas served in this way, because at that spatial scale dispersal may become a vital trait of a successful biocontrol agent. Moreover, predators less efficient in clearing pest arthropods from a plant (milkers) may produce more dispersers and therefore promote their chances to reach distant sites. Clearly, there is every reason to reconsider carefully the criteria for a good biocontrol agent when the aim is to achieve control over a large spatial scale. Fourth, there is the long-standing question of where best to collect candidates for biological control. We expect to find milker-type predators near the borders of the geographical range (where predator densities are low) and killer-type predators in the center (where predator densities are high; see Pels and Sabelis 1999), but much more (theoretical and empirical) work is needed to substantiate this claim. A more elaborate discussion of these four implications is given in Chapter 32, albeit based on predictions of how selection molds predator virulence only and not on how it acts on virulence and defense in systems with three trophic levels.

Appendix 22.A Evolutionarily Stable Herbivore Emigration Rate

Assume a metapopulation of patches with local interactions and global migration. As in Van Baalen and Sabelis (1995c), all the patches start with exactly the same number of predators and prey (metapopulation-wide equilibrium and no stochastic variation in the number of colonizers), that is, $N_0 = N(0)$ and $P_0 = P(0)$, and the redistribution of predators and prey occurs after completion of the local interaction (sequential interaction rounds). To assess the evolutionarily stable emigration rate (given r_N, k, and r_P), a measure of prey fitness is defined as the total number of herbivore emigrants per herbivore foundress, $w = m_N(N_{\text{tot}}/N_0)$, where N_{tot} is the total number of prey produced in a patch. Defined in this way, the fitness measure reaches an optimum between no migration ($m_N = 0$; extinction of all patches) and an emigration rate equal to the per capita growth rate ($m_N = r_N$; no offspring produced in the patch).

Further, assume that the two types of herbivore clones are attacked in proportion to their relative abundance, but that the mutant is so rare that $N_{\text{res}} + N_{\text{mut}} \approx N_{\text{res}}$. Then the herbivore dynamics in the patch are entirely dependent on the resident population and the rare mutant's presence does not influence the growth of the resident herbivore population. As a result of this last assumption the model yields an analytical solution for the fitness of the rare mutant. The dynamics of the rare mutant are given by

$$\frac{dN_{\text{mut}}}{dt} = \left(r_N - m_{N,\text{mut}}\right) N_{\text{mut}} - kP\frac{N_{\text{mut}}}{N_{\text{res}}} . \tag{22.7c}$$

Solving Equations (22.7a), (22.7b), and (22.7c) gives an explicit description of how the number of mutant herbivores changes with time,

$$N_{\text{mut}}(t) = \frac{N_{0,\text{mut}}}{N_0}\left\{ N_0 e^{(r_N - m_{N,\text{mut}})t} - P_0\frac{k}{r_P - \left(r_N - m_{N,\text{res}}\right)} \right.$$
$$\left. \left[e^{(r_P + m_{N,\text{res}} - m_{N,\text{mut}})t} - e^{(r_N - m_{N,\text{mut}})t} \right] \right\} , \tag{22.7d}$$

with $N_{0,\text{mut}} = N_{\text{mut}}(0)$.

The fitness of the rare mutant w_m is the sum of dispersing mutants divided by $N_{0,\text{mut}}$,

$$w_m = m_{N,\text{mut}} \int \frac{N_{\text{mut}}(t)}{N_{0,\text{mut}}} , \tag{22.8a}$$

or

$$w_m = \frac{m_{N,\text{mut}}}{N_0}\left\{ N_0 \frac{e^{(r_N - m_{N,\text{mut}})\tau} - 1}{r_N - m_{N,\text{mut}}} - P_0\frac{k}{r_P - \left(r_N - m_{N,\text{res}}\right)} \right.$$
$$\left. \left[\frac{e^{(r_P + m_{N,\text{res}} - m_{N,\text{mut}})\tau} - 1}{r_P + m_{N,\text{res}} - m_{N,\text{mut}}} - \frac{e^{(r_N - m_{N,\text{mut}})\tau} - 1}{r_N - m_{N,\text{mut}}} \right] \right\} . \tag{22.8b}$$

The evolutionarily stable emigration rate m_N^* can be found as the emigration rate of the resident population for which no mutant with a different emigration rate has higher fitness,

$$\left.\frac{dw_m}{dm_{N,\text{mut}}}\right|_{m_{N,\text{mut}}=m_{N,\text{res}}} = 0 , \tag{22.9}$$

Equations (22.8b) and (22.9) yield an implicit function, from which a solution for m_N^* can be obtained,

$$m_N^* = \frac{\left(N_0 + \frac{kP_0}{r_P - \Delta}\right)\left(\frac{e^{\Delta\tau} - 1}{\Delta}\right) - \left(\frac{kP_0}{r_P - \Delta}\right)\left(\frac{e^{r_P\tau} - 1}{r_P}\right)}{\left(\frac{kP_0}{r_P - \Delta}\right)\left(\frac{e^{r_P\tau} - 1}{r_P^2} - \frac{\tau e^{r_P\tau}}{r_P}\right) - \left(N_0 + \frac{kP_0}{r_P - \Delta}\right)\left(\frac{e^{\Delta\tau} - 1}{\Delta^2} - \frac{\tau e^{\Delta\tau}}{\Delta}\right)} , \tag{22.10a}$$

where $\Delta = r_N - m_N^*$ and τ is the total time of the predator–prey interaction,

$$\tau = \frac{1}{r_P - \Delta} \ln\left(1 + \frac{r_P - \Delta}{k}\frac{N_0}{P_0}\right) . \tag{22.10b}$$

Part F

Vaccines and Drugs

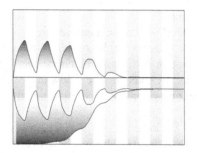

Introduction to Part F

Evolutionary virulence management necessarily takes a long-term perspective and concentrates on population-level characteristics. Yet, in practice we also have to interfere with diseases on a short-term basis and, especially in the case of humans, the welfare of individual patients is an additional concern. This establishes the need to evaluate the longer-term effects of short-term protection measures. Only on this basis can we understand which compromises can be made or, even better, whether it is possible to devise practices that allow both satisfactory short-term and long-term disease control.

The main individual-level protection measures are drug treatment and vaccination. Both have consequences for public health by affecting the population dynamics of the disease, though in different manners: the former by shortening the infectious period, the latter by changing the inflow of fresh susceptible hosts. And both tend to have evolutionary consequences in terms of resistance evolution and vaccine escape.

After drug resistance develops, we are basically back at square one in terms of the control effort, since resistant types have at best a very slightly reduced fitness. If the drug-based selective regime is maintained long enough, resistant and nonresistant types even tend to become equally fit in a drug-free environment because of the incorporation of genetic modifiers.

In the case of drug resistance we appear to be rapidly running out of alternative options, whereas the prospects are considerably better for vaccination as, in principle, we can keep adapting the vaccine type. In addition, vaccine-escape mutants tend to have a lower basic reproduction ratio as they do not gain a foothold before implementation of the vaccination scheme.

Chapter 23 discusses the problems that arise from the rapid evolution of antibiotic resistance. Bonhoeffer addresses this issue by means of a mathematical model for the dynamics of infection that tackles the question of how to use existing antibiotics with maximal effect for the treatment of bacterial infections, while simultaneously delaying, and possibly even reversing, the emergence of resistance. Within this framework he assesses the effects of different strategies, like cycling antibiotic therapy, combination therapy, and others. The chapter also comments on how to define optimal treatment policies, and, in this context, how to weigh long-term against short-term benefits.

In Chapter 24 McLean develops a simple model framework to evaluate the effect of vaccination schedules on the emergence of vaccine-escape mutants. The depression of a competitively superior strain by a vaccine that confers little cross-immunity may change the competitive balance, setting off an outbreak of an earlier, competitively inferior strain. An unexpected finding is that such effects take much longer to occur than might be guessed naively. The message to virulence managers is that we should not allow ourselves to be lulled into a false feeling of security. The good news is that the escaping strain should have a smaller basic reproduction ratio than the one that dominated earlier on.

Chapter 25 considers pathogen populations that exhibit a vast diversity in antigen types from the start. These are maintained by immune selection as a stable, discrete set of independently transmitted types without overlap in antigenic repertoires. Gupta argues that when antigenic types can provide cross-protection, the dynamics of each antigenic type may change dramatically, and so will the dynamics of pathogen virulence. Therefore, to understand changes in pathogen virulence in the absence or presence of interventions such as vaccination, one should consider the underlying composition of the pathogen population.

Chapter 26 takes the argument in Chapter 25 a step further. One move in the combat against the diversity of circulating antigenic types has been the development of conjugate vaccines that simultaneously offer protection against several serotypes of the pathogen. With the increased use of such vaccines the concern now is that these vaccinations vacate niches for other serotypes. Lipsitch critically reviews the empirical data sets that have been used to identify serotype replacement, outlines better ways of tracing such processes, and argues that serotype replacement can be, but is not necessarily always, bad. This is because serotype replacement may augment the effectiveness of a vaccination program in a community when nonvaccine serotypes outcompete vaccine serotypes.

Taken together, the chapters in Part F give an overview of the current scientific insight into the potential evolutionary implications of the major measures of short-term disease control.

Managing Antibiotic Resistance: What Models Tell Us?

Sebastian Bonhoeffer

23.1 Introduction

The rapid ascent of antibiotic resistance is cause for great concern in public health. Resistant bacteria not only compromise the success of treatment, but can also be transmitted and cause an epidemic of resistant organisms. In hospitals, antibiotic resistance is commonly observed in organisms such as *Streptococcus pneumoniae* (Doern *et al.* 1996), *Neisseria gonorrhoeae* (Cohen 1994), and *Mycobacterium tuberculosis* (Bloom and Murray 1992), as well as in nosocomial pathogens including *Staphylococcus aureus* (Swartz 1994), *Enterococcus* spp. (Swartz 1994; Arthur and Courvalin 1993), and *Klebsiella* spp. (Jacoby 1996). At the same time, the development of new antibiotic agents is becoming increasingly difficult and costly.

Antibiotics, therefore, must be seen as a limited resource in our efforts to control and cure bacterial infections. This raises the question of how to use existing antibiotics with maximal effect to treat bacterial infections while simultaneously reversing or delaying the emergence of resistance. A number of measures have been proposed to counteract the evolution of resistance, including the improvement of hospital hygiene (Murray 1994), the possible use of vaccines (Jernigan *et al.* 1996), tighter controls on antibiotic use in clinical practice as well as agriculture (Anonymous 1995), and alternative patterns of antibiotic use such as sequential cycling of antibiotics or combination therapy (Swartz 1994; McGowan 1986).

These measures are not mutually exclusive, and an effective plan for the future use of antibiotics must consider how they should be combined. In this chapter, however, I focus only on how different patterns of antibiotic use may affect the evolution of resistance. I present and analyze a series of mathematical models to investigate the strengths and weaknesses of different treatment strategies. The purpose of such models is to move the discussion from a verbal to a more formal basis of reasoning. Clearly, mathematical models are no substitute for clinical trials, but they may help to provide theoretical guidelines for the design of efficient treatment strategies.

23.2 Evaluation of Drug Treatment Strategies

Before addressing the question of how to make optimal use of an antibiotic, we must first define a precise criterion according to which different treatment strategies can be ranked. The simplest criterion is to rank treatment strategies according to the time taken until a given percentage of patients are infected with resistant bacteria. Clearly, the timeframe in which resistance emerges is an important factor in the evaluation of a treatment strategy. However, it is also clear that it cannot be used as the sole ranking criterion, as its maximum value occurs when the antibiotic is never used.

Therefore, we need a criterion that balances the value of preserving a drug's effectiveness with the value of curing infections. Such a criterion should rank treatment policies according to the extent to which they reduce the number of infected individuals and increase the number of uninfected individuals over time before the antibiotic fails because of resistance (Bonhoeffer *et al.* 1997).

Mathematically, this amounts to maximizing the number of uninfected hosts or minimizing the number of infected hosts over a period of time long enough for the antibiotic to lose its effectiveness in the patient population through resistance. (It turns out that both criteria are mathematically equivalent.) In this chapter, I refer to this criterion as the "maximum benefit criterion."

23.3 Dynamics of Infection: A Simple Model

To illustrate some points that are central to this chapter, I first present a simple model of the dynamics of resistance in the presence of treatment, and extend this later on. Consider three populations: S, the density of uninfected individuals; I_w, the density of individuals infected with nonresistant wild-type bacteria; and I_r, the density of individuals infected with resistant bacteria. Uninfected individuals enter the population at a rate B, and die at a rate d (accounting for the natural death rate through causes unrelated to infection). Uninfected individuals are infected (either by wild-type or resistant bacteria) at a rate proportional to the product of their density, S, the density of infected hosts, $I_w + I_r$, and a transmission rate parameter, β. Infected individuals die at a rate μ, which includes natural and disease-associated mortality. Infected individuals may recover spontaneously from infection at a rate θ_s. Patients infected with nonresistant bacteria may also recover from infection through antibiotic therapy. This occurs at a rate, $q\theta_t$, where q is the fraction of patients who receive therapy, and θ_t is the rate at which patients recover when treated. Once an infection is cleared, the patient returns to the pool of uninfected hosts. Altogether, we obtain the following for the dynamics of infected and uninfected individuals,

$$\frac{dS}{dt} = B - dS - \beta S(I_w + I_r) + \theta_s(I_w + I_r) + q\theta_t I_w \,, \tag{23.1a}$$

$$\frac{dI_w}{dt} = \beta SI_w - \mu I_w - \theta_s I_w - q\theta_t I_w \,, \tag{23.1b}$$

$$\frac{dI_r}{dt} = \beta S I_r - \mu I_r - \theta_s I_r .\tag{23.1c}$$

This model makes several assumptions:

- There is no fitness cost associated with resistance. Sensitive and resistant bacteria are identical with respect to all bacteria-specific parameters (β, θ_s, μ) for the spread in the patient population in the absence of antibiotics. (This assumption is later relaxed.)
- Individuals who are treated and cured (or recover) become immediately susceptible again. Hence, the possibility of temporary or life-long immunity is ignored. It can be shown, however, that including temporary or permanent immunity does not affect the conclusion drawn here concerning optimal treatment policies based on the criterion of maximum benefit.
- The possibility of superinfection of wild-type-infected individuals by resistant bacteria (or vice versa) is not considered. This can be safely ignored if there is cross-immunity during infection or if the incidence of both types of infection is low in the population.
- This model describes directly transmitted bacterial infections, because transmission is contact dependent.

23.4 The Steady State

In the absence of therapy ($q = 0$), this system converges to two different steady states depending on the basic reproduction ratio. The basic reproduction ratio, commonly denoted by R_0, is defined as the number of secondary infections caused by a single infected individual placed into an entirely susceptible population (Anderson and May 1991). Clearly, if $R_0 > 1$, then the infection is capable of spreading in the population. If $R_0 < 1$, then the infection cannot spread and will disappear from the population. For this model the basic reproduction ratio is given by $R_0 = B\beta/[d(\mu + \theta_s)]$. If $R_0 > 1$, then the system converges to the endemic steady state given by $S^* = (\mu + \theta_s)/\beta$ and $I = I_w + I_r = B/\mu - d(\mu + \theta_s)/(\beta\mu)$. If $R_0 < 1$, then the system converges to the uninfected steady state given by $S = B/d$ and $I = 0$. An understanding for R_0 can be obtained by noting that B/d is the density of uninfected individuals in the absence of the infection, $1/(\mu + \theta_s)$ is the average duration of infection, and β is the rate at which susceptible individuals are infected.

Provided $R_0 > 1$, then in the presence of treatment ($q > 0$), the system converges to the resistant steady state given by $S^* = (\mu + \theta_s)/\beta, I_w = 0, I_r = B/\mu - d(\mu + \theta_s)/(\beta\mu)$. Hence, in the presence of treatment, the resistant bacteria eventually replace the nonresistant ones.

23.5 Gauging Antibiotic Therapy

Suppose the system is in the infected steady state. Suppose also that when therapy is initiated, the majority of patients are infected by nonresistant bacteria and only a small fraction of patients, p_0, harbor resistant bacteria. Let us calculate the total

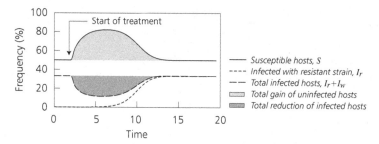

Figure 23.1 A numerical simulation of the Equations (23.1a) to (23.1c). The total gain of uninfected hosts is given by the light shaded area. The total reduction of infected hosts is given by the dark shaded area. Both the total gain and the total reduction are equivalent as criteria for the evaluation of antibiotic treatment policies.

increase in uninfected patients gained by antibiotic therapy before the antibiotic fails through resistance. Mathematically, this amounts to calculating the integral of $S - S^*$ over the time during which the antibiotic is used, where $S^* = (\mu + \theta_s)/\beta$ is the density of individuals in the infected steady state. This integral is illustrated graphically by the light shaded area in Figure 23.1. (Strictly speaking, this integration gives the total gain in uninfected patients multiplied by time. So the success of a treatment policy would be measured in a unit such as "uninfected patient years.")

To calculate this, I divide Equation (23.1c) by I_r and integrate over the duration of treatment, T. Thus, we have

$$\int_0^T \frac{1}{I_r} \frac{dI}{dt}\, dt = \beta \int_0^T \left(S - \frac{\mu + \theta_s}{\beta} \right) dt$$

$$\ln[I_r(T)/I_r(0)] = \int_0^T \frac{1}{I_r} dI_r\, dt = \beta \int_0^T (S - S^*)\, dt \,. \tag{23.2a}$$

Hence, we obtain for the gain in "uninfected patient years" after a time T,

$$G(T) = \int_0^T (S - S^*)\, dt = \frac{1}{\beta} \ln \frac{I_r(T)}{I_r(0)} \,. \tag{23.2b}$$

If treatment is continued for a sufficiently long time, the system approaches the resistant steady state. In that case, $I_r(T) = I_r(0) + I_w(0)$. Then, for the total gain which can be achieved with an antibiotic until it loses its effect on the patient population, we obtain

$$G = G(\infty) = -\frac{\ln(p_0)}{\beta} \,, \tag{23.2c}$$

where $p_0 = I_r(0)/I_r(T)$.

Interestingly, the total gain of uninfected hosts is independent of the rate $q\theta_t$ at which patients are treated and cured – that is, it is independent of the treatment strategy. (Note that we are considering treatment strategies at the population level

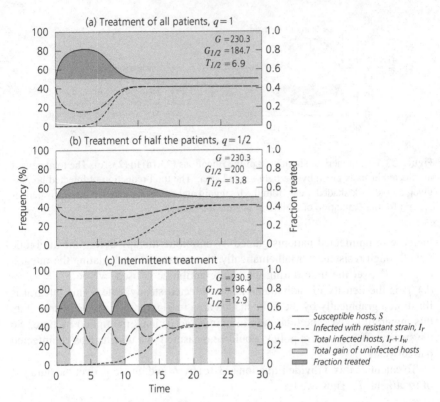

Figure 23.2 The total gain G of uninfected hosts is independent of the treatment strategy. However, the time until half of the patients are infected with resistant bacteria and the cumulative gain of uninfected hosts during that time depend on the treatment strategy.

and not at the level of the individual infected patient.) Whether a large or small fraction of the population receives treatment, or whether treatment is continuous or intermittent, the eventual gain of uninfected hosts is always the same regardless of the chosen treatment strategy (see Figure 23.2).

The conditions under which the results for G and $G(T)$ were derived are important. In particular, I assumed that at the time when the antibiotic is applied to the population some patients are already infected with resistant bacteria. Although this may often be a reasonable assumption, it will not generally be true. Secondly, as pointed out before, I assumed that there is no fitness cost associated with resistance. In as far as these fitness costs have been measured, they are often small (Bouma and Lenski 1988; Schrag and Perrot 1996); but this may not generally be the case. I relax these assumptions further below.

It is interesting that the derivation of the total gain of uninfected hosts depends only on Equation (23.1c). The result is therefore independent of the particular form of the dynamics of the wild-type-infected and uninfected hosts [i.e., Equations (23.1b) and (23.1c)]. Hence, the result is only dependent on the assumptions affecting Equation (23.1c), which strengthens the generality of the result. Using Equations (23.1a) and (23.1b), one can show that the total reduction of infected hosts is also independent of the fraction of patients who receive therapy, q. Therefore, the total reduction of infected hosts and the total gain of uninfected hosts are equivalent criteria for the benefit of an antibiotic treatment policy.

Furthermore, one can show that including a class of temporarily or permanently immune individuals does not affect the conclusion that the gain of uninfected hosts is independent of the treatment strategy. However, importantly, if a class of immune individuals is included, the criterion then is to maximize the number of individuals who have never been infected rather than the number of individuals who are not infected at a given time.

More interesting properties of this simple model are worth pointing out. For example, we can calculate the time taken for resistant bacteria to grow from a fraction, p_0, to a fraction, p_1, in the presence of treatment. To do this, I transform Equations (23.1a) to (23.1c) and express the dynamics of resistance in terms of the total number of infected patients, $I = I_w + I_r$, and the fraction of patients infected with resistant bacteria, $p = I_r/I$. We then obtain

$$\frac{dS}{dt} = B - dS - \beta SI + \theta_s I + q\theta_t I(1-p),\tag{23.3a}$$

$$\frac{dI}{dt} = \beta SI - \mu I - \theta_s I - q\theta_t I(1-p),\tag{23.3b}$$

$$\frac{dp}{dt} = q\theta_t p(1-p).\tag{23.3c}$$

The solution for the fraction of resistant infected hosts is

$$p(t) = \frac{p_0}{(1-p_0)e^{-q\theta_t} + p_0}.\tag{23.4a}$$

Hence, we obtain for the time to increase from p_0 to p_1

$$T_p = \frac{1}{q\theta_t} \ln\left(\frac{1-p_0}{p_0}\frac{p_1}{1-p_1}\right).\tag{23.4b}$$

Interestingly, the time for the rise of resistance therefore depends inversely on $q\theta_t$, the rate at which patients are treated and cured, but only depends logarithmically on the initial frequency of resistant infections, p_0. Hence, large changes in p_0, even of orders of magnitudes, have only a small effect on the time for resistance to rise to high frequencies. However, even small changes of $q\theta_t$ have a strong effect on the rise of resistance.

23.6 Treatment with Two Antibiotics: An Extended Model

To extend this simple model, I first introduce the possibility that resistant bacteria have a selective disadvantage compared to the nonresistant bacteria in the absence of antibiotic therapy. Such a fitness cost to resistance could be manifest in any of the parameters β, μ, or θ_s that describe rate of infection, mortality, and spontaneous recovery, respectively. For simplicity, I assume, however, that the fitness cost is manifest by a faster rate of recovery from resistant infection. Hence, patients who harbor resistant bacteria are on average infected for a shorter period of time than patients infected with wild-type bacteria, resulting in a decreased overall transmission of resistant bacteria.

Secondly, I distinguish two different pathways by which patients may become resistant to treatment. First, patients may be infected by a resistant carrier. This pathway is already included in the simple model and is called "primary resistance." Alternatively, a small subpopulation of resistant bacteria may be present in a patient predominantly infected by nonresistant wild-type bacteria and may outgrow the nonresistant bacteria when the patient receives antibiotics. This is called "acquired" or *"de novo"* resistance. Hence I include a new parameter that describes the fraction of patients who are initially predominantly infected by wild-type, but who develop resistance when receiving antibiotics.

Thirdly, I extend the model to describe treatment with several antibiotics. For simplicity, I present the model for two antibiotics A and B, but the model and the results can easily be generalized for more than two antibiotics.

The extended model is

$$
\begin{aligned}
\frac{dS}{dt} = {} & B - dS - \beta(I_w + I_A + I_B + I_{AB})S \\
& + \theta_{sw}I_w + \theta_{sA}I_A + \theta_{sB}I_B + \theta_{sAB}I_{AB} + \theta_t(1 - \phi_2)q_{AB}I_w \\
& + \theta_t(1 - \phi_1)[(q_A + q_B)I_w + q_AI_B + q_BI_A \\
& + q_{AB}(I_A + I_B)] \,,
\end{aligned}
\tag{23.5a}
$$

$$
\frac{dI_w}{dt} = [\beta S - \mu - \theta_{sw} - \theta_t(q_A + q_B + q_{AB})]I_w \,,
\tag{23.5b}
$$

$$
\frac{dI_1}{dt} = [\beta S - \mu - \theta_{sA} - \theta_t(q_B + q_{AB})]I_A + \theta_t\phi_1 q_A I_w \,,
\tag{23.5c}
$$

$$
\frac{dI_2}{dt} = [\beta S - \mu - \theta_{sB} - \theta_t(q_A + q_{AB})]I_B + \theta_t\phi_1 q_B I_w \,,
\tag{23.5d}
$$

$$
\begin{aligned}
\frac{dI_{12}}{dt} = {} & (\beta S - \mu - \theta_{sAB})I_{AB} + \theta_t\phi_1[q_{AB}(I_A + I_B) + q_AI_B + q_BI_A] \\
& + \phi_2\theta_t q_{AB}I_w \,.
\end{aligned}
\tag{23.5e}
$$

The extended model takes the following form. There are five variables: S, the susceptible hosts; I_w, the individuals infected with nonresistant wild-type bacteria; I_A, individuals infected with bacteria resistant to antibiotic A; I_B, individuals infected with bacteria resistant to antibiotic B; and I_{AB}, individuals infected with bacteria resistant to both antibiotics. The parameters θ_{sw}, θ_{sA}, θ_{sB}, and θ_{sAB} are the recovery rates of wild-type, A-resistant, B-resistant, and AB-resistant infected hosts, respectively. The parameters q_A, q_B, and q_{AB} are the fraction of patients treated with A, B, or AB. Since these parameters reflect fractions, their sum is ≤ 1. (If the sum is smaller than 1, then a fraction of patients does not receive any treatment.) Finally, the parameters ϕ_1 and ϕ_2 describe the fraction of patients that develops *de novo* resistance when treated with a single antibiotic or with both simultaneously. (Both ϕ_1 and ϕ_2 are smaller than 1, since they are fractions.)

Before we consider treatment strategies using both antibiotics, we first return to single antibiotic therapy to see how the introduction of a fitness cost for resistance and the possibility of *de novo* resistance change the finding that the gain of uninfected hosts is independent of the treatment strategy. We assume that only antibiotic A is used, and that when therapy is first applied to the population, all patients are infected by nonresistant bacteria. A calculation similar to that for the simple model in the previous section yields for the fraction of A-resistant hosts as a function of time,

$$p(t) = \frac{\phi_2(e^{q_A \theta_t t} - 1)}{\phi_2(e^{q_A \theta_t t} - 1) + 1},$$ (23.6a)

and for the gain of uninfected hosts,

$$G(T) = \frac{1}{\beta}\left[-\ln \phi_1 + \Delta\theta_s T + \ln z + \ln\left(1 - \phi_1 - \frac{\Delta\theta_s}{q_A\theta_t}\right)\right],$$ (23.6b)

where T is the time past the start of therapy, $\Delta\theta_s = \theta_{sA} - \theta_{sw}$ is the cost of resistance to antibiotic A, and z is the ratio between total density of infected hosts at time T and at equilibrium before the start of therapy.

If there is no cost to resistance ($\Delta\theta_s = 0$), then the equilibrium densities before treatment and during treatment are the same (i.e. $z = 1$), and the gain of uninfected hosts is given by

$$G = \frac{1}{\beta}\ln\frac{1 - \phi_1}{\phi_1}.$$ (23.6c)

Hence, in this case, the gain of uninfected hosts is again independent of $q_A\theta_t$, the rate at which patients are treated and cured with antibiotic A. Therefore, including the possibility of *de novo* resistance does not affect the finding that the gain of uninfected hosts is independent of the treatment strategy, provided that there is no cost to resistance. However, if there is a cost to resistance, then the gain of uninfected hosts $G(T)$ increases with increasing $q_A\theta_t$. However, the numerical effect is small, because G only depends logarithmically on $q_A\theta_t$.

The main numerical effect of a cost of resistance to the total gain is represented by the term $\Delta\theta_s T$, which reflects that the equilibrium of infected hosts is lower during therapy than before therapy, because the resistant bacteria have a larger rate of recovery and, therefore, a smaller net transmission rate than the nonresistant bacteria in the absence of antibiotic therapy.

23.7 Multiple Antibiotic Therapy

Let us now consider treatment strategies using both antibiotics and represented by two scenarios: TR, the majority of resistant infections are caused by transmission of resistance, and DN, the majority of resistant infections initially result from *de novo* resistance, and only in later stages of the epidemic does transmission of resistance become numerically important.

For scenario TR, let us assume that the problem of AB-resistance has progressed to a point such that the large majority of AB-resistant patients have acquired the resistant bacteria by infection from a resistant carrier. By comparison, the number of cases with *de novo* acquisition of AB-resistance is numerically negligible. Mathematically, this amounts to $\phi_2, \phi_1 \approx 0$ and $I_A, I_B, I_{AB} > 0$ at the start of antibiotic therapy in Equations (23.5a) to (23.5e). Solving for the gain of uninfected hosts we get

$$G(T) = \frac{1}{\beta}[\ln(z/\tilde{\rho}_0) + \Delta\theta_s T] , \tag{23.7}$$

where $\tilde{\rho}_0$ is the frequency of AB-resistant bacteria at the start of therapy, $\Delta\theta_s = \theta_{sAB} - \theta_{sw}$ is the cost of AB-resistance, and z is, as above, the ratio of the steady states of $I = I_w + I_A + I_B + I_{AB}$ before and after therapy. Hence, provided the initial incidence of primary resistance is considerably greater than the incidence of *de novo* resistance, the long-run benefit of treatment (as measured by the criterion of maximum benefit) is independent of the treatment protocol. All multiple antibiotic therapies result in the same gain of uninfected hosts. This finding is not surprising given our discussion of the simple model above, in which we also assumed a small subpopulation of patients is infected with resistant bacteria before therapy is applied to the patient population. We conclude that if the resistance problem progresses to a point at which most cases of resistance result from the epidemic spread of resistance, then there is little that can be achieved by changing patterns of antibiotic use.

For scenario DN, we assume that, at least initially, the majority of resistant cases result from *de novo* resistance. Later on, however, once resistant infections have become frequent, primary resistance may outweigh *de novo* resistance.

With two antibiotics, we have new options for treatment. The two antibiotics can be cycled periodically, which I call cycling antibiotic therapy (CAT). Alternatively, the patient population can be fractionated into two groups each receiving one antibiotic; this I call simultaneous single antibiotic therapy (SSAT). Finally, all patients who receive treatment may be given both antibiotics together, which I call combination therapy (COT).

If there is no cost to resistance ($\theta_{sw} = \theta_{sA} = \theta_{sB} = \theta_{sAB}$), then it can be shown analytically that CAT and SSAT result in the same total gain of uninfected hosts given by $G \approx -1/\beta \, \ln\phi_1^2$ for $\phi_1 \ll 1$. The gain of uninfected hosts for COT ($q_A = q_B = 0$, $q_{AB} > 0$) is $G \approx -1/\beta \, \ln\phi_2$ for $\phi_2 \ll 1$. Hence, COT is better than CAT or SSAT according to the criterion of maximum benefit, if $\phi_1^2 > \phi_2$. That is, if the square of the probabilities of inducing single resistance by single antibiotic therapy is larger than the probability of inducing AB-resistance by AB-therapy. Otherwise, if $\phi_2 > \phi_1^2$, then any antibiotic therapy in which a patient receives only one antibiotic at a time would outperform combination therapy.

There is a plausible argument for $\phi_2 \ll \phi_1^2$, at least for mutation-borne resistance. Assume that there are n bacteria in an infected individual. If we assume that a fraction Q of these carry a mutation that confers resistance to A, and another fraction Q carry a mutation that confers B-resistance, then a fraction Q^2 should carry mutations conferring resistance against both antibiotics. The expected probability that single antibiotic therapy will result in the emergence of resistance is approximately given by $\min\{1, nQ\}$. Similarly, the probability that AB-treatment will result in AB-resistance is approximately $\min\{1, nQ^2\}$. To a first approximation, Q is given by the mutation rate of the bacteria for which biologically realistic values lie between 10^{-10} and 10^{-7}. Therefore, $\phi_2 = nQ^2 \ll \phi_1 = nQ$ and, hence, COT is likely to result in a considerably greater gain of uninfected hosts than both SSAT and CAT.

It is important to keep in mind that this line of reasoning applies to mutation-borne resistance. However, in many cases, resistance is conferred by the horizontal transfer of accessory elements such as plasmids or prophages. Under these conditions, simultaneous acquisition of multiple resistance is common (Falkow 1975). This would increase ϕ_2 considerably, and, therefore, it is more problematic to predict the relative merits of COT, SSAT, and CAT for horizontally transferred resistance. However, there is also no clear argument against COT, and it is equally unclear whether there is any benefit to single antibiotic therapies.

If there is a cost of resistance, then analytical results are more difficult to obtain. Numerical simulations, however, show that CAT is generally worse than SSAT. The intuitive reason for this is that the optimal policy at any given point of time is to treat with that antibiotic for which there is least resistance in the patient population. Imagine we start with antibiotic A. Treating with A increases the frequency of resistance to A in the patient population. Eventually the resistance to A exceeds that to B, and from this time on it is better to use B. In doing so, resistance to B increases and therefore requires a switch back to A. Hence, the best policy is to switch back and forth between A and B, which is essentially equivalent to SSAT.

Provided $\phi_2 < \phi_1^2$, COT is also better than SSAT or CAT when there is a cost of resistance. Hence, altogether, COT is the therapy of choice according to the maximum benefit criterion for directly transmitted bacteria in which resistance is mutation borne. For an illustration of these treatment strategies for scenarios TR and DN, see Box 23.1.

23.8 Discussion

We can draw the following conclusions from the mathematical models:

- The evaluation of alternative treatment protocols requires a precise definition of what constitutes an optimal treatment policy. Ranking treatment policies according to the time taken until a certain fraction of patients is infected with resistant bacteria is inappropriate because the best result occurs when antibiotics are not used at all. A better criterion to quantify the success of a treatment policy is to measure the total gain of uninfected hosts until the antibiotic therapy fails through resistance.

Box 23.1 Comparing treatment policies using two antibiotics

Consider three treatment strategies for the use of two antibiotics in a patient population (see figure):

- The antibiotics are cycled periodically in the patient population, but at at any point in time only a single antibiotic is used in the patient community, i.e., cycling antibiotic therapy (CAT);
- Of all the patients who receive therapy, a fraction receives the first antibiotic, whereas the rest receives the other, but at any point in time each patient receives only a single drug, i.e., simultaneous single antibiotic therapy (SSAT);
- All patients who receive therapy receive both antibiotics in combination, i.e., combination therapy (COT).

Two scenarios are distinguished:

- The majority of resistant cases result from the transmission of resistant bacteria (scenario TR);
- The majority of resistant cases result from *de novo* resistance (i.e., the selective outgrowth of a resistant subpopulation in a patient predominantly infected with nonresistant bacteria) and only later during the epidemic does transmission of resistance become numerically important (scenario DN).

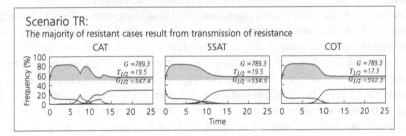

Scenario TR:
The majority of resistant cases result from transmission of resistance

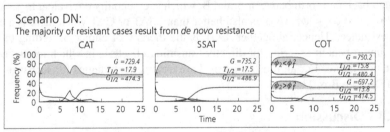

Scenario DN:
The majority of resistant cases result from *de novo* resistance

For scenario TR, all three strategies result in the same total gain of uninfected hosts, $G = 789.3$, as measured by the shaded area under the curve. More generally, one can show analytically that all possible strategies eventually result in the same gain. However, the strategies differ with respect to the gain $(G_{1/2})$ or the time $(T_{1/2})$ taken until half of the cases are resistant to treatment. For scenario DN, SSAT

continued

Box 23.1 *continued*

outperforms CAT. It can be shown analytically that SSAT always results in a larger gain of uninfected hosts than CAT if there is a fitness cost to resistance. If there is no such cost, then both strategies result in the same gain. Whether COT outperforms both single drug treatment strategies depends on the parameters ϕ_1 and ϕ_2, which describe the probability for the selective outgrowth of a resistant subpopulation in a predominantly sensitive infected patient in response to single or combination therapy, respectively. If $\phi_1^2 > \phi_2$, then COT is generally better, and if $\phi_1^2 < \phi_2$, then COT is generally worse than the other two strategies. For an argument as to why, in many cases, $\phi_1^2 > \phi_2$ holds, see the main text. The numerical difference in the gain of uninfected hosts between these strategies is rather small, because the choice of strategy only matters up to the point at which resistance is so frequent that most resistant cases result from transmission. The numerical difference between COT and the other two strategies increases the larger ϕ_1^2 is in relation to ϕ_2.

- For directly transmitted bacterial infections, the long-term benefit of using a single antibiotic from its introduction to the time when resistance precludes its use is largely independent of the pattern of use. Hence, all treatment strategies result in a similar total gain of uninfected hosts.
- When two or more agents are used, SSAT is generally superior to CAT, unless there is no cost to resistance, in which case both strategies are equivalent.
- For mutation-borne resistance, we expect COT to outperform SSAT and CAT significantly. However, if resistance is conferred by plasmids or prophages, the situation is less clear.

In the light of the predictions of these models, two of the diseases for which the model is most appropriate provide an interesting contrast. For tuberculosis, combination therapy is widely applied (CDC 1989; Blower *et al.* 1996), while gonorrhea is usually treated with a single antibiotic (Kam *et al.* 1995; Moran and Levine 1995). Although problems of nonadherence and the rise of tuberculosis among immunocompromised persons complicate the picture, the consistent use of multi-drug therapy and its general success (until recently) in stemming the spread of tuberculosis in developed countries is in accordance with the predictions of the model. With gonorrhea, there has been considerable spread of resistance to a number of antibiotic classes, which might have been preventable with the more widespread use of combination therapy.

In a strict sense, the model considered here applies directly only to those bacterial infections, such as tuberculosis, gonorrhea, and some diarrheal diseases, in which the recovery from the infection coincides with the termination of carriage and transmission of the infectious organisms. Many of the organisms that cause nosocomial infections are not obligate pathogens of this kind, but are organisms that colonize the nose, nasopharynx, or gut of healthy patients and cause disease when they enter and proliferate in normally sterile sites (Fekety 1964). As a result, infection, colonization, and shedding (transmission) are distinct states, and treatment of an infection may or may not terminate colonization or transmission. For such organisms, a different model may be more appropriate (Levin *et al.* 1997). It is not straightforward whether the conclusions of this model about the general inferiority of cycling will extend to these pathogens (or to the commensal bacterial flora or sexually recombining pathogens). However, in the absence of a specific reason, cycling of antibiotics should be undertaken with caution.

Finally, this chapter exclusively considers the effects of antibiotic-use policies at the population level, not the consequences for each individual patient. It is well known that in various public health contexts, particularly vaccination and chemotherapy, the interests of individuals and the community do not coincide. Ethical considerations dictate that physicians treat their patients in ways that maximize the patients' own health. Nonetheless, within the limits dictated by duties to individual patients, antibiotic treatment policies can be modified in the interests of public health (McGowan 1986; Anonymous 1995). Models of the kind presented here can be used to predict and evaluate the efficacy of such modifications.

The goal of the mathematical models presented here is to describe the effect of different treatment strategies in a very general fashion. Future models will have to address the specific properties of particular pathogens. There are many ways in which these models can be extended. In particular, it would be interesting to investigate the spread of resistance in pathogens that are not directly transmitted and to study the effect of horizontal transfer of resistance by plasmids.

Acknowledgments   I am very grateful to Bruce Levin and Marc Lipsitch, with whom I have collaborated on the work that forms the basis of this chapter.

Evolution of Vaccine-resistant Strains of Infectious Agents

Angela R. McLean

24.1 Introduction

Vaccination is one of the most notable successes of modern medicine. Smallpox has been eradicated, and many serious infectious diseases of childhood have been brought under control, with a vast reduction in the associated morbidity and mortality. To achieve this required placing a huge selection pressure upon the associated pathogens. Despite this pressure, there has been little evolution of the pathogen strains that escape from vaccine-induced immunity.

In this chapter, I first present a modeling approach that allows consideration of competition between strains of pathogens and their responses to changes in the balance of competition that are imposed by a vaccination campaign. This framework allows the calculation of conditions that would allow the emergence of a vaccine-resistant strain. The numerical simulation of the evolution of vaccine resistance gives interesting insights into the time scale over which it might occur. Finally, I discuss four case studies from infectious diseases of humans.

24.2 Theoretical Framework

This section describes the basic theoretical framework on which the discussion in this chapter is built.

Basic reproduction ratio

The community-level impact of vaccines is best considered within the context of the basic reproduction ratio R_0, which is defined as the number of secondary cases caused by one infectious individual introduced into a community in which everyone is susceptible. R_0 can be generalized to R_p, the number of secondary cases caused by one infectious individual introduced into a community where a fraction p have been vaccinated and everyone else is susceptible. An infectious agent becomes eradicated when a vaccination campaign renders R_p less than 1. The larger the value of R_0, the more difficult it is to eradicate an infectious disease (Anderson and May 1991; Macdonald 1952). R_0 can be calculated from age-stratified incidence or serological data, and, thus, p_c (the critical vaccination proportion for eradication) can be inferred (see Box 24.1). For a perfect vaccine (one that completely protects all recipients forever), $R_p = pR_0$, where p is the proportion of the community that is vaccinated. For a vaccine that is anything less than perfect, the

---

Box 24.1 Determining basic reproduction ratios from serological data

To a good approximation,

$$R_0 = \lambda L$$

where the force of infection λ is the annual rate of infection per susceptible host and L is the life expectancy of the population. R_0 is calculated from serological data by first calculating λ and then inferring R_0.

An age-structured serological profile yields the proportion susceptible at each age, a. This can be expressed as $p_{susc}(a) = \exp(-\lambda a)$. Thus, λ can be estimated from the slope in a plot of the natural logarithm of the proportion susceptible, $\ln[p_{susc}(a)]$, against age a.

Table 24.1 Basic reproduction ratios and critical vaccination proportions for different childhood infectious diseases in different countries.

Infection	Location	Date	R_0	p_c (%)
Measles	UK	1950s	15	93
Measles	Senegal	1964	18	94
Smallpox	India	1960s	4	75
Polio	USA	1955	6	83

fraction vaccinated must be discounted by the vaccine efficiency ψ. p_c is calculated by finding the value of p such that $R_p = 1$. When comparing values of R_0 and p_c for different infectious diseases in different settings (Table 24.1), it is easy to see why smallpox has gone, polio is going, and measles is still with us.

The honeymoon period

Since host–parasite interactions are nonlinear, an apparently straightforward intervention can engender an unexpected response. One example is the "honeymoon period": the period of very low incidence immediately following the introduction of a mass vaccination program. The honeymoon period occurs because susceptible hosts accumulate much more slowly in a vaccinated community, so it takes a long time to reach the threshold number required for an epidemic (McLean and Anderson 1988; Clements *et al.* 1992). Such patterns were predicted using mathematical models in the 1980s and have since been observed in communities in Asia, Africa, and South America (Chen *et al.* 1994b; Cutts and Markowitz 1994; McLean 1995a).

Competition: An inevitable consequence of cross-reactivity

Directly transmitted infectious diseases are obligate parasites of their hosts; for them, hosts are a substrate over which they must compete – either for internal resources or to avoid immune recognition. Any two pathogens that share cross-reacting epitopes are inevitably in competition to be the first to infect susceptible

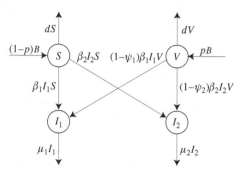

Figure 24.1 A simple model for the evolution of vaccine resistance. Four groups of hosts are represented: susceptible, S; vaccinated, V; infected with wild type, I_1; infected with vaccine-resistant type, I_2.

hosts. When the pathogen has strongly immunogenic conserved epitopes, competition leads to the simple dominance of a single strain. The dominant strain is the one with the largest basic reproduction ratio R_0. However, when conserved epitopes are only weakly immunogenic, competition can result in a shifting balance of strains with a complex antigenic structure (Gupta *et al.* 1994a, 1996).

Modeling the emergence of vaccine-resistant strains

What might happen when vaccination is imposed upon such competitive interactions? Consider a simple case in which a single strain of pathogen circulates before vaccination. Suppose that strain constantly generates less fit (i.e., lower R_0) mutants, some of which are vaccine-escape mutants. In the absence of vaccination, a vaccine-resistant strain is outcompeted if it has a lower R_0 than the wild type. Vaccination acts to shift the competitive balance between the wild-type and resistant strains.

A simple model for such an infection is presented in Figure 24.1 (McLean 1995b). In brief, four types of people are represented in the model: susceptible, vaccinated, infected with strain 1, or infected with strain 2. Any one individual can only be infected with one of the two strains, and recovery from one confers total immunity against the other. In this situation, only one of the two strains can persist in the population, and it will be the one with the higher basic reproduction ratio. The model in Figure 24.1 is described by the equations

$$\frac{dS}{dt} = (1 - p)B - (\beta_1 I_1 + \beta_2 I_2) - dS \,, \tag{24.1a}$$

$$\frac{dV}{dt} = pB - (1 - \psi_1)\beta_1 I_1 V - (1 - \psi_2)\beta_2 I_2 V - dV \,, \tag{24.1b}$$

$$\frac{dI_1}{dt} = \beta_1[S + (1 - \psi_1)V]I_1 - \mu_1 I_1 \,, \tag{24.1c}$$

$$\frac{dI_2}{dt} = \beta_2[S + (1 - \psi_2)V]I_2 - \mu_2 I_2 .\tag{24.1d}$$

Equations (24.1c) and (24.1d) show that, in the absence of vaccination, population I_1 is at equilibrium when $I_1 = 0$ or when $S = \mu_1/\beta_1$. Equally, I_2 is at equilibrium when $I_2 = 0$ or when $S = \mu_2/\beta_2$. Unless $\mu_1/\beta_1 = \mu_2/\beta_2$, either I_1 or I_2 must be absent at equilibrium. The condition $\mu_1/\beta_1 = \mu_2/\beta_2$ is equivalent to the condition in which the two strains have the same basic reproduction ratio, $R_{0,1} = R_{0,2}$. In the general case, the strain with the higher basic reproduction ratio competitively excludes the other. For convenience, let us suppose that $R_{0,1} > R_{0,2}$, so that before vaccination strain 2 is outcompeted. Now suppose that we vaccinate against strain 1, and that the vaccine gives immunity that is weaker and less cross-reactive than naturally acquired immunity. The assumption that vaccine-induced immunity is weaker than naturally acquired immunity is represented in the model by the term $(1 - \psi_1)$, where ψ_1, the vaccine's efficacy, is less than 1. The assumption that vaccine-induced immunity is less cross-reactive than naturally acquired immunity is represented by setting ψ_2 (the vaccine's efficacy) against strain 2 to be smaller than ψ_1 (its efficacy against strain 1). For example, we could set the efficacy against strain 1 at 95%, and that against strain 2 at 50%. The vaccine shifts the competitive balance between the two strains. There is a level of vaccine coverage above which the second strain will emerge as a result of the vaccination campaign. This situation is illustrated in Figure 24.2. Vaccination begins at time 3 years. There follows a period of very low incidence (the honeymoon period) before epidemics of strain 1 restart. Notice in Figure 24.2b that vaccine efficacy remains at 80% during these post-honeymoon epidemics. The post-honeymoon epidemic that starts at time 15 years is a result of the slow accumulation of unvaccinated susceptible hosts. A small number of those who have been vaccinated are also infected because of the incomplete protection conferred by the vaccine. Several decades later, a much larger epidemic occurs, and, at the same time, vaccine efficacy plummets. The second strain has achieved competitive dominance as a result of the growing number of vaccinated individuals. These vaccinated people are well protected against strain 1, but have only minimal protection against strain 2. The vaccinated reproduction ratio for strain 2 is larger than that for strain 1. It takes several decades of accumulation of vaccinated people before this shift in competitive advantage manifests itself in epidemics of strain 2. However, for this combination of parameters, the effect is inevitable. It is not, however, an unavoidable consequence of vaccination. Highly cross-reactive and immunogenic vaccines can eradicate both strains at coverage levels below those at which the second gains the competitive advantage. Alternatively, low levels of vaccination (or very high fitness costs to a mutant able to escape from vaccine-induced immunity) leave the pre-existing strain as the competitive superior. Figure 24.3 illustrates these possibilities with plots of the vaccinated reproduction ratio R_p against proportion vaccinated p. Figure 24.3a has the same parameter values as Figure 24.2. Here, the vaccine is only weakly immunogenic against the second strain, and the vaccinated reproduction ratio R_p for strain 2 falls rather slowly as the proportion

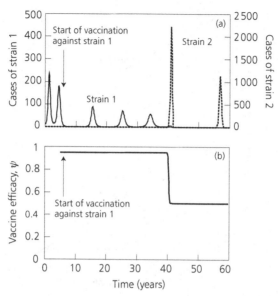

Figure 24.2 Emergence of a vaccine-resistant strain. (a) At time 3 years, a vaccination campaign is introduced that reaches 80% of newborns. The vaccine is 95% effective against the circulating strain (strain 1) but only 50% effective against the vaccine-resistant strain (strain 2). A 10-year honeymoon period ensues, followed by post-honeymoon outbreaks of strain 1. Almost 40 years after introduction of vaccination, there is an outbreak of the vaccine-resistant strain. (b) During the post-honeymoon outbreaks, vaccine efficacy is unchanged. These outbreaks are a natural consequence of the nonlinear nature of interactions between susceptible and infectious individuals. At time 40 years, when strain 2 emerges, vaccine efficacy falls from 95% to 50%, signaling the arrival of the new strain.

vaccinated increases. With 80% vaccination coverage, R_p for strain 2 is greater than R_p for strain 1, so eventually an outbreak of the vaccine-resistant strain is inevitable. If vaccination coverage had been much lower, strain 2 would not have gained the competitive advantage. A vaccine with greater cross-reactivity does not encounter these problems. Figure 24.3b illustrates an example in which both strains are eliminated before the second strain can gain the competitive advantage.

Thus, there are three possible explanations as to why we have not seen outbreaks of vaccine resistance in response to the major vaccination campaigns against childhood infectious diseases. The first explanation (Figure 24.2a) is that the outbreak simply has not occurred yet. The second (Figure 24.3a) is that coverage is too low to give the competitive advantage to resistant strains. The third (Figure 24.3b) is that current vaccines give enough cross-immunity to prevent resistant strains ever emerging.

What if a resistant strain were to emerge? The good news is that it would have lower R_0 than the pre-existing strain. This means that the vaccination coverage levels that allowed the control of the first strain should enable the control of the second. However, this conclusion is dependent upon the assumptions of this

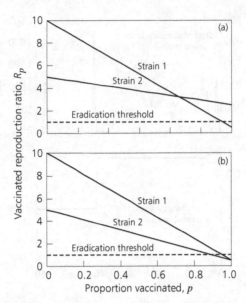

Figure 24.3 Vaccinated reproduction ratios predict the outcome of competition under different vaccination regimens. The strain with the higher vaccinated reproduction ratio eventually dominates. (a) Vaccine with low cross-reactivity. At high vaccination coverage, the vaccine-resistant strain has the competitive advantage. (b) Vaccine with high cross-reactivity. If a vaccine is so cross-reactive that the vaccine-resistant strain only gains competitive advantage at coverage levels above that at which the wild-type strain is eradicated, vaccine-resistant strains cannot emerge.

model, and is not necessarily true for a situation in which multiple strains were able to coexist before vaccination (McLean 1995b).

24.3 Case Studies from Infectious Diseases of Humans

Given the basic framework, the next question is how to put it to work; this I do by describing how the ideas can be applied in four concrete cases.

Measles

Measles have been subjected to one of the most intense control efforts of all the infectious diseases of childhood. When sequences of currently circulating measles virus are compared with historical samples, an increase can be observed in the rate of nucleotide change in the measles' hemagglutinin gene from the point in time when vaccination became widespread (Rota *et al.* 1992). Furthermore, this sequence variation translates into antigenic differences between currently circulating strains and the strains that make up the vaccine (Tamin *et al.* 1994). Serum from individuals infected with current wild-type strains reacts four to five times more effectively with wild-type strains than it does with the vaccine strain. Fortunately, the reverse is not true. Serum from people who have recently been vaccinated has an equally strong antibody response to either strain. Thus, for the moment, there is no evidence that vaccine-escape mutants of measles are about to emerge.

Hepatitis B

Antigenic subtypes of hepatitis B occur naturally, and hepatitis B vaccine-escape mutants have been identified (Carman *et al.* 1990; Wallace and Carman 1997). Since the vaccine is relatively new, there is no large pool of vaccine recipients to act as fuel for an epidemic of vaccine-resistant hepatitis B. However, as the number of people vaccinated against hepatitis B grows, the transmission of the variant hepatitis B virus must be considered. It has already been suggested that the variant sequence should be included in future vaccines.

Pertussis

The Netherlands had an effective pertussis control program in place. In 1989–1995, the number of annually reported cases was less than 500. Then, in the summer of 1996, a pertussis outbreak occurred in which, for the year as a whole, over 2 500 cases were reported (Figure 24.4a). There was no discernible shift in the age distribution of cases (Figure 24.4b). A post-honeymoon period epidemic would show a clear signature increase in the average age at infection – after 7 years of low incidence, this would have been large enough to discern even with cases reported in age bands of 5 years. It has been suggested, but not yet confirmed, that the resurgence of pertussis resulted from the emergence of strains of *Bordetella pertussis* that are less sensitive to the immune protection provided by the vaccine (de Melker *et al.* 1997).

Smallpox and monkeypox

The eradication of smallpox and the consequent cessation of vaccinia vaccination is often held up as the Holy Grail of vaccine strategies. The patterns of competition among strains discussed here apply to any group of infectious agents that share cross-reactive antigens – not just different strains of the same pathogen. Monkeypox, smallpox, and vaccinia give an intriguing example. Before the eradication of smallpox, infection of humans by the monkeypox virus was rare, and human-to-human transmission rarer still. Vaccinia immunization protects against monkeypox virus infection, and so, presumably, did immunity to smallpox. A recent outbreak of monkeypox virus in Zaire was characterized by large numbers of human cases (mostly among smallpox-naive individuals) and long chains of human-to-human transmission. Thus, it may be that first smallpox infection and then vaccinia immunization were protecting exposed individuals from infection by monkeypox virus. Now that smallpox has been eradicated and vaccination has ceased, a pool of individuals susceptible to monkeypox infection has accumulated and appears to have fueled an epidemic. Reintroduction of vaccinia immunization is being considered (Anonymous 1997a, 1997b).

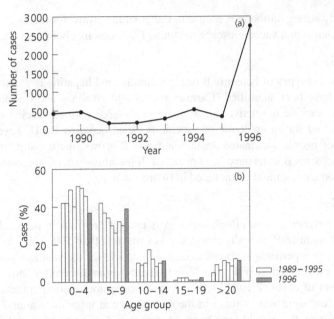

Figure 24.4 Pertussis outbreak in the Netherlands. (a) Annual incidence 1989–1996. After many years of low incidence, the Netherlands experienced a major outbreak of reported cases of pertussis in 1996. (b) The age distribution of cases in the outbreak was indistinguishable from that of earlier years.

24.4 Discussion

This chapter set out to ask why vaccination has been so successful, and, in particular, why infectious agents have been unable to evolve resistance to vaccine-induced immunity. The first possibility is that vaccine-resistant strains simply have not arisen yet. In the models investigated here, it can take a very long time for enough vaccine recipients to accumulate for a vaccine-resistant strain to achieve competitive dominance. The second, and, I believe, most likely possibility, is that current-generation vaccines are effective enough and cross-reactive enough that a vaccine-resistant strain cannot become the competitive superior. A third, and unlikely, possibility is that so few individuals are vaccinated that the vaccine-resistant strain remains competitively inferior. This last point is, however, worth bearing in mind for new vaccines that might be targeted at particular risk groups, while leaving the rest of the community unvaccinated. Subject to limits on the aggregation and accumulation of vaccine recipients, this might prove to be a useful method for continuing to avoid the emergence of vaccine-resistant strains in the future.

Pathogen Evolution: The Case of Malaria

Sunetra Gupta

25.1 Introduction

The rapid development of molecular methods for use in epidemiological studies has led to the accumulation of vast amounts of data on pathogen diversity. Whether or not the observed diversity is relevant to the management of virulence depends on the functional status of the polymorphic elements, either as factors of virulence or in the fundamental processes that underlie the population dynamics of these systems. In this chapter, we focus on antigenic diversity in pathogen populations, and its implications for virulence management.

To understand how antigenic diversity has evolved, it is first necessary to develop a reasonable system of classification. The term "strain" is used widely to designate different parasite types. In the case of certain pathogens that do not exchange genetic material, the term "strain" is synonymous with "clone" (Maynard Smith *et al.* 1993). However, whether they are clonal or not, pathogens may be usefully classified into strains with respect to a certain set of relevant markers. For example, the population may be divided into drug-resistant and drug-susceptible strains on the basis of their response to a particular chemotherapeutic agent. This response may be determined by variation at a single locus or a combination of genetic loci. Similarly, a pathogen population may be categorized into antigenic types with respect to variation at a set of genetic loci encoding these antigens. When these antigens induce immune responses that have an effect on the transmission success of the pathogens, the associated antigenic types, or strains, may be seen as epidemiological units. It is in this sense that we use the word "strain" in this chapter.

25.2 Maintenance of Pathogen Diversity in Single-locus Systems

Simple multistrain models of infectious disease systems can be used to investigate the interaction between antigenic types circulating within a host community. By making the assumption that cross-immunity only affects the transmission of other strains, we may specify a simple model with host compartments that correspond to the proportion immune to a given strain j, r_j; the total proportion of hosts with immunity, r_+; and those infectious for strain j, i_j. Individuals who are not immune to strain j may be infected by j with a force of infection given by λ_j. Nonimmune individuals, the proportion of which is given by $1 - r_+ = s$, will invariably become infectious. However, individuals who are immune to a strain other than j (i.e., $1 - s - r_j$) will only become infectious with a probability of σ, where the parameter $1 - \sigma$ measures the degree to which infection with a given strain limits the transmission of strains. If $\sigma = 1$, then the strains do not interact, whereas if $\sigma = 0$, then there is total cross-protection between the strains. This system may be described by the following set of differential equations,

$$\frac{dr_j}{dt} = \lambda_j(1 - r_j) - dr_j , \tag{25.1a}$$

$$\frac{ds}{dt} = -\sum_k \lambda_k s - d(1 - s) . \tag{25.1b}$$

$$\frac{di_j}{dt} = \lambda_j[s + (1 - \sigma)(1 - s - r_j)] - \gamma i_j . \tag{25.1c}$$

We assume, in this simple example, that immunity to any particular strain is life-long. Thus, losses from the immune classes occur at the same rate as mortality, d. The loss of infectiousness, at rate γ, is assumed to occur on a much faster time scale than the loss of immunity through death at rate d. Since the strains do not interact except through the host immune system, the force of infection, λ_j, is simply $\beta_j i_j$, where β_j is a combination of parameters affecting the transmission of strain j. It can be shown, following conventional lines of analysis, that the basic reproduction ratio of strain j, $R_{0,j}$, is β_j / γ.

Figure 25.1 demonstrates the behavior of a two-strain system under various levels of cross-immunity. In the absence of cross-immunity, the strains independently equilibrate at the values they would have achieved in the absence of the other, as shown in Figure 25.1a. The latter records the changes in the proportions immune to two strains, where strain A (dashed curve) has an R_0 of 3 and strain B (continuous curve) has an R_0 of 2. In general, when $\sigma < 0$, two strains 1 and 2 may only coexist if the following conditions are satisfied (Gupta *et al.* 1994b),

$$\sigma > \frac{R_{0,1} - R_{0,2}}{R_{0,2}(R_{0,1} - 1)} , \tag{25.2a}$$

$$\sigma > \frac{R_{0,2} - R_{0,1}}{R_{0,1}(R_{0,2} - 1)} . \tag{25.2b}$$

It is clear from Equations (25.2a) and (25.2b) that if there is total cross-immunity (i.e., $\sigma = 0$), then no solution to the system of equations is possible unless the basic reproduction ratios of the strains are equal. This result is demonstrated in the numerical simulation of the model shown in Figure 25.1c, which shows that strain A (with the higher R_0) will displace strain B in the course of time. The eventual displacement of the strain of lower R_0 by a more successful strain is known as the principle of competitive exclusion (May and Anderson 1983a; Bremermann and Thieme 1989). If cross-immunity is partial (Figure 25.1b), strains may be maintained within a certain range of differences in basic reproduction ratios as

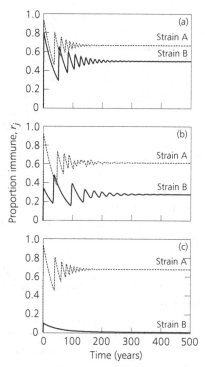

Figure 25.1 Temporal changes in the proportions of the host population immune to strain A ($R_0 = 3$: dashed curve) and strain B ($R_0 = 2$: continuous curve). The degree of cross-immunity between the strains is given by (a) $\sigma = 1$, (b) $\sigma = 0.5$, and (c) $\sigma = 0$.

defined by Equations (25.2a) and (25.2b). Figure 25.1b records the proportions of the host population immune to strains A and B when $\sigma = 0.5$. Both strains A and B exhibit damped oscillations around a stable equilibrium, which increases with the degree of coupling introduced by cross-immunity. As the degree of cross-immunity wanes, the systems uncouple until they are independent, as shown in Figure 25.1a.

25.3 Multilocus Antigenic Diversity with Genetic Exchange

Analysis of single-locus systems indicates that levels of cross-immunity between strains have to be low for them to be stably maintained. If a single genetic locus is responsible for antigenic diversity, this condition is essentially satisfied if the conserved determinants (i.e., antigens or epitopes shared by the strains) are weak. However, for most infectious disease systems, more than one locus is likely to be involved, either because there is more than one dominant antigen or because the dominant antigen is encoded by different loci. This is particularly relevant for those organisms that undergo antigenic variation in the host, such as *Plasmodium falciparum* and trypanosomes.

If multiple loci are involved in defining a strain, competition is mediated not only by cross-immunity (because of conserved determinants), but also through cross-protection resulting from sharing variants of the polymorphic determinants. In the first case, the relevant immune response is one that is elicited by an immunological determinant shared by all strains or by an immune response that acts nonspecifically against all variants. By contrast, cross-protection (as we define it) involves strongly variant-specific immune responses, and results from sharing one or more of these variants with a strain that has previously infected the host.

The simplest model for such a system involves two immunologically dominant loci, each with two alleles or variants. For example, where A and B are alleles at one locus, and X and Y are alleles at the second locus, the four possible types of strains are AY, AX, BX, and BY. Strains that do not share any alleles do not interfere with each others' transmission (i.e., $\sigma = 1$ for that particular pair), since the immune responses directed against one strain, say AX, are ineffective against the other, say BY, since neither anti-A nor anti-X responses recognize either B or Y. By contrast, for strains that do share alleles, cross-protection may range from none to complete protection ($0 < \sigma < 1$), depending on the fate of a particular strain, say AX, within a host who has protective immune responses either against A or X, but not both (because of exposure to either AY or BX). These features may be included in the single-locus model described above, Equations (25.1), by designating each genotype as a strain (i.e., strain j is defined by n_L loci), and modifying the variable $1 - s$ such that it is specific to each strain, $1 - s_j$, and represents those hosts immune to any strain in the subset M_j of strains sharing alleles (at the relevant polymorphic loci) with strain j. This subset M_k includes strain j itself. The dynamics of $1 - s_j$ can be described by

$$\frac{d(1 - s_j)}{dt} = \sum_{k=1}^{M_j} \lambda_k s_j - d(1 - s_j) \, . \tag{25.3a}$$

Individuals who have not been infected either by j or any strain sharing alleles with j, that is, a proportion s_j of individuals, are completely susceptible to strain j. However, those that have been exposed to a strain sharing alleles with j, but not exposed to strain j itself, that is, a proportion $1 - s_j - r_j$ of individuals, will become infectious with a probability of σ when they are infected by strain j. The dynamics of the proportion of the population infectious for strain j may thus be represented as

$$\frac{di_j}{dt} = \lambda_j[s_j + \sigma(1 - s_j - r_j)] - \gamma i_j \, . \tag{25.3b}$$

The resultant multilocus model is described by Equations (25.1a), (25.3a), and (25.3b). Figure 25.2 schematically represents the relationship between the proportions immune to strains that share alleles in the 2-locus 2-allele case. We focus on the strain AX. Those immune to strain AX (r_{AX}) are shown in white. The area outside r_{AX} is shown in gray. Within this area, the dark gray region represents individuals who are not immune to AX, but who do have immunity to a strain that

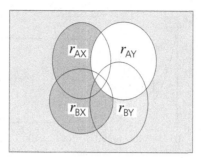

Figure 25.2 The relationships between the proportions of a host population immune to four different strains, AY, AX, BX, and BY, of a 2-locus 2-allele system in which A and B are alleles at one locus, and X and Y are alleles at the second locus. *Source*: Gupta *et al.* (1997).

shares alleles with AX (i.e., AY or BX). Individuals in this category transmit AX with probability σ, where $1 - \sigma$ is the degree of protection against a new strain occurring. When cross-protection is complete, only individuals in the gray region are able to transmit AX. Conversely, in the absence of cross-protection, the entire hatched region (i.e., outside the white circle) is susceptible to AX.

Another important feature of multilocus systems concerns the possibility of genetic exchange through, for instance, sexual processes in the case of *P. falciparum* (Walliker 1989), coinfection of the same host cell for human immunodeficiency virus 1 (HIV-1; Robertson *et al.* 1995), or transformation in bacteria (Feavers *et al.* 1992). The impact of genetic exchange on the population structure of infectious disease agents may be examined within this framework by modifying the force of the infection term λ_j to include the assumption that the progeny of parasites within hosts infectious for two or more strains will consist of defined fractions of the various combinations of all the different strains, k and l, that may generate strain j through recombination.

Figure 25.3 illustrates the behavior of this model system for the 2-locus 2-allele case in the extremes at which cross-protection is complete ($\sigma = 0$). We have assumed that all four strains have the same basic reproduction ratio R_0. In the limit where cross-protection is absent, all strains independently equilibrate at the same level $(1 - 1/R_0)$, as in the single-locus case. However, in the limit at which cross-protection between strains sharing alleles is complete (Figure 25.3) – although they are all identical in every respect – one set of strains strongly dominates, in terms of abundance, over the other strains. It is invariably the case that these strains do not share alleles with each other. The principle of competitive exclusion thus operates within a multilocus system by selecting a group of strains that are distinguished by unique antigenic repertoires. The latter can coexist because they do not create any immunological interference with each other. Other antigenic types, although they may be generated frequently by recombination, have a low transmission success because they share alleles with the dominant antigenic types.

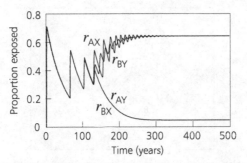

Figure 25.3 Temporal changes in the proportions of the host population exposed to the four strains, AX, BY, AY, and BX, where AY and BX encode for strongly immunogenic antigens ($\sigma = 0.15$) and recombination occurs randomly between the loci. The strains have the same R_0 ($= 3$), but the proportion initially infected varies slightly, i.e., $i_{AX} = i_{BX} = i_{AY} = 0.0001$; $i_{BY} = 0.0001001$.

Thus, those antigens that elicit the strongest immune response, which in turn have the strongest impact on transmission success, may be organized by immune selection, acting within the host population, into sets of nonoverlapping variants. The pathogen population may exhibit such a discrete strain structure despite frequent recombination, where one set of nonoverlapping variants exists at much greater frequency than the other (Gupta *et al.* 1996). However, for this pattern to emerge and to be stable over time, the intensity of acquired immunity to a specific variant antigen (encoded by a given allele) within the host population must considerably reduce the fitness of all genotypes that possess that allele. This is reflected in the analytical result that a stable, discrete strain structure only occurs if the level of cross-protection exceeds an upper threshold, $\sigma_T = 1/2R_0$. Conversely, above an upper threshold σ_L no strain structure exists. Between these two thresholds, pathogens may exist as a set of strains that exhibit cyclical or chaotic fluctuations in frequency over time (Gupta *et al.* 1998). They may still be organized by immune selection into discrete groups of variants in which all the members in a given group have different alleles at every locus, but the dominancy of the group, relative to that of other groups, may fluctuate widely over time, either cyclically or chaotically. Figure 25.4 shows an example of cyclical behavior in a 2-locus 3-allele system and chaotic behavior in a 3-locus 3-allele system. It is clear that in the latter case, stochastic effects in small populations may interfere with persistence (Bolker and Grenfell 1995; Ferguson *et al.* 1998). Nonetheless, we may expect to see the following broad patterns in longitudinal molecular epidemiological data. For polymorphic antigens that elicit weak immune responses, the abundances of the different strains (as defined by different combinations of alleles) will be determined by their respective transmission successes, or fitnesses. In the cyclical or chaotic region, in which the degree of cross-protection associated with the antigen is moderate, irregular epidemic cycles will be observed. For antigens that elicit very strong protective immune responses, a stable discrete strain structure will prevail.

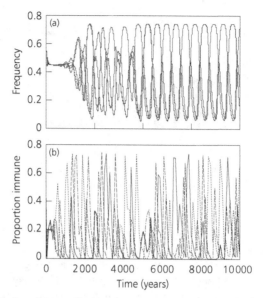

Figure 25.4 Examples of (a) cyclical behavior in a 2-locus 3-allele system, and (b) chaotic behavior in a 3-locus 3-allele system for genes associated with weak-to-moderate immune selection. Each curve represents a set of discrete antigenic types.

25.4 *Plasmodium falciparum*: A Case Study

P. falciparum is one of four species of malaria that infects humans. In most endemic areas, a large majority of individuals are infected with *P. falciparum* but do not exhibit any clinical symptoms. The cases that exhibit symptoms are largely confined to young children and show a spectrum of severity – most cases would be classified as chronic "mild" illness, and a minority would be classified as severe disease requiring hospitalization. Severe malaria can be broadly classified into two syndromes: severe malarial anemia and cerebral malaria (CM). Figure 25.5 shows the typical age distribution of these two syndromes among children in Gambia. The figure shows a disjunction between the average ages of presentation for the two syndromes; the same observation has been made over a wide range of malaria endemicity in sub-Saharan Africa (Snow *et al.* 1994). In this section, we discuss how a theory of antigenic structure in pathogen populations, in conjunction with epidemiological data, can help us to understand the virulence characteristics of *P. falciparum*, and the potential impact of control on malaria mortality.

Infection with *P. falciparum* occurs when sporozoites are injected into a human host from an infectious mosquito. Polymorphic antigens exist on the surface of the sporozoites, but these are only exposed to the immune system for an average of 30 minutes before taking up residence in the liver. Infected liver cells are subject to destruction by cytotoxic lymphocytes (CTL). Those cells that are not destroyed eventually burst and release merozoites into the bloodstream. Polymorphic antigens exist on the surface of the merozoites, but these antigens are a target

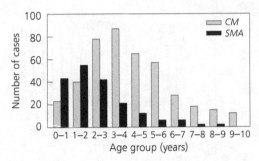

Figure 25.5 Age distribution of severe malarial anemia (SMA) and cerebral malaria (CM) from a large hospital-based case-control study in Gambia, recorded as number of cases in each age group. *Source*: Hill *et al.* (1991).

of immunity for only an average of 20 seconds before they invade erythrocytes. Parasites that grow within erythrocytes induce antigens on the surface of the cell, collectively known as PfEMP. These antigens elicit a strong variant-specific agglutinating antibody response (Newbold *et al.* 1992), which may be related to their location on the surface of the erythrocyte, which causes them to be exposed to the immune system for a very much longer period of time (at least 18 hours) than various other polymorphic antigens. A recent study in Kenya (Bull *et al.* 1998) demonstrated a significant difference in frequency with which a disease-causing isolate was recognized by the index case by comparison with sera from age-matched controls from the same community, implying that antibody responses against PfEMP1 play a major role in protection against disease. PfEMP1 is also known to be involved in the process of adherence of infected blood cells to host endothelium (Magowan *et al.* 1988; Gardner *et al.* 1996), which is believed to be important to the survival of the parasite by removing it from circulation through organs such as the spleen, from which they might be cleared. Sequestration of parasites through cytoadherence in postcapillary venules may also provide a hypoxic environment favorable to parasite growth (Trager and Jensen 1976). It has also been proposed that reinvasion of erythrocytes may be facilitated by the crowding of infected cells sequestered in postcapillary venules (Berendt *et al.* 1994). Antibodies that block cytoadherence may thus have a major effect on the maintenance of parasitemia. For these reasons, the host immune response against PfEMP1 may be seen as a major selective force on *P. falciparum*, and it may be postulated that the latter organisms are organized into distinct nonoverlapping subsets of PfEMP1 variants that cocirculate within a population as independently transmitted antigenic types. Such a hypothesis is consistent with age-specific patterns of seroconversion, which suggests that antibody responses to different isolates are acquired independently (Gupta *et al.* 1994c; Bull *et al.* 1998).

The effects of self-organization into discrete antigenic types can be directly related to its virulence characteristics, in which the dominant antigens have some effect on the virulence of the pathogen. In the case of *P. falciparum*, such a link is

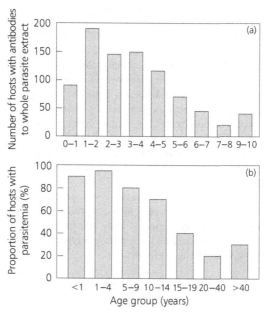

Figure 25.6 Typical age distributions of (a) chronic mild malarial disease and (b) *P. falciparum* infection. The former are recorded as the number in each age class presenting with mild malaria at the Medical Research Council clinic in Fajara, Gambia, during the period October–December 1992. *Data source*: A.M. Greenwood, adapted from Gupta *et al.* (1994a). The latter shows the proportion in each age class with asexual parasitemia in a region of the southern Cameroons. *Source*: Gibson (1958).

provided because PfEMP1 is the ligand for cytoadherence of the infected erythrocytes to endothelium. It has been observed that serious complications of *P. falciparum* malaria are associated with the sequestration of parasites in various critical organs, which leads to organ damage through reduced perfusion, high metabolic demands of infected cells, and concentrated release of toxic wastes. Microvascular obstruction through cytoadherence in organs, such as the brain, is believed to be a critical step in the development of CM. Thus, a simple explanation for the maintenance of the spectrum of virulence is that *P. falciparum* is a construct of independently transmitted strains the virulence characteristics of which are defined by their unique PfEMP1 profiles. In particular, we postulate that certain of these are specifically associated with CM. That the average age of CM appears to be consistently higher than other manifestations of malarial disease may thus be explained by proposing that CM-causing strains of *P. falciparum* have a lower force of infection.

Differential trends have been noted in protection against mild and severe disease among carriers of the hemoglobin S variant (i.e., sickle; Allison 1964), for certain human leukocyte antigen (HLA) associations, and also with regard to the impact of impregnated mosquito nets (Alonso *et al.* 1993). A large case-control study

Box 25.1 Maintenance of pathogen diversity despite cross-immunity

Antigenic determinants are pathogen-specific molecules that elicit an immune response in the host. For the sake of simplicity, three antigenic determinants per pathogen are distinguished in the diagrams below: these are shown as ellipses, triangles, and squares along vertical lines that represent pathogen types or strains. Antigenic determinants may be either conserved (shown as monomorphisms of a black ellipse variant) or polymorphic (shown as dimorphisms of black and white triangle and square variants). The antigenic determinants that dominate the effect on the immune response (strong determinants) are indicated by large ellipses, triangles, or squares, whereas those with little effect (weak determinants) are represented by small shapes.

If the conserved antigenic determinants are stronger (large black ellipses versus small black and white other shapes), infection with a given pathogen type (the one indicated by the dashed rectangle) renders a susceptible host (represented by a continuous rectangle) immune to all those pathogen types that share the conserved determinant (in this case, the large ellipses):

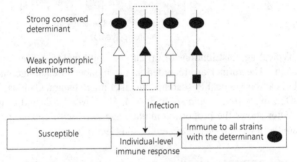

By contrast, if the polymorphic antigenic determinants dominate the effect on the immune response (large black and white triangles and squares versus small black ellipses), an individual host becomes immune only to those other pathogen types that share at least one of the polymorphic antigenic variants (in this case, either a black square or a white triangle):

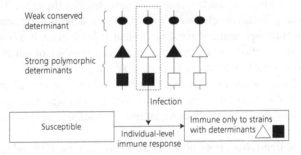

In both cases, competitive exclusion operates. In the first case (considered again in the left panel below), this results in the single pathogen type that possesses the
continued

Box 25.1 *continued*

highest R_0 value (here a combination of large black ellipses, small white triangles, and small black squares) prevailing in the pathogen population (see Box 2.2). In the second case (right panel below), the immune response always leads to a polymorphism at the level of the pathogen population: any host occupied by the single pathogen type with highest R_0 can still be coinfected by variants with alternative antigenic determinants.

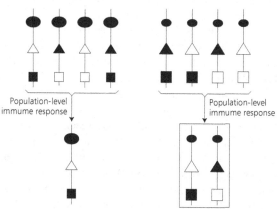

Thus, in the example for the second case, if the pathogen type with a white triangle and a black square has the highest R_0, hosts (represented by a continuous rectangle at the bottom of the right panel above) infected by this pathogen type can still be shared by a pathogen type with a black triangle and a white square. These two pathogen types therefore circulate in the host population as independently transmitted strains. If the polymorphic determinants correspond to different virulence levels, a polymorphism of virulence levels can thus be maintained in the pathogen population.

---

performed in Gambia (Hill *et al.* 1991) indicated that the HLA class I antigen, HLA-B53, was associated with an approximately 40% protection against severe disease, but was not significantly associated with protection against mild disease or infection. These patterns may be explained by assuming severe disease is caused by independently transmitted parasite strains. The effect of host genotypes may be explained within a static framework by assuming that the probability of disease is nonlinearly related to the number of effective infections by a certain strain (Gupta and Hill 1995). The effect of mosquito nets may also be explained in this manner, or, alternatively, by considering that the community-wide use of mosquito nets affects the overall dynamics of the system, and thus may decrease the transmissibilities of mild and severe strains. This may disproportionately reduce the rate of disease associated with the rarer strain (Gupta *et al.* 1994a).

The hypothesis that strain-specific antibodies to PfEMP1 may be effective in preventing disease (as a consequence of reducing the reproductive success of the parasite) is consistent with the age distribution of chronic, mild malarial disease. The incidence of mild malaria tends to decline uniformly with age (Figure 25.6a), such that children 5–9 years of age experience significantly fewer episodes than those who are 1–4 years of age (Greenwood *et al.* 1987; Marsh *et al.* 1989; Cox *et al.* 1994). By contrast, the prevalence of parasitemia tends to remain at high levels throughout childhood, often peaking at 5–9 years of age (reviewed in Gupta and Day 1994a, 1994b). In most hyperendemic areas, the prevalence of infection falls after the age of 15 years, and may be maintained at quite low levels in adults, as shown in Figure 25.6b. These patterns suggest antibodies to PfEMP1 may be effective against disease (and transmission), but cannot prevent reinfection by the same strain. The fall in prevalence of infection in adults implies that infection-blocking immunity is achieved against all parasite strains after repeated exposure to a different antigen or set of antigens. For instance, a conserved response, such as antibodies to the C-terminus of MSP-1, may reach densities at which mero-zoite invasion is blocked, as suggested by immunization experiments (Daly and Long 1993). Similarly, immunity to sporozoite antigens (Hofmann *et al.* 1987) or CTL against liver stages (Hill *et al.* 1992) may, after many exposures, achieve densities that render such responses effective against infection by all strains. Fur-thermore, recent data collected at hospital sites across Africa (Snow *et al.* 1998) imply a strong component of nonvariant specific (or strain-transcending) immunity involved in protection against severe, noncerebral illness that requires hospitaliza-tion (Gupta *et al.* 1998), as distinct from the mild chronic symptoms that may be prevented by strain-specific immunity. However, these conserved responses col-lectively correspond to low levels of cross-immunity, which are therefore unable to disrupt the strain structure of *P. falciparum* (see Box 25.1).

25.5 Impact of Vaccination

A stable, discrete set of antigenic types that do not share alleles or variants at the loci that influence transmission cocirculate within a host population in a manner analogous to totally unrelated pathogens such as mumps and measles. Thus, even though they may exchange genetic material, these antigenic types effectively be-have as independently transmitted strains. An infectious disease that is a construct of many such strains has a very low average age at first infection, since "first in-fection" records the event of becoming infected by any one of a large number of strains. It has been shown (Gupta *et al.* 1994c) that the average age of first infec-tion with any one of *n* strains, a_n, is given by

$$a_n = \frac{H}{\sum_j R_{0,j}} \,, \tag{25.4}$$

where H is the average duration of strain-specific immunity. *P. falciparum* malaria may be such an example where the average age of first infection with the parasite as a whole is very low, even though that of its constituent strains is considerably

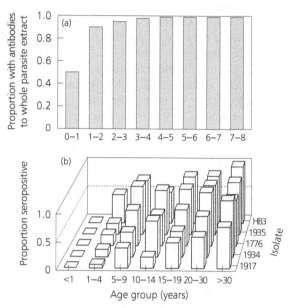

Figure 25.7 The proportion of individuals in each age class with antibodies against (a) whole *P. falciparum* extract, and (b) specific strains of *P. falciparum* as represented by the five different isolates 1917, 1934, 1776, 1935, and HB3, within a population in the Madang region of Papua New Guinea. *Source*: Gupta *et al.* (1994c) based on data from K. Trenholme and K. Day.

higher. In most endemic areas, the prevalence of individuals with antibodies to whole parasite extract rises very sharply with age, as shown in Figure 25.7a. By contrast, the proportion with antibodies to a specific strain (here defined by the characteristics of its PfEMP1 type) rises much more gradually, as shown by data collected in the Madang region of Papua New Guinea (Figure 25.7b). The low average of infection with *P. falciparum* has traditionally been interpreted as a manifestation of its high basic reproduction ratio. The strain-specific serological age profiles shown in Figure 25.7b suggest, however, that the actual magnitude of R_0 for malaria may be very much lower than previously assumed (Gupta *et al.* 1994c). This result has important implications for the control of *P. falciparum* malaria, as the level of vaccine coverage required to make a significant impact is directly related to the basic reproduction ratios of the constituent strains.

Antigenic diversity within a parasite population can only be stably maintained under conditions in which the immune responses to antigens conserved between the strains are relatively ineffective – otherwise, the repertoire would be reduced through competitive exclusion (May and Anderson 1983a; Bremermann and Thieme 1989; Gupta *et al.* 1994b). In preparing vaccines for use against organisms that exhibit extensive antigenic diversity, the question arises as to whether we should target the more immunogenic polymorphic determinants or the less immunogenic conserved determinants. While an ideal vaccine would be capable of artificially boosting a conserved immune response that is only weakly protective under natural conditions, it may not be possible to create. The alternative is to arrange a cocktail of the dominant polymorphic antigens. It is unlikely, however, that such a cocktail would include the entire range of relevant epitopes. We have explored the epidemiological consequences of using vaccines based on polymorphic antigens that only partially cover the spectrum of antigenic types or strains circulating within a community, in which the latter frequently exchange genetic material (Gupta et al. 1997).

It is intuitively obvious that for the case in which there is no cross-protection, vaccinating against a particular strain has no effect on the others. As cross-protection increases, the parasite types that share alleles with those included in the vaccine are diminished in prevalence, while those that do not share these alleles increase in prevalence. Thus, if cross-protection is strong enough to cause competition between all the parasite types, but not strong enough to generate strain structure, vaccinating against a particular set of antigenic types greatly reduces the selection pressure against strains that do not contain any of the vaccine variants, and causes them to increase in frequency (McLean 1995b; Lipsitch 1997). The increase in prevalence of these strains becomes more and more exaggerated as cross-protection increases. If cross-protection is high enough to cause one set of discrete antigenic types to dominate in the pathogen population, then vaccinating with antigenic types that are not part of the dominant subset may cause a precipitous increase in the prevalence of a strain that was previously only present at low levels through immune selection. This second case precisely underscores the importance of conducting appropriate population studies before proceeding with a vaccine that has been proved effective in laboratory studies and Phase I trials against only a subset of the antigenic types circulating in a defined population. It is conceivable that the epitopes of less-prevalent antigenic types will have been identified in vitro, as they are likely to be more immunogenic than the dominant combinations and may be singularly effective in a vaccine. Altering the distribution of antigenic types in this manner carries the danger of greatly increasing the burden of disease should the selected types be even slightly more virulent than the types displaced by vaccination. On the other hand, vaccinating with a range of subdominant strains may be more useful than would appear by simply considering the frequencies of these strains within a parasite population. This may have the effect of reducing the frequencies of the dominant strains considerably. Where the latter are the more virulent types, this may have a beneficial effect on the community.

25.6 Discussion

The management of virulence must be viewed in the context of the fundamental processes that shape pathogen population structure. Immune selection by the host is particularly important as virulence factors are often vital targets for immune attack.

P. falciparum is an example in which a critical polymorphic antigen (PfEMP1) is also implicated as a major virulence factor. Age distributions of clinical symptoms of *P. falciparum* malaria may be explained by proposing that the different syndromes are caused by independently transmitted antigenic types or strains, which have evolved through immune selection by the host.

In general, host immune responses can cause populations of infectious pathogens to self-organize into a stable collection of independently transmitted strains with nonoverlapping repertoires of antigenic variants. Such a discrete structure is stable for cases in which the antigens elicit strong, protective responses; however, over a large range of intermediate levels of protection, discrete subsets of strains may change in prevalence in either a cyclical or chaotic manner. Changes in pathogen virulence may therefore occur as a natural adjunct to the unstable dynamics of a particular system, and may follow a nonlinear pattern in the aftermath of intervention through vaccination, chemotherapy, or ecological measures such as mosquito nets in the case of *P. falciparum*.

26

Vaccination and Serotype Replacement

Marc Lipsitch

26.1 Introduction

Vaccination has been an undisputed success in the control of many infectious diseases, both viral and bacterial. In recent decades, researchers have attempted to extend these successes to the development of vaccines against a variety of other infectious agents, ranging from long-standing public health threats like typhoid, gonorrhea, and malaria, to newly emerging or newly discovered organisms, such as human immunodeficiency virus (HIV) and hepatitis B virus. While some of these vaccine development efforts have succeeded quite rapidly – the hepatitis B vaccine is a good example – many have not yet produced highly effective vaccines. The presence of substantial antigenic diversity is a common feature that characterizes many of the infections for which vaccines have proved elusive. This diversity can take either, or both, of two forms. Within a single infected host, the expression of particular antigens may change during the course of an infection by a variety of mechanisms, including intragenomic recombination, phase variation through changes in the lengths of oligonucleotide repeats, and simple mutation. This process of antigenic variation may disrupt antigen-specific immune responses, with important consequences for the maintenance of infection and pathogen virulence. Antigenic diversity can also occur at the population level; in this case, the pathogens of a particular species circulating in the host population are characterized by polymorphism in one or more antigens. Each of these forms of polymorphism may increase the number of antigenic variants that a vaccine must "cover" to give strong protection, thereby increasing the difficulty of vaccine development.

In this chapter, I discuss population-wide polymorphism in a single antigen targeted by vaccination in the context of two closely related vaccines: polysaccharide–protein conjugate vaccines against *Streptococcus pneumoniae* (pneumococcus) and *Haemophilus influenzae*. Both of these bacteria exhibit considerable diversity in their capsular polysaccharide – the antigen against which presently available vaccines are directed. As vaccination may have complex effects in changing the patterns of diversity in these organisms, the design of vaccines against *H. influenzae* and *S. pneumoniae* involves not only the usual considerations of immunology, cell biology, and biochemistry, but also questions of population biology. As such, this problem of vaccine design is an excellent example of the sorts of issues raised by a real-world problem of "virulence management." The mathematical models that are the stock-in-trade of population biologists can offer assistance in controlling disease from these organisms, but useful application and interpretation of these models require that they incorporate a rather large amount of specific biological information about the organisms in question. Since the questions raised by such vaccines are relatively new, one of the most important functions of these models is to identify areas in which further empirical (immunological and epidemiological) research would be particularly helpful in advancing our ability to predict the effects of vaccine use, and, thus, designing vaccines that will be more effective at the population level.

This chapter considers the current generation of vaccines – known as conjugate vaccines – against *H. influenzae* and *S. pneumoniae* to illustrate the ways in which knowledge of basic immunology and microbiology can be combined with the techniques of population biology to address an applied question in virulence management. The chapter begins with a brief description of the common and distinctive features of the biology of these two organisms, the strategies adopted to address the problem of vaccination in the presence of antigenic polymorphism, and the effects observed from these vaccines. I then discuss some predictions of mathematical models of vaccination against such polymorphic bacterial pathogens, and describe how the models' predictions can aid in the interpretation of data and the design of clinical vaccine trials. I conclude the chapter by emphasizing the other role of models: to identify some areas of relative biological ignorance in which further research would be particularly fruitful to help predict and evaluate the effects of vaccination.

26.2 Biology, Diversity, and Impact of Two Pharyngeal Pathogens

H. influenzae and *S. pneumoniae* are both obligate colonizers of the human oro- and nasopharynx. Most people who carry these organisms remain healthy, and most transmissions of the organisms occur between asymptomatic carriers (Austrian 1986; Moxon 1986). Nonetheless, as a result of the large number of people who carry these organisms, even a relatively low rate of disease among carriers translates into a substantial threat to public health. Pneumococci are presently responsible for an estimated 7 million cases of otitis media, 500 000 cases of pneumonia, 50 000 cases of bacteremia, and 3 000 cases of meningitis annually in the USA (CDC 1997). *H. influenzae* is a frequent cause of otitis media and other relatively minor infections, but the greatest concern with this organism arises from its role in invasive disease – particularly meningitis and bacteremia. Virtually all invasive disease associated with *H. influenzae* was caused by serotype b (Hib) until the advent of effective vaccination against this serotype. In the prevaccine period, Hib caused invasive disease in an estimated one in 200 children under 5 years of age in the USA (CDC 1995).

H. influenzae and *S. pneumoniae* are characterized by extensive antigenic diversity. Both organisms are classified by serotype, which is determined by the capsular polysaccharide. *H. influenzae* has six known serotypes, a–f, in addition to so-called "nontypeable" strains, which are nonencapsulated. As stated earlier, Hib is responsible for virtually all invasive disease caused by this species (in the absence of vaccination); however, all of the capsulated types are relatively rare in surveys of isolates from healthy carriers and from noninvasive disease, such as otitis media; in both of these groups, nontypeable strains are by far the most common.

The diversity of serotypes is considerably greater in pneumococci (with 90 known serotypes). And although "rough" (unencapsulated) variants of pneumococci are viable in the laboratory, they are rarely found in isolates from either healthy carriers or pneumococcal disease patients. Among the 90 serotypes, the variations in prevalence (rates of carriage) and in virulence are considerable (Smith *et al.* 1993); virulence is defined here to mean the probability that a particular type will cause disease in an individual who carries it. In contrast to *H. influenzae*, no single pneumococcal serotype is responsible for the majority of serious disease; however, a loosely defined group of so-called "pediatric serotypes" is commonly found both in carriage and in disease isolates, especially in children, in many parts of the world.

26.3   Conjugate Vaccines

The present generation of vaccines, known as polysaccharide–protein conjugate vaccines, consist of capsular polysaccharide covalently bound to a protein carrier. As the vaccines are based only on a single, highly polymorphic antigen, the protection offered by them is specific to the capsular polysaccharide serotypes included in the vaccine.

In the case of *H. influenzae*, the vaccine includes only one capsular serotype, Hib, because this type plays such a disproportionate role in invasive disease. Conjugate vaccines against Hib are now part of routine childhood immunizations in developed countries, where they have reduced the incidence of invasive Hib disease by 90% or more (Booy and Kroll 1997). Hib conjugate vaccines offer protection not only against invasive disease, but also against carriage of the organism. As a result, these vaccines induce herd immunity – a decline in the prevalence of the organism that offers indirect protection to unvaccinated individuals by reducing their exposure to infection or colonization by the organism (Barbour *et al.* 1995). Herd immunity may explain why the reduction in invasive Hib disease in some populations has exceeded the fraction of the population that received the vaccine (Booy and Kroll 1997). Indeed, the apparent herd effects of Hib vaccination have led to speculation that the eradication of Hib might be possible if vaccination could be used on a global scale. Unfortunately, the current price of Hib vaccines places them out of reach of many developing countries (Booy 1998).

Pneumococcal conjugate vaccines have been designed with a variety of formulations, each of which includes between seven and 11 of the serotypes most commonly associated with disease in particular populations. These vaccines are still in clinical trials, but the results of these trials so far have been promising: all major studies have shown protection against carriage of those pneumococcal serotypes included in the vaccine (Dagan *et al.* 1996a, 1998; Obaro *et al.* 1996; Mbelle *et al.* 1999; O'Brien *et al.* 2001). More recently, the vaccine has shown efficacy against invasive disease (Black *et al.* 2000) and otitis media (Eskola *et al.* 2001) in major clinical trials.

26.4 Serotype Replacement

Although the reduction in carriage achieved by conjugate vaccines is beneficial from the perspective of herd immunity, it has also raised concerns about the possibility of "serotype replacement." Since the protection offered by conjugate vaccines is specific to the capsular type(s) included in the vaccine, it has been suggested that reducing carriage of these vaccine types may leave open an ecological niche that will be filled by serotypes not included in the vaccine (Greene 1978; Farley *et al.* 1992; Wenger *et al.* 1992; Nitta *et al.* 1995; Obaro *et al.* 1996).

In light of this concern, the design of vaccines against *H. influenzae* and *S. pneumoniae* presents a well-defined, practical problem in virulence management: given a bacterial species with a range of serotypes (and nonserotypeable strains) that vary in their tendency to cause disease, how can the composition of a vaccine be optimized to achieve the maximum reduction in disease from that species? As described above, the strategy adopted in *H. influenzae* vaccination (inclusion of only one serotype, which was responsible for virtually all serious disease) is different from that adopted in the design of pneumococcal vaccines currently being evaluated (inclusion of multiple serotypes). While both vaccines have been effective in reducing the carriage of serotypes they target (Barbour *et al.* 1995; Dagan *et al.* 1996a; Obaro *et al.* 1996; Mbelle *et al.* 1999), they have had quite different results with respect to serotype replacement.

So far, serotype replacement has not been detected following vaccination against Hib. Studies of *H. influenzae* carriage in Finland (Takala *et al.* 1991) and the UK (Barbour *et al.* 1995; Booy *et al.* 1995) both failed to find any evidence of increased carriage of nontype b *H. influenzae* as a result of vaccination. Although there have been reports of increases in invasive disease from other nasopharyngeal bacteria in the period since Hib vaccination began (Baer *et al.* 1995; Urwin *et al.* 1996), there is no evidence of a causal link to Hib vaccination (Booy *et al.* 1995; Urwin *et al.* 1996).

In contrast, there is considerable evidence of serotype replacement in pneumococcal conjugate vaccine studies. Three studies, in The Gambia (Obaro *et al.* 1996), Israel (Dagan *et al.* 1998), and South Africa (Mbelle *et al.* 1999), demonstrated statistically significant increases in nasopharyngeal carriage of nonvaccine serotypes in children who received the conjugate vaccine, compared to control children. Most recently, serotype replacement has been demonstrated in otitis media, probably the most common, albeit least severe, of the diseases caused by pneumococci. Children in Finland were randomized to receive the pneumococcal conjugate vaccine or a control vaccine (hepatitis B), and followed for otitis media. When otitis media was diagnosed, the causative organism was isolated when possible and, if pneumococcal, it was serotyped. In this study, there was a 34% increase ($p = 0.053$) in nonvaccine-type otitis media among vaccinees compared to controls, although total pneumococcal otitis media was lower in vaccinees. This study was the first report of increased disease from nonvaccine types through pneumococcal conjugate vaccination (Eskola *et al.* 2001).

These contrasting experiences raise a number of questions for future trials and possible adoption of conjugate vaccines. Under what circumstances is serotype replacement most likely, and what factors influence the extent of such replacement? How does the possibility of serotype replacement affect the optimal design of vaccines for an organism like pneumococcus (pneumococcus vaccines may incorporate many, though not all, serotypes)? If replacement does occur, how is it likely to affect the total amount of disease? How can clinical trials be optimized to detect replacement if it occurs? Finally, what do the results of these trials tell us about the dynamics of bacterial populations and the interactions between bacteria of different types in colonizing vaccinated and unvaccinated hosts?

26.5   Role of Mathematical Models

As clinical trials of pneumococcal conjugate vaccines continue and new ones are planned, it is important to learn as much as possible about the potential for serotype replacement and its likely effects. Mathematical models can be useful in defining the likely extent of serotype replacement in various contexts, optimizing the design of clinical trials to discern whether such replacement occurs, and interpreting the results of these trials. With these goals in mind, I recently constructed and analyzed a compartmental mathematical model (Anderson and May 1991) that represented the transmission dynamics of colonizing bacteria with multiple serotypes, and the effect of vaccination on these dynamics (Lipsitch 1997).

This model simultaneously considers the transmission of two (or more) strains of the same organism, and it is designed to analyze the effects of competitive interactions between (among) these strains, in which carriage of one serotype reduces the probability that a host will be colonized with another serotype. If such competitive interactions occur, then serotype replacement is possible because vaccine-induced reductions in some serotypes will increase the opportunities for others to spread in the population. Recently, we showed that such competitive interactions are present in a mouse model of pneumococcal intranasal carriage: colonization with one pneumococcal strain can inhibit the acquisition of additional strains (Lipsitch *et al.* 2000). Epidemiological studies provide some indirect evidence of such competitive interactions (Reichler *et al.* 1992), while laboratory studies suggest mechanisms by which different species of streptococci (Johanson *et al.* 1970; Sanders *et al.* 1976) or different strains of *H. influenzae* (Venezia and Robertson 1975) might compete in the nasopharynx. At present, however, little is known about the precise nature of these interactions, and perhaps the most compelling evidence that competition occurs comes from the replacement observed in pneumococcal conjugate vaccine studies.

Figure 26.1 shows a diagram of the model; the assumptions and structure of the model are as follows. In the absence of vaccination, it is assumed that individuals are born into the susceptible compartment S at a rate B and are removed from that compartment (and all other compartments) at a specific per capita death rate.

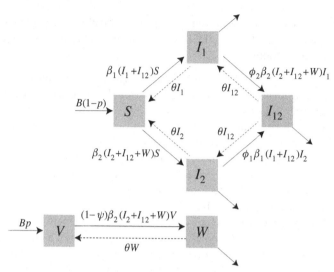

Figure 26.1 Compartmental model of transmission dynamics of colonizing bacteria with multiple serotypes, and the effect of vaccination on these dynamics. Solid lines between compartments indicate colonization; dashed lines indicate loss of colonization. *Source*: Lipsitch (1997).

There are two pneumococcal serotypes present, designated 1 and 2, and susceptible hosts may be colonized by either of these two types, moving them into the I_1 or I_2 compartment, respectively; the rate constants are β_1 and β_2, respectively. Colonization with each type is proportional to the total number of individuals carrying that type. Colonization has an average duration of $1/\theta$. While carrying one serotype, a host may be colonized by the other type, moving the individual into the dually colonized compartment I_{12}. This secondary colonization also occurs at a rate proportional to the prevalence of the colonizing type, but a rate that is ϕ_j (j = 1 or 2) times the rate at which a susceptible individual would be colonized by the same type. Thus, ϕ_j is an inverse measure of the competitive inhibition of type j by the resident type in a host.

When vaccination begins, a fraction p of all individuals is assumed to be vaccinated at birth. In the model, these individuals are born into the vaccinated compartment V. It is assumed that vaccination completely protects an individual against carriage of type 1. To consider the effects of including more than one bacterial serotype in the vaccine, the model can accommodate vaccines that are effective only against type 1 (monovalent vaccines), as well as those that give either partial or full protection against type 2 (bivalent vaccines). The parameter $1 - \psi$ represents the degree of protection offered by the vaccine against serotype 2; the compartment W contains vaccinated individuals colonized with serotype 2.

By varying the parameters of the model, one can compare the effects of different levels of vaccine coverage (fractions of the population vaccinated), different assumptions about the competitive interactions among pneumococcal serotypes, and different types of vaccines (monovalent versus bivalent). Lipsitch (1997) describes the predictions of the model in detail. In summary, the major predictions of the model are as follows:

- If there is competition between different serotypes to colonize hosts, then vaccination against serotype 1 alone increases the prevalence of serotype 2. The extent of replacement, measured as the increase in the prevalence of serotype 2, will be greatest when vaccine coverage is high and for situations in which serotype 2 is strongly inhibited from colonizing individuals who carry serotype 1. Serotype replacement may take either of two forms: an increase in prevalence of a type already present in the population, or the appearance and spread of types that were previously absent from the population because they were unable to compete with the vaccine type(s).
- Bivalent (or polyvalent) vaccines can also cause replacement if the protection offered against different serotypes is uneven. In particular, if a vaccine has relatively low efficacy against serotype 2 but very high efficacy against serotype 1, then use of a bivalent vaccine may increase the prevalence of type 2.
- If there are only two serotypes interacting in a population, there is a limit to the amount of replacement that can occur. Specifically, the increase in the prevalence of serotype 2 will always be less than (or at most equal to) the decrease in the prevalence of serotype 1. Thus, for example, if the prevalences of serotypes 1 and 2 prior to vaccination are 15% and 20%, respectively, then the prevalence of serotype 2 following vaccination will be no more than 35%.
- If more than two serotypes are competing to colonize hosts, then this limitation need not hold. In the presence of more than two serotypes, it is possible for vaccination to increase the prevalence of a single, nonvaccine type more than it reduces the prevalence of the vaccine type.
- Although replacement is a source of concern, it may also be beneficial. If serotypes compete to colonize hosts, then increases in the prevalence of the nonvaccine types will help to reduce the prevalence of the serotypes included in the vaccine. Thus, replacement augments the effects of herd immunity in reducing the exposure of all members of the population to vaccine serotypes. This results in a trade-off between the breadth of coverage of a vaccine (number of serotypes covered) and the effectiveness of the vaccine in reducing carriage of each serotype at the population level.

The model's predictions have several implications for the interpretation of existing data from the use of conjugate vaccines, for the design of vaccine trials, and for the choice of vaccine composition. The next three sections describe some of these implications.

26.6 Pneumococcal Conjugate Vaccines versus Hib Vaccines

As noted above, the absence of serotype replacement observed with the use of Hib in developed countries contrasts with the findings of considerable serotype

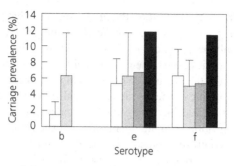

Figure 26.2 Carriage of three serotypes of *H. influenzae* in children vaccinated against serotype b (white bars) compared with controls (light gray bars). Error bars (half shown) indicate 95% confidence interval (binomial approximation). Dark gray bars show the maximum carriage of serotypes e and f in vaccine recipients that could result from replacement in a population in which only a small proportion of susceptible hosts are vaccinated (as in the study). Black bars show the equivalent figures in a hypothetical study in which virtually all susceptible hosts were vaccinated. *Data source*: Barbour *et al.* (1995); calculations, Lipsitch (1999).

replacement in two studies of pneumococcal vaccines. What might account for this difference?

The mathematical model suggests an explanation. As noted above, the model predicts that, in a pairwise interaction between two serotypes, the increase in prevalence of a nonvaccine type will be no more than the reduction in prevalence of a vaccine serotype. This principle is illustrated in Figure 26.2, which presents data from a study of Hib conjugate vaccine in the UK (Barbour *et al.* 1995). In the figure, the white bars show the prevalence of each of three *H. influenzae* serotypes – b, e, and f – in vaccinated individuals, and the light gray bars show the prevalence of each of these serotypes in controls. If one assumes that Hib interacts independently with each of the two nonvaccine serotypes (e and f), the two-serotype model can be used to calculate the maximum prevalence of these nonvaccine types that would be expected in vaccinees if these serotypes compete very strongly with serotype b. The dark gray bars show the maximum prevalence of types e and f expected in the study, in which only a small fraction of the community was vaccinated; the black bars indicate the equivalent figure if the whole community had been vaccinated. As is clear from Figure 26.2, the increase in nonvaccine-type carriage in vaccinees would be minuscule and statistically undetectable in a study of this kind. [Indeed, the study from which these data are drawn was not designed to detect replacement; data on the prevalence of types e and f were used to control for general changes in the prevalence of *H. influenzae* that could have been attributable to factors other than vaccination (Barbour *et al.* 1995).] The reason is that the prevalence of Hib was so low before vaccination that even its complete removal by widespread vaccination would have little effect on competing bacteria. The prevalence of Hib carriage in other developed countries is similar to that measured in the UK study. Therefore, the model suggests that the lack of replacement, even following widespread use of the Hib conjugate vaccine in developed countries, may simply be a result of the low prevalence of Hib.

If this interpretation is correct, then strain replacement would be more likely to occur in areas in which the prevalence of Hib is higher, or for vaccination against other organisms whose prevalence is higher. This difference could account for the contrasting outcomes of vaccination against Hib and pneumococci. Differences in the biology of colonization or in the interactions between bacterial types may also have a role in these contrasting outcomes. Distinguishing the relative importance of these two explanations requires further research into the biological interactions of bacterial populations in the nasopharynx, as well as studies of the effects of conjugate Hib vaccination in areas where Hib's prevalence is higher.

26.7 Detection of Replacement: Design of Clinical Trials

If a conjugate vaccine is used by a large fraction of the human population in a community, it may alter the composition of the bacterial population in that community, not only in vaccinated individuals, but also in unvaccinated individuals. Vaccination may reduce the prevalence of those serotypes included in the vaccine, thereby protecting unvaccinated individuals against exposure to these serotypes (herd immunity). Similarly, if serotype replacement occurs and vaccinated individuals become more likely to carry nonvaccine serotypes, then the exposure of unvaccinated individuals to these serotypes increases. As a result of these indirect effects, strain replacement will be magnified in communities in which large numbers of individuals are vaccinated. This process is also evident from Figure 26.2. There, the shaded bars show the model's prediction of the maximum increase in nonvaccine-type carriage in vaccine recipients in a community in which the vaccine is used for only a very small proportion of individuals, while the striped bars show the same increase in a community in which everyone is vaccinated. As is clear from Figure 26.2, replacement will be most easily observed in communities in which the level of vaccine coverage is high.

Therefore, if one wants to answer the question as to whether the use of a vaccine will induce serotype replacement, clinical trials that randomize vaccination to a small fraction of individuals within a community will be less effective than those that use whole communities as the units of randomization. Results of studies of pneumococcal vaccines in which communities among Native Americans in the southwestern USA are the units of randomization have been published recently (O'Brien *et al.* 2001).

26.8 Is Serotype Replacement Always Bad?

For an organism like the pneumococcus, in which a number of serotypes are capable of causing disease, the choice of serotypes for inclusion in a conjugate vaccine is very important. One possibility is to include as many serotypes as possible, to achieve the broadest possible protection. In addition to some technical limitations on the number of serotypes that can be included in a single vaccine, there are other reasons why such a strategy would not be ideal. As noted above in the last prediction from the model, serotype replacement can augment the effectiveness of a vaccination program in a community. This occurs because increases in the prevalence of nonvaccine serotypes competitively inhibit carriage of vaccine serotypes. Ideally, then, one would like to design a vaccine that maximizes these beneficial effects while minimizing the risk of added disease from increased carriage of nonvaccine serotypes. A variant of this strategy has been proposed by Ewald (1996), who refers to it as the "virulence–antigen" strategy.

The question is how such a balance can be accomplished. As described so far, the model deals only with carriage of various serotypes; it does not directly address the problem of disease. The effect of vaccination on disease depends both on changes in patterns of carriage of different serotypes, and on the propensity of the different serotypes to cause disease. It is clear that serotypes of both *H. influenzae* and *S. pneumoniae* vary considerably in their pathogenicity (Wenger *et al.* 1992; Smith *et al.* 1993). If these serotype associations are known to be stable, the ideal stated in the previous paragraph could be achieved by including in a vaccine the most pathogenic serotypes, but excluding those that tend to be avirulent, thereby taking advantage of any increases in the prevalence of the avirulent serotypes to augment the effect of the vaccine.

There are several potential problems with this approach. First, the model predicts that widespread use of a vaccine may result in the appearance of bacterial types that were previously absent from the population because of competition from vaccine types. It would obviously be difficult to predict the virulence of these novel types, since competitive inferiority to existing types need not correlate with low virulence (Topley 1919; Lipsitch and Moxon 1997). Second, both species discussed here are highly transformable. Although the capsular type seems to be very tightly associated with virulence in *H. influenzae* (Moxon and Vaughn 1981), transformation studies in pneumococci revealed complicated interactions between capsular type and other genes in determining virulence (Kelly *et al.* 1994), so the existing associations between virulence and capsular type (Barnes *et al.* 1995; Takala *et al.* 1996) may be short-lived. If pneumococcal conjugate vaccines are used on a widespread scale, it will be important to maintain surveillance for shifts in the serotype associations of invasive disease.

Thus far, the problem of serotype replacement has been discussed mostly as it applies to serotypes not included in the vaccine. However, if the vaccine is only weakly effective in immunizing against carriage of some of the serotypes included in it, then these serotypes may actually increase in prevalence following vaccination. This can occur if the efficacy of the vaccine against these serotypes is outweighed by its effect in removing competing serotypes. It is clear from the results obtained in trials published thus far that the protection offered by the vaccine against the combined serotypes included within the vaccine is less than 100%. It therefore is important to monitor the results of future trials to determine whether prevalence of any of the individual vaccine serotypes increases in vaccinated hosts.

26.9   Limitations of the Models and Areas for Future Work

The mathematical models described here, like all such models, involve a number of simplifications. In some cases, these simplifications are deliberately introduced to make the model more tractable and to focus attention on fundamental processes of transmission and competition of different serotypes. In other cases, the simplifications are necessary because much remains unknown about the biology and, especially, the immunology of the carriage of these organisms. The biological assumptions and limitations of the model are discussed in more detail by Lipsitch (1997). In this section, we describe some of the presently unanswered biological questions that are relevant to our understanding of the effects of vaccination.

Very little is known at present about the immune response to natural exposure (either asymptomatic carriage or disease) to *S. pneumoniae*. Several studies show that carriage of pneumococci can elicit serum anticapsular antibody responses in children (Gwaltney et al. 1975; Gray and Dillon 1988) and adults (Musher et al. 1997). Furthermore, the success of conjugate vaccines in reducing pneumococcal carriage suggests that acquired immune responses (presumably antibody-mediated) can affect carriage. However, the effect of antibodies acquired from natural exposure on subsequent acquisition and carriage of pneumococci is poorly understood. One study showed that possessing a pre-existing antibody did not prevent colonization, but did, however, reduce the duration of carriage (Gwaltney et al. 1975). Few other data, however, are available to address this issue. Similarly, little is known about the extent to which naturally acquired antibody responses protect against invasive disease (Smith et al. 1993).

Another area of relative ignorance concerns the interactions between different species, strains, or serotypes of bacteria in the nasopharynx. As noted above, various possible mechanisms for competitive interactions between bacterial types have been found, but the significance of these mechanisms *in vivo* remains unknown. The results of vaccine trials (Obaro et al. 1996; Mbelle et al. 1999) suggest that carriage of one pneumococcal serotype inhibits carriage of additional serotypes, but the mechanisms of this inhibition remain unknown. The mathematical model described above assumes that different serotypes inhibit one another directly – that is, the presence of one type reduces the chance of acquiring an additional type. We recently showed that inhibition occurs in animal models (Lipsitch et al. 2000). However, other mechanisms of inhibition are possible. In particular, stimulation of cross-reactive antibodies (to antigens other than the capsular polysaccharide) by one serotype might inhibit future carriage of other serotypes, even after carriage of the first type has ceased. Simulations of mathematical models that incorporate such effects show that vaccination could have rather different effects from those predicted in the model described here, which includes only direct inhibition. Finally, no data are available (to my knowledge) regarding the possibility that carriage of one bacterial population in the nasopharynx may modulate the virulence of other bacteria to which the host is subsequently exposed. Further research into the microbiology and immunology of the host–bacteria relationship in the nasopharynx is important to understand and predict the effects of conjugate vaccination.

Box 26.1 Models of vaccination effects on pathogen diversity

Several mathematical models have been designed to explore the consequences of vaccination on heterogeneous populations of pathogens. Perhaps the first such model was that by McLean (1995b; see also Chapter 24), who sought to describe the conditions under which vaccination against a pathogen that was initially homogeneous could lead to the emergence and spread in the host population of vaccine-resistant escape mutants. More recent efforts focused on the question of how an already heterogeneous pathogen population would respond to vaccination.

Gupta and colleagues have used a series of models to study the way in which host immunity, both natural and vaccine-induced, structures the populations of pathogens that present several important antigens to the immune systems of infected hosts. Their work in this area arose from the question of whether stable, multilocus genotypes (strains) can persist in pathogen populations that undergo frequent sexual recombination. The models indicate that the answer to this question is positive, because host immunity can act as a strong, epistatic selective force in favor of particular, nonoverlapping combinations of antigenic types and against any recombinants that could be generated by these types (Gupta *et al.* 1996; see also Chapter 25). Subsequent modeling focused on the ways in which vaccination would affect such structured parasite populations. The results of these models are complex, but they show, in general, that vaccination against one or several strains in such a system may either result in elimination of all strains or in a restructuring of the population, such that the strains not affected by the vaccine increase while those affected by the vaccine decline (Gupta *et al.* 1997; compare also Chapter 25).

The major challenge in modeling heterogeneous parasite populations is to simplify the model sufficiently to permit meaningful analysis. If one considers a population with n strains, a full epidemiological model that tracks only infection with each strain includes 2^n equations (assuming that individuals can be infected with more than one strain at a time, which is usually true biologically). If the model includes a "recovered" or "immune" class for each strain, then the number of equations grows to 3^n. Various assumptions are necessary to reduce this combinatorial complexity to a manageable level. Lipsitch (1997) considers only two strains and ignores acquired immunity to carriage. Gupta *et al.* (1997), by contrast, include immunity, but make the simplification that the probability of immunity to each strain is independent of immunity to other strains. Additional work, using a combination of mathematical simplifications and more complex simulations, is needed to deal with these complicated systems. As described in the main text of this chapter, extrapolation from two-strain systems to systems of more than two strains is not always possible; the need to study multiple-strain systems amplifies the importance of developing better techniques for doing so.

As the answers to such biological questions become clearer, mathematical models will need to be modified to reflect the expanded understanding.

26.10 Discussion

Two further points about the effects of conjugate vaccines on pneumococcal populations are worthy of mention.

First, the choice of serotypes for inclusion in conjugate vaccines has been different in different locations, but has generally been designed to cover those serotypes that are most often implicated in invasive disease. Often, these types coincide with those serotypes that show the greatest levels of antibiotic resistance (Dagan *et al.* 1996b; Butler 1997). As a result, conjugate vaccination has led to a reduction in the percentage of antibiotic-resistant pneumococci carried by vaccinees in two studies (Dagan *et al.* 1996a; Mbelle *et al.* 1999).

Second, it is important to note that capsular polysaccharide is not the only possible target for vaccination. Several pneumococcal vaccines based on protein antigens are in various stages of testing (Paton 1998). Since these protein antigens show considerably less variation among pneumococcal isolates, vaccines based on them should be less vulnerable to serotype replacement, and they may be useful as complements or alternatives to polysaccharide conjugate vaccines.

The occurrence of serotype replacement in several studies of pneumococcal conjugate vaccines confirms the validity of the concerns expressed in anticipation of trials of both Hib and pneumococcal conjugate vaccines. As the results of further clinical trials are made available, it will become clearer how general this phenomenon is. Mathematical models are useful in suggesting ways to improve the design of these trials and the interpretation of their results (Box 26.1). Furthermore, the epidemiological findings of these studies should be the impetus for further research into the role of serotype and other factors in determining the variation in pneumococcal virulence, the nature of immune responses to organisms like the pneumococcus at the nasopharyngeal mucosal surface, and other questions in the biology of bacterial carriage.

Acknowledgments   Marc Lipsitch was supported by grant # GM19182 from the US National Institute of Health. The author thanks Prof. E.R. Moxon for raising the problem of serotype replacement and for many helpful discussions.

Part G

Perspectives for Virulence Management

Introduction to Part G

The authors of this book have been encouraged by the editors to stick their necks out and dream up strategies for virulence management, phrased as concretely as possible. As editors, we believe that good science proceeds by making definite predictions so that they can be rigorously put to the test. Phrasing predictions in the form of recommendations forces a healthy definitiveness; no one is allowed to hide under a slightly woolly phrasing. It must nevertheless be understood that not all the recommendations outlined here can as yet be taken at face value; many of the issues raised require additional theoretical and experimental research.

The stress on management aspects is the defining feature of this last part of the book. Whereas earlier parts review particular mechanisms of virulence evolution with a perspective on potentially ensuing options for virulence management, for this part of the book the authors were invited to focus on the following questions:

- For which specific empirical settings can the various possible options of virulence management strategies be expected to apply?
- For each given context, which options appear to be particularly promising?
- What are the open research questions that have to be addressed before measures of virulence management can be recommended for implementation?

After an introductory chapter that is meant to summarize what has been achieved so far, each chapter in this part covers one of the main potential arenas for virulence management: human, wildlife, and livestock diseases, crop protection, and pest control.

Given the variety of these arenas we must recall the range of virulence concepts highlighted in the Introduction – the common denominator is the damage wrought to individual health, to Darwinian fitness, to economic return, or to the pest to be controlled. In particular, the notion of virulence considered in the chapters on human and animal diseases captures various forms of damage to host individuals. In the plant pathology chapter, virulence refers to a pathogen's ability to obtain access to host individuals, a type of virulence that is of great practical importance in plant disease control. The chapter on pest control focuses on the ability of a disease to spread between host populations and thus to produce differential damage at the metapopulation level.

Chapter 27 sets the stage for the later chapters by placing the theory developed in the previous parts in a wider context, with an emphasis on the interface between the theoretical and experimental literature. After giving a methodologically oriented overview of the field, stressing restrictions and caveats, Sabelis and Metz attempt to summarize the main results on virulence evolution gleaned from the previous chapters and the literature. From that perspective the authors identify what they see as gaps in our current knowledge that need to be filled to transform the study of virulence evolution and management into a mature science.

Chapter 28 considers those opportunities for virulence management of human diseases that result from influencing the modes of pathogen transmission from one host to another. After reviewing the available evidence for a diarrheal disease (cholera) and a vectorborne disease (malaria), Ewald concludes that the predicted effects of transmission route manipulation are broadly, although preliminarily, corroborated by comparative studies. For cholera the evidence stems from water purification efforts in South America, and for malaria from mosquito-proofing measures in the USA. Such practices offer short-term as well as long-term benefits, appear ethically uncontroversial, achieve high levels of cost effectiveness, and are expected to be evolutionarily stable – they therefore constitute prime candidates for virulence management initiatives.

In Chapter 29, De Leo and Dobson point out that the goal of preserving charismatic wildlife biodiversity in our human-dominated world makes virulence management of feral populations an issue. Since the time it takes for drug resistance to develop tends to be very short and containment is absent, drug treatments are generally not advisable. There is a dearth of empirical data at the population level, which is especially problematic because wildlife systems are far less controllable than systems in other arenas of virulence management. One strategy to overcome this difficulty is to estimate the disease-induced reduction in host population density from the relative prevalence of the disease in corpses and live animals to gauge where efforts can be invested most usefully. At present, realistic measures aim at the containment of potential disease sources; these options are of two types. First, uncontrolled "spill-overs" from reservoir populations to endangered populations must be minimized. Owing to their higher densities, the former populations can harbor more virulent strains. Second, we should try to reduce the risks of disease transport in controlled wildlife translocations.

Chapter 30 considers livestock that are kept under such controlled conditions and for which diseases are so costly that there is a strong incentive to go beyond the simplifications made in more generally applicable models. It is against this background that de Jong and Janss discuss a verbal model, in which the body of an animal is subdivided into two compartments by the action of the evolving immune system: one in which the virus multiplies and from which it spreads, and one in which it rarely enters but is at its most harmful. Their considerations lead to a number of recommendations for stocking schemes that enable an infection to be stamped out quickly and, at the same time, prevent the disease evolving to escape from the standard control measures.

Whereas Chapter 30 focuses on effects of the host defense system, it is the total absence of systemic defenses in plants that underlies the analysis by Jarosz in Chapter 31. In such a setting, the relative rate of increase of a plant disease depends mainly on the available amount of free tissue. This engenders a dominant impact of seasonality through the dramatic changes in amount of leaf tissue, which lead to different selective pressures the year round. In particular, in winter the amount of leaf tissue is so small that stochastic and metapopulation effects may kick in. The resultant mechanisms affect the genetic structure of the populations in different manners and therefore open up different avenues for virulence management. Inferring the relative importance of these mechanisms from the observed genetic structure of a population makes it possible to devise management strategies accordingly.

Chapter 32 considers biological pest control from a metapopulation perspective. From this vantage point it becomes mandatory to distinguish the virulence of a biocontrol agent (pathogen, parasitoid, or predator) toward its individual victim (individual-level virulence) from its virulence toward local populations of its victim (patch-level virulence). Whenever the aim is to achieve establishment, spread, and long-term persistence of the biocontrol agent over a large geographical area, it becomes important to consider whether the biocontrol agent exploits local pest populations so as to maximize the number of dispersers. That aim is not necessarily best achieved by a high rate of killing of pest organisms. Elliot, Sabelis, and Adler argue that the key to understanding changes in patch-level virulence is to assess the degree to which biocontrol agents monopolize exploitation of local pest populations. Thus, any means by which this monopoly can be broken will increase the virulence of the biocontrol agent. This results in a suite of measures that can be taken to increase the effectiveness (i.e., the patch-level virulence) of biocontrol agents.

A few of the recommendations in this part have already passed more or less rigorous tests so that they indeed are entering the realm of practical application. Other practical measures suggested here appear to have a good chance of surviving further empirical testing. Generally speaking, however, we are dealing with the very initial stages of a fascinating and hopefully, in the long run, practical science. The following chapters are meant to provide a sense of direction.

27

Taking Stock: Relating Theory to Experiment

Maurice W. Sabelis and Johan A.J. Metz

27.1 Introduction

This book is concerned with the way natural selection affects the virulence of disease agents, here loosely defined as damage to the host, and with how we can use this knowledge to design strategies for managing virulence. These questions are rooted in Darwinian thinking about evolution (Poulin 1998; Stearns 1999). If it were possible to resolve these questions at the level of evolutionary storytelling only, this book would not exist. The impetus behind this book came from recent advances in mathematical evolutionary theory, in particular the ongoing merger of the theories of population dynamics and natural selection. This merger enables quantitative, and therefore testable, predictions of the outcome of selection for a given ecological setting. As so often, applied problems form an ideal testbed for the new tools.

Since disease agents have short generation times and usually harbor considerable genetic variation, natural selection can potentially cause rapid changes in the genetic make-up of pathogen or parasite populations. Therefore, the evolution of parasite virulence is an obvious area to test new evolutionary theories. Another matter is whether such tests promise immediate applications. The theory of evolutionary dynamics is not at a stage that can produce lists of management strategies to solve any particular problem with certainty; to be fair, is any theory able to? However, any measure, even the considered absence of action, is guided by some theory in whatever verbal or mathematical form. Given that measures will be taken anyway, we do better to evaluate their effects by comparing them against the predictions of the theory of evolutionary dynamics. Of course, one hopes for more direct applications in a not too distant future, but some scepticism is not altogether unwarranted (e.g., Bull 1994). However, within the scientific approach such scepticism should not bar us from making and testing predictions, and, clearly, one of the most difficult ways of making predictions is to devise management strategies. Moreover, at the applied end, to devise management strategies without incorporating as best as possible the available plethora of evolutionary predictions would be an ostrich-type strategy, since changes in virulence are an undeniable feature of the epidemiology of disease agents.

In this chapter we place the theory discussed in the previous chapters in a wider perspective, take stock of what has been achieved in our field with an emphasis on the interaction between theory and experiments, and touch on a number of

conceptual issues that still require further resolution. This sets the scene for the
chapters on virulence management that follow in this part.

27.2 Panoramic View of Virulence Evolution

Before presenting a survey of how the theory relates to experiments at the process
level, we first take a grander view and consider the prerequisite for any evolutionary theory to work, the genesis of sufficient genetic variation. This leads us
naturally to say a few words about the popular concept of the ecology and evolution of virulence, which is largely shaped by the publicity that surrounds emerging
diseases. We argue that from a theoretical viewpoint there is little new to disease
emergence, and that work on more mundane aspects of virulence evolution can
make a greater contribution to public health improvement. After this, we consider
the time scales on which the various relevant processes act and how this is relevant
to the questions a prospective virulence manager might ask. This perspective then
leads to a few general caveats that a manager should keep in mind.

Sources of genetic variation in virulence-related traits

A prerequisite for natural selection to influence parasite virulence is the existence
of genetic variation in traits that affect this virulence. If – to keep things simple
– we temporarily equate virulence with the extra host death rate induced by the
disease agent (additional to the death rate from other causes), then the differences
in virulence among disease agents are striking, be it between species or between
genetic variants within species.

Changes in virulence may arise from simple mutations. For example, a single
base-pair change in a fungal gene that encodes a product that elicits programmed
cell death in a plant (the so called hypersensitive plant response, which creates
a "scorched earth" around the infection site) suffices to cause a dramatic change
in the virulence of *Cladosporium fulvum* to tomato plants (De Wit 1992; Joosten
et al. 1994). Such changes often relate to traits involved in evasion from the host's
defense system, but they may also concern traits that help the host cell to recover
after pathogen invasion, as shown for *Salmonella* bacteria (Fu and Galán 1999).
Bacteria, such as *S. enterica* and *Escherichia coli*, have been shown to harbor mutator phenotypes (LeClerc *et al.* 1996; Heithoff *et al.* 1999). Given that their environment continuously changes, mutator alleles may be important for pathogens to
evolve new ways to enter the host, find their niche, avoid competition with other
microbes, and circumvent host defense barriers (Moxon *et al.* 1994; Taddei *et al.*
1997b). Among the viruses, the human immunodeficiency virus (HIV) is the best
known example of a mutator that continually evades immune surveillance (Nowak
et al. 1991).

Other variations result from genetic recombination between pathogen races and
genetic exchange between related species. When together in the same host cell,
viruses with a segmented genome (e.g., influenza-, arena- and bunyaviruses) may
exchange structural genes in a modular fashion (Koonin and Dolja 1993). Bacteria may swap genes via transformation and exchange of mobile genetic elements

(e.g., plasmids, transposons) and fungal networks may come into genetic contact via anastomosis. A recent example of the latter is the interspecific hybridization between two species of a plant fungus (*Phytophthora* spp.) that led to a new aggressive pathogen of alder trees in Europe (Brasier *et al.* 1999; Brasier 2000). Also, bacteria and fungi themselves can be infected by pathogens that alter the virulence of their host. For example, a bacterial virus (bacteriophage) encodes the gene clusters (so-called pathogenicity islands) that determine the colonization success of cholera bacteria (*Vibrio cholerae*), but another different virus carries the cholera toxin genes (Karaolis *et al.* 1999). Indeed, such converting bacteriophages may act as efficient vehicles for horizontal gene transfer and may explain quantum leaps in virulence through the transfer of blocks of genetic material, rather than through the accumulation of single nucleotide mutations, as shown for *Salmonella* spp. (Mirold *et al.* 1999). All these remarkable ways of generating genetic diversity cause some microbes to become (a)virulent to their host and/or to jump from one host to another.

Emerging or re-emerging diseases?

Some so-called emerging diseases have had devastating effects on their host as well as on their host's populations. An often-heard argument is that these newly emerging diseases prove the existence of major genetic changes in virulence. This is not necessarily true. For one thing, the emergence of a new disease may reflect newly emerging knowledge rather than anything else. For another, the emergence may actually be a re-emergence purely as a result of the mechanisms of population dynamics (i.e., natural selection does not play a role). For example, bubonic plague, caused by the bacterium *Yersinia pestis*, kills its host in a matter of days. It carved a path of death through Europe, once Genoese tradesmen, who contracted the disease in the Crimea, introduced it in 1346. Within 4 years it had moved to Scandinavia and even entered Greenland, killing a third of the European population on its way, according to the medieval chronicler Froissart (McNeill 1976). This 14th century Plague was not a unique event: other Great Plagues have occurred over the past 2 500 years in Asia, the Middle East, North Africa, and Europe and the disease continues to cause 200 reported deaths per year worldwide (Tikhomirov 2001). In some areas bubonic plague has even re-emerged as a significant health concern (Kumar 1995).

How can a disease kill its host so fast, and yet maintain itself and re-emerge? For bubonic plague the answer is that rodents represent a reservoir of the bacteria, and that flea parasites can serve as a vector not only between rodents, but also from rodents to humans. Thus, whenever rodents such as rats become abundant near urban centers, there is the risk of a pest outbreak. Once the bacterium enters the human host, the disease may progress and its transmission may take place via the handling of infected tissues (septicemic plague) and via pulmonary droplets (pneumonic plague), but historical evidence suggests that these transmission routes are less important than that from rats to humans. Stochastic, spatially structured population models predict that bubonic plague can persist in relatively small rodent populations from which occasional human epidemics emerge (Keeling and Gilligan 2000). Thus, bubonic plague is driven by its dynamics in the rat population. Consequently, the isolation of infected hosts and/or vaccination are ineffective as methods to eradicate the disease. Rat culling is the solution, provided it takes place at a time of low prevalence of infection among the fleas, since, in the absence of rats, hungry fleas will try their luck among humans.

Plague is just one example showing that a disease outbreak after a long disease-free period does not necessarily demonstrate a newly emerging disease. It may reflect the re-emergence of a disease as a result of the dynamics intrinsic to the parasites and their hosts. However, this is unlikely to be the explanation for all disease outbreaks, because new hypervirulent types of parasites inevitably arise. For example, single point mutations in plasmids of *Yersinia* spp. have been shown to change virulence dramatically (Rosqvist *et al.* 1988; Galyov *et al.* 1993).

Natural reservoirs probably play an important role in the persistence of disease agents (Reid *et al.* 1999; Osterhaus *et al.* 2000), but genetic change is likely to be involved in species jumps followed by outbreaks in populations of the new host. Examples are HIV-1, HIV-2, and the human T-cell lymphotropic virus (HTLV), which have a simian origin (SIV and STLV, respectively; Myers *et al.* 1992; Koralnik 1994; Gao *et al.* 1999). Also, the canine distemper virus was originally not pathogenic for lions, but later entered the lion population in the Serengeti Park (Roelke-Parker *et al.* 1996; Carpenter *et al.* 1998).

Evolutionary time scales

The emergence of new diseases and, to a lesser extent, the generation of variability are the main topics in popular discussions of virulence evolution. Clearly, their importance must not be disregarded. However, the arrival of new genetic material does not drive epidemics on its own. Hypervirulent disease agents are bound to emerge every so often, but they spread only if the appropriate selective conditions arise. The previous chapters in this book take the existence of variation in the virulence of parasites and pathogens as a starting point from which to analyze the processes that ensue: population dynamics of parasite and host genotypes, and the impact of natural selection on genetic variation in virulence and resistance.

The major part of this book deals with interactions between microparasites (viruses, bacteria, fungi) and their hosts. These are characterized by much shorter time scales of dynamics and evolution for the parasite than for the host. So, in evolutionary time emerging microparasites have a head start. In the long term, the host population will respond to parasite-imposed selection pressure by developing resistance. Indeed, for pathogens that are a significant public health burden, such as *Mycobacterium tuberculosis*, *Plasmodium* spp., HIV, hepatitis-B virus, and *V. cholerae*, there are hints of human resistance genes, which hopefully, with the completion of the Human Genome Project, will soon be further substantiated (McNicholl *et al.* 2000). To cope with the parasite's head start, hosts have been naturally selected to produce variable offspring through sexual recombination and enhance offspring resistance by choosing mates with reduced parasite load or indirect signs of resistance to parasites. These are consequences solely of the time-scale differences between microparasites and their hosts. It is not immediately clear how these consequences change for systems characterized by less extreme time scales between parasite and host evolutionary rates (e.g., macroparasites, such as parasitic nematodes, and arthropods) or by nearly equal time scales (e.g., hymenopterous parasitoids and predatory arthropods).

Much evidence indicates that evolutionary change in parasite–host systems can take place quite rapidly. First and foremost, there is the emergence of multiple resistance of commensal and pathogenic bacteria against well-established and new antibiotics applied in medical and veterinary situations (Kristinsson *et al.* 1992; Austin *et al.* 1999). Key elements behind this rapid increase are pre-existing resistant mutants and the exchange of transposable elements between bacteria and between hosts. Bacterial plasmids that carry resistance and virulence factors can be transferred to sensitive strains of bacteria by cell-to-cell contact, and thereby also enable the transfer of resistance from one host to another. A second example comprises laboratory experiments in which the genetic composition of bacterial populations is recorded. Whereas in chemostats and serial transfer of batch cultures populations are usually taken over by a single genotype, this is not the case in a constant batch culture system without nutrient input or cell removal (Finkel and Kolter 1999). Instead, highly dynamic changes in genetic composition are observed, even within a period of just a few months.

Finally, there is the famous data set on changes in virulence of the myxoma virus and resistance in rabbits in Australia between 1950 and 1985 (Fenner and Fantini 1999). After the release of highly virulent virus strains, resistance increased and virulence dropped to intermediate levels. During the past decennium, however, virulence increased again. These empirical examples show that evolutionary change can take place at a time scale close to that of the ecological dynamics. Thus, evolution cannot be ignored when studying ecological dynamics, and vice versa.

A management viewpoint

Let us now consider the central problems for a prospective virulence manager from the time-scale perspective. On the time scale given by the parasite's generation time, these are:

- How and why virulence levels change as the parasites continue to persist in the host population; and
- How the changed properties of the parasites feed back to the dynamics of the parasite–host system.

Extending the time scale to the usually much longer generation time of the host, the additional problems are:

- How the parasite–host dynamics influence selection for resistance in the host; and
- How this in turn changes the selective environment of the parasite.

Previous chapters answer these questions in the form of theoretical statements, giving conditions under which parasites become gradually benign to their host, evolve high virulence, or vary in virulence over time. These theoretical statements help us order our thinking. However, with regard to their practical implications they should be interpreted as no more than hypotheses, as the extent to which the theoretical conditions are fulfilled in a concrete situation is not fully clear in advance.

Once the hypotheses are formulated correctly, the next step, before even considering the management applications, is to formulate consistent hypotheses on how to interfere with or even redirect the evolution of virulence. Among the desired goals such things as low virulence in the parasites of crops, cattle, or humans, and high virulence in the parasites of weeds and pests could be considered. However, the formulation of management measures does not imply that their application be recommended immediately. Clearly, the science of virulence evolution is not yet at a stage where the recipes in this book can be recommended to one's doctor or local politician. In general, physicians should not rush to apply untested evolutionary predictions on how to combat disease outbreaks. Such advice should not be uncritically accepted in the absence of experimentally established fact, unless the measures fit in neatly with existing practice to abate harm. Yet, the phrasing of measures to manage virulence stimulates critical tests in practice, whenever reasonable, possible, and ethical.

Theoretical caveats: what sort of extensions are needed?

The main message of the simplest models of virulence evolution is that, under certain assumptions, diseases will become relatively benign, and even more so when the general living conditions of the host are improved. Do we then have an all-encompassing recipe for eternal bliss? Of course we do not. The reasons why the recipe often fails are hidden in the model's assumptions:

- First, for many disease agents it is not realistic to assume that they do not compete with each other within hosts. Even when hosts are isolated from contact with others soon after disease symptoms first appear, multiple-strain infections can still occur during the period between actual infection and the emergence of the first symptoms. However, to account for the possibility of multiple infections rapidly causes bookkeeping to become a complex procedure, which poses one of the major challenges to the development of the theory. Part C of this book is devoted to models of within-host dynamics, as well as to alternative approaches to the analysis of how multiple infections affect the evolution of virulence.
- A second assumption in need of further theoretical exploration is the structure of the host population. Populations in the real world are rarely homogeneously mixed. Instead, they may exhibit an externally imposed metapopulation structure or develop fancy patterns in space through the host–parasite dynamic itself. Such spatially extended models are discussed in Part B.

▪ The third critical assumption that requires more research is that natural selection acts only on the disease agents. In this way, the evolutionary responses of the host and of other interacting organisms in the food web are altogether ignored. These coevolutionary responses and their consequences for the evolution of parasite virulence are dealt with in Parts D and E.

▪ Finally, to complete the list of potentially modifiable assumptions, models may be extended through alternative modes of transmission, such as vertical instead of horizontal transmission or indirect instead of direct transmission (via vectors or intermediate hosts).

Practical caveats: how well can we understand a concrete case?

All the extensions that result from relaxing one or another assumption discussed above can lead to a suite of alternative explanations for empirical observations. For example, the decline to intermediate virulence observed in the rabbit myxoma virus can be explained in at least seven different ways:

1. The observed decline represents a snapshot during the long-term approach toward avirulence, a prediction consistent with the simple models provided the per host disease-induced death rate evolves independent of other parameters.
2. The observed intermediacy is the end result of an evolutionary process constrained by a trade-off between virus-affected parameters (e.g., the per host transmission rate, the per host recovery rate, and the per host disease-induced death rate).
3. The observed decline is a time series of evolutionary end states and the critical variable under gradual change is the disease-free death rate of the rabbits.
4. The introduction of the myxoma virus caused a decline in rabbit density such that the probability of multiple infections drastically declined, as did, concomitantly, the virulence level favored by selection (multiple infections select for higher virulence than single infections).
5. The rabbit population is structured in family groups with high internal contact frequency, yet very low external contact frequency, and so virus strains that wipe out family groups too quickly hamper their own propagation through the rabbit population as a whole.
6. The originally virulent strain of the virus triggered selection for resistance in the rabbit population, which caused the virus prevalence to decline and the probability of multiple infection to decrease. In the longer run this will lead to an arms race between the virus and the rabbit, which potentially could give rise to co-evolutionary cycling of resistance and virulence levels. In this respect it is interesting to note that recent assessments (1992–1994) showed an upsurge of the virulence level (Fenner and Fantini 1999).
7. And, as if this list were not already long enough, there is another hypothesis based on indirect transmission of the virus, first formulated in a recent authoritative ecology textbook (Begon *et al.* 1996). The idea is that blood-feeding lice take up strains of the virus that reside in or near the rabbit's skin and that these strains have lower rates of multiplication than others that exploit more

profitable tissues in the rabbit. Selection for vectorborne transmission therefore favors less virulent strains of the myxoma virus.

This example of rabbit myxomatosis shows that changes in virulence can potentially be explained by many different hypotheses, each of which draws attention to particular, realistic features of the ecological setting of the microparasite–host system. To develop new hypotheses and discriminate between them, mathematical models are an indispensable tool to enable careful reasoning and generate precise, and therefore testable, predictions. Empirical tests are needed to assess which of these hypotheses survive scrutiny. These scientific activities are expected to become a breeding ground for further analyses and for the design of experiments on virulence management.

27.3   Conceptual Issues

Theoreticians can simply make the assumptions they need and define concepts on the basis of their convenience in deriving conclusions. The empiricists thus have to deal with the problems of how to check the assumptions made, or rather of how to identify the systems for which these assumptions may hold at least approximately, and of how to make workable the concepts so conveniently dreamt up. The task of checking assumptions is examined in Section 27.4. First, however, we assess in some detail the two grand concepts central to the analysis of virulence evolution, to wit virulence and fitness. In the world of simple models these are simple concepts. However, the world often is far from simple, in which case the twin questions arise as to how these concepts can best be extended so that conclusions based on them generalize, and as to whether suitable operationalizations, or usable substitutes, can be derived that are relatively accessible experimentally.

The monster of fitness

The concept of fitness has all the essential features of a monster. It is a grand idea that is permeating virtually every field of scientific thought. Just as we picture the monster of Loch Ness as a marine dinosaur, we have a pretty good idea of what fitness should be like. Yet, when it comes to obtaining tangible proof, a precise and quantifiable measure of fitness, the specific ecological setting within which the focal organism functions and interacts becomes crucial. Whereas, in principle, theorists can derive fitness measures given any ecological setting, experimental observers face the formidable task of first establishing a complete understanding of their ecological system before they can select a suitable fitness measure and assess how a given organismal trait affects it. Therefore, the real monster exists in the eye of the observer (and not in that of the theorist). Given the large number of sightings in the ecological literature, we may hope the fitness monster will fare better than the monster of Loch Ness (Sheldon and Kerr 1972), and eventually we may even be able to derive a fairly complete picture.

For a better understanding of the problem, take pen and paper and write down a list of candidate representative measures for the fitness of a microparasite. These could include:

- Parasite's population growth rate;
- Net per generation reproduction of the parasite;
- Rate at which new infective hosts are produced per infected host;
- Number of new infective hosts produced over the lifetime of an infected host (i.e., the basic reproduction ratio R_0);
- As well as a number of variants that all differ in the way these measures are averaged over generations.

This exercise shows that it is not at all self-evident which measure is most appropriate.

Now, consider a rare mutant that invades a homogeneously mixed population. Local linearization of the mutant's dynamics at near-zero mutant densities yields a dynamics in which all individuals (disease cases) reproduce effectively independently. The dominant Lyapunov exponent of this linear dynamics provides the correct fitness concept, and, if the coefficients in the linear equations are constant, this reduces to the dominant eigenvalue (Metz *et al.* 1992). In biological terms, this is directly related to the long-term, time-averaged per capita growth rate, but when the population is at equilibrium, R_0, as formalized for the general case by Diekmann *et al.* (1990, 1998), can be shown to provide an equivalent measure. This is the fitness measure used in most of the preceding chapters. For $R_0 > 1$ ($R_0 < 1$), fitness is positive (negative). Thus, evaluation of the value of R_0 suffices to pinpoint the conditions under which the resident population resists invasion by any conceivable mutant. This has led to the use of pairwise invasibility plots (Geritz *et al.* 1998), introduced in Chapter 4.

Recently, the concept of invasion fitness has been extended to metapopulations. Metz and Gyllenberg (2001) and Gyllenberg and Metz (2001) defined a quantity R_m as the average number of new mutant dispersers (infective particles) produced by one newly released mutant disperser. They showed that this quantity qualifies as a substitute fitness measure provided the (resident) metapopulation is at equilibrium. However, much work is still required to identify fitness measures under equilibrium and nonequilibrium conditions in spatially extended systems and in food webs with two or more interacting organisms.

Virulence, what's in a word?

In the medical world virulence is most commonly equated with disease severity. In many models this is abstracted as the per capita disease-induced host death rate. This is a very convenient choice, because this death rate often represents a significant aspect of disease severity as it is commonly perceived, and is also a factor of great importance for a disease agent that – for its own transmission – depends on a living host. Thus, the parasite has to balance within-host growth against host survival, since death of the host halts further transmission of the parasite. However, this definition of virulence is clearly oversimplified, because disease severity may involve many negative effects on the host other than increased mortality, and several of these other effects may not hamper, and may even promote, transmission. As an example of the second point, some parasites castrate their host, which is detrimental to the host's fitness, but not necessarily negative to the parasite, provided it is transmitted horizontally. If castration prolongs the lifetime of the host, the average number of parasite transmission events to new hosts may even be increased.

Thus, discrimination should be made between the effects of disease severity that reduce the fitness of the host and those that alter the fitness of the parasite. This distinction is at the heart of various attempts to clarify the definition of virulence (Shaner *et al.* 1992; Andrivon 1993; Poulin and Combes 1999). We resist the urge to give our own all-encompassing definition to supersede the rest. Instead, we feel that with the present stage of our knowledge it is better to use the word in a blanket fashion and to concentrate on the particulars of specific cases. Earlier chapters all deal with parasite strategies to exploit the host for multiplication and for transmission, and the authors refer to virulence to indicate the detrimental effect of parasitic exploitation on the host, just as resistance indicates the detrimental effect of the host's defense on the parasite. Clearly, the parasite may incur costs because of its virulence toward its host, just as the host may incur costs because of its resistance against the parasite. Whenever it is necessary to single out those effects on the host that are important for parasite fitness, this has been stated in words rather than by introducing new terms.

The Encyclopedia of Ecology and Environmental Management (Calow 1999) also refers to another definition of virulence: the effect of the parasite on the host population level. In this definition, virulence also includes aspects of the parasite's ability to be transmitted to other hosts in the population. Virulence is then equivalent to R_0, a usage that must be avoided. Virulence and transmission should be distinguished from each other to separate the effects of the parasite's exploitation strategy on the host from those on the ability of the parasite to transmit itself or its offspring to new hosts. It is clear, however, that virulence does not necessarily relate to the damage inflicted on an individual host only. For example, in a metapopulation context virulence may be defined as the impact of the parasite on a local population, and transmission may be defined as the ability of the parasite to reach new *local* populations of hosts. Thus, virulence can be discussed at the level of the local population (or, more briefly, the patch level) as opposed to the individual level (see Chapters 22 and 32). In fact, local populations can then be envisaged as individual hosts, much like individual hosts represent a collection of cells. Clearly, how virulence is defined depends on the issue at stake.

Confusion sometimes arises from the summary statement that parasites have genes *for* (a)virulence. Virulence is a byproduct of the way the parasite exploits its host; it is certainly not its function or goal. Sometimes the severity of a disease is to a large extent the result of the immune reaction against it, in other cases it is a consequence of the growth of the pathogen population inside the host. Pathogens certainly do not express virulence genes to trigger such defense responses. The primary function of these genes is not clear, but they probably confer some advantage to the pathogen instead of specifically eliciting its destruction (Poulin and Combes 1999). Avirulence means that the pathogen is eliminated relatively quietly when its gene products are recognized by the immune system of the host. Note, however, that the same genes would lead to disease and virulence in hosts not capable of recognizing the pathogen's gene products. Thus, whether the pathogen is avirulent to its host or not depends on the manner in which resistance genes in the host and virulence (determining) genes in the pathogen interact.

In plants the situation is a little different because the defense responses are less graded. What largely matters is whether the pathogen can gain entrance to the host or not, but once it is has gained entrance, there is little difference, at least on the time scale of agricultural practice, in the amount of damage inflicted. In addition, phytopathology started from a concern about crop losses (i.e., damage to the local population), which led to the use of the term virulence gene for those genes that allow the pathogen to gain entrance, with resistance genes as the matching term on the host side. In this book the authors who deal with plant systems (Chapters 17 and 31) chose to use the term "matching" virulence to distinguish virulence through nonrecognition by the host from virulence as a byproduct of the parasite's exploitation strategy, which they call "aggressive virulence".

27.4 The Dialogue between Theorists and Empiricists

Here we take stock of our current understanding of the process of virulence evolution as treated in this book and the recent literature. As in any field of science, evolutionary epidemiology gains its momentum from the dialogue between theorists and empiricists. Here, we attempt to identify the key elements of this dialogue and also try to identify the points on which theorists and empiricists seem to misunderstand each other. We hope to foster communication by rephrasing theoretical issues in more general terms, by scrutinizing assumptions that underlie the theory, and by comparing explicit theoretical predictions against what can be gleaned from the literature on empirical tests.

Trade-off relationships prevent evolution toward avirulence

Parasitologists long held the view that all parasites evolve to become mild for their host. This view was rejected on theoretical grounds because it assumes that the disease-induced host death rate (i.e., virulence) is independent of the parasite transmission per infected host. It seems more likely that there is a trade-off between transmission and virulence. The underlying rationale is that higher replication rates of the disease agent lead to higher virulence as well as higher transmission, and that both these relationships have a genetic basis. Empirical support for these relationships and assumptions is provided by a study of the malaria parasites of mice (Mackinnon and Read 1999a). Increased virulence as a result of increased pathogen replication is observed for the case of African trypanosomes (Diffley *et al.* 1987). That the increased pathogen load correlates with increased transmission is substantiated by studies on:

- Myxoma virus of rabbits (Fenner and Fantini 1999);
- Phages of *E. coli* bacteria (Bull and Molineux 1992);
- Malaria parasites of humans (Dearsly *et al.* 1990; Day *et al.* 1993);

■ Microsporidian parasites of water fleas (Ebert 1994); and
■ HIV (Quinn *et al.* 2000).

Thus, all in all, there is good reason to think that parasites generally do not evolve to avirulence, because of the trade-offs with, for example, transmission.

Virulence management is not simply about transmission intervention

Empiricists often take the view that measures to reduce pathogen virulence are a matter of reducing their transmission. This is true insofar as reduced transmission causes both per host infection and within-host competition among pathogen strains to decrease. However, surprising as it may seem, theory shows that in the long run it may be more effective to base measures of virulence management on how they influence the degree to which a pathogen "is in control of exploiting its host". If the pathogen is not in control because of competition with other pathogens or erratic mortality of the host, then it pays for pathogens to multiply rapidly at the expense of the host. But if the pathogen controls the resources contained in the host, even when delaying its use of them, then virulence may be expected to decrease.

For example, antibiotic therapy in itself reduces the pathogen population and thereby its transmission, but this may actually favor increased virulence. The pathogens lose their host anyway, so their only opportunity for transmission is during the time between infection and the administration of antibiotics. Thus, even though the parasite population decreases and fewer hosts become infected, it is possible that pathogen virulence increases. As another example, eradication of mild strains that could monopolize exploitation of their host (e.g., by triggering an immune response that harms later invaders more than the initial ones) might create opportunities for more virulent strains that lack this ability. In this situation it would be advisable to tolerate rather than combat mild strains. Thus, it is somewhat misleading to focus on transmission intervention in the design of virulence management strategies. Instead, it is more appropriate to develop strategies that "give a pathogen control over the exploitation of a host's resources"; as long as pathogens do not have to compete with other pathogens and/or can exploit their host without interference from environmental causes, the well-being of the host is in the evolutionary interest of the parasite.

Virulence increases when ordinary transmission is bypassed

Theory predicts that selection favors fast-replicating strains of the disease agent when transmission is guaranteed (i.e., when it does not depend on host survival). These conditions are met in so-called serial passage experiments (SPEs), in which pathogens or parasites are transferred from one host to another, either artificially by injection or through natural transmission in dense host cultures, and thereby the constraints on real-world infectious processes are relaxed. Hosts usually have low genetic diversity (clonal or inbred lines), but the pathogens and parasites are genetically variable and undergo mutation. The striking general outcomes of these

SPEs are a rapid increase in the parasite-induced reduction of host fitness and that the passaged parasites outcompete the ancestral strains used to initiate the SPEs, which indicates parasite adaptation to the hosts used in SPEs (Ebert 1998a).

The rate at which virulence increases is most rapid for ribonucleic acid (RNA) viruses, slower for deoxyribonucleic acid (DNA) viruses and bacteria, and slowest for eukaryotes, which suggests that generation time and mutation rate determine the rate of change. Albeit obtained in the laboratory, SPEs taught an important lesson: disease agents are less virulent than they can be because they depend on host survival for their transmission. Outside the laboratory, the very same lesson applies when transmission is facilitated by the long-term survival of propagules, mediated by water or vectors. In all these cases disease agents tend to be more virulent, a view championed by Ewald (1983, 1991a, 1994a, and Chapters 2 and 28).

Vertical transmission lowers virulence

Another theoretical prediction supported strongly by empirical evidence is that low virulence is favored by selection when transmission is vertical [i.e., from mother to offspring (Ebert and Herre 1996)]. All else being equal, a parasite transmitted exclusively vertically should not harm its host, because the number of new infections depends on the fecundity of the host. For example, Bull *et al.* (1991) propagated populations of phage-infected *E. coli* in two ways. In one treatment, phage replication was wholly dependent on host reproduction. In the other, phages could be transmitted both vertically and horizontally. The two selection regimes had the expected effects on host fitness: infected bacteria in the vertically transmitted lines increased in density much more quickly than did those in the lines in which horizontal transmission could occur. Genetic changes in both host and parasite were responsible for the evolved differences in virulence. Further support for low virulence with vertical transmission comes from observations on an initially virulent parasitic bacterium in an amoeba. After 5 years of mainly vertical transmission, this bacterium evolved to be mild to its host (Jeon 1972, 1983).

A striking example of the role of vertical versus horizontal transmission is provided by the cytoplasmatic endosymbionts that are transmitted from mother to daughter and sons, but enter a dead-end in the latter (O'Neill *et al.* 1997). These symbionts have little impact on the survival of their female host, but promote their transmission by causing the female host to produce more daughters. If the symbionts end up in male hosts, they render mating with uninfected females incompatible – an effect referred to as spiteful because it reduces the fitness of uninfected hosts relative to infected hosts. Alternatively, they cause the males to die. The latter occurs only when male death allows the symbiont to be transmitted horizontally (e.g., via consumption of the lethargic male). Thus, virulence is sex specific: symbiont-induced mortality is much higher in the horizontally transmitting sex than in the vertically transmitting sex (Hurst 1991; Ebert and Herre 1996).

Multiple infection increases virulence

There is good support for the prediction that virulence increases when horizontal transmission leads to the multiple infection of hosts and consequently to within-host competition. A striking interspecific correlation between nematode virulence, multiple infection, and opportunities for horizontal transmission was observed in the nematode parasites of fig wasps (Herre 1993, 1995). Within fig wasp species, the reproductive success of individual females can be related to the presence and/or absence of nematode infections, which thereby provides an estimate of nematode virulence. For some wasp species, typically a single wasp pollinates a fig inflorescence so that she only carries the nematodes that enter the inflorescence, and the only nematodes that leave are carried by her offspring (vertical transmission). In other wasp species (which pollinate different species of figs) many wasps pollinate a single inflorescence, which allows nematodes from different host lineages to mix before dispersal (horizontal transmission). The virulence caused by nematodes increases with the degree of within-fig competition and concomitantly with the potential for horizontal transmission of nematodes between wasps within a fig.

Conflicting trends often result from poor experimental design

Several reports on evolutionary change in virulence seem at first sight to contradict theory, but actually do not because of pitfalls in the experimental design. For example, contrary to expectation, experiments with the gut parasites of water fleas showed a higher virulence under the vertical transmission regime (Ebert and Mangin 1997). This, however, was because the experimental manipulations, inadvertently, created higher within-host competition under a vertical transmission regime. In other cases, virulence did not change [e.g., conjugative plasmids (Turner *et al.* 1998)] or exhibited only small increases when switched from horizontal to vertical transmission [e.g., bacteriophages (Messenger *et al.* 1999)]. Possibly, this resulted from low within-host competition despite horizontal transmission and a shallow trade-off relation between transmission and virulence. As a final example, artificial selection for high (or low) weight loss in malaria-infected mice – a measure correlated with the death rate in mice – did not give rise to the expected low (or high) transmission. Instead, it gave rise to increased replication rates of the malaria parasites. Possibly, this effect arose because fast replication was the best strategy for the parasite to reach the syringes used for transmission under either selection regime (Mackinnon and Read 1999b).

Evolutionary dynamics takes place at multiple levels

It is generally agreed that virulence evolves by selection that acts both within hosts and between hosts. Variants of the disease agents compete for resources within hosts and, in addition, they compete to infect new hosts. Within-host competition favors the fast-growing strains, which may reduce the infectious period for the host, either through host mortality or by triggering a stronger immune response. Between-host competition of pathogens favors those strains that balance their within-host growth rate and per host transmission rate against the consequences for the infectious period. Relative to the single-infection case, virulence is predicted to increase when hosts become infected by multiple strains.

However straightforward this prediction may seem, reality may be more complex because natural selection and population dynamics may interact. The critical point is whether density-dependent feedback brings the pathogen–host system to states in which multiple-strain infections prevail or become rare. Hence the evolution of virulence cannot be predicted without taking the population dynamics of host and pathogen into account (and vice versa). To ignore the feedback of population dynamics when formulating general predictions can be very dangerous. Theorists are usually well aware of this problem, but empirical researchers sometimes are not; this may lead to the rejection of hypotheses for the wrong reasons. Much the same pitfalls may arise if pattern formation and coevolution in parasite–host systems are ignored. Spatial patterns may create additional selection levels and reciprocal selection on parasites and hosts may enrich the dynamic repertoire of virulence (and resistance) evolution.

27.5 Gaps in Current Knowledge

Not only should a dialogue between theorists and empiricists result in a mutual agreement as to which predictions stand up to scrutiny, but also it should identify the gaps in our present knowledge. Below we argue that current progress in evolutionary epidemiology hinges crucially on a better empirical insight into the underlying processes. We emphasize three areas of empirical research in need of more in-depth work: within-host competition, between-host transmission, and responses of the host to infection. We conclude this section by highlighting some major gaps in our current theoretical understanding.

Within-host processes

No doubt it is overly simplistic to model within-host competition between multiple strains of parasites as a race of fast replication rates. For within-host interactions between parasites, all the competition mechanisms known to occur in ecological communities can be expected, such as:

- Resource limitations may trigger exploitation competition;
- Parasite clones may interfere with each other by producing toxins that kill off their competitors or block entry to later infections (interference competition);
- An increase in density of one parasite clone may trigger an immune response that affects all the clones in the host (apparent competition).

Indeed, replication of a disease agent like a virus is much like a predator–prey–resource system. By infecting a cell, the virus preys on its resource, while it itself is being subjected to predation by T-cells that are part of the host's immune system (Levin *et al.* 1999).

Although there are examples of independently transmitted strains [e.g., variant surface antigen serotypes of malaria parasites (Gupta *et al.* 1996; see Chapter 25)], reality probably offers a plethora of competition outcomes, such as:

- Priority phenomena ("first come, first serve" irrespective of replication rate);
- Improved or reduced transmission under mixed versus single infection;
- Evasive specialization to different resources within the host or to different host genotypes (Taylor *et al.* 1997a, 1998b; Taylor and Read 1998).

Biological research into the mechanisms that underlie these competitive interactions is still in its infancy. No doubt, the consequences for the evolution of virulence are decisive. For example, if parasite clones invest more energy in interference than in replication, multiple infections may lead to reduced instead of increased virulence (Chao *et al.* 2000).

Apart from competition, parasites may also interact within hosts in other ways. They may gain by cooperation, despite the potential to cheat that such an interaction always entails (Brown 1999). For example, infectious bacteria need to reach a critical density to overcome host defenses and establish the infection. Many different bacterial pathogens are now known to regulate virulence-determining processes in a manner dependent on cell density through cell–cell communication via a diffusible signal molecule [e.g., N-acylhomoserine lactone (Williams *et al.* 2000)]. This phenomenon, referred to as "quorum sensing", may be an important determinant of the population variance observed in virulence assessments. It clearly requires an evolutionary explanation of its own (Brookfield 1998; Brown 1999), and its potential role in the evolution of virulence is virtually unexplored.

Between-host replication

Whether within-host interactions are competitive or cooperative, within-host selection in itself is unlikely to maximize parasite replication rates. This is because fast multiplication rates *per se* do not promote between-host transmission. Many parasites have specialized transmission stages and, thus, invest in traits that increase the per capita success of transmission, even when this is at the expense of their replication rates. It may well be that the increased replication rates favored by selection in SPEs actually result from the loss of investment in transmission-related traits (Ebert 1998a). Clearly, there is a need for selection experiments to elucidate the key traits for transmission, rather than just the traits that determine the numerical outcome of within-host competition.

More insight is also required into the costs and benefits involved in the various modes of transmission, as this may help us understand why certain transmission routes evolve in the first place. Parasites are necessarily involved in a struggle for transmission and they may employ one transmission route or a combination of these from a range of possibilities:

- Transmission via long-lived propagules versus exclusive reliance on transmission from a living host;

- Direct versus indirect transmission;
- One form of indirect transmission versus another (e.g., waterborne versus vectorborne);
- Horizontal versus vertical transmission;
- Control over the host as a transmission vehicle or control over the vector (parasite-induced host behavior).

How transmission routes evolve together with virulence has been explored for some special cases [e.g., horizontal and vertical transmission to hosts that cannot be coinfected by other parasites (Yamamura 1993; Lipsitch *et al.* 1995b, 1996)]. The results of these analyses do not indicate a virulence–avirulence continuum between horizontal and vertical transmissions, but a more comprehensive theory is needed to allow for multiple infections of hosts and a wider variety of transmission routes.

Although pathogens may maximize the per host production of transmission stages, they can also promote transmission by expanding their host range. Many medically and veterinarily important pathogens can infect more than one species of host. However, such a generalist strategy probably carries a cost to the pathogen. Each host species will impose a different selection regime on the pathogens, so that adaptation to one host occurs at the expense of adaptation to another. Multi-host pathogens may therefore tend to be less virulent to their hosts. For example, SPEs rapidly lead to pathogen lines that are much milder to their original host than the ancestor pathogens (Ebert 1998a) and nematode parasites seem to be less virulent to fruit flies when they attack various host species (Jaenike 1993). However, multi-host pathogens may also be less dependent on the survival of their host by virtue of a larger overall host population. Exactly how this influences the evolution of virulence in multi-host pathogens is not yet clear.

In some cases, multi-host pathogens express unusually high virulence. This may occur in hosts that do not contribute to the transmission of the pathogen. For example, rodent-associated hantaviruses are extremely virulent to humans, which represent a dead end for transmission of the virus. Unusually high virulence may also result from short-sighted evolution within hosts (Lipsitch and Moxon 1997). Pathogens mutate and experience severe selection within hosts, which leads to invasion of vital host tissues other than those needed for the pathogen's transmission (e.g., bacterial meningitis). However, the persistence of such mutator pathogens in evolutionary time still requires an evolutionary explanation (Moxon *et al.* 1994; Taddei *et al.* 1997b), and their role in the evolution of virulence is only now beginning to be explored (Bergstrom *et al.* 1999).

Reactions of the host

Changes in disease severity do not result solely from genetic changes in the pathogen. Damage to the host results from pathogen aggressiveness and host resistance. Given a sufficiently tight coupling between the interacting populations, reciprocal selection may drive parasite–host coevolution.

There is now good evidence that genotype–genotype interactions are of over-riding importance in the expression of virulence, especially in the pathogens of plants (Thompson and Burdon 1992; Clay and Kover 1996), but also in:

- Trypanosomes that infect bumblebees (Shykoff and Schmid-Hempel 1991; Schmid-Hempel and Schmid-Hempel 1993);
- Microsporidia that infect waterfleas (Little and Ebert 2000);
- Webworms that feed on wild parsnip (Berenbaum and Zangerl 1998).

Moreover, it has been shown for various invertebrates and vertebrates that host re-sistance has a genetic basis and carries a cost (Toft and Karter 1990; Henter and Via 1995; Kraaijeveld and Godfray 1997; Sorci *et al.* 1997; Webster and Wool-house 1999).

How reciprocal selection affects the coevolutionary dynamics of parasite–host systems is little explored empirically (but see Henter and Via 1995; Burdon and Thrall 1999; Fenner and Fantini 1999), which leaves an expanding body of the-ory (e.g., on maintenance of polymorphism and coevolutionary cycling) largely untested (Frank 1992b, 1993b, 1994a, 1996a, 1996c; Dieckmann *et al.* 1995; Haraguchi and Sasaki 1996; Sasaki and Godfray 1999; Sasaki 2000; see Chap-ters 4 and 14 to 17).

Parasitic relationships may evolve to become mutualistic (Michalakis *et al.* 1992), and mutualistic relationships may vary between parasitism and mutualism depending on the time scale and the spatial scale under consideration (Bronstein 1994a, 1994b). To understand why reciprocal exploitation provides net benefits to each partner or otherwise requires a sharp insight into the common interests shared by the partners, their private interests, and the conflicts of interests that may arise (Herre *et al.* 1999).

Whether a symbiont evolves to become mutualistic, commensalistic, or para-sitic also depends on the degree to which it can control exploitation of the host and on the opportunities for transmission to new hosts (e.g., Yamamura 1993; Lipsitch *et al.* 1995b, 1996). Similarly, the host opens or closes transmission routes avail-able to the symbionts and thereby influences the evolution of virulence, benevo-lence, or cooperation. Although it is clear that vertical transmission can promote the evolution of mutualism and that vertical transmission may evolve depending on the symbiont-induced benefits that accrue to the host (Yamamura 1993; Law and Dieckmann 1998), the full set of conditions has not yet been established. This is because host control over symbiont transmission and the full range of competitive interactions among symbionts within hosts have not yet been taken into account.

Recent molecular phylogenetic studies showed that bacterial symbionts belong to deeply branching clades that are strictly parasitic or strictly mutualistic (Moran and Wernegreen 2000). This pleads against the theoretical notion that transitions from parasitism to mutualism occur frequently. Thus, symbiotic interactions be-tween bacteria and animal hosts may be more constrained than assumed in evolu-tionary models so far. Such constraints may result from major genomic changes associated with a certain bacterial lifestyle, which implies a massive loss of genes that code for a variety of functional capabilities and thereby limits the evolutionary options for bacterial lineages.

Gaps in the theory

The most critical gap in our knowledge is the lack of good biological insight into the role of within-host competition and within-host evolution. Once this gap begins to close, we may begin to explore the theoretical outcomes of virulence evolution in more realistic settings. The models needed for such explorations have to be based on a framework that allows complex book-keeping procedures, especially when it comes to modeling multiple infections and the ensuing competitive interactions within hosts. Physiologically structured models phrased in terms of partial differential equations offer such a framework (Metz and Diekmann 1986; see also Diekmann and Heesterbeek 1999; Metz and Gyllenberg 2001; Gyllenberg and Metz 2001), and their application to the analysis of adaptive dynamics offers much promise.

In addition, we are only beginning to understand the role of spatial processes in population dynamics and evolution (Dieckmann *et al.* 2000). Studies of excitable media have shown the potential for self-organization, and parasite–host systems have all the essential properties to exhibit this behavior. A major future task is to understand how contact networks develop in space, determine the spread of disease agents, and influence the evolution of their virulence.

27.6 Discussion: Toward Virulence Management

"It is time to close the book on infectious diseases." This statement was made in 1969 by the USA surgeon general after inspecting a successful disease control campaign. It reflects the widespread optimism that good sanitation, vaccines, and antimicrobial agents would conquer infectious diseases (Binder *et al.* 1999). However, the public health successes of the 1960s and 1970s were followed in the 1980s and 1990s by ominous developments, such as the resurgence of diseases (e.g., tuberculosis) and the emergence of the HIV pandemic.

Just as with the current HIV, both the medieval Plague and the 1918 epidemic of Spanish Flu had major impacts on human populations and significantly altered the course of human history (McNeill 1976). Increased contacts with faunal elements (e.g., rodents), tradesmen, colonizers, and soldiers were at the root of new diseases that had ravaging effects on endemic populations (Diamond 1997). Such epidemics have even caused landscapes to metamorphose. When the Asian rinderpest virus entered Eritrea via cattle brought by Italian invaders, it took only 5–10 years for the virus to spread through Africa with devastating effects on cattle and pastoral tribes. This changed vast areas in Africa from grassland into bushy Savannah and woodland, thereby giving way to wildlife and tsetse. The tsetse flies, in turn, prevented humans and their cattle from regaining the areas formerly occupied. So, the rinderpest outbreak at the end of the 19th century caused irreversible changes in the African landscape (Anonymous 2000a; see Chapter 3). Diseases can cause dramatic mortality and ecosystem changes, and they can also leave traces in the genetic make-up of human populations. These include the polymorphism in genes involved in discriminating self from nonself [the major histocompatibility complex (MHC)] and the polymorphism for sickle-cell anemia maintained by heterozygote advantage in areas with high malaria incidence (Hamilton and Howard 1997). The importance of diseases is also reflected in that resistance against diseases is shown to be a target for mate preferences and hence subject to sexual selection (Penn and Potts 1999).

The fact that several of these dramatic historical events took place over a time scale of 10–100 years shows that disease agents can, in principle, impose new selection pressures over vast areas and evolve on a time scale shorter than or near to the generation time of their host. To ignore natural selection in designing campaigns to combat (re-)emerging diseases is therefore (as already stated) a serious mistake. The same conclusion can be drawn from the development of microbial resistance against antibiotics (Kristinsson et al. 1992), the resistance of malaria mosquitoes against dichlorodiphenyltrichloroethane (DDT; Coetzee et al. 1999; Roberts et al. 2000), the resistance of malaria parasites against chloroquine (Peters 1985; Conway and Roper 2000), and from many other examples. The challenge is to develop sustainable ways to control disease agents, and thereby take selection into account (Kolberg 1994). For example, the World Health Organization (WHO) advocates the use of pyrethroid-impregnated bednets to reduce the incidence of malaria (the so-called Roll Back Malaria campaign). However, this has already led to selection on the mosquitoes for resistance against pyrethroids (e.g., in French-speaking West Africa). In addition, selection for mosquito behavior is expected to circumvent the use of bednets, for example by modifying the diurnal pattern of biting activity (Anonymous 2000b). Deeper biological insight into mosquito behavior may well provide clues as to how to "attract and kill", and there is much potential for the control of mosquitoes by using their natural enemies (Takken and Knols 1999). However, in developing sustainable control strategies it is important to ask which new selective pressures will be created and how the development of resistance to control methods can be slowed or even prevented. From this viewpoint, the central question is whether evolutionarily stable control strategies can be designed.

The idea of an evolutionarily stable control strategy is based on the assumption that the world is constant at least on the time scale under consideration. Yet the world does not remain constant on a larger time scale. Population growth, increased mobility, and possibly also climatic change will create new opportunities for diseases (Epstein 2000), whereas genetic change will help the disease agents to evade the barriers put in their way. Therefore, instead of chasing the facts it may pay off to think ahead and ask how to steer epidemiological changes brought about by population dynamics and natural selection. This requires that our control strategies not only be evolutionarily stable, but also be robust. In this chapter we have summarized what we consider the main rules of thumb that have surfaced so far, as well as the main provisos. The following chapters examine specific fields of application in further detail.

28

Virulence Management in Humans

Paul W. Ewald

28.1 Conceptual Basis for Virulence Management

Studies of virulence evolution have attracted attention in part because of the potential for practical applications to health problems. One of these potential applications involves forcing disease organisms to evolve toward low virulence (Ewald 1988, 1991a, 1994a). If evolutionary interventions to control virulence are feasible they may be especially cost-effective solutions, because they may control entire classes of diseases through interventions considered valuable for other reasons. In particular, interventions that favor reduced frequency of infection may simultaneously favor evolutionary reductions in the inherent harmfulness of infectious agents (referred to hereafter as "pathogen virulence").

Evolutionary considerations and the current state of empirical investigation suggest that one class of such interventions, potentially capable of tipping the competitive balance in favor of milder pathogen strains, involves alterations in the mode of pathogen transmission. This chapter assesses the current state of investigation into options for the virulence management of diarrheal diseases, as well as of other categories of disease for which transmission may be relatively independent of host mobility. For each category of disease I discuss interventions that may foster virulence management through the alteration of transmission mode. Finally, I broaden the argument to include virulence management through vaccination strategies. A thorough evaluation of these ideas requires a long series of hierarchically organized tests, so the current state of evidence represents work in progress, which may require decades to fully develop.

28.2 Virulence Management of Diarrheal Diseases

Principles of virulence management seem especially applicable to diarrheal disease for several reasons. First, routes of transmission that depend on host mobility (e.g., direct contact) are distinct from those that do not (e.g., waterborne transmission). Second, the hypothesized effect – the reduction of virulence in response to reduction in the potential for waterborne transmission (see Chapter 2) – is at a relatively advanced stage of testing. Third, the critical final stage of empirical testing – experimental improvement in water quality – is particularly feasible. Finally, if the evolutionary effect is demonstrated, implementation could proceed expeditiously; logistical, financial, and social barriers to reducing waterborne transmission are surmountable, because people generally prefer their water to be free of

fecal contamination, the technology to provide uncontaminated water is available
and relatively inexpensive, and the provision of uncontaminated water gives other
health benefits, such as reduced incidence of infection, which help justify the intervention.

Feasibility criteria

The theory presented in Chapter 2 predicts that waterborne transmission should
favor evolutionary increases in the virulence of diarrheal pathogens (Figures 2.5
and 2.6 in Chapter 2). The positive association between the lethality of diarrheal
bacteria and waterborne transmission across the entire spectrum of diarrheal bacteria of humans (Figure 2.2 in Chapter 2) supports this idea. This lends credence to
the possibility of reducing the virulence of diarrheal pathogens and raises the possibility that reductions in opportunities for waterborne transmission could cause
evolutionary reductions in pathogen virulence. If such reductions in virulence occur rapidly, virulence management could serve as an alternative or complement to
standard interventions.

For reductions in waterborne transmission to serve as a tool for virulence management, the target pathogens must fulfill several criteria:

- Sufficient genetic variation in pathogen virulence must be present or readily
  generated to provide the raw material for an evolutionary shift toward benignity.
- The advantage of mild strains must be sufficiently great to allow them to displace virulent strains in an acceptably short period of time.
- The reduction in disease that results from this displacement must be comparable
  to the reduction generated from nonevolutionary interventions.

Interspecific studies: *Shigella*

Geographical and temporal comparisons within genera of diarrheal bacteria reveal
trends consistent with the association between lethality and waterborne transmission. The composition of *Shigella* species, for example, shifted regularly toward
the more mild species within a decade or so after water supplies had been purified
in different countries over the past century (Ewald 1991a, 1994a). As strong as
these trends are, the possibility that they are caused by some difference between
the *Shigella* species, other than that of inherent virulence, cannot be excluded.
Also, do such interspecific trends underestimate the effect of water purification
as a virulence management tool because they do not reveal intraspecific changes?
Immune-mediated competition, for example, tends to be stronger within species
than between species; if a mild variant within a species is favored by reductions
in the potential for waterborne transmission, it tends to have a stronger immune-
mediated suppressive effect on other severe variants within the species than that
between severe species within the same genus. (This argument does not require
the concept of species to be unambiguous, but only that those variants within a
"species" are more closely related to each other than they are to the other species
in a genus.)

Intraspecific studies: *Vibrio cholerae*

One key to the generation of stronger tests is the ability to assay directly the virulence of pathogens. Such assays are difficult, because virulence of infections depends on host resistance in addition to the inherent virulence of the pathogen, two influences that are often difficult to disentangle.

In this regard the bacterial agent of cholera, *V. cholerae*, appears to be a particularly valuable study organism because its virulence depends largely on the production of cholera toxin. This toxin generates an efflux of fluid into the small intestine, which appears to provide two benefits to *V. cholerae*:

- The efflux flushes out competitors in the intestinal tract, allowing viable *V. cholerae* to pass down and out of the tract. (*V. cholerae* can prosper in this environment because it can swim fast in the intestinal tract, and also adhere to the intestinal lining to avoid being washed out with the other bacteria.)
- The efflux creates a fluid stool that probably facilitates transmission through contamination of the external environment and dissemination in water supplies.

For the pathogen, the costs of toxin production include the following:

- The metabolic costs of producing the toxin;
- The negative effect of the toxin on host mobility and an increased probability of host death.

Death from cholera results primarily from dehydration, which occurs with loss of fluid through the intestine because of the effects of the toxin. Recently initiated laboratory studies on *V. cholerae* strains isolated from Latin America are designed to evaluate whether toxigenicities of the *V. cholerae* biotype eltor are associated with variation in water quality (Ewald *et al.* 1998). The focus is on the cholera outbreak in South America, because its distinct onset in Peru at the beginning of 1991 allows an assessment of whether evolutionary effects of waterborne transmission can occur over a time period comparable to the interval necessary for other categories of interventions (e.g., vaccination or hygienic improvements to reduce the frequency of infection). Within 2 years of entering Peru, the descendants of the Peruvian *V. cholerae* had spread from this epicenter throughout most countries of South and Central America (Tauxe *et al.* 1995). This spread set up a series of natural experiments that may allow detailed testing of the proposed association between waterborne transmission and the toxigenicity of *V. cholerae*.

Cholera and waterborne transmission in Brazil. We first focused on Brazil and neighboring countries because substantial variation in water quality has been quantified by the ministries of health of these countries and by the World Health Organization. The toxigenicity of *V. cholerae* in this region is negatively associated with access to potable water (Figure 28.1).

These data are preliminary in several respects. For example, the degree to which the different data points are independent is not known. Use of molecular approaches to clarify phylogenetic associations should allow independent pairwise comparisons in statistical tests. Additional strains need to be obtained to make the

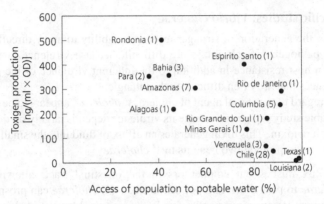

Figure 28.1 Toxigenicities of *Vibrio cholerae* biotype eltor from South America and the USA. Toxigenicities are expressed per optical density (OD) unit of bacteria. Names of states or countries are given next to the data point that corresponds to the geometric mean toxigenicity of the strain(s) isolated from the area. Numbers in parentheses refer to the number of different strains tested from each location. About 20 separate measurements of toxigenicity were made for each strain. *Source*: Ewald *et al.* (1998).

accuracy of each data point similar. (As indicated in Figure 28.1, some data points are based on multiple isolates whereas others are based on only one.) Perhaps most importantly, changes in toxigenicity need to be followed over time to determine whether harmful strains that enter areas with relatively pure water become less harmful over time.

   Although the data in Figure 28.1 suggest that *V. cholerae* has evolved toward a lower level of virulence, they do not specify how mild *V. cholerae* could eventually become in response to cycling in areas with clean water supplies. To provide such an indication, Figure 28.1 also shows the rate of toxin production of strains isolated from the southern USA. Zymodeme analysis indicates that these US strains cluster with the eltor strains of *V. cholerae* (rather than with strains of the classic biotype), but are only distantly related to the other eltor strains (Salles and Momen 1991). They therefore appear to have been present in the USA for decades, perhaps being the remnant of a global outbreak of cholera that occurred many decades ago. Their low toxigenicity thus provides an indication of how low *V. cholerae* toxigenicity could become in an area with low contamination of drinking water. Accordingly, although the frequency of seropositivity to *V. cholerae* in local populations can be substantial, cases of cholera in this Gulf area are extremely rare (Ewald *et al.* 1998).

   Cholera and waterborne transmission in Chile. To obtain a more direct assessment of whether *V. cholerae* evolves reduced virulence in response to reduced potential for waterborne transmission, we are assessing the toxigenicities of *V. cholerae* as a function of time for various countries in South and Central America. Our most complete data set comes from Chile. Chile is a particularly

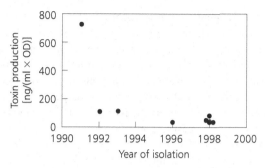

Figure 28.2 Toxigenicities of *V. cholerae* biotype eltor isolated from Chile, from the onset of the South American outbreak in 1991 to the beginning of 1998. Each data point corresponds to a different isolate. Other details are as described in Figure 28.1. *Source*: Ewald *et al.* (1998).

important country for the evaluation of this hypothesis because its drinking water supplies are among the least contaminated of those South American countries in which *V. cholerae* infections have become endemic. *V. cholerae* entered Chile from Peru at the onset of the South American epidemic in 1991; the first case was reported in Chile at about the same time as the first case was reported in Brazil, about 2.5 months after the first case reported in Peru (Tauxe *et al.* 1995). Peru's water supplies were substantially more contaminated than Chile's, particularly at the onset of the South American epidemic. Our preliminary data suggest that toxigenicity has declined as a function of time since the introduction of *V. cholerae* from Peru into Chile (Figure 28.2).

These data are interesting in terms of the toxigenicities of strains isolated in recent years, which are nearly as low as those isolated from the Gulf coast of the USA (compare the US isolates in Figure 28.1 with the 1998 isolates in Figure 28.2). Accordingly, the number of reported cases in Chile dropped to one in 1994 (Tauxe *et al.* 1995). In that year about 50 000 cases were reported in Brazil and 24 000 in Peru (Tauxe *et al.* 1995). These figures, coupled with similarities between the toxigenicities of the US Gulf strains and the most recent Chilean strains, support the idea that the evolutionary management of virulence is feasible for *V. cholerae* if water quality can be raised sufficiently.

Reasons for reduced toxigenicity of Vibrio cholerae. At least four hypotheses could be advanced to explain the evolution of reduced toxigenicity of *V. cholerae* in response to a reduction in waterborne transmission:

- The reduction in toxigenicity could result from the increased costs and decreased benefits of toxin production, as outlined above.
- The reduction could result if the growth of *V. cholerae* in external environments disfavors toxigenicity. The culture of parasites outside hosts causes attenuation as a result of relaxed selection for genes that contribute to virulence. Growth of *V. cholerae* in external environments might similarly disfavor virulence.

■ The decline of toxigenicity in Chile could be interpreted as a result of the duration of the outbreak. Theory suggests that, as an outbreak becomes endemic, pathogen virulence may decrease because the decline in the density of susceptible hosts favors the more benign competitors among the pathogen strains (Lenski and May 1994). To evaluate this hypothesis analogous data are needed from "control" countries invaded by *V. cholerae* at the same time, but for which water quality has remained low. If the reduction in toxigenicity of *V. cholerae* in Chile is attributable at least in part to its water quality, its reduction should be stronger than that found in such control countries with poor water quality. To test this idea we are measuring toxigenicities of *V. cholerae* from countries such as Guatemala, Peru, and Ecuador.

■ The decrease in waterborne transmission might favor decreased virulence by reducing the genetic heterogeneity of the population of pathogens within a host. Although this hypothesis is probably generally applicable across a broad range of disease organisms, it does not appear to be particularly applicable to *V. cholerae* because its pathogenicity does not involve the direct use of host tissues as resources, but instead involves the secretion of a product that benefits other *V. cholerae* bacteria in the intestinal lumen (as described above for toxin action). Under such circumstances, increased variation within hosts could lead to decreased virulence, because those variants that do not produce toxin can obtain the fitness benefits of toxin production without paying the price.

Of course, it is also possible that some unknown variable, correlated with waterborne transmission, causes a change in toxigenicity over time. Experimental reductions in the potential for waterborne transmission are needed to resolve this issue.

28.3 Virulence Management of Vectorborne Diseases

Other categories of disease transmission involve entirely different mechanisms, but the various mechanisms share fundamental similarities with regard to selective pressures that shape virulence. One such similarity involves the importance of host mobility for transmission (see Chapter 2).

Dependence of vectorborne pathogens on host mobility

Like waterborne pathogens, vectorborne pathogens are less dependent on host mobility and tend to be more virulent than directly transmitted pathogens (see Chapter 2). Vectorborne transmission can be subdivided, however, into transmission events that occur from largely mobile infected hosts and those that occur from largely immobile infected hosts. People immobilized by their illness are more likely to be at home or in a hospital than those who do not feel ill, so transmission from homes and hospitals should tend to involve relatively virulent variants of the pathogen population. Analyses of virulence, therefore, need to separate transmission that occurs in dwellings of sick individuals from transmission outside such dwellings.

Consideration of mosquito-proof housing makes apparent the importance of this distinction for virulence management. When dwellings such as houses and hospitals are mosquito-proof, those people who stay inside do not transmit pathogens to others via the vectors that live outside. In analogy to the modification, introduced in Chapter 2, that accounts for the distinction between waterborne and direct transmission of diarrheal pathogens [see Equations (2.2) to (2.4)], models for vectorborne transmission can be modified to account for the distinction between transmissions inside and outside dwellings. Specifically, the equation for transmission to vectors [Equation (b) in Box 2.3] can be replaced by two such equations, one that includes transmission from hosts within dwellings used for sleeping and resting, and one that excludes such transmission. As in the case of waterborne pathogens, this separation enables distinction between modes of transmission that are mostly dependent on and those that are mostly independent of host mobility. An increase of virulence, illustrated for waterborne transmission in Chapter 2, should therefore also occur when transmission involves vectorborne transmission from dwellings. That is, when vectorborne transmission occurs freely from indoor environments as well as outdoor environments, a higher level of virulence evolves than when transmission occurs only in the outdoor environment; the latter is expected to require relatively higher levels of host mobility. The vector-proofing of dwellings should therefore shift the transmission options from the former to the latter situation, and should therefore favor decreased virulence.

Assumptions about hosts and pathogens

The logic leading to this conclusion is based on several assumptions:

- When people are severely ill, they tend to stay indoors (i.e., at home or in a hospital) rather than go outdoors.
- The illness is associated with a significant period of transmissibility of the pathogen from the infected person to the vector.
- A sufficient amount of variation in pathogen virulence exists or could be generated (e.g., by mutation) to allow for evolutionary reductions in virulence.

The first assumption seems so obvious from common experience that it hardly needs to be validated in most areas. However, it still needs to be considered explicitly because it might not apply to some groups of people who, for example, do not live in houses. Similarly, it might not be valid at some times of the year if, for example, vector-proof houses are so stifling that people prefer to rest outside.

The second assumption appears to be true for most vectorborne pathogens (e.g., for yellow fever and dengue, see Faust *et al.* 1968 and Vaughn *et al.* 1997), but it has been a source of concern particularly in malaria, because in some circumstances the density of gametocytes (i.e., the *Plasmodium* stages that are infective to mosquitoes) in the blood of patients tends to peak after treatment. The available data suggest, however, that this lack of coincidence between gametocyte density and illness is an artificial consequence of treatment (see below). Studies on patients who were not given antimalarial drugs show a concordance between the presence of gametocytes and the timing of severe symptoms (Kitchen 1949). Although gametocytes are often present after the cessation of malaria symptoms (Kitchen 1949), gametocytes generated late during infection (i.e., after the symptomatic period) tend to be less infective for mosquitoes (Sinden 1991).

Gametocytes tend to be present after treatment among patients given antimalarial drugs (Garnham 1966). A set of controlled experiments recently demonstrated that among mice infected with *P. chabaudi*, the delay in gametocyte increase is caused by chloroquine treatment (Buckling *et al.* 1997). Effects of treatment on the timing of transmissible stages therefore need to be considered in attempts to cause evolutionary changes in virulence by mosquito-proofing. If infective gametocytes are present during recuperation from illness, the retarding effect of treatment on gametocyte emergence may have little negative effect on virulence management as long as recuperation is associated with rest in mosquito-proof dwellings. If infective gametocytes appear after a domiciliary recuperation period, then the delay in gametocyte appearance because of treatment may restrict the effectiveness of mosquito-proofing as a virulence management tool in those areas for which prompt treatment is not readily available.

The third assumption is supported by the variation in virulence of malaria infections. *P. falciparum* infections are mild where the potential for vectorborne transmission is low and sporadic (Elhassan *et al.* 1995; Gonzalez *et al.* 1997). This tendency occurs more generally along the northern edge of *P. falciparum*'s range in sub-Saharan Africa (D.S. Peterson, personal communication), where the parasite's distribution may be limited by the restricted abundance of mosquitoes. Similarly, *P. vivax* strains tend to be milder in geographical areas associated with low and sporadic mosquito transmission. This variation is partly attributable to differences in the parasite's tendency to form dormant resting stages, called "hypnozoites," which are especially apparent in more northerly latitudes where transmission is seasonally sporadic and infections are mild (Krotoski 1985; Ewald 1994a). In tropical areas where transmission can occur more continuously throughout the year, illness caused by *P. vivax* more closely resembles that of *P. falciparum*, so much so that it is sometimes difficult to distinguish the two on the basis of symptoms alone (Ewald 1994a; Luxemburger *et al.* 1997). Low rates of transmission do not appear to be associated with benignity when transmission is relatively continuous, perhaps because children have lower levels of acquired immunity as a result of the lower frequency of infections (Snow *et al.* 1997).

Implications of geographical variation

The preceding geographical considerations suggest that virulence management could occur relatively quickly if mosquito-proofing programs are enacted at the edges of *P. falciparum*'s distribution, where transmission is sporadic, and then progressed inward toward the center of *P. falciparum*'s distribution. This approach capitalizes on the mild strains that are apparently present in the *P. falciparum* gene pool. Such a progression, however, might not be necessary, as variations in pathogen virulence are present even in areas with intense and relatively continuous transmission (e.g., Kun *et al.* 1998).

The influences of exposure to infection on host resistance is a potential confounding variable in any efforts to control malaria through alteration of transmission. One hypothesis that has recently attracted attention proposes that reductions in entomological inoculation rates (EIRs) have little effect on overall mortality and morbidity in areas with moderate-to-high bite frequencies; in such situations the benefits of reduced EIRs might be offset by reductions in acquired resistance (Snow and Marsh 1995). Indeed, a recent study showed that increased frequencies of infection during early infancy are associated with decreased risks of severe malaria later in life (Snow *et al.* 1998). With regard to evolutionary effects, however, it is important to realize that in such areas of high transmission, mosquito-proofing is expected to cause an evolutionary shift toward benignity with relatively little effect on frequency of infection, and hence with little effect on the benefits of acquired immunity. The primary effect would be to have milder infections early in life, which are more like vaccines than life-threatening experiences. At low EIRs, mosquito-proofing should lower frequencies of infections to the point of eradication (Watson 1949). At some intermediate (and difficult to specify) EIR, an important trade-off might occur: infection rates might be lowered sufficiently by mosquito-proof housing to cause the benefits of evolutionary reductions in virulence to be offset by a lowered acquired immunity. To resolve these issues, the effects of mosquito-proof housing on acquired immunity, reduced frequency of infection, and reduced virulence all need to be assessed. By gradually introducing mosquito-proof housing from areas of sporadic transmission toward more central areas, these variables could be monitored to detect any negative effects that might outweigh the positive effects of screening on the frequency and virulence of infection.

Mosquito-proofing in the Tennessee Valley

The most thorough experimental test of the effectiveness of mosquito-proof housing on malaria transmission was conducted by the Tennessee Valley Authority (TVA) during the 1930s and 1940s in a large section of northern Alabama (Watson 1949). In 1939, the TVA began a campaign to mosquito-proof all houses in the area. They divided the area into 11 zones and completed the mosquito-proofing of each zone at different times. Before the mosquito-proofing began, about half of the people in the area tested positive. Mosquito-proofing virtually eradicated malaria from the area within 7 years (Figure 28.3). No other intervention was enacted prior to the decline (Watson 1949).

These results demonstrate several important points. First, they support a central assumption of the virulence management argument: *Plasmodium* populations can be affected by mosquito-proofing. Without such data, skeptics could argue that mosquito-proofing may have little effect on *Plasmodium* populations because much transmission may occur outside of houses.

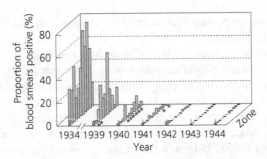

Figure 28.3 Seropositivity of blood samples for *Plasmodium* presented as a function of year during the mosquito-proofing program carried out in Alabama by the Tennessee Valley Authority. Each horizontal row corresponds to one of the 11 geographical zones within the study. The asterisk designates the year in which mosquito-proofing was completed for all houses in the zone. *Source*: Watson (1949).

Second, the results demonstrate a traditional epidemiological benefit – reduction in the prevalence of infection – which may justify the large-scale experiments needed to assess effects of mosquito-proofing on the evolution of *Plasmodium* virulence. To justify such experiments from ethical and economic perspectives, new areas could be selected on the basis that they have a slightly more difficult control problem than that encountered in Alabama. If the *Plasmodium* is eradicated, the evolutionary experiment fails but the intervention is a success in the context of disease prevention. If the *Plasmodium* is not eradicated, the evolutionary experiment can take place, and the intervention may be a success through reductions in pathogen prevalence and virulence.

Third, the results show that the experiment is socially, economically, and logistically feasible. The costs of mosquito-proofing (in 1944 dollars) was about $100 per house for the area with the poorest quality of housing; the costs of maintaining the mosquito-proofing was about $12 per house per year (Watson 1949). Modern technology has generated materials that are more effective, more durable, easier to apply and maintain, and more pleasant to live with than those used in the TVA study. Costs, therefore, should not increase as much as indicated by a simple adjustment for inflation of the TVA costs.

Mosquito-proof housing should guard against other vectorborne pathogens as well. The effectiveness of mosquito-proof housing against transmission of dengue, for example, is indicated by the resistance to dengue invasion when mosquito-proof housing is generally present. Over the past two decades thousands of cases of dengue fever have occurred on the Mexican side of the US/Mexico border along the Gulf of Mexico. Dengue has been introduced repeatedly into Texas, but has failed to spread in spite of the ubiquitous presence of *Aedes* vectors. For every reported case acquired on the Texas side of the border there are about 1 000 reported cases on the Mexican side (CDC 1996). The pervasiveness of mosquito-proof dwellings on the Texas side appears to be responsible for this difference.

Similarly, malaria has been introduced on numerous occasions in recent years to areas in the USA and Europe, where it has been endemic previously. Appropriate vectors are present, yet little secondary transmission occurs; when it does, outbreaks have been self-limited and localized (Wyler 1993; Dawson *et al.* 1997; Kun *et al.* 1997; for an analogous example involving severe diarrheal disease in areas with protected water supplies see Weissman *et al.* 1974).

28.4   Virulence Management in Dwellings

Other categories of transmission may involve even more subtle distinctions. Transmission of respiratory tract pathogens, for example, may be dependent on the mobility of infected hosts or on that of uninfected susceptible hosts. If the options for transmission by the latter route are reduced, the remaining options for transmission are more dependent on host mobility, and the competitive balance is again tipped in favor of milder variants. As durability of respiratory tract pathogens in the external environment increases, transmission is expected to depend increasingly on the mobility of susceptible hosts, and therefore decreasingly on the mobility of infected hosts (see Chapter 2). Interventions that restrict transmission to the latter route of transmission should therefore favor evolution of reduced virulence.

An example of such interventions is negative pressure, modular ventilation systems that rapidly remove airborne pathogens from locations such as homeless shelters, homes for the elderly, or prisons, which would otherwise allow prolonged transmission of respiratory tract pathogens, such as *Mycobacterium tuberculosis*. Such systems provide effective control of transmission in hospital settings. The evolutionary perspective suggests that an additional benefit may occur. The long-term transmission from such areas, which depends on the mobility of susceptible hosts, is restricted; the variants in such areas therefore depend more strongly on the mobility of infected individuals and should therefore evolve toward mildness. Evaluating the validity of this hypothesis has become more urgent now that antibiotic resistant strains of *M. tuberculosis* are more widespread.

28.5   Discussion: The Intervention Spectrum

In this chapter the potential value of virulence management through manipulation of transmission routes is stressed. Manipulation of other potential influences, particularly within-host competition, may also affect the evolution of virulence, but the emphasis here is on transmission routes for three reasons. First, as reviewed here and in Chapter 2, the hypothesized influence of transmission routes on the virulence of human diseases now has a broad, though preliminary, base of empirical support generated through comparative studies. Second, transmission routes are easily manipulable with current technology. Third, the theory predicts that the transmission route would influence the evolution even if within-host variation were the same for the pathogens under consideration; for example, even if each host is infected with a clone, parasites that do not depend on host mobility for transmission would still achieve a greater net benefit from host exploitation than would parasites that depend on host mobility. In this case, however, the competitive advantage involves reaching susceptible hosts first and precluding the entry of competitors, either by destroying the host or by stimulating a protective immune response. The resolution of the various evolutionary influences on virulence depends on the critical evaluation of models that incorporate various combinations of the different hypothetical influences. This evaluation needs to encompass models that simplify the system by excluding hypothetical influences that make intuitive sense, but may have negligible effects in real systems.

The practical value of this emphasis on transmission route rests on its cost effectiveness, its ethical acceptability, and its stability against evolutionary neutralization (in contrast, for example, with the neutralization of antibiotic treatment through the evolution of antibiotic resistance). Regarding cost effectiveness, for every category of disease there exists some intervention that in theory provides both the traditional (nonevolutionary) benefit of reducing the frequency of infection and the evolutionary benefit of reducing the harmfulness per infection. For most if not all of these interventions the traditional benefit alone brings the intervention into the realm of cost effectiveness in some settings. Any evolutionary benefit merely increases the cost effectiveness. If the preliminary analysis of *V. cholerae* in Chile is a valid indicator, this evolutionary benefit may virtually eradicate severe disease without eradicating infection.

Regarding evolutionary neutralization of an intervention, attempts to eradicate pathogens place selective pressures on the pathogen to evolve characteristics that inhibit eradication (such as antibiotic resistance or antigenic characteristics that permit vaccine escape; see Chapters 23, 24, and 26). The shift in emphasis from eradicating the pathogen to controlling the disease is therefore important in terms of achieving both the intervention goal and stability of the resolution if the goal is approached. The goal to eradicate the pathogen creates greater selective pressures on the pathogen with more intense interventions. Eradication is a realistic goal only if the genetic variation in the pathogen population does not contain (or cannot generate) variants that persist in the face of the intervention. This has occurred for vaccination against smallpox and seems feasible in at least a few other vaccination programs, such as those against polio and measles. However, the rapid evolution of antibiotic resistance and the antigenic variation present among many problematic pathogens (such as *P. falciparum* and human immunodeficiency virus) indicates that the susceptibility to pathogen eradication is very variable, and that our attempts to eradicate these more flexible adversaries are less likely to succeed. At a minimum, for these pathogens we need fundamentally new eradication techniques, which require substantial time to develop; thus there will be long periods during which these pathogens will be, at least partially, uncontrolled.

With regard to ethical acceptability, as long as the evolutionary interventions are restricted to those that have a positive effect on disease control even in the absence of an evolutionary effect, ethical considerations will favor the investments required. This conclusion is especially true if the intervention also provides an aesthetically preferred outcome besides disease control, such as better housing,

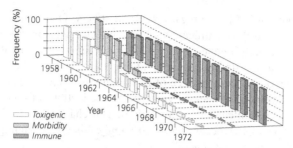

Figure 28.4 Changes in the incidence of diphtheria, toxigenicity of *C. diphtheriae*, and immunity during an extensive vaccination effort in Romania. *C. diphtheriae* evolved toward benignity, as evidenced by a decrease in the percentage of isolates that were toxigenic. "Toxigenic" refers to the percentage of isolates that produced toxin. "Morbidity" refers to the incidence of diphtheria relative to the incidence during the first year of intervention. "Immune" refers to the percentage of the population that was immunologically positive for diphtheria toxin. *Source*: Pappenheimer (1982).

cleaner water, and better ventilation. Some interventions have inherent difficulties. The use of antibiotics, for example, often pits the individual against the group because they help the patient being treated, but impose a cost in terms of antibiotic resistance on others in the group who become infected at a later time (see also Chapter 5). Vaccination similarly can pit individual against group if such a high frequency of vaccination is necessary for eradication that it is forced on individuals who do not wish to be vaccinated. This problem generally becomes more acute as eradication is approached, because the negative effects of the vaccine are often perceived to outweigh its positive effects as the risk of infection approaches zero.

Nevertheless, even with vaccination and antibiotic usage there are strategies that should result in the evolutionarily enhanced management of virulence (Ewald 1994a, 1996, 1999). For example, vaccines that are formulated from virulence antigens should selectively suppress the virulent variants that express those antigens, and leave in their wake mild variants that function as free live vaccines. Although this strategy has only recently been suggested, its primary criterion – formulation of the vaccine based on virulence antigens – was met by chance in two vaccination programs: those designed to control *Corynebacterium diphtheriae* and *Haemophilus influenzae* type b. These programs have proved to be two of the most successful vaccination programs in history. In both cases the available evidence indicates that the target organisms evolved toward reduced virulence (Ewald 1996). Figure 28.4 presents the most complete data on the association between diphtheria vaccination and toxigenicity of *C. diphtheriae*.

The more general argument is that disease organisms can evolve rapidly in response to changes in their environment. Interventions that humans enact to control human pathogens often change the pathogen's environment in ways that dramatically alter pathogen survival and reproduction. If this alteration differentially affects the fitness of pathogen variants, evolutionary change occurs. Throughout the history of health sciences, humans have enacted policies of epidemiological intervention with little (if any) regard for these effects. One of the goals of virulence management is the identification of these effects so that interventions can be better chosen. Though fragmentary, current evidence suggests that this approach is on the right track, with both theory and empirical studies indicating that evolutionary reductions in the virulence of human diseases have already occurred in response to controllable human activities such as reductions in waterborne transmission and use of virulence antigen vaccines. Our challenge is to determine exactly how such processes occur and how to apply this understanding to improve intervention strategies more generally.

29

Virulence Management in Wildlife Populations

Giulio De Leo and Andy Dobson

29.1 Introduction

Historically, control of virulence in wild animals has only been attempted when the disease threatened humans or their livestock. However, as populations of some wild animals have become increasingly rare, public demand to protect endangered species has lead to an increasing effort to control disease in wildlife. Habitat fragmentation and the ensuing edge effects have further exposed wildlife populations to exotic species and livestock that may act as vectors for infectious and parasitic diseases to which the wildlife population has not been exposed previously. Small populations are at greater risk, because the loss of individuals can reduce genetic diversity, make the population more sensitive to the natural fluctuations of the environment, and trigger a population crash as a consequence of high predation pressure or the disruption of social structure (May 1988; Hutchins *et al.* 1991). Moreover, these negative effects may be enhanced by the loss of immunity through the natural elimination of the disease at low population densities. If the disease is then accidentally reintroduced into the now immunologically naive population, hosts may suffer a higher level of mortality with respect to epidemics of previously endemic diseases (Cunningham 1996).

In this chapter, we first present a simple formula to estimate the time for virulence or resistance to evolve. Then, after a brief consideration of the potential consequences of vaccination programs on the evolution of resistance, we briefly review some of the reasons why wildlife virulence management is still a science in its infancy. We present some simple methods to detect the impact of infectious diseases on wildlife populations, and discuss the impact of parasites and pathogens on reservoir hosts. We analyze the power and limits of infection manipulation at the population level – that is, experiments in which a part of the population is treated for the disease. We briefly outline the problem of local adaptation and how this can alter the effectiveness of biological control programs. Finally, we review the possible adverse effects of wildlife translocations that result from disease transmission, and we propose some guidelines to minimize the risk of disease translocation.

29.2 Time Needed for Resistance to Evolve

Asymmetries in generation time between pathogen and host have a fundamental effect on the success of antiparasitic measures. For management purposes, it is important to assess how much time is necessary for a host population to evolve a significant degree of resistance to an introduced pathogen or parasite. Although this subject has been studied most intensively in terms of agricultural insect pests and the evolution of pesticide resistance within parasite populations, the results can be generalized to a fairly broad range of host–parasite associations. Let us consider a homozygote host population that is at a dynamic equilibrium with its parasite. Let us assume that a rare resistant mutant is introduced into the population, and that resistance against the parasite is controlled only at one loci. As summarized by May and Dobson (1986) for a deterministic case in which selection force is not too strong, the expected time T_R taken for resistance to appear is approximately

$$T_R \approx T_g \ln(p_f/p_0)/\ln R_{0,\text{mut,res}*} , \qquad (29.1)$$

where T_g is the cohort generation time (the average time from birth to reproductive maturity) of the host, $R_{0,\text{mut,res}*}$ is the average number of offspring produced by a mutant heterozygote relative to the resident at the population dynamic equilibrium, p_0 is the initial frequency of the resistance allele, and p_f is the frequency of the resistance allele when resistance is first recognized. Equation (29.1) shows that T_R depends directly on the parasite's generation time T_g, but only logarithmically on the other two factors: the threshold at which resistance is recognized, and a measure of the selection strength $R_{0,\text{mut,res}*}$. The frequency of initially rare resistant alleles p_0 typically ranges from 10^{-2} to 10^{-13}, but once the logarithmic transformation is applied, this potentially enormous range collapses to a mere factor of six. Similarly, values of $R_{0,\text{mut,res}*}$ that range from 10^{-1} to 10^{-4} also make a reduced contribution to the expression after logarithmic transformation. The dominant factor is thus the generation time T_g. The same argument applies when we compute how much time is necessary for a pathogen to evolve the ability to bypass host immune defenses, or to develop a significant degree of resistance against pesticides or antibiotics. The net effect is that resistance to pesticides and antibiotics typically appears within 10–100 pest or pathogen generations, despite enormous variations in other factors (May and Dobson 1986). This explains why resistance takes so much longer, when measured in absolute time, to appear in mammals than in weeds, worms, bacteria, viruses, and insects.

29.3 Drugs and the Development of Resistance

The main objective of virulence management in wild animal populations is straightforwardly simple: to control or eradicate parasitic and infectious disease in endemic areas and prevent spread of the diseases beyond these areas. Traditionally, control has been achieved by removing the perceived source of infection and reducing transmission by culling. Using drugs and vaccines to control disease in wildlife species has been fairly uncommon, with some notable exceptions (e.g., for foxes and African hunting dogs against rabies, cattle against rinderpest, and gorillas against measles). However, the potential consequences of different culling and vaccination programs on the host–parasite relationship are still poorly understood. The widespread use of vaccines and drugs may potentially boost the evolution of resistance of targeted pathogens. While little is known about free-living populations, much theoretical and empirical evidence suggests that the evolution of resistance can, indeed, occur in human diseases. The development of resistance of *Plasmodium falciparum* to antimalaria drugs is probably the best-studied case. Malaria parasites have developed mechanisms of resistance against virtually every drug that has been used against them (Peters 1982, 1987; Björkman and Phillips-Howard 1990). Several hybrid transmission–genetics models for the development of resistance by *P. falciparum* have been derived to simulate the effects of different drug dosage (Cross and Singer 1991). These studies show that a strategy of maximally effective treatment for all infected individuals in the short term can be deleterious – in terms of rapid development of drug resistance – in the long term.

Of course, this conclusion cannot be simply extended to wildlife diseases, because the development of resistance depends upon the specific biology and epidemiology of the host and the infective agent in consideration. However, the potential for drug resistance remains (at least for intensive rearing systems such as those for livestock and aquaculture) where a pervasive use of antibiotics can accelerate the selection process, and accidental contacts with free-living individuals can spread the disease in wildlife populations.

29.4 Problems in Managing Virulence in Wildlife

Besides some striking examples, such as the myxoma virus in Australia, quantitatively tracking the evolution of host–parasite associations in wildlife presents daunting obstacles mainly because it is difficult to gather data other than parasite mean prevalence and incidence in the field. Additionally, measures of genetic variability in both the pathogen and the host are rarely available. This is unfortunate, because parasites may play a crucial role in maintaining host genetic diversity, population abundance, and community structure, as discussed in Chapter 3 (see also Grenfell and Gulland 1995). The interplay between parasite virulence and the costs of host resistance is an important mechanism responsible for maintaining polymorphism in these systems. "Boom and bust" cycles, where new resistant races of a crop are attacked by new virulent strains of a pathogen, have long been documented in agricultural practice (Burdon and Jarosz 1991). Several studies have detected genetic variations in disease incidence for host populations of plants (Burdon 1987; Simms and Fritz 1990; Marquis and Alexander 1992), but there is a paucity of data for wild vertebrates (Read 1995). The greater emphasis on the study of plant genetics compared with animal epidemiology is partially because of the greater empirical base for plants, which is a consequence, presumably, of the relative ease of genetic and parasitological screening in plants (Read 1995). Variation in disease resistance in domesticated and laboratory animals has been documented (Wakelin and Blackwell 1988; Kloosterman *et al.* 1992). As for wildlife diseases, it is well known that individual hosts differ in their response to many infections and that the differential response to infections may well have a genetic basis (McCallum 1990; Grenfell *et al.* 1995)

Nevertheless, despite some empirical evidence, and an impressive amount of theoretical considerations, our understanding of the evolution of virulence in wildlife is still insufficient to allow effective management of virulence in free-living populations (Read 1995). Experimental data are too sparse to support quantitative predictions on the evolution of virulence and/or resistance.

One of the main problems is that much of the attention in wildlife diseases focused, until recently, on the effects of pathogen and parasite at the individual level, rather than at the population level, with almost no attempt to understand the evolutionary consequence of the host–parasite interactions. The standard veterinary approach often focuses on individuals, although it recognizes that the pathogen indeed affects the population if a substantial number of individuals have the same symptoms. Unfortunately, pathological abnormalities are not always evident, particularly in the case of macroparasites (Gulland 1995). Wild animals tend to manifest few recognizable signs of disease, and often isolate themselves from others when affected (Young 1969). In practice, detecting the impact of a macroparasite on its host population may be substantially harder than detecting the impact of microparasites, because macroparasites are generally enzootic (fairly stable within host population), in contrast to the epizootic (outbreaking) behavior of microparasites. Estimating the effects of the pathogens at the individual level in terms of increased host mortality or reduced host fertility in the wild may be technically difficult and very expensive, whereas laboratory tests may often underestimate the pathogenicity of the disease. Finally, the problem remains of how to translate the effect of the pathogen on the individual host to the population level. In this scenario, the identification of genetic differences may present paramount difficulties, and may be simply impractical.

This should not be surprising: counting individuals (or corpses) is so much easier than assaying genotypes! On the other hand, wildlife epidemiologists have already fought against problems that are conceptually simpler (but equally crucial) than assessing genetic traits. For instance, the definition of virulence in the wild and its estimation are hampered by many theoretical and experimental difficulties. Originally, there was a tendency to overlook the potential ability of parasite diseases to determine wildlife abundance or affect community structure – an approach probably inherited from the epidemiology of human diseases, in which factors other than parasitic agents usually regulate population density. As a consequence, parasite-induced mortality was often assumed to be compensatory, rather than additive (Holmes 1982). This assumption reflects the underlying problem of comparing host–population dynamics without disease agents to dynamics with disease agents, without altering other factors that influence the dynamics of the host–parasite system. The problem is exacerbated because diseases can influence many aspects of the host's lifecycle, physiology, and ecology (Cunningham 1996): They can cause death, increased susceptibility to predation or further disease through physiological and behavioral modifications, increased susceptibility to other stress factors (such as drought or food shortage), reduced reproduction capacity, or a combination of the above effects (Scott 1988; Gulland 1995).

Several statistical and graphical methods have been developed to measure parasite-induced mortality from data on prevalence and intensity in the field

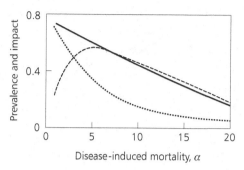

Figure 29.1 The prevalence of disease in dying hosts (continuous curve) compared with the prevalence of disease in the total host population (dotted curve) and the extent of impact on the host population, D (dashed curve), all shown as a function of disease pathogenicity α. A large difference in prevalence between dying hosts and the total population suggests that the disease is having an impact on the host population. *Source*: McCallum and Dobson (1995).

(Crofton 1971; Anderson and Gordon 1982; Pacala and Dobson 1988; Gulland 1995). Observations of the disease status of the population may be achieved through suitable sampling plans of animals captured and then released in radio-tagging and mark-and-recapture experiments (McCallum and Dobson 1995). For diseases of low pathogenecity, serological tests, whenever applicable, may help to identify individuals that harbor the infective agents. Methods based on analysis of life tables can provide estimates of the likelihood of survival of infected animals, with respect to disease-free individuals. However, these methods are less effective in assessing the effects of parasites on the fecundity of their hosts.

29.5 Detecting the Impact of Infectious Diseases

To check whether the disease is indeed having an impact on the host population, it may be useful to estimate the difference between the prevalence of disease in dead and dying hosts and that of the whole population (McCallum and Dobson 1995). Figure 29.1 shows disease prevalence in dead and dying hosts, the prevalence of the disease in the total population, and the extent that a host population is depressed below its disease-free carrying capacity as a function of pathogenicity. Prevalence in dying hosts declines monotonically with pathogenicity, although not as rapidly as does prevalence in the total population. A large difference in prevalence between dying hosts and the total population suggests that the disease, indeed, has an impact on the host population. This difference may be even greater if a third factor, such as stress, causes morbidity and low resistance to infection.

In the case of single-pathogen and single-host infections, simple models can be used to quantify the impact of a parasite or pathogen on a host population, provided that data are available for the effects on both host individuals and the extent of infection within the host population. If the host–pathogen system is at equilibrium, the extent to which the disease depresses the host population D

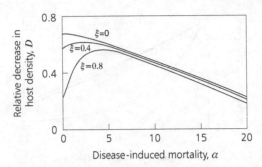

Figure 29.2 The effect of disease pathogenicity α on the extent D that a host population is depressed below its disease-free carrying capacity for three different values of ξ, the impact of the disease on host fecundity ($\xi = 0$, infected animal is sterile; $\xi = 1$, no impact). If pathogens primarily affect mortality ($\xi = 0.9$ in this figure), those with intermediate levels of pathogenicity have the greatest impact on their host population. If the major effect is on fecundity ($\xi = 0$ in this figure), then the greater the decrease in fecundity, the larger the impact on the host. *Source*: McCallum and Dobson (1995).

(1 − the ratio between host density at equilibrium with infection present and the disease-free carrying capacity) can be computed (McCallum and Dobson 1995) as

$$D = i^*[\alpha + b(1 - \xi)/r_0] , \tag{29.2a}$$

where α is the disease-induced mortality, r_0 is the instantaneous rate of growth of the population, b is the birth rate of uninfected hosts, i is the equilibrium prevalence of the pathogen, and ξ is the impact of the disease on host fecundity ($\xi = 0$, infected animal is sterile; $\xi = 1$, no impact).

An analogous equation for macroparasites (McCallum and Dobson 1995) is

$$D = M\alpha/r_0 , \tag{29.2b}$$

where M is the mean parasite burden in the population and α is the increase in host death rate per parasite. Pathogenicity can be estimated approximately as the inverse of the life expectancy of an infected host's microparasites, and as the gradient of the relationship between parasite burden and the inverse of life expectancy for macroparasites. The demographic rates r and α can be estimated from literature, by using life-table analysis, or, if these data are not available, through allometric relationship with body weight. The parameters i and M need to be measured through suitable epidemiological surveys.

If the disease primarily increases mortality, high pathogenicity does have a minor effect on host population (Figure 29.2). Therefore, if a disease occurs in most hosts, it is probably mild and unlikely to be a major problem. If the disease is proved to be highly pathogenic in laboratory tests, it is also unlikely to cause problems, particularly in low-density populations, because infected animals die before the disease can spread. On the contrary, if the main effect of the pathogenic agent is on host fecundity, diseases present at high prevalence may have a major impact on host populations.

29.6 Pathogens and Parasites with Reservoir Hosts

It is important to remember that these results apply only to single-host infections that can be assumed to be at equilibrium. Conversely, a pathogen that infects a range of host species can cause a variety of problems for endangered species. A simple model may clarify the issue. Let S and I be the densities of susceptible and infected individuals, respectively, of an endangered host species. Assume also that:

- A second host species harbors the disease (the reservoir host);
- This second species is at its endemic equilibrium;
- The prevalence (infected fraction) in the reservoir population is high (e.g., because of the low pathogenicity in that species);
- The demography or epidemiology of the reservoir species is not affected by the first species.

Let $\lambda_{\text{reservoir}}$ be the *force of infection* exerted by the reservoir species onto the first species – that is, $\lambda_{\text{reservoir}}$ denotes the probability that a susceptible individual of the first species is recruited to the infective class through an infective contact with an individual of the reservoir species. Then, the classic SI model should be modified as

$$\frac{dS}{dt} = r_0 S \left(1 - \frac{S}{K} \right) - \beta I S - \lambda_{\text{reservoir}} S , \qquad (29.3a)$$

$$\frac{dI}{dt} = \beta I S + \lambda_{\text{reservoir}} S - (\alpha + d) I , \qquad (29.3b)$$

where, as usual, K is the disease-free carrying capacity of the endangered species, r_0 is the instantaneous rate of population growth, d is the disease-free mortality rate, β is the infection rate constant, and α is the disease-induced mortality rate. Let us further assume that in the absence of the reservoir species, the carrying capacity K of the endangered species is under the critical level K_c required for the maintenance of the disease,

$$K < K_c = \frac{\alpha + d}{\beta} . \qquad (29.4a)$$

Or, equivalently, the basic reproduction ratio is smaller than one,

$$R_0 = \frac{\beta K}{\alpha + d} < 1 . \qquad (29.4b)$$

If the endangered population does not have any interaction with the reservoir species ($\lambda_{\text{reservoir}} = 0$), the disease quickly dies out. But as soon as $\lambda_{\text{reservoir}} > 0$ (i.e., the probability of infective contacts between the two species is not negligible), then the disease can spread in the endangered population, even if the infective agent is so virulent ($\alpha \gg d$) that an infected hosts always dies before being able to transmit the disease to other susceptible individuals. In fact, in this case, it is the reservoir species that provides the necessary infective contacts to maintain the disease in the population of the endangered species. Moreover, if the force of infection $\lambda_{reservoir}$ is sufficiently high (i.e., $\lambda_{reservoir} \geq r_0$, which is likely to occur since endangered populations are typically characterized by low growth rates), the disease can drive the population to extinction. This is probably what happened with the canine distemper virus, which is assumed to have caused a large decline in the last remaining free-living colony of black-footed ferrets. The population of black-footed ferrets, strongly depressed by over-hunting, was probably too small to sustain the pathogen. When the pathogen was introduced into the colony from prairie dogs, or feral canids, it quickly spread, causing the deaths of around 70% of the ferrets.

29.7 Manipulation of Infection at the Population Level

When managing host populations in the wild, we should be aware that an increase or a reduction of host density can potentially alter the competitive outcome of different parasitic strains that exploit the same host. It is well known that the occurrence of multiple infections can foster the selection for increased virulence (Van Baalen and Sabelis 1995a; see also Chapter 3). A reduction of host density (e.g., through selective removal of the host) decreases the number of infective contacts per unit time, which, in turn, can favor the selection of less virulent but more persistent strains (De Leo and Guberti, unpublished). In fact, a virulent strain cannot survive in a very sparse population of hosts, because it is likely to kill its host (and thus die also) before infecting a novel healthy one. In contrast, life expectancy of a mildly infected host may be long enough to allow successful transmission of the infective agent. It is clear, anyway, that to quantify authoritatively the effect of disease on free-living individuals, infection should be manipulated at the population level – that is, from experiments in which a part of the population is treated for the disease (McCallum and Dobson 1995).

Unfortunately, no treatment for disease is without risk, and the capture and handling of animals required to administer treatment may involve substantial hazards to the animals, as exemplified by reports of increased mortality detected in treated African wild dogs in the Serengeti. Moreover, manipulation is often impractical because it requires many replications and several control populations to be statistically significant. As stressed by McCallum and Dobson (1995), the problem of estimating survival rates is similar to that of estimating a proportion, and a 95% confidence interval for a proportion is approximately $\pm 1/\sqrt{n}$, where n is the sample size. Thus, at least 100 individuals must be marked to estimate survival to within 10%. Moreover, if we want to test the difference in survival between the control and treated subpopulations, it may be necessary to handle 400–500 animals to detect a 10% difference in survival between the two groups with 95% confidence (McCallum and Dobson 1995). As a consequence, there are almost no examples of population-level manipulations of infection using an entirely natural host population. Several endangered species have been vaccinated against diseases – such as the African hunting dog (*Lycaon pictus*) against rabies (McCallum and Dobson 1995) and the mountain gorilla (*Gorilla gorilla beringei*) against measles

(Hutchins *et al.* 1991) – but comparison of survival with untreated control animals has not been undertaken.

Laboratory experiments may certainly provide some clue, but the results cannot always be straightforwardly transferred to populations in the wild. For instance, Scott (1987) detected a substantial effect of *Nematospiroides dubius* in laboratory colonies of mice. But the results were much less dramatic when he repeated the experiment using *Apodemus sylvaticus* in seminatural conditions.

However, some manipulative studies of parasite impact on nonendangered wildlife populations were very successful. Hudson *et al.* (1992), for instance, analyzed the impact of the parasitic nematode *Trichostrongylus tenuis* on the red grouse (*Lagopus lagopus scoticus*) in the north of England and in Scotland. By suitably treating part of the free-living population and using radiotelemetry aerials to monitor both control and treated individuals for up to 6 months, they showed that treated birds (those with a lower parasite burden) have a higher survival rate and significantly better fecundity than the control birds with natural parasite burden. Hudson *et al.* were also able to prove that high parasite burden increases the susceptibility of host to predations.

Other manipulative experiments on the impact of parasites in the field include studies on white-footed mice (*Peromyscus leucopus*; Munger and Karasov 1991), ectoparasites on gerbils (*Gerbillus andersoni allenbyi*; Lehmann 1992), and nematodes in Soay sheep (*Ovis aries*; Gulland *et al.* 1993).

29.8 Disease Risks of Wildlife Translocations

In conservation biology the increased concern with efforts to counteract wildlife habitat destruction and species extinction has led to countless wildlife translocations of individual rescued animals and captive breeding programs of endangered species for their eventual reintroduction to the wild (Cunningham 1996). Translocation in the form of large-scale movement of wildlife has become very common also for sporting purposes. New parasites imported with the translocated animals can lead to undesirable consequences for the managed species (either the translocated animals or resident animals of the same species), for other resident species at the site of translocation, or for both (Hutchins *et al.* 1991; Davidson and Nettles 1992; Ballou 1993; Viggers *et al.* 1993).

Many documented cases of wildlife intervention result in the transfer of disease (Davidson and Nettles 1992; Viggers *et al.* 1993; Woodford and Rossiter 1993; Gulland 1995). It is debatable, though, whether an introduced parasite will be mildly or strongly virulent for the new host species. Some well-known cases show that invading parasites may be more harmful to newly acquired host species than they are to hosts with which a long-evolved relationship has likely been established. The African trypanosomes (*Trypanosoma brucei*), for instance, produce mild infections with insignificant mortality when they infect indigenous ruminants, while they are highly virulent in recently introduced species of ruminant livestock. The African rinderpest pandemic and the myxomatosis pandemic in Australia described in Chapter 3 seem to confirm this view. Of course, newly

422   *G · Perspectives for Virulence Management*

introduced parasites can be more virulent simply because the resident host has not yet developed an effective immune response to counteract the parasites (Ebert and Herre 1996).

Nevertheless, experimental evidence indicates that the opposite is usually true. Novel parasites are usually less harmful (Ballabeni and Ward 1993; Ebert 1994), less infectious (Lively 1989; Ebert 1994), and less fit (Edmunds and Alstad 1978; Ebert 1994) than the same parasite species infecting the host to which it is adapted. In each of two reciprocal cross-infection experiments, Lively (1989) showed that a digenetic trematode (*Microphallus* spp.) was significantly more infective to snails (*Potamopyrgus antipodarum*) from its local host populations. This gives strong evidence for local adaptation by the parasite and indicates a genetic basis to the host–parasite interaction. It suggests that the parasite should be able to track common snail genotypes within populations and, therefore, that it could be at least partially responsible for the persistence of sexual subpopulations of the snail in those populations that have both obligately sexual and obligately parthenogenetic females. By analyzing geographical patterns of virulence of a horizontally transmitted microparasitic disease in *Daphnia* (a planktonic crustacean), Ebert (1994) showed that a parasite's ability to infect and exploit a novel host may decrease with increasing geographical and, presumably, genetic distance from the host to which the parasite is adapted. This finding indicates local adaptation of the parasite, but contradicts the hypothesis that long-standing coevolved parasites are less virulent than novel parasites. In this case, local adaptation may be seen as the extension of host specificity on a microevolutionary scale (Thompson 1994; Ebert and Herre 1996). Virulence can be explained as the consequence of balancing the positive genetic correlation between host mortality and strain-specific spore production.

The issue of local adaptation of parasitic agents to their host is particularly important even when we approach this problem in terms of biological control. An interesting example of the importance of the source of the controlling organisms is given by the introduction to Australia of a herbivore to consume the floating weed *Salvinia molesta* (Room 1990). After failing to control the weed with herbicide, a beetle species from Trinidad was introduced that was known to consume a closely related species, *S. auriculata*. The first attempt was completely unsuccessful, but, subsequently, other beetles of the same species were collected from a Brazilian site where they were known to feed on *S. molesta*. In this case, the introduction was successful. This example shows that local adaptation is possible, and that the source of material used in biological control programs can influence the outcome of an introduction.

29.9 Minimizing Disease Risks in Wildlife Translocations

Although some effort has been undertaken to assess the disease risks involved in animal translocation (Davidson and Nettles 1992), our current knowledge of the parasites of wild animals and, specifically, of their biology and pathogenicity, is still too poor to allow for a quantitative risk evaluation. However, some rules for sound management can been derived in the effort to reduce disease risk

(Cunningham 1996). For example, reintroduced animals should be screened for known pathogens, all animals that die in captivity should be necropsied, and, if possible, animals that die after release should also be necropsied (Viggers *et al.* 1993). Moreover, it may be useful to assess the parasite status of the resident populations of the same or related species to the one being introduced, even though this operation may be expensive and, at times, impractical. Clinically healthy animals should not be regarded as parasite-free, but should be subject to the most effective (species-dependent) test to detect potential infective agents. The translocation of animals to areas free of related species may decrease the risk of interspecific transmission of the disease. Moreover, managers responsible for wildlife translocation should remember that, if no diseases have been detected in a particular population for many years, this does not mean that that population is not susceptible to any disease. Further guidelines for hygiene and quarantine procedures, and for monitoring the parasite status of both captive and free-living animals are given in an excellent article by Cunningham (1996).

It is not always wise, however, to eliminate completely all the parasites from a targeted host species, particularly if the host and the parasite share a long history of coadaptation. The importance of rinderpest in structuring the ungulate community in the Serengeti (Chapter 3) indicates that the removal of a pathogen from a system may have a major impact on the structure of the entire community. Moreover, the removal of parasites from animals to be reintroduced may be unnecessary if they harbor parasites to which they will be exposed in their natural habitat.

29.10 Discussion

One important goal of wildlife reserves is to preserve attractive biodiversity in our human-dominated world. Such reserves are much smaller than the original ranges of the species contained within them. Moreover, the level of human intervention is much higher than for truly wild populations. It is therefore essential to consider virulence management. The problem of virulence management in wildlife is generally more complex than that in, for example, a veterinary context, because of the effects of natural population dynamics and because the community comprises multiple species. We have provided a few thoughts on how such problems can be approached on a general level. The impact of a disease can be assessed from the relative prevalence of the disease in corpses and live animals. This information can be used to assess where management efforts can be invested most usefully. In addition, we discuss an uncontroversial elementary type of virulence management, to wit the containment of potential disease sources by minimizing spill-overs from reservoir populations to endangered populations, and by reducing the risks of disease transport during enforced wildlife translocations.

There is much anecdotal evidence for the evolution of virulence in wildlife. However, the paucity of data for wildlife populations and our poor knowledge of the parasites and their biology and pathogenicity mean that general statements cannot be made about quantitative risk assessments or about the effects of introducing a new pathogen into a novel host. Currently, we can do little more than make gross qualitative statements and general exhortations. In particular, wildlife managers should be made aware that virulence and transmission may change, or evolve, in direct response to traditional management attempts to control pathogens and their hosts. The general ideas sketched in this chapter, and elsewhere in this book, should give at least some idea of those aspects that should be considered. In particular, we strongly recommend:

■ Carrying out small-scale experiments before pathogen treatment is applied on a larger scale;
■ Taking into account the potentially wider consequences of pathogen removal or wildlife management and translocations.

To quote McCallum and Dobson (1995), an experimental approach to handling disease threats to wildlife population "is not only good science, it is good management practice as well."

30

Virulence Management in Veterinary Epidemiology

Mart C.M. de Jong and Luc L.G. Janss

30.1   Introduction

In northern Europe, there is currently a tendency toward more "natural" livestock production. The proponents of this view believe that animals should be kept in herds and able to move about freely, that the (regular) administration of drugs should be reduced, and that it is wrong to destroy whole flocks to eradicate disease. In addition, there remain requirements for economically viable livestock production that yields safe food for human consumption. These aims can be reached in two ways: by complete eradication of certain agents (e.g., by mass vaccination), or by avoidance of disease caused by other agents. In writing about infectious diseases, the terminology often causes confusion. We adhere to the following definitions: agent – the microorganism that can sometimes cause disease (because disease does not always occur, and because the effect may even be favorable, the term pathogen is avoided); infection – colonization of the host (i.e., some replication of the agent in the host); and disease – clinical symptoms of the host caused by the agent. Eradication is preferable when the disease causes severe economic damage (e.g., foot-and-mouth disease, classic swine fever) or when there are serious consequences for human health (e.g., bovine tuberculosis, bovine spongiform encephalopathy).

When there is a high rate of reintroduction (e.g., from other animal species), the monetary costs to restore the infection-free status again and again can be higher than the losses caused by the disease. Thus, when the agent is common in many animal species and causes disease only occasionally, measures that lead to fewer clinical symptoms (disease) in the host population but do not require eradication of the agent are preferred. Efforts should then focus on the control (of the spread) of the infection instead of on complete eradication. Livestock and plant production systems are probably the areas in which such disease control can be used most extensively. In these production systems, vaccination and special hygienic measures can be applied and, additionally, spatial structures of hosts can be manipulated, movement of the hosts can be limited, and genotype frequencies in host populations can be manipulated.

In addition, disease prevention can be achieved by intervening in risk factors that are responsible for the development of clinical symptoms after an infection, or by managing the infection itself. Management of the infection by influencing the timing of the infection, the amounts of the infectious agent received, and

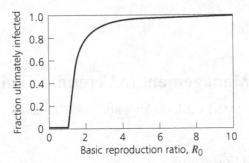

Figure 30.1 The important nonlinear relationship between the basic reproduction ratio, R_0, of a disease and the fraction of infected individuals. When R_0 is near the critical value 1 (say, between 1 and 2), measures to reduce transmission can have large effects on disease spread, while, in other cases, similar measures may have little effect.

the ability of the agent to cause disease may be used to avoid disease. Managing the infection requires a quantitative approach because infection always entails some nonlinear interaction (transmission), and, consequently, less exposure to a risk factor may have either a dramatic effect or almost no effect (Figure 30.1). Mathematical and statistical models help us understand the mechanisms of disease spread and allow us to quantify disease transmission (Kermack and McKendrick 1927, 1932, 1933, 1937, 1939; Anderson and May 1979; May and Anderson 1979; Becker 1989; Diekmann *et al.* 1990, 1995; de Jong and Kimman 1994; de Jong *et al.* 1995). For example, epidemiological models predict that herds with more than a critical fraction of animals completely protected by vaccination experience no major outbreaks, whereas vaccinating less than the critical fraction has almost no effect (Figure 30.1).

Other models have been used to estimate the effect of vaccination on transmission when vaccination does not lead to complete protection (Halloran *et al.* 1992; de Jong and Kimman 1994; Longini *et al.* 1996). For spread of the infection, the overall rate of transmission is quantified by the so-called reproduction ratio R_0 of the infection, which is calculated on the basis of parameters that describe susceptibility, infectivity, and the contact structure (Koopman and Longini 1994; de Jong 1995).

These parameters can be estimated in transmission experiments, as shown for the Pseudorabies virus in pigs that are subjected to vaccination against this infection (Bouma *et al.* 1997a, 1997b). Such a quantitative approach helps to determine an acceptable level of protection against infection spread ($R_0 < 1$), without aiming at the improbable goal of "zero" transmission. However, the methods tested and the control programs designed on the basis of these experiments can be used to eradicate an infectious agent, because $R_0 < 1$ implies that the infectious agent will vanish from a finite population.

Alternative methods, which might result in the control of a disease but not complete eradication of the agent, could be used to reduce the infection pressure. For this to work, either the chances of disease following infection should depend disproportionally on the infectious dose received (e.g., Medema *et al.* 1996), or reduction of infection pressure should result in selection (evolution) toward less virulence (Ewald 1994a). Therefore, in our view, demands for more sophisticated approaches to disease control will ultimately also lead to a demand for virulence management and control. These more sophisticated approaches require the application of quantitative methods to transmission control. Studies (either theoretical or experimental) on virulence evolution are particularly important because of, on the one hand, concerns about host–agent coevolution and, on the other hand, the new potentials to interfere with this evolution by quickly changing hosts' genotypes through the use of DNA tools. As a prelude to the study of virulence evolution, we argue that simple models generally do not seem to apply to host–agent systems when the host is an animal, and that models should be used that account for the hosts' immune response as an important interfering mechanism. In general, we describe the study of virulence evolution and virulence management as it relates to our experience on disease transmission.

30.2   Virulence Evolution Made Simple

Several chapters of this book discuss models on the evolution of virulence; for earlier work on virulence evolution, see the references in these chapters. In the simplest models, greater replication of the agent within the host was assumed to lead simultaneously to greater damage and to more production (and shedding) of the agents – that is, a direct relationship between virulence (damage to the host possibly results in death) and infectivity (shedding) is assumed. This standard model sees the host as one "bag" of cells, like a tissue culture in virus research, or a container filled with growth medium for a bacterium. Competition between agents within a host is then merely competition for a limited amount of food, which is known as "scramble competition" in ecology. For the agents, use of resources (i.e., cells) is directly linked to replication and shedding. Therefore, in the "one-bag" model, there is indeed a direct link between virulence and infectivity. The one-bag model predicts that the within-host evolution of agents leads to higher virulence. However, when secondary infections are negligible, the overall evolutionary dynamics will ultimately lead to R_0 maximization and, therefore, to medium virulence – at least when mass action is assumed.

This one-bag model can only be true if there is no evolution of the host's defenses, as explained further below. In the case of a host without a defense mechanism, the agent would move, on an evolutionary time scale, toward exploiting all the compartments (e.g., organs, tissues) of the host. (Although on a shorter time scale the agent may not be able to use certain cells or tissues – e.g., because it lacks the right receptor molecules.) When a host possesses a defense mechanism, containment of the agent to some compartments and exclusion from others might evolve. It is therefore questionable whether the simple one-bag model holds in general, as it is known that hosts do not passively undergo an infection, but generally respond actively via their immune system. Such an active response may limit, on an evolutionary time scale, the replication of the agent to certain compartments, as is explained in Section 30.3.

We now discuss an example to show the consequences of the presence or absence of an immune response. Bovine virus–diarrhea virus (BVDV) is a pestivirus that transmits mainly from cows to their unborn calves (Moerman *et al.* 1993, 1994). For BVDV, both immune-tolerant ("passive") and immune-competent ("active") hosts exist. Immune-tolerant hosts appear through a fetal infection (via their mother) by a completely avirulent noncytopathogenic (NCP) strain. Infection of a pregnant cow by an NCP strain can cause infection of the fetus also, and can induce immune-tolerance in the fetus when infection occurs before full development of the immune system – that is, the calf learns to recognize BVDV as part of its own antigen system. When infected immune-tolerant calves are born, they can spread the NCP virus to other pregnant cows, which causes immune-tolerance in other newborn calves. Moreover, when immune-tolerant females reproduce, they give birth to BVDV-infected immune-tolerant calves. These mechanisms can lead to persistent (avirulent) BVDV infections and to large numbers of BVDV immune-tolerant animals on farms.

Cytopathogenic (CP) strains of BVDV are virulent and can spread in populations of immune-tolerant hosts (each host being envisaged as a passive "bag of cells"), but do not spread in populations of immune-competent hosts. The effects of infection by a virulent CP strain are markedly different on farms with a persistent NCP-strain infection than on farms with no (other) BVDV infection. On farms with a persistent NCP infection (i.e., with large numbers of immune-tolerant animals), a CP strain is very virulent, and causes a disease known as mucosal disease, which quickly kills the animals. However, in herds with no previous contact with BVDV, CP strains cause little harm: an adequate immune response means the viral attack is quickly stopped and generally does not spread. As a result of these dynamics, CP strains do not persist in the cattle population. Hence, they must arise *de novo* through mutation in immune-tolerant animals in which a CP strain takes over from an NCP strain. Further evidence for the *de novo* creation of CP strains is that the CP strain newly arising in an immune-tolerant animal is similar to the NCP strain. Infection of an immune-tolerant animal with a very different CP strain is not possible: the animal is not immune-tolerant for that strain.

The observation that virulent CP strains of BVDV can outcompete avirulent NCP strains within single hosts and can spread in immune-tolerant populations fits with the simple one-bag model, which predicts that within-host competition leads to increased virulence. However, this conclusion applies only to systems of hosts that do not have an immune response to the agent. In hosts that have an immune response, the agent causes very limited clinical symptoms (Moerman *et al.* 1994). The case of BVDV shows that immunity can be an important element in the evolution of the agent–host relationship, and thus in virulence evolution. The one-bag model probably applies only to very restricted cases in which there is no immune response. As shown, the one-bag model applies to BVDV immune-tolerant cattle, and it may also apply to other infections for which there is no immunity, such as prion diseases and opportunistic (nosocomial) infections in immune-compromised patients.

30.3 Two-bag Model of Virulence Evolution

Darwinian selection acting on the host normally results in an immune system capable of stopping the multiplication of disease-causing agents as soon as the agent is detected by the host's immune system. In such a case, the one-bag model, in which all agents are involved in scramble competition with each other, is not sufficient to understand the coevolution of agent and host. A more realistic model consists of at least two bags:

- A first bag in which the agent can replicate and spread, but in which the replication does not cause much damage, and, hence, virulence is limited when the agent is constrained to that bag;
- A second bag in which replication does cause damage resulting in (severe) clinical disease.

This two-bag model assumes that the active immune reaction of the host can restrain agents to some local tissue and not allow the agent to reach other more vital tissues and thus inflict severe damage to the host (i.e., be virulent). The two-bag model thus incorporates two aspects that we consider important to understand virulence evolution: one is the independence of disease (clinical symptoms) and infection (replication), and the other is the role of the immune system in defending the host from disease, but not necessarily from infection.

The independence of disease and infection is brought about in the first place by compartmentalization of the host: replication in some tissues (e.g., epithelium) may be less harmful than in other tissues (e.g., neural tissues). The immune response of the host is selected to fight off infections at the point of entry to such an extent that the total damage by the agent (and by the immune response) is minimal. Containment of the infection is important to limit the damage: replication in more vital tissues has to be precluded, because fighting the infection in vital tissues always causes damage from the agent and the immune response (e.g., by killing virus infected cells). Compartmentalization and immune response are therefore jointly important in the host's defense. Some models of the immune system do not take compartmentalization into account and model the host as one bag. However, at least a two-bag model is needed to distinguish between effects of the immune system on disease and on agent replication.

For the agent, a situation can arise in which there are two compartments in the host because the selection on the host differs from that on the (pathogenic) agent. On the one hand, there is selection on hosts to have an immune system that protects the host against clinical disease. On the other hand, there is selection on the agent to maximize its replication and to possess traits that enhance transmission. The combination of selection on the host and on the agent may result in agents that can readily replicate and spread when restrained to a local tissue, without causing (severe) disease.

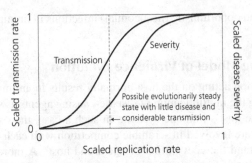

Figure 30.2 The two separate effects of the replication speed of the causative agent on both the severity of the disease and the transmission rate. The replication rate has been scaled relative to its largest feasible value, and the transmission rate has been scaled relative to the maximum value that it takes (at the largest replication rates).

In many diseases, such a situation does seem to have evolved (e.g., in bacterial and viral meningitis; see also Levin and Bull 1994). Meningitis is caused by an agent that mostly spreads in the host population without any clinical symptoms, and only occasionally do tissues surrounding the brain and spinal cord become infected, which leads to severe disease (e.g., *Haemophilus influenzae*). Another example is pseudorabies in pigs, which spreads through replication of the virus in the mucosal tissues, and only causes severe disease in those cases in which it also invades neural tissues.

Figure 30.2 depicts a conceptual model for the conflict between the evolution of the host's immune system and the evolution of the agent's virulence. There are two responses to the replication of the agent in the host: one is the infectivity of the host as a function of the amount of replication, and the other is the amount of disease of the host as a function of the replication. In the one-bag model, these two responses are completely correlated, so the two boundary curves in Figure 30.2 coincide. If higher replication goes hand-in-hand with severe disease, the result for the virus is an evolutionary steady state characterized by the balance between the detrimental effect of the disease and the beneficial effect of its replication. However, in the two-bag model, the two responses shown in Figure 30.2 are themselves subject to evolutionary change. The host immune response evolves so that with sufficient replication of the agent, there is little disease. The agent evolves to avoid the immune response as much as possible while obtaining maximum replication and sufficient infectivity in the host. Both evolutionary processes together create an evolutionary steady state in which there is much infectivity but little disease. This predicted evolutionary steady state corresponds to agents of medium virulence, which are constrained to local tissues, where they succeed in replicating.

In the arms race between evolution of the host's immune system and that of the replication (virulence) of the agent, there are two important borders. Beyond one border, agents replicate so much (or the immune reaction is too weak) that they cause too much clinical disease in the host. The chance that the host suffers clinical symptoms during its life is then high, and there is a strong natural selection for adequate immune responses in the hosts (we assume that increasing host immunity or decreasing agent replication is interchangeable). Beyond the second border, the host immunity becomes too strong (or conversely, agent replication too little), so that agents no longer sufficiently replicate to transmit from host to host. In this case, agents do not spread that well, the chance for the animal to become diseased is small, and, thus, there is less selection for the host to have an adequate immune response. Assuming that in the absence of disease-causing organisms there is a cost to immune responses, then the immune response will become less effective, and the possibilities for the agent to spread increase. As depicted in Figure 30.2, the two borders can create a window in which agents of medium virulence (which usually do not cause clinical disease) remain existent, provided that replication without clinical disease is sufficiently effective to maintain the agent in the population.

It is important in the above assumption that both the replication characteristics of the agent and the immune response repertoires of the host are not one-dimensional: there are different ways to increase replication in the host, and there are also different ways for the host to respond. The various ways to achieve greater replication may differ in the amount of disease associated with that replication, and the various ways for the host to respond to the agent may differ in the amount of replication of the agent that they still allow. Given such differences, Darwinian selection tends to favor those characteristics of the host that reduce disease and increase transmission.

The above model, which is based on an evolutionary conflict between effective immune reaction and agent virulence, predicts the exact opposite of Ewald's (1994a) one-bag model. With the one-bag model, evolution is predicted to lead toward lower virulence when transmission is reduced; with the two-bag model, evolution is predicted to lead toward (relatively) increasing virulence when transmission is reduced. In fact, the argument that evolution acts to uncouple disease and replication predicts that the evolutionarily steady state occurs when agent virulence is just at the border of causing some clinical disease, and host immunity is just at the border of blocking some replication of the agent. Such borders are not sharp: the genetically or phenotypically weakest individuals (i.e., by chance circumstances) will be the ones that become clinically diseased, and the strongest individuals (genetically or phenotypically) will be able to prevent replication of agents completely.

The difference between the one-bag and two-bag models lies in what is assumed about the selection effect on hosts. For those agents that have been confronting a particular host population for a longish time, the selection pressure on agents and hosts as described for the two-bag model will indeed have occurred. In contrast, for those agents that are newly emerged in a host population, evolutionary adaptation of agents and hosts may not yet have occurred. The outcome of a new encounter between a host and an agent depends on whether the immune system, which in vertebrates appears to be very versatile, can effectively deal with the new agent. The usual outcome is that the agent has little opportunity to replicate, and, if it does replicate, does not cause disease. In exceptional cases, an agent new to the host causes disease and selection to begin within the host. Until the selection effect on the host becomes apparent and uncoupling of disease and transmission occurs, the one-bag model applies – as happens, for example, in some newly emerging influenza strains.

In the veterinary setting, we can change the genetic make-up of host populations very quickly by artificial selection (breeding). This genetic selection may seek to improve the immune reaction of the host to attempt to restrain the agents to the first bag. It is, however, crucial to consider whether breeding leads to a smaller chance of becoming clinically diseased, or to a reduction in infectivity of the subclinically diseased animals (reduction in transmission). A change in the host population that leads to less transmission, but does not bring the reproduction ratio R_0 below 1 (and immediately eradicates the agent), leads to selection for more virulent agents. Those changes in the genetic make-up of the host population that do not influence transmission – but only the chances of clinical disease – would not lead to more virulence. The exact quantitative relations are therefore crucial and should be determined in transmission experiments.

30.4   Toward Virulence Management

In this section, we consider the implications of two models for the evolution of virulence in terms of the analysis of virulence management.

One-bag model

The first of the two models that we consider, a simple one-bag model, predicts a favorable evolution of virulence when the transmission of agents is reduced in some way. It is likely, however, that this model is not applicable to organisms with actively interfering immune systems. Exceptions for which the one-bag model appears to apply to higher host organisms occur in cases where the immune system is not functional, as in our example of cattle that are immune-tolerant to BVDV. Other examples in which the simple one-bag model holds include scrapie in sheep; one of the transmissible spongiform encephalopathies (TSEs) such as bovine spongiform encephalopathy (BSE, or mad-cow disease); and Creutzfeld–Jacob disease in humans. For the TSEs, it is assumed that there is no immune response, as the infective agent is just a host protein in another conformational state (Prusiner 1982, 1995). One possible strategy to eradicate scrapie would be selection to favor resistant genotypes of sheep (Smits et al. 1997). The one-bag model predicts that selection would, in any case, be successful in lowering the number of clinically diseased sheep. If transmission is not affected, more resistant genotypes in the population will result in less disease. If transmission is affected, selection is toward less virulence, as described above. Unfortunately, the current plans are to eradicate the scrapie agent, and this is only possible if transmission is lowered sufficiently to bring the reproduction ratio to below 1. It is therefore important to monitor whether this goal is met. Methods developed by Becker (1989) and extensions thereof (especially those suited for the analysis of small farm population dynamics; Moerman *et al.* 1994; de Jong *et al.* 1996) make it possible to quantify the transmission separately from clinical disease.

From the two-bag model – in which evolution of the host's immune system was incorporated – it was concluded that the effects on virulence evolution caused by changing transmission rates are opposite to those of the one-bag model. In the evolutionarily steady state of the two-bag model, there was medium virulence with some clinical disease, and most of the replication occurred in nondiseased hosts. Therefore, for veterinary application, it should at least be considered possible that models other than Ewald's (1994a) model may apply, and that selection for less-susceptible hosts (in the sense of reduced transmission) could also lead to increased virulence.

The previous considerations lead us to question how virulence can be managed and how the described risks of virulence evolution in veterinary practice can be dealt with.

Two-bag model: highly virulent diseases

For extremely virulent and contagious diseases, such as classic swine fever in pigs and foot-and-mouth disease in cattle, we foresee realistic possibilities for disease control and virulence management. Since government regulations require any observed infection to be stamped out, virulence management should seek to reduce the risk of a major epidemic and keep the size of the minor outbreak limited. The host animal populations in countries that are free of these agents for an extended period are not adapted to the agent, and the agent is manifested in the host population normally as a virulent agent. In regions where foot-and-mouth disease virus (FMDV) and classic swine fever virus (CSFV) are endemic, these viruses seem less virulent.

Consider now that vaccination can largely reduce the transmission of agents such as CSFV. This implies that major outbreaks among pigs in a vaccinated herd are effectively prevented, but minor outbreaks are possible. Infected herds will be detected, but minor outbreaks are very difficult to detect because many animals need to be tested. A separate issue is the evolution toward avirulence during an epidemic, which can occur with and without vaccination. Nonvirulent strains cannot be detected easily by farmers and veterinarians, as the primary basis for detection is the observation of diseased animals.

To meet all these demands, we suggest that herds be stocked as shown in Figure 30.3 (a majority of vaccinated animals are kept together with a minority of unvaccinated animals). This strategy results in a population with the following characteristics:

- An infected herd is easily recognizable because susceptible (unvaccinated) animals are more likely to catch the infection and become ill; for CSFV, this requires eradication of the animals on the farm.
- When animals are mixed in such proportions that the reproduction ratio for the disease is below 1, the infection does not spread within the herd at large, and

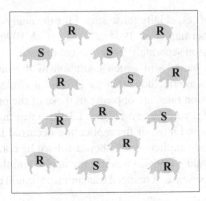

Figure 30.3 A farm on which most individuals have become resistant through vaccination (R), but some are susceptible (S).

there is low risk that the infection will spread to other herds in the time between infection and eradication.

- Since replication of the agent is mainly through susceptible animals, there is little, if any, evolutionary pressure on agent virulence.

The appropriate levels of immunity and the appropriate proportions for the mix of vaccinated and unvaccinated animals can be determined with transmission experiments (de Jong and Kimman 1994). In general, at the herd level, the reproduction ratio of the disease should be below 1. For pseudorabies virus, the necessary levels of immunity (in terms of dose and repetition of vaccination) were determined through transmission experiments (Bouma *et al.* 1997a, 1997b). In those experiments, a standard SIR model was used for the analysis. For more general studies, the SIR model could also be extended to a model with four classes: susceptible, (subclinically) infected, clinical, and removed (in this model, subclinical and clinical individuals are assumed to have a different level of infectivity).

Two-bag model: less-virulent endemic diseases

For less-virulent agents, virulence management appears to be more difficult. Agents of medium virulence are in the evolutionarily steady state, and are often endemic. According to the two-bag model, increasing host immunity to reduce the occurrence of such endemic infections could eventually result in increased virulence of the agent. Still, for many endemic agents, especially those harmful to humans, the aim is to reduce the infection levels. Long-term suppression of an endemic infection, without risk of virulence adaptation, could be achieved by the use of two or more types of "resistant" individuals. Instead of stocking herds with a mix of both types as in Figure 30.3, here it would be more sensible to uniformly stock each herd with one type (Figure 30.4). When this method is combined with the continuous turnover of the resistant types in each herd, an adapted, more virulent strain dies out as soon as the herd is completely stocked with the other resistant type.

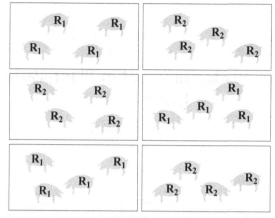

Figure 30.4 Stocking of herds with two types of resistant animals, R_1 and R_2; when the resistant types are continuously turned over in each herd, it should be possible to suppress endemic infections over the long term without increasing agent virulence.

30.5 Discussion

We argue that for the application of virulence management in veterinary epidemiology it is crucial to recognize the effects of the active immune response of animals toward most agents. This creates a conflict between the evolution of the agent's virulence and that of the host's immunity, which results in an evolution of virulence that is contrary to predictions from Ewald's (1994a) model:

- For the application of virulence management, a distinction should be made between exotic virulent agents that are only occasionally introduced and endemic agents of medium virulence;
- For virulent agents, vaccination that reduces transmission could be an effective solution;
- The evolution of nonvirulent strains that cannot be detected quickly enough, which makes eradication difficult, will not occur;
- To reduce the transmission of less-virulent agents without increasing the risk of greater agent virulence, it is necessary to use at least two resistant types of animals. The suggested mixtures of different hosts apply to acquired immunity (by vaccination) as well as to genetic immunity (genetic resistance).

The above considerations should provide an adequate conceptual foundation to deal with selection for genetic resistance while preventing evolution toward more virulent agents.

Acknowledgments The authors appreciate and acknowledge the important constructive comments of Michiel van Boven.

31

Virulence Management in Plant–Pathogen Interactions

Andrew M. Jarosz

31.1 Introduction

Theoretical models have been much improved in their ability to identify factors that are likely to affect the evolution of virulence. However, empirical tests of model predictions are surprisingly few, despite the fact that some factors can be measured and sometimes manipulated under field conditions (e.g., superinfection rates and the relative importance of horizontal versus vertical transmission). Many plant–pathogen interactions appear at first glance to be excellent systems for empirical testing of theoretical models, but a closer examination often reveals features that complicate interpretation. One striking example is the seasonal nature of epidemics of many plant–pathogen interactions (Figure 31.1). Epidemics often display distinct spikes during the plant's growing season followed by a sharp crash in pathogen population size. The crash results from the restricted climatic conditions under which most fungal and bacterial pathogens can infect plants (Agrios 1988). As a consequence, epidemics are not simply limited in occurrence by the availability and density of susceptible hosts, but also by climatic conditions conducive to infection. Indeed, some disease-management strategies are designed to take advantage of these climatic constraints by requiring that crops be planted either in areas not conducive to disease development (spatial escape from disease) or during seasons when epidemics are unlikely (temporal escape). This seasonal fluctuation suggests that plant and pathogen are rarely, if ever, at numerical equilibrium. It is also unlikely that virulence is under intense selection during the decline or crash phase of an epidemic (Figure 31.1). Further, the numerical dynamics of plant–pathogen interactions suggest that the strength and perhaps direction of selection change throughout the year, and that the strength of other evolutionary factors also varies seasonally. For example, genetic drift is important during the crash phase because of reduced population size. In this chapter, I discuss how numerical dynamics of epidemics affect the population genetics of plant–pathogen interactions, which in turn affect the evolution of virulence. I also attempt to provide some insight on how population genetics may be used to direct efforts to manage pathogen virulence.

Another complication of plant–pathogen systems is that plants are constructed in a modular fashion and have many repeated structural elements. For diseases that cause localized lesions, this means a plant is composed of numerous infection

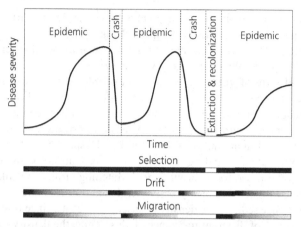

Figure 31.1 The relationship between pathogen population size, estimated as disease severity, and the evolutionary forces of selection, genetic drift, and migration. The pathogen population becomes extinct locally during the second crash phase and is recolonized by migration from another site. *Source*: Burdon (1992).

sites, be they roots, leaves, or flowers. When the number of infections per plant is small, pathogens may not be competing within the plant. Only after the number of infections increases to the point at which the whole plant is affected is there competition among infection sites. Therefore, the effect of multiple infections per plant may not be the straightforward process envisioned in some models of the evolution of virulence (see Chapter 9). Further, plants often have the ability to shed infections via senescence of infected modules. This can happen during the growing season or, more commonly, at the end of the growing season when modules are shed as part of the dormancy process. Plants that recover from infection in this manner are not immune to reinfection, as is often assumed to be the case for vertebrates that recover from microparasitic diseases. Plants are often fully susceptible to reinfection, unless the initial disease episode systemically induces resistance [see Hammerschmidt and Dann (1997) for a review of systemically induced resistance in plants]. Thus, the density of susceptible individuals changes in a manner that differs from models in which recovered individuals are immune from infection (see Chapter 9).

31.2 Two-faced Virulence

Virulence is addressed in two ways within plant–pathogen systems. The first, matching virulence, has received a great deal of attention in agricultural systems and, to a lesser extent, within natural systems. Matching virulence is part of the classic gene-for-gene interaction discovered by Flor (1956), in which specific pathogen gene products trigger resistance reactions in the plant (see also Chapter 17). Virulence alleles code for the absence or altered forms of the product that allow the pathogen to avoid triggering plant defenses. Matching virulence is under negative frequency-dependent selection, which causes host populations to display high variability. The theoretical dynamics have been modeled extensively (e.g., Leonard 1994; Frank 1997; Kirby and Burdon 1997).

The second form of virulence is the effect of infection on host fitness, and it essentially corresponds to the damage inflicted on the host following infection. For convenience, I refer to it as aggressive virulence. It has been much less studied in plants, which contrasts with the emphasis it has received in general models (see Chapter 9) and in animals (see Chapter 30). Despite the rich theoretical tradition for both matching and aggressive virulence, I do not know of any work that combines these two forms of virulence. Matching virulence is likely to affect the evolutionary trajectories of aggressive virulence. For example, the presence of matching virulence reduces the density of susceptible hosts, since host populations are often mixtures of different resistance alleles. Thus, the density of susceptible hosts changes as new matching virulence alleles enter a pathogen population via mutation or migration.

31.3 Epidemiology, Genetics, and Evolution of Virulence

In the remainder of this chapter, I assume that the pathogen causes a local lesion disease. With this type of disease, the pathogen infects a plant part, causing a localized lesion or canker on leaf, stem, meristem, fruit, flower, or root tissue. A single infection event does not have the potential to systemically ramify through the whole plant. The increase in the amount of diseased tissue on an individual plant results from the accumulation of numerous infection events. Many fungal and bacterial species of pathogens conform to this general pattern (Agrios 1988); and because of their explosive rate of increase when the plant and environment are conducive to infection, these pathogens are particularly important for agricultural applications. The seasonal fluctuations in pathogen population size influence the strength of selection, migration, genetic drift, and recombination that operate within the pathogen population. Figure 31.1 depicts 3 years of a typical epidemic cycle. The pathogen displays logistic increases in population size during the early portions of the growing season. As autumn approaches, plants begin to senesce and environmental conditions become less favorable for new infections, causing pathogen population size to decrease. Further population decreases occur during the winter through mortality of the dormant pathogen. At the beginning of the next growing season, pathogen population sizes are often thousands of times lower than they were at the end of the previous growing season.

An example of how the ecology of a pathogen can affect evolutionary dynamics was demonstrated in a model by Barrett (1987) is used to analyze the evolution of matching virulence for a local lesion pathogen in crop mixtures. In this model, the evolution of pathogen genotypes that could infect all resistance genotypes in the mixture (i.e., commonly called a super-race) was influenced by ecological parameters. The proportion of spores reinfecting the same plant (termed autoinfection) was important: a low autoinfection rate increased the probability that a super-race would increase in frequency. A second determinant was the length of the growing season. Longer growing seasons favored super-races.

Seasonal changes in selective regimes

Since the seasonal fluctuation of an epidemic is caused by abiotic factors, the evolutionary interaction between plant and pathogen can be maintained in a state of nonequilibrium. Early in the season when pathogen numbers are low, pathogen genotypes with matching virulence and high fecundity are favored. The amount of plant material is not a limiting factor at this stage: selection favors genotypes that can readily overcome the resistance found in the plant population and spread very efficiently through the population. If fecundity and aggressive virulence are positively correlated, then pathogen genotypes with high aggressive virulence are also favored.

Later in the epidemic, when the amount of plant material becomes a limiting factor, competitive ability among infection sites is important. Since active infection requires matching virulence, this form of virulence is still favored, but the selective regime for aggressive virulence is not clear. Theory predicts that multiple infections per host should favor highly aggressive pathogen genotypes (Chapter 9). This prediction should be true generally for plant pathogens, but the dynamics may be modified by the form of competition between pathogen genotypes and by the effect of infection on plant senescence. Genotypes that are efficient at extracting resources from the host are favored during this phase of the epidemic. While this seems to indicate strong selection favoring increases in aggressive virulence, the plant's ability to lose infected parts through differential senescence may actually favor reductions in aggressive virulence [plant pathogens are known to alter senescence patterns in their host; e.g., senescence is delayed in soybeans infected with soybean dwarf virus (Hewings 1989)]. If a pathogen can delay senescence of an infected plant part, then it could gain some fitness advantage by maintaining an existing infection when the probability of infecting new plant material is low. If conditions favor the option to maintain an existing infection site, then reduced aggressive virulence could be favored during this phase of the epidemic.

During the crash phase, the pathogen population dynamics are influenced by the pathogen's ability to either grow as a saprophyte or survive as dormant mycelium or spores. The crash phase has largely been ignored in numerical epidemiology (Campbell and Madden 1990), and its significance on population structure has rarely been appreciated (Burdon 1992). However, the evolutionary dynamics of pathogen populations are influenced because the crash alters the selective regime and can enhance the importance of genetic drift and migration through reduced population size. Since the pathogen does not act as a parasite during this phase, the selective regime should be altered considerably. Traits that improve either saprophytic growth or dormancy are favored over traits that improve parasitic ability. Thus, matching and aggressive virulence genes are not under direct selection.

Both forms of virulence may be neutral, or may actually lower fitness if these virulence loci adversely affect survival during the crash. A negative effect on survival is entirely possible, especially for matching virulence alleles. The virulent phenotype is conditioned by the alteration or loss of the gene product that triggers a plant's defense apparatus. Since the normal gene products are known to have functionality unrelated to the gene-for-gene interaction (Dangl 1994; Laugé and de Wit 1998), it seems reasonable that the virulence allele may have a fitness cost that affects its survival during the crash phase.

Metapopulation structure

Even if virulence is a neutral trait, it is still subject to the effects of genetic drift when the population size is reduced by a factor of several thousands during the crash. Thus, significant year-to-year variation in virulence structure may occur, unrelated to selection that acts during the growth phase of an epidemic. Another consequence of the crash is that migration events that occur early in the next growing season have a significant effect on the resident population structure. The combined effects of genetic drift and early season migrations have the potential to radically alter the virulence structure of a pathogen population.

If the crash phase results in local extinction of the pathogen, then population level coevolution between the plant and pathogen is negated. Subsequent pathogen populations must be reestablished via migration. These migrants may have been under very different adaptive regimes, and their population structure may be entirely different from that found locally before the extinction event.

The role of local extinctions can be extremely important in predicting the numerical and genetic dynamics of plant–pathogen systems, because the scale of the interaction is increased. Thus, the interaction between plant and pathogen is increasingly influenced by metapopulation dynamics (Burdon 1992; Thrall and Antonovics 1995, Thrall and Burdon 1997, 1999). For example, metapopulation dynamics are the most likely explanation for the pattern of matching virulence found in *Melampsora lini*, a rust fungus that infects *Linum marginale* in Australia (Burdon and Jarosz 1991, 1992; Jarosz and Burdon 1991). Rust populations at ten sites covering a distance of 95 km were sampled for up to 4 years. All *M. lini* populations across all years were dominated by some combination of four matching virulence genotypes. Of all the samples, 82% were characterized as belonging to one of these four common races. If selection operated at the population level, we would expect a predominance of matching virulence genotypes that could overcome the host resistance genotypes found at a site. With the possible exception of one very large population site on the Kiandra plain, pathogen population structure did not appear to be shaped by the *L. marginale* resistance structure at a site. Further, *M. lini* populations at all sites displayed considerable year-to-year variation in matching virulence, which again was not in response to any rapid turnover in the local plant population.

The annual crash phase of the epidemic was severe even at large sites (Jarosz and Burdon 1992), which led to frequent local extinctions of *M. lini* populations

(Burdon and Jarosz 1992). These numerical and genetic patterns are consistent with the idea that *M. lini* populations are shaped by metapopulation dynamics that involve frequent local extinction events and high migration rates among sites. The four common matching virulence genotypes appear to be the best adapted across the *M. lini* metapopulation in the Snowy Mountains. The high spatial and temporal variation is caused by the crash phase, which results in either high genetic drift or local extinction–recolonization events.

If local extinctions are fairly common, then the evolutionary trajectory of a pathogen population is shaped by an interaction between local and metapopulation effects. This may be particularly important for many plant–pathogen interactions in which metapopulation dynamics may be expressed on a time scale similar to that of local dynamics. As a consequence, many metapopulation models may not be applicable for describing plant–pathogen interactions, because they treat local dynamics as instantaneous relative to metapopulation dynamics (Levins 1969, 1970; Hanski 1997). When local and metapopulation dynamics interact on the same time scale, they can produce evolutionary trajectories that would not be predicted by either single population or metapopulation models (Thrall and Antonovics 1995; Thrall and Burdon 1999).

Sexual versus asexual reproduction

In evolutionary dynamics, a pathogen is also influenced by the mode of reproduction. Pathogens vary greatly in their dependence on sexual versus asexual reproduction. We expect that sexually reproducing pathogens will display more diversity than asexual forms. Testing this prediction is difficult, but some fungal plant pathogens offer an opportunity to make within-species comparisons. For example, *Puccinia graminis tritici* is an obligate pathogen of wheat that causes a local lesion leaf disease (commonly called stem rust). The pathogen has an alternating life cycle in which the asexual phase occurs with wheat, and the sexual phase occurs with barberry plants (*Berberis* spp.). Early in the 20th century, attempts were made to disrupt the stem rust life cycle by eradicating barberry plants from the plains in the USA (Roelfs 1982). While the eradication program was largely successful, the disease still persists in the Plains States as an asexually reproducing pathogen that migates yearly from Mexico (see Roelfs 1986 for a review of this disease). This asexual population was sampled in the 1970s for electrophoretic variability, and it was found to be consistently less diverse than a sexual population of *P. graminis* found to the west of the Rocky Mountains in the USA (Burdon and Roelfs 1985).

The absence of meiotic recombination also alters the target of selection. This raises the possibility of the formation of coadapted gene complexes that is not possible in sexual species because of the disruptive action of recombination. I am not aware of any study that implicates coadapted gene complexes in the evolution of either matching or aggressive virulence, but other work suggests that it is possible. An elegant study by Bouma and Lenski (1988) found that the genome of *Escherichia coli* could respond to the presence of a plasmid that confers resistance to tetracycline. Initially, the plasmid exacted a fitness cost, since the plasmid-free strain of *E. coli* was competitively superior in environments that did not contain tetracycline. However, this fitness cost proved to be transient. The plasmid-containing strain was allowed to evolve in the presence of tetracycline for 500 generations, and was put in competition against the original plasmid-free strain. The evolved strain was competitively superior in environments with and without tetracycline. The fitness increase in the evolved strain resulted from the accumulation of favorable mutations within the *E. coli* genome that compensated for the presence of the plasmid.

This result may have implications for the evolution of matching virulence, since virulence alleles may have associated fitness costs when they are not needed to avoid plant defenses. If true, the frequency of a virulence allele should decline when the matching resistance allele is absent or at a low frequency within a plant population. The Bouma and Lenski data suggest that fitness costs may be transient, and virulence-allele frequency does not decline if beneficial compensating mutations become fixed in a pathogen population. From an agricultural standpoint, this means that ineffective matching resistance alleles must be withdrawn from a crop before the beneficial mutations accumulate. If this is done correctly, then the matching virulence allele is eventually purged from the pathogen population, and the resistance allele may be recycled into the crop at some later date.

Genetic analyses of population structure

Interest in the genetic population structure of plant pathogens increased in the 1990s. Many studies investigated genetic structure from the standpoint of a pathogen's mode of reproduction (see Milgroom 1995 for a review), and the results are surprising in some cases, especially for pathogens that are not known to undergo meiosis. For example, many bacterial pathogens of both plants and animals have population structures that deviate significantly from the clonal lineage stereotype of asexual species (see Maynard Smith *et al.* 1993 for a review). The plant pathogen *Pseudomonas syringae* has distinct lineages, but the pattern of variability within a lineage suggests that considerable recombination is occurring. Other pathogenic bacteria (e.g., *Neisseria gonorrhoeae*) have population structures that approximate panmictic population. Thus, the rate of recombination in these species may be considerably greater than that seen in *E. coli*. Plant viruses also have the ability to genetically recombine through homologous and heterologous template-switching by polymerase enzymes (Simon and Bujarski 1994). I am not aware of any study that examines population structure for plant pathogenic viruses. However, the ability to genetically recombine and the high mutation rate found in these life forms based on ribonucleic acid (RNA) present the possibility that genetic population structure may approach that of panmixia.

Many fungal plant pathogens produce both sexual and asexual spores, and it is often unclear which spore type is involved in the development of an epidemic. Studies on the genetic structure of pathogen populations have provided useful information with regard to this question. Populations of *Stagonospora nodorum* (syn. *Septoria nodorum*) – the cause of glume blotch of wheat – were found to be at or near linkage equilibrium for seven restriction fragment length polymorphism (RFLP) loci (McDonald *et al.* 1994). Prior to this study, the founding inoculum was assumed to come from asexual conidia, but these data suggest an important role for sexually produced ascospores. Another pathogen, *Rhynchosporium secalis*, a leaf blotch pathogen of cereals, has a population structure that approximates the equilibrium predicted by random mating (McDonald *et al.* 1989). However, a sexual cycle is not known for *R. secalis*. These data suggest that *R. secalis* can either undergo parasexual recombination or has a yet to be discovered sexual cycle.

A number of excellent studies by Michael Milgroom and his laboratory characterized *Cryphonectria parasitica* populations – the cause of chestnut blight in the USA. Considerable between-population variation was found across the eastern USA (Milgroom and Lipari 1995), but genetic distances between populations were not correlated with geographical distance. This may be a consequence of the chestnut blight's recent invasion onto the continent and its subsequent spread. Within-population studies suggest that sexual and asexual spore types disperse differently. Genetic analysis among trees within a population from western Virginia indicated a randomly mating population, while clones were often found within a tree (Milgroom *et al.* 1992). This finding suggests that the sexual ascospores are largely responsible for spread between trees, while asexual conidia are important for spread within a tree.

In contrast to the population structure of *C. parasitica* in the eastern USA, two populations in Michigan were largely clonal within and among trees (Liu *et al.* 1996). The clonal structure can be explained by the spread of double-stranded RNA (dsRNA) hyperparasites that infect *C. parasitica* (Fulbright *et al.* 1983; see Chapter 21 for review of dsRNA hyperparasitism). Double-stranded RNA infections inhibit or severely reduce sexual reproduction in *C. parasitica* (Anagnostakis 1988). Since it is dispersed within the asexual conidia but not in the sexual ascospore (Jaynes and Elliston 1980), this inhibition is adaptive from the dsRNA's standpoint.

31.4 Population Structure and Virulence Management

Although an evolutionary perspective of virulence management is still in its infancy, population genetics has great potential as a disease management tool. However, the full pay-off will not be attained until we begin to bridge the gap between theoretical and empirical work. Mathematical models must begin to generate predictions that can be rigorously tested with empirical data. Once this integration occurs, it will be possible to use empirical data to predict the dynamics of a system *and* determine how the system can be changed to manage virulence. Currently, only a handful of studies have led to concrete management strategies. While the findings are not extremely robust, they begin to demonstrate the utility of an evolutionary plus population genetics perspective for disease management.

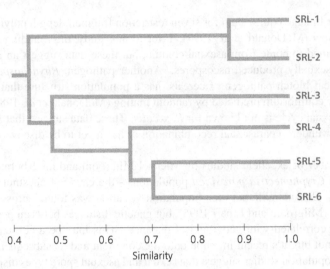

Figure 31.2 Phenogram of the six lineages of *Magnaporthe grisea* at Santa Rosa, Colombia. Relationships based on data for MGR-DNA haplotypes. *Source*: Levy *et al.* (1993).

Three examples are used to illustrate how population genetic data and an evolutionary perspective have been utilized to assist in disease management, and how this work can lead to practical control recommendations. Keller *et al.* (1997) utilized a hierarchical sampling strategy to investigate the population structure of *Phaeosphaeria nodorum*, an ascomycete fungus that causes glume blotch on wheat. Across Switzerland, the population of *Ph. nodorum* was found to be nearly panmictic, since it was not differentiated across three spatial scales (regional, field, and within-field) and there was no evidence that particular pathogen genotypes were specialized on specific crop varieties. Thus, for this pathogen, disease management strategies must be targeted at a regional scale.

In the second example, a coevolutionary perspective was utilized in the management of two foliar diseases of the common bean *Phaseolus vulgaris*. Common beans are thought to have two centers of origin, Mexico and the Andes Mountains (Gepts and Bliss 1985; Singh *et al.* 1991), and commercial varieties can be traced to one or the other center. *Phaeoisariopsis griseola*, which causes an angular leaf spot, is differentiated into two subspecies, one that specializes on varieties with a Mexican ancestry, and the other on the Andean germplasm (Guzmán *et al.* 1995). The bean rust pathogen *Uromyces appendiculatus* displays a pattern of cross infertility that suggests it is also differentiated into Andean and Mexican subspecies (Martinez *et al.* 1994, 1996; Maclean *et al.* 1995).

Guzmán *et al.* (1995) propose a resistance breeding strategy that takes advantage of this pattern of specialization and intermixes resistance genes between the two centers of origin. The goal is to overcome matching virulence in the specialized pathogens by incorporating alien resistance genes to which they are not adapted. It is important, though, that each plant–pathogen pair be evaluated in its

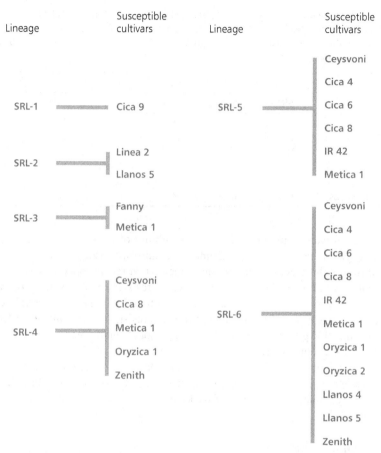

Figure 31.3 Susceptibility of rice cultivars to the six lineages of *Magnaporthe grisea* at Santa Rosa, Colombia. *Sources*: Hamer *et al.* (1993), Levy *et al.* (1993), Zeigler *et al.* (1994).

own right, as this specialization pattern is not universal among all common bean pathogens. For example, the pathogen *Colletotrichum lindemuthianum*, which causes an anthracnose disease on the common bean, was not subdivided into Andean and Mexican subspecies (Balardin *et al.* 1997).

In the third example, the population structure of the rice blast pathogen *Magnaporthe grisea* was used to design a novel breeding strategy (Zeigler *et al.* 1994). *Ma. grisea* isolates from a breeding nursery in Santa Rosa, Colombia, fell into one of six distinct lineages (Figure 31.2), with each lineage being variable with regard to the rice cultivars that could be infected (Figure 31.3). While each rice cultivar was known to carry some resistance, all of them were susceptible to at least one *Ma. grisea* lineage. Consequently, all cultivars appeared to be susceptible when grown agronomically in Colombia. However, no *Ma. grisea* lineage was able to

Table 31.1 Disease reactions of the five parental lines used in a breeding scheme to produce Oryzica Llanos 5. Reactions are to the six *Magnaporthe grisea* lineages found at Santa Rosa, Colombia (R = resistant and S = susceptible). *Sources*: Hamer *et al.* (1993) and Correa-Victoria *et al.* (1994).

Cultivar	*Magnaporthe grisea* lineage					
	SRL-1	SRL-2	SRL-3	SRL-4	SRL-5	SRL-6
Colombia 1	R	R	R	S	R	R
5685	R	R	R	R	S	S
Cica 7	R	R	S	S	R	S
IR 36	R	R	R	R	S	S
Cica 9	S	S	R	R	R	R
Oryzica Llanos 5	R	R	R	R	R	R

infect all cultivars. This fact raised the possibility of using susceptible cultivars in a lineage-exclusion breeding scheme. The cultivar Oryzica Llanos 5 was generated from a three-generation crossing scheme that incorporated resistance from five cultivars (Table 31.1). This new cultivar is resistant to all six *Ma. grisea* lineages, while each parent cultivar is susceptible to at least one *Ma. grisea* lineage. It has been grown commercially since 1989, and resistance has remained effective through the 1998 growing season (M. Levy, personal communication). The significance of this exclusion breeding scheme is that it allows old resistance genes to be recombined in a manner that extends their usefulness. Another practical advantage is that resistance genes are already in an agronomically useful genetic background. This greatly reduces the number of backcrosses needed to produce a new, useful cultivar.

31.5 Discussion

These examples illustrate how an evolutionary perspective can aid in the control of disease. The current work represents only a small fraction of the potential contribution to disease and virulence management. However, the full potential will not be realized until empirical and theoretical studies are synthesized. Those who carry out theoretical work are in an excellent position to produce models that can be evaluated directly by field researchers. From the perspective of plant–pathogen interactions, these new models need to include seasonal numerical dynamics so their effect on evolutionary dynamics can be evaluated. Specifically, the models should be used to evaluate the effects of nonequilibrium situations brought on by the seasonal epidemics. Models should also be used to evaluate the effect of fluctuating selection regimes, genetic drift, and metapopulation interactions on overall evolutionary dynamics. Some researchers have begun to use models to evaluate metapopulation dynamics, and have found that they can alter expected outcomes of interactions (Thrall and Antonovics 1995).

Unfortunately, the importance of an evolutionary perspective is not fully appreciated within agriculture. A goal of this chapter is to highlight how the ecology, or numerical dynamics, of a pathogen affects population genetic structure. The recent drive for sustainable agriculture has stressed an ecological perspective to improve disease management, but the role of population genetics is still not appreciated (Anonymous 1996). This chapter illustrates that the ecology and evolution of plant–pathogen systems are inextricably intertwined, and that disease management strategies will need to integrate population genetics if there is to be any hope of attaining long-term sustainability with these strategies. Indeed, ecological disease management strategies, such as crop mixtures, can also influence pathogen virulence in a manner that could enhance sustainability (Barrett 1980).

32

Virulence Management in Biocontrol Agents

Sam L. Elliot, Maurice W. Sabelis, and Frederick R. Adler

32.1 Introduction

Although biological control is founded upon the virulence of natural enemies to the targeted pests, there has been little effort to understand how this might change, let alone to manage it. Frank Fenner and colleagues can be credited with being the first (and last!) to monitor changes in virulence of a biological control agent, namely the myxoma virus used to control rabbits in Australia (Fenner and Fantini 1999). This is despite a body of literature showing that the virulence of natural enemies can and does change in response to selective forces, either natural or artificial. These studies cover a wide taxonomic range of organisms, including fungal pathogens of plants (Burdon and Thrall 1999; Brasier *et al.* 1999; see also Chapter 31), microsporidian parasites of daphnids (Ebert 1994), pathogens and parasites of humans (Ewald 1994; Chapters 2 and 28), malarial parasites of rodents (Chapter 12), pathogens and parasites of social *Hymenoptera* (Schmid-Hempel 1998; Boot *et al.* 1999; Oldroyd 1999), nematode parasites of fig wasps and fruit flies (Herre 1993, 1995; Jaenike 1996, 1998), and hymenopteran parasitoids of aphids (Henter 1995; Henter and Via 1995). The results of this work have suggested that the course of virulence change can be predicted and possibly even manipulated. In biocontrol, a predator, parasitoid, or pathogen is used to control a pest population, a pest being defined as an animal, plant, or microorganism that is perceived to be damaging to some human activity. In this setting, the practical aim of virulence management would usually be to increase virulence. This contrasts with the management of virulence of pathogens or parasites of humans, domestic animals, or crop plants, in which low virulence is the aim.

So how does one attempt to manage the virulence of a biocontrol agent? Theory predicts that increased virulence is selected for when a natural enemy is limited in its control over the resources contained within a victim. This occurs through competition with other natural enemies (including other strains of the same species) or through external sources of victim mortality (Van Baalen and Sabelis 1995a, 1995b; see Chapters 7, 9, 11, and 22). Both of these factors are potentially manipulable and in this chapter we present a framework in which this might be attempted.

The field of biological control employs a range of natural enemies to control pest organisms. Workers using pathogens as biocontrol agents have traditionally recognized virulence (commonly termed "aggressiveness" in plant pathology, see, e.g., Jarosz and Davelos 1995) as an important attribute to assess, but we hope to demonstrate that virulence is also a trait possessed by arthropod natural enemies (herbivores of weeds, predators and parasitoids of arthropod pests, and fungivorous arthropods). Various biocontrol strategies can be considered, for the sake of the arguments developed in this chapter, to differ principally in the degree to which the natural enemy exploits the pest resources and converts them into new natural enemies. We use the term *inundative biocontrol* for the release of a natural enemy as a biological pesticide with a rapid and short-term effect with no reliance on subsequent generations of the biocontrol agent, *inoculative biocontrol* for the release of an agent that does not exert immediate control but acts over more than one generation, and *classical biocontrol* for the introduction of an exotic natural enemy to a new area to achieve control over many generations. These strategies form a continuum, with a further strategy being the manipulation of the environment to foster naturally occurring enemies of the pest (*conservation biocontrol*).

We begin by proposing that the virulence of a natural enemy is not necessarily assessed adequately by a consideration of the effect on individual victims, but requires consideration of how victims are distributed and exploited in space. From this standpoint, we question whether high virulence to an individual victim really is a desirable trait in a natural enemy. We then ask how stable the virulence of a biocontrol agent is likely to be in the field, how it might be manipulated in the field, and how it is affected by the mode of mass rearing (the production of large numbers of the natural enemy prior to release). We expand our treatment to include other components of pest management systems, asking why insect pathogens appear to attack herbivores more than the predators of those herbivores when they could, in principle, infect both. Throughout the chapter, our focus is on virulence from the point of view of the natural enemy under the assumption that the victim cannot adapt. The ecological and evolutionary response of the pest to its enemy's virulence is briefly discussed in the penultimate section.

32.2   At What Level Should Virulence be Considered?

The archetypal parasite is an organism that lives in association with another individual organism (the host or victim) which it exploits, to the latter's detriment. The negative effect on the host (ideally measured as a loss in the host's fitness) is the parasite's *virulence* and depends to a large degree upon the exploitation strategy that the parasite adopts. For an organism such as a pathogen, its virulence is clearly definable as the effect it has on an individual host, and this we term "individual-level virulence." Arthropod herbivores are often considered as parasites of plants (e.g., Crawley and Pacala 1991; Begon *et al.* 1996) and can therefore be ascribed an individual-level virulence. We argue that this also applies to predators. In the simplest terms, a given predator–prey interaction can be lethal or nonlethal, a binary form of virulence. Intermediate shades can occur, because attacks may be nonlethal but still damaging and because the mere detectable presence of a predator can have fitness consequences for its prey, for example through antipredator behavior (Heads 1986; Dixon and Baker 1988; Lima and Dill 1990; Stamp and Bowers 1993; Pallini *et al.* 1998).

It is at this individual level that the notion of virulence is normally applied, but we argue that selection on virulence acts at different spatial scales (Boerlijst *et al.* 1993; Miralles *et al.* 1997; Van Baalen and Sabelis 1995c; Sabelis *et al.* 1999c, 1999d). For example, many insect pests of plants have a tendency to form discrete infestation foci that cover an area ranging from part of a plant up to several neighboring plants. Such a patch of herbivores is, to an organism which invades that patch (and will reside there for more than a short period), just as much of a host as an individual is to the archetypal parasite. In this case we use the term "patch-level virulence." While we expect the properties of a host patch to have stark differences to those of an individual host – individual organisms may emigrate whereas cells cannot – there are striking similarities (Levin *et al.* 1999). For either an individual victim or a patch of victims, parallels can be found in:

- The quantity and quality of invading propagules;
- The natural enemies' ability to reproduce within and to exploit the victim(s);
- The probability of encountering other natural enemy genotypes;
- The role of host defenses;
- The amount of natural enemy propagules produced by the end of the interaction with an individual victim or with the victims in a patch.

Such analogies have been used to study the virulence strategy of a predatory mite invading a patch of herbivorous prey mites (Van Baalen and Sabelis 1995c; Pels and Sabelis 1999; Sabelis *et al.* 1999c, 1999d). Again, there is no need to restrict this to a particular class of natural enemy. To a pathogen or parasite invading a colony of eusocial arthropods, the colony is akin to an individual host (Schmid-Hempel 1998, p. 263) and the host population need not be social for this to apply (Oduor *et al.* 1997). The critical point is that the individual-level virulence in such a setting is only one of a range of variables which will contribute to an overall patch-level virulence. For example, the importance of the production of new infective propagules within a local population can be critical in determining the overall, patch-level virulence, as with the fungal pathogen *Metarhizium flavoviride* in orthopteran hosts (Thomas 1999). At this patch level of virulence, there may well be a trade-off between virulence and transmission, just as at the individual level.

At which of these (or other) levels we should consider virulence depends upon the patchiness and viscosity of the populations of the target organism and the biocontrol agent, as well as on the biocontrol strategy being employed. Individual-level virulence is more important in a purely inundative approach, as there is no dependence on reproduction, or in a nonpatchily distributed pest population. In contrast, in inoculative and classical biocontrol or in a more patchy pest population, patch-level virulence comes to the fore. It is important to realize, however, that even with the inundative use of a biocontrol agent, it may be at the patch level that selection on virulence is most influential in the natural populations from which that agent has been taken.

32.3   Is High Virulence Always Desirable?

Rather than a highly exploitative strategy, a natural enemy may adopt a strategy of restraint in exploiting its host or host patch, thereby preserving for longer its resource, exploiting its victims' capacity for growth or reproduction, and so ultimately producing more propagules than it would otherwise. This is incorporated into trade-off functions like black boxes when considering individual-level virulence (but see Nowak *et al.* 1991 for an exception), but it is only at the patch level that a mechanistic explanation for this trade-off has been attempted. The latter case considers local predator–prey dynamics in which high and low virulence strategies are termed "killer" and "milker," respectively. The trade-off revolves around the emigration of predator individuals, which relieves pressure on the victims in the patch, and thereby allows prey reproduction to be "milked" in the low virulence strategy (Van Baalen and Sabelis 1995c; Chapter 22). By producing more propagules per patch, the milker may ultimately better control metapopulations of a pest organism, which is usually the aim of biocontrol. In other words, high virulence may even compromise biocontrol (Te Beest *et al.* 1992; Thomas 1999), especially where a trade-off exists between virulence and transmission (see Bull 1994; Messenger *et al.* 1999). This is likely to become more of a concern as we move across the spectrum from inundative to classical biocontrol, that is, as the colonization of new pest patches becomes more important.

This point can be illustrated by referring to the example of the use of predatory mites (Acari: Phytoseiidae) in the Africa-wide control of the exotic cassava green mite (*Mononychellus tanajoa* Bondar, Acari: Tetranychidae). A survey of neotropical natural enemies resulted in ten candidate species of predatory mite of which three established successfully in Africa upon release (Yaninek *et al.*, unpublished). It appeared that the two predators with the fastest predation and population growth rates hardly spread between cassava fields, whereas *Typhlodromalus aripo* DeLeon, with lower growth and predation rates, readily spread beyond the release fields and still had an impact on the pest population. The latter species stands out not only because of its lower predation rate, but also because it forages actively only at night on the upper leaves, thereby passing up opportunities for predation (A. Onzo, personal communication). During the day it sits in the growing tips of the cassava plants and waits for the few prey that walk up to feed on leaf primordia. As a result of its lower functional and numerical response, this predator does not overexploit its prey as fast as the other two. While this predator does not exterminate a patch of prey as quickly as would a more virulent one, this strategy allows it to reach larger numbers. It then has a stronger foothold from which to spread to other prey patches. Thus, the successful establishment of *T. aripo* may well represent a case in which low virulence promotes biocontrol.

Another example of low virulence being desirable in a (classical) biocontrol agent is found in weed control. Many of the successful biocontrol agents of weeds castrate their hosts, but do little harm otherwise. Castration is usually partial or temporary, which is considered the result of a host escaping from its castrator

(Minchella 1985; Hurd 1998; Hsin and Kenyon 1999). Some castrators are insects, such as pre-dispersal seed-predating weevils (Crawley 1989). Other examples are eriophyoid mites that make galls or cause other growth deformations in buds and flowers of plants, thereby blocking reproduction of the plant (Rosenthal 1996; Sabelis and Bruin 1996). Such parasites may have little immediate effect on the vigor of the host, which has been cited as a potential drawback for their use as biocontrol agents (Cromroy 1979). Such a view, however, seems shortsighted when it is remembered that the main goal is the control of *populations* of weeds, not individual plants *per se*. In their review, Te Beest *et al.* (1992) found that pathogens used in classical biocontrol of weeds are most successful where they reduce host reproductivity or cause low mortality. Simulation modeling by these authors predicted optimal weed control from an introduced pathogen when the pathogen caused 66% mortality or sterility. Increased pathogen virulence led to oscillations and increased weed biomass, but the effects on weed population dynamics are unlikely to be straightforward (Crawley 1989) and one must consider carefully what is meant by virulence. In terms of the host's fitness, the impact of castration is catastrophic and an evolutionary biologist might label this as high virulence, but in terms of host survival the parasite's impact is low and a biocontrol worker may characterize it as low virulence. By castration, a parasite may use resources that would otherwise be directed to reproduction (Baudoin 1975; Obrebski 1975; Kover 2000), thereby minimizing its effect on the longevity of the host and gaining a foothold for dispersal through a population of hosts (García-Guzmán *et al.* 1996).

Further evidence of the potential for castrators to better regulate populations of their hosts comes from a concordance of predictions from modeling with experimental evidence in the case of insect–parasitic nematodes (Jaenike 1998). Castrating parasites of animals may even prolong the host's life span, as demonstrated by artificial castration of a nematode (Hsin and Kenyon 1999) and, similarly, castration of plants by fungal pathogens has been linked to increased viability, vegetative vigor, or preferential survival under grazing (Bradshaw 1959; Clay 1990; Burdon 1991). Ebert and Herre (1996) state that it is a mystery why castration as a parasitic strategy is not more common or even universal. However, this strategy probably evolves only when single infections prevail: it is expected to be highly vulnerable to invasion by a more virulent parasite genotype that uses the resources left available by the castrator. When this happens, the best strategy for the parasite would be to switch to higher virulence as the advantage of the prolonged life span of the host is lost.

32.4   Is High Virulence a Stable Trait in Biocontrol Practice?

The classic case of a change in virulence in a pathogen is the introduction of the myxoma virus to control the European rabbit, *Oryctolagus cuniculus* (L.), in Australia, where subsequent evolution toward a more benign state was reported (see Kerr and Best 1998; Fenner and Fantini 1999). This observation has been criticized somewhat in terms of the bioassays used (Parer *et al.* 1994; Parer 1995) and the possibility that virulence may subsequently have increased (Dwyer *et al.* 1990; Ewald 1994a; Fenner and Fantini 1999), but there is little doubt that virulence has changed. This is the critical feature for our discussion of the evolution of virulence, whatever the direction of the change.

For a biocontrol agent, it is highly unlikely that virulence will remain unaltered during mass rearing (see Section 32.6 below) and after introduction in the field, unless the original sample of biocontrol agents contained little relevant genetic variation and the processes creating such variation are relatively slow. This is because the evolutionarily stable virulence under these conditions is unlikely to match that under the natural conditions in which they were moulded by selection. We therefore expect a shift in virulence toward a new evolutionarily stable state. The remaining questions are then whether the changes will be toward higher or lower virulence, how far along this trajectory the natural enemy will evolve, and what the consequences are for the agent's efficiency. As we have stated, selection may not occur at the individual level, so patch-level virulence may need to be considered. Predicting the direction of change relies on an understanding of the theory of the evolution of virulence. Thus, virulence should increase by selection when there is horizontal transmission (Anderson and May 1991), transmission by vectors (Power 1992; Ewald 1994a), a high background mortality (Anderson and May 1979; May and Anderson 1983b), a high probability of multiple infections (Nowak and May 1994; Van Baalen and Sabelis 1995a, 1995b), or when parasites produce propagules for long-term survival outside the host (Hochberg 1989; Bonhoeffer *et al.* 1996; Gandon 1998). It is important to realize, however, that these factors may interact with one another. For example, low host mortality may allow more time for parasite genotypes to reproduce within the host and so compete, thereby leading to selection for higher virulence than expected if only the effect of mortality under single infections is considered (Ebert *et al.* 1997; Gandon *et al.* 2001).

Three determinants of the degree of change in virulence are the genetic variation (extant or novel) of the natural enemy, the magnitude of the selection pressure, and (of critical practical relevance) which biocontrol strategy is to be used. In classical biocontrol, the relevant time scale covers many generations of the natural enemy, so there is plenty of time for virulence to evolve. In inoculative biocontrol the time scale is shorter, but we can still expect at least some transient changes over generations. For inundative biocontrol, however, there is only one natural enemy generation and the biocontrol agent has no opportunity to evolve. The exception to this occurs when pathogens are used and they reproduce inside host individuals, in which case there will be many generations of the pathogen and there may be substantial changes in the pathogen's virulence (Sokurenko *et al.* 1999). Environmental manipulation (or conservation biocontrol) may incidentally alter the pattern of selection for virulence of naturally occurring biocontrol agents, for example by increasing background mortality of the host. An intriguing but little explored area is the extent to which the virulence expressed by an exploiter is a plastic response to a changing environment. The expression of virulence factors by pathogenic bacteria within their animal or human host may depend on the detection of a sufficiently large population of the same species of bacterium, a phenomenon known as "quorum sensing" (Williams *et al.* 2000). Whether this sort of phenomenon also occurs in biocontrol systems is an open question. However, it is now well established that insect pathogens, especially viruses, can be more virulent when their host is stressed. It is also common for pathogens to remain at such low levels in their host that they are very difficult to detect, but under some circumstances are still lethal (e.g., Marina *et al.* 1999). Is the reason increased susceptibility of a stressed host, or within-host evolution, or could it be that the pathogen alters its strategy of host exploitation because it detects the increased likelihood of the host dying?

32.5 How Can Virulence be Manipulated in the Field?

If we have some understanding of what determines virulence we are a little way down the road to manipulating it. We have so far discussed the spatial scale at which virulence should be assessed, whether individual-level or patch-level virulences are desirable features of a biocontrol agent, and how likely they are to change in the field. The theory on the evolution of virulence provides pointers as to how virulence management may be undertaken in a given system. The first step is to elucidate the principal transmission route(s) of the natural enemy (horizontal versus vertical, the role of vectors). Then whether there is a trade-off between transmission and disease-induced mortality must be determined (otherwise avirulence of the natural enemy is expected to evolve). Having established these basic features, the key point is to consider the degree to which a given genotype is able to keep control over the exploitation of the victim's resources in the face of competition with other genotypes or other sources of victim mortality. Thus virulence can be increased in the field by any measure by which the background mortality or the probability of multiple infections is increased (Nowak and May 1994; Van Baalen and Sabelis 1995a, 1995b), or by any measure that reduces the cost to the natural enemy of overexploiting the victim's resources.

One possible route to manage virulence is via vectors of the biocontrol agent, if such exist. Alternatively, chemical SOS signals of the plant, supplementary food (extrafloral nectar), or refuges (domatia) that attract or maintain natural enemies may be manipulated to increase the chances of multiple infection of the herbivores and so promote the virulence of natural enemies (Sabelis *et al.* 1999a, 1999c, 1999d; Chapter 22). The ability of a plant to foster such "bodyguards" may also provide an explanation for the high virulence of the insect pathogen and saprophyte *Bacillus thuringiensis* to insects. The bacterium is commonly found on the plant phylloplane. If it is the plant which maintains this population, independent of potential insect hosts, then there is no cost to the bacterium of overexploiting the victim, and so no constraint on evolving higher levels of virulence (Elliot *et al.* 2000). In this case, breeding for mutualistic plants will not only directly benefit the insect pathogen but may also indirectly trigger selection for increased virulence of these pathogens.

The integration of different pest control strategies may also allow management of virulence. Thus, selective pesticides or, most interestingly, other natural enemies will provide an alternative source of background mortality of the pest, which causes the biocontrol agent to "compete" with them and generates selection for high virulence. If high virulence is desired, then the clear implication is that an integration of control strategies should maintain high virulence in the biocontrol agents. When strains of biocontrol agents are screened for high virulence to the target pest, factors can be included that will contribute to the maintenance of that virulence, such as having long-lived propagules (Hochberg 1989; Bonhoeffer *et al.* 1996; Gandon 1998). For any biocontrol agent, the potential for its virulence to be a plastic trait should be borne in mind. We have discussed the possibility of pathogens displaying plasticity within their host (Marina *et al.* 1999; Williams *et al.* 2000), but it is also known that organisms exploiting a patch of victims can adjust their strategy according to the availability of victims or the presence of competitors (Janssen *et al.* 1998). It is possible that the response of an exploiter to potential competitors is to reallocate its resources from within-host reproduction to some form of interference competition, such as the production of toxins (Chao *et al.* 2000). In this case, multiple infections would lead to the capacity for virulence of the biocontrol agents to be "wasted", from the point of view of effective biocontrol.

32.6 Does Mass Rearing Affect Field Virulence?

A common concern in the mass rearing of biocontrol agents is the maintenance of their effectiveness in biocontrol (Hopper *et al.* 1993; Thompson 1999; van Lenteren and Nicoli 1999). The selective pressures in a mass rearing can be quite different from those in the field and routine procedures are usually in place to limit the loss of virulence. Pathogens are normally cultured on artificial media, which can lead to a loss in virulence to its original host as the trade-off between virulence and transmission rate is removed or even reversed. This effect can be ameliorated by using a media that more closely resembles the host nutritionally (e.g., Hayden *et al.* 1992). It is known that small populations of parasitoids may contain sufficient genetic variation to allow selection for higher or lower virulence (Henter 1995). In mass rearings, in which parasitoids are commonly reared in factitious hosts (i.e., different hosts from the target hosts, and ones which can themselves be reared more easily or cheaply), the loss of virulence toward the target pest is a practical problem. A routine procedure with pathogens or parasitoids is to pass them through the target host species to restore virulence, but this is done with little scientific understanding of the mechanisms which restore virulence. Such periodical selection is for high virulence at the individual level and one may question the desirability of selection for this single trait. Even for some pathogens used inundatively, survival in the field is usually as important for effective pest control as is virulence. For arthropod natural enemies, behavioral traits are also important and it is unlikely that single traits measured in the laboratory will predict the efficacy of the biocontrol agent in the field (Bigler 1994).

Biocontrol agents may be reared on the target pest, as with the phytoseiid mite *Phytoseiulus persimilis* Athias-Henriot, a predator of the two-spotted spider mite. In this instance, virulence of *P. persimilis* to a patch of prey is determined by its rate of conversion of prey into eggs and its retention in the prey patch (Sabelis and van der Meer 1986). Both of these traits are selected for in commercial mass rearings because there is selection for increased growth rate and dispersing mites are lost. Thus, the high patch-level virulence observed in the field (Pels and Sabelis 1999; Chapter 22) is expected to be conserved in a mass rearing. A converse example can also be found among predatory mites: *Hypoaspis aculeifer* (Canestrini) is used in the biocontrol of the bulb mite *Rhizoglyphus robini* Claparède, a pest of lily and freesia corms, but is reared on a nontarget mite, *Tyrophagus putrescentiae* (Schrank). In the rearing, a genetic polymorphism in preference for these two prey and associated reproductive success is maintained by hybrid advantage (Lesna and Sabelis 1999). This means that those genotypes with a preference for the target pest are maintained, but are diluted in a mixture of other genotypes that perform less well on the target pest. Thus, the mass-reared predators have a lower virulence with respect to the target pest than does the specialist genotype.

32.7 Pathogen Virulence Toward Herbivores and Their Predators

A quick glance at the field of invertebrate pathology shows that many more pathogens are known from arthropod herbivores than from arthropod predators or parasitoids (their natural enemies). There is a clear bias as the majority of invertebrate pathologists are interested principally in controlling herbivorous pests using microbes. However, this may also be a genuine biological pattern that begs an evolutionary explanation. Carnivores may vector the pathogen between local herbivore populations (see Brooks 1993), for example following ingestion of infected prey and passage through the gut of a predator (Vasconcelos *et al.* 1996) or following the external pick-up of propagules (Pell *et al.* 1997; Roy *et al.* 1998). Ewald's (1987b, 1994a) explanation is that high virulence to the carnivore would be counterproductive for the pathogen. Assuming that predator and prey can be infected only once, recent modeling work showed that the pathogen should evolve to be relatively mild to the predator, provided it is sufficiently more mobile than its prey (Elliot *et al.*, unpublished).

Multiple infections may also play a role in explaining lower virulence in carnivores than in their victims. Carnivores may well be able to avoid infected prey, so reducing the likelihood of multiple infections, while it is hard for a prey patch to prevent invasions by infected carnivores. The long-term consequence of this is a lower relative virulence to the carnivore. A parallel can be seen with arthropod vectors of animal parasites, for which the conventional explanation of low virulence to the vector is that of Ewald (1987b, 1994a; Koella 1999). A more powerful explanation, however, is that the vector takes only a few blood meals in its life (and so is more likely to be infected singly), whereas the host may receive much unwelcome attention from vectors (and so is more likely to be multiply infected; Macdonald 1957; Molineaux and Gramiccia 1980; Anderson and May 1991). Thus, pathogens and carnivores are expected to be compatible biocontrol agents where multiple infections of the carnivore are sufficiently rare that the pathogen's virulence to them does not impede their acting as vectors.

32.8   Ecological and Evolutionary Response of the Pest

So far we have only considered evolution of the natural enemy, ignoring the consequences of decreased pest density and the pest's evolution of resistance to its enemies. Introducing a biocontrol agent ultimately leads to a decrease in pest density and a natural enemy density lower than the initial one, which in turn decreases the probability of multiple infection and thereby the optimal virulence. This is an explanation for the observed reduction in virulence of the myxoma virus used for the classical biocontrol of rabbits (Fenner and Fantini 1999). It represents an alternative to the explanation given by Anderson and May (1991), which is also based on trade-offs but assumes single infections.

The use of biocontrol agents also generates selection pressure on the target pest to evolve resistance. This has been observed in the case of rabbits that developed resistance to myxoma (Fenner and Fantini 1999). Such a pest response opens the possibility for coevolution between exploiter and victim. The end result of this process is not immediately obvious, because pest density may increase and the mean susceptibility may decrease, so it is unclear how all of this affects the probability of multiple infections. Clearly, population dynamics determines densities and the probability of multiple infection, whereas the direction of evolution is determined by the probability of multiple infection. Thus, the interplay between evolution and population dynamics determines the outcome.

Holt and Hochberg (1997) highlighted an apparent discrepancy in the persistence of classical biocontrol versus chemical control. Under chemical control pests may rapidly become resistant (Roush and Tabashnik 1990; Gould 1991) whereas under classical biological control such resistance has not been reported (Croft 1992). This discrepancy may result from genetic constraints on selection, from differences in selection pressure, or from differences between the control agent and the pest in their capacity to respond evolutionarily. It remains an open question why the inundative use of pathogens has led to resistance (Tabashnik 1994; Moscardi 1999), whereas inundatively released arthropod natural enemies have not, as in the case of greenhouse biocontrol. This pattern is confirmed by an elegant experiment carried out with pea aphids *Acyrthosiphon pisum* (Harris), and a parasitoid *Aphidius ervi* Haliday (Henter 1995; Henter and Via 1995). Here, increased host resistance to parasitization and increased parasitoid virulence were obtained by selection in the laboratory, but increased host resistance did not arise under strong parasitization pressure in the field. This suggests that resistance to parasitoids (and predators) is more costly than resistance to pathogens or chemical pesticides. This is understandable, since barriers to chemicals, or their breakdown (and pathogens are quite reliant on chemical means of overcoming host defenses), are often based on changes in the expression of a small set of genes (Roush and Tabashnik 1990; Sayyed *et al.* 2000). This contrasts with the costs of polygenic traits involved in establishing morphological defenses to arthropod natural enemies (Tollrian and Harvell 1999) and perhaps to the costs of immune responses (Kraaijeveld and Godfray 1997; Fellowes *et al.* 1998). We suspect a gradient in costs from relatively cheap biochemical defenses through immune responses to costly morphological defenses. As the development of resistance is likely to depend upon the costs to the pest organism, it should be more rapid when defenses are less costly (Sasaki and Godfray 1999), for example against chemicals or pathogens. For models incorporating more costly resistance, Sasaki and Godfray (1999) showed that resistance may either not arise or may develop and break down in cycles. This may explain why Henter and Via (1995) did not find an immediate development of resistance after a sudden increase in parasitization pressure and possibly also why biological control with arthropod predators has rarely resulted in resistance when compared with control by pesticides or pathogens.

32.9 Discussion

In this chapter we question the assumption that high individual-level virulence is the Holy Grail of biological control. We argue that virulence at the individual level may be but one component of virulence toward a patch of victims. It may be at the patch level that selection on virulence has occurred in the natural setting, and it may be at this level that virulence will be most relevant in biocontrol. After all, it is in the control of (meta)populations of a pest that biocontrol workers are ultimately interested. Whenever the agent is expected to produce more than one generation, its patch-level virulence is composed of its ability to find and attack new victims, convert these into offspring, and disperse these locally. Thus, a suite of traits becomes important as one moves from inundative via inoculative to classical biocontrol. In the first case, a rapid kill of pests in a localized area is desired, whereas in the last case the aim is to achieve establishment, spread, and long-term persistence of the biocontrol agent over a large geographical area.

When interacting with an individual pest or with a local population of pests, a biocontrol agent faces the possibility that its resource will become unavailable because of incidental pest mortality or competing exploiters. The key to understanding changes in virulence is to understand the degree to which the exploiter is able to maintain control over the victim's resource. Any means by which this control can be reduced increases the virulence of the natural enemy and perhaps its effectiveness for biocontrol.

As a final word of caution, changes of virulence have both population dynamical effects and effects on victim resistance. These will in turn alter the selective pressures on virulence. If virulence increases, pest numbers are likely to decrease and pest resistance increase. As these processes feed back into the population dynamics of the biocontrol agents, they will further affect the probability of multiple infections and so alter the pattern of selection on virulence. Thus, the processes and options for virulence management discussed here must be put into a broader coevolutionary and population dynamical context.

Acknowledgments   We thank Arne Janssen and his graduate student discussion group, Bas Pels, and Minus van Baalen for their helpful comments on the manuscript of this chapter. Sam L. Elliot was supported by the Netherlands Foundation for the Advancement of Tropical Research (WOTRO).

33

Epilogue

Ulf Dieckmann, Karl Sigmund, Maurice W. Sabelis, and Johan A.J. Metz

Far from conquering infectious diseases through good sanitation, vaccines, and antimicrobial agents, populations of humans – as well as those of other animals and plants – continue to be harassed by an onslaught of pathogens. Complex processes of host–pathogen adaptation are responsible for the perennial persistence of this threat.

To develop sustainable control strategies, it is important to ask which new selective pressures on virulence will thus be created, and how resistance against control measures can be slowed, prevented, or even reversed. On the one hand, population growth, increased mobility, and climate change create new opportunities for diseases, while on the other hand adaptations allow disease agents to overcome the current transmission barriers.

Can epidemiological changes be steered in the desired directions and can they be prevented from veering off course in detrimental ways? That is what this book is about. Its aims are

- To show how evolutionary epidemiology as a science can profit from modeling techniques that take both population dynamics and natural selection into account;
- To explore the design of strategies for virulence management based on models of the evolutionary dynamics of pathogen–host systems;
- To highlight important unresolved research questions that need to be addressed before evolutionary predictions and management options are to be trusted; and
- To foster the dialogue between theorists and empiricists in the field of evolutionary epidemiology.

What are the general predictions regarding the evolution of virulence traits, as they have emerged throughout this book? An overarching principle appears to be the following: the more control the pathogen has over the host, the smaller the likelihood that the pathogen will become virulent. The basis for this prediction is that whenever exploitation by a pathogen cannot be interfered with by other pathogens or environmental causes, the well-being of the host is in the evolutionary interest of the parasite. However, under unclear conditions of "ownership", restrained exploitation is less likely. This can best be viewed as an instance of the Tragedy of the Commons. Whenever the resource (the host) is not safely monopolized, consideration of long-term benefits loses importance. The following points can be viewed as special cases of this principle.

- Under conditions of guaranteed transmission, selection favors the strains that replicate faster. This highlights why it is important to analyze alternative transmission modes, be they actual or potential.
- Under vertical (horizontal) transmission, low (high) virulence is favored. A vertically transmitted pathogen is generally much more closely tied to the host than an invader can ever be. There are relatively few theoretical analyses on this aspect, but it is likely to play an essential role in long-term evolution.
- The larger the multiplicity of infections, the higher the virulence. This is the result of arguments based on game theory (rather than optimization arguments) and the core of the Tragedy of the Commons: the more players, the less interest each has in safeguarding the common resource.
- Compared with well-mixed systems, in socially or spatially structured systems less virulent parasites are favored.

From these considerations, important options for virulence management emerge. We list them succinctly, with cross-references to the corresponding chapters for the essential caveats.

- Evolutionary optimality principles should be used with caution when managing virulence evolution under frequency-dependent selection (Chapter 4).
- In the presence of multiple infections, the evolutionary stability of biological control strategies must be assessed in the light of multiple levels of selection (Chapters 9 to 12, 22, and 32).
- Routes of pathogen transmission that function independently of host health, or that even intensify for sick hosts, should be at the very focus of management measures (Chapters 2 and 28).
- Altering transmission networks so that virulent pathogens are exposed to the detrimental consequences of their aggressive exploitation strategy selects for decreased virulence (Chapter 7).
- In animal husbandry and crop management, enhancing the relatedness of infected hosts is expected to select for decreased virulence (Chapters 7 and 11).
- Influencing the likelihood of horizontal versus vertical transmission can select for decreased virulence (Chapters 20 and 21).
- Models based on subdividing the host organism into compartments may help us to prevent the disease from escaping standard control measures (Chapter 30).
- By preventing hosts from acquiring multiple infections, decreased virulence can be selected for (Chapters 9 to 11).
- In the presence of multiple infections, long-term benefits arise from sanitation and vaccination that would otherwise be absent (Chapter 11).
- Tolerating relatively benign parasites, rather than trying to eliminate them, may often be advisable from an evolutionary perspective (Chapter 5).
- Strengthening the "conspiracy" between plants and arthropod predators in tritrophic interactions can improve the prospects for controlling the herbivores sandwiched in between (Chapter 22).

- In the context of biological control efforts, it must be kept in mind that pathogens genetically engineered to have a high virulence may be at a selective disadvantage (Chapters 6 and 32).
- Management-induced supply of novel genetic material from source to sink populations provides the genetic variation needed for local responses in the sink, but can also swamp the local adaptation of pathogens and hosts (Chapter 8).
- Uncontrolled spillover of pathogens from reservoir populations to endangered populations must be minimized, since the former pathogens tend to be more virulent than those in the latter populations (Chapter 29).
- By fostering host evolution and mate choice, it is possible to diminish disease losses in breeding programs for endangered species and livestock production (Chapters 15 and 18).
- To keep virulence at bay in systems with gene-for-gene interactions between parasites and their hosts, fostering the genetic diversity of hosts is essential (Chapter 17).
- If gene-for-gene interactions prevail, breeding schemes can be devised to select for hosts that have simultaneous resistance against multiple pathogen strains (Chapter 31).
- If more than one antibiotic is available to treat a bacterial infection, they should be administered, in a population-wide campaign, to individual hosts through combination therapy (Chapter 23).
- The suppression of a competitively superior pathogen strain by a vaccine that confers little cross-immunity may set off outbreaks of earlier, competitively inferior strains (Chapter 24).
- When nonvaccine serotypes outcompete vaccine serotypes, serotype replacement may augment the effectiveness of a vaccination program in a community (Chapter 26).

It hardly needs to be emphasized that many theoretical and empirical questions remain still unanswered. As with all good engineering, the development of techniques for virulence management requires a process of stepwise scaling up from small-scale predictions and controlled experiments toward applications of realistic complexity. In this context, our impression is that the following problems need to be addressed most urgently.

- A deeper understanding of the genetic basis of virulence traits is needed to better predict virulence evolution in the short term.
- More information is needed on the evolution of mutation rates.
- More experimental clues to the mechanisms of intraspecific competition of pathogens within hosts are needed.
- Actual patterns of competitive and cooperative interactions between different pathogenic strains need to be better understood.
- The existence of alternative transmission routes and their evolutionary implications should be explored in greater depth, both empirically and theoretically.
- Evolutionary implications of spatial structure are as yet imperfectly understood.

- The role of ecological networks in shaping pathogen–host interactions is still largely unexplored, especially if the hosts harbor several pathogen species and parasites can use multiple hosts.
- The richness of patterns in pathogen–host coevolution driven by reciprocal selection warrants further analysis.

If readers feel that this leaves them with more unanswered questions than they had before, the editors will be perfectly satisfied. The main goal of this book is to set a research program for evolutionary virulence management firmly on the road.

References

Page numbers of reference citations in this volume are given in square brackets.

Abrams PA & Matsuda H (1997). Fitness minimization and dynamic instability as a consequence of predator–prey coevolution. *Evolutionary Ecology* **11**:1–20 [*199, 234*]

Abrams PA, Matsuda H & Harada Y (1993). Evolutionarily unstable fitness maxima and stable fitness minima of continuous traits. *Evolutionary Ecology* **7**:465–487 [*45*]

Adler FR & Brunet RC (1991). The dynamics of simultaneous infections with altered susceptibilities. *Theoretical Population Biology* **40**:369–410 [*124, 140*]

Adler FR & Mosquera J (2000). Competition, biodiversity, and analytic functions in spatially structured habitats. *Ecology* **81**:3226–3232 [*138–139*]

Adviushko SA, Brown GC, Dahlman DL & Hildebrand DF (1997). Methyl jasmonate exposure induces insect resistance in cabbage and tobacco. *Environmental Entomology* **26**:642–854 [*316*]

Agrios GN (1988). *Plant Pathology*. San Diego, CA, USA: Academic Press, Inc. [*436, 438*]

Agur Z, Abiri D & Van der Ploeg LHT (1989). Ordered appearance of antigenic variants of African trypanosomes, explained in a mathematical model based on a stochastic switch process and immune selection against putative switch intermediates. *Proceedings of the National Academy of Sciences of the USA* **86**:9626–9630 [*246*]

Allison AC (1964). Polymorphism and natural selection in human populations. *Cold Spring Harbor Symposia on Quantitative Biology* **29**:137–149 [*355*]

Alonso PL, Lindsay SW, Armstrong Shellenberg J, Keita K, Gomez P, Shenton FC, Hill AG, David PH, Fegan G, Cham K & Greenwood BM (1993). A malaria control trial using insecticide-treated bednets and targetted chemoprophylaxis in a rural area of The Gambia, West Africa. *Transactions of the Royal Society of Tropical Medicine and Hygiene* **87**(Suppl. 2):37–44 [*355*]

Al-Yaman F, Genton B, Reeder JC, Anders RF, Smith T & Alpers MP (1997). Reduced risk of clinical malaria in children infected with multiple clones of *Plasmodium falciparum* in a highly endemic area: A prospective community study. *Transactions of the Royal Society of Tropical Medicine and Hygiene* **91**:602–605 [*168–169*]

Ameisen JC (1996). The origin of programmed cell death. *Science* **272**:1278–1279 [*194*]

Ameisen JC, Idziorek T, Billautmulot O, Loyens M, Tissier JP, Potentier A & Ouaissi A (1995). Apoptosis in a unicellular eukaryote (*Trypanosoma cruzi*): Implications for the evolutionary origin and role of programmed cell death in the control of cell-proliferation, differentiation, and survival. *Cell Death and Differentiation* **2**:285–300 [*193*]

Anagnostakis SL (1982). Biological control of the chestnut blight. *Science* **215**:466–471 [*291–292, 294–295*]

Anagnostakis SL (1988). *Cryphonectria parasitica*, cause of chestnut blight. In *Advances in Plant Pathology*, Vol. 6, eds. Sidhu GS, Ingram DS & Williams PH, pp. 123–136. London, UK: Academic Press [*443*]

Anagnostakis SL & Day PR (1979). Hypovirulence conversion in *Endothia parasitica*. *Phytopathology* **69**:1226–1229 [*292–293*]

Anagnostakis SL & Kranz J (1987). Population dynamics of *Cryphonectria parasitica* in a mixed-hardwood forest in Connecticut. *Phytopathology* **77**:751–754 [*293*]

Anagnostakis SL, Hau B & Kranz J (1986). Diversity of vegetative compatibility groups of *Cryphonectria parasitica* in Connecticut and Europe. *Plant Disease* **70**:536–538 [*293*]

Anderson RM (1982). *The Population Dynamics of Infectious Diseases: Theory and Applications*. London, UK: Chapman & Hall [*11, 16*]

Anderson RM & Crombie JM (1985). Experimental studies of age-intensity and age-prevalence profiles of infection: *Schisto-soma mansoni* in snails and mice. In *Ecology and Genetics of Host–Parasite Interactions*, eds. Rollinson D & Anderson RM, pp. 11–145. Linnean Society Symposium Series, London, UK: Academic Press [*32*]

Anderson RM & Gordon DM (1982). Processes influencing the distribution of parasite numbers within host populations with special emphasis on parasite-induced host mortalities. *Parasitology* **85**:373–398 [*417*]

Anderson RM & May RM (1978). Regulation and stability of host–parasite population interactions. I. Regulatory processes. *Journal of Animal Ecology* **47**:219–247 [*31, 75*]

Anderson RM & May RM (1979). Population biology of infectious diseases: Part I. *Nature* **280**:361–367 [*150, 426, 453*]

Anderson RM & May RM (1981). The population dynamics of microparasites and their invertebrate hosts. *Proceedings of the Royal Society of London B* **291**:451–524 [*10, 14, 112, 288–289*]

Anderson RM & May RM (1982). Coevolution of hosts and parasites. *Parasitology* **85**:411–426 [*10–11, 35, 39, 48, 65, 74, 86, 106*]

Anderson RM & May RM (1991). *Infectious Diseases of Humans: Dynamics and Control*. Oxford, UK: Oxford University Press [*15, 32–34, 39, 48, 60, 62, 69, 87, 91, 113, 124, 162, 247, 328, 339, 366, 453, 456–457*]

Anderson RM, May RM, Medley GF & Johnson A (1986). A preliminary study of the transmission dynamics of HIV, the causative agent of AIDS. *IMA Journal of Mathematics Applied in Medicine and Biology* **3**:229–263 [*78*]

Anderson RM, May RM & Gupta S (1989). Non-linear phenomena in host–parasite interactions. *Parasitology* **99**(Suppl.):S59–79 [*176*]

Anderson RM, Donnelly CA & Gupta S (1997). Vaccine design, evaluation, and community-based use for antigenically variable infectious agents. *Lancet* **350**:1466–1470 [*60*]

Andersson M (1994). *Sexual Selection*. Princeton, NJ, USA: Princeton University Press [*251*]

Andersson DI & Hughes D (1996). Muller's ratchet decreases fitness of a DNA-based microbe. *Proceedings of the National Academy of Sciences of the USA* **93**:906–907 [*248*]

Andreasen V & Christiansen FB (1993). Disease-induced natural selection in a diploid host. *Theoretical Population Biology* **44**:261–298 [*229–230*]

Andreasen V & Christiansen FB (1995). Slow coevolution of a viral pathogen and its diploid host. *Philosophical Transactions of the Royal Society of London B* **348**:341–354 [*231*]

Andreasen V & Pugliese A (1995). Pathogen coexistence induced by density-dependent host mortality. *Journal of Theoretical Biology* **177**:159–165 [*124, 138, 231*]

Andrivon D (1993). Nomenclature for pathogenicity and virulence: The need for precision. *Phytopathology* **83**:889–890 [*388*]

Anonymous (1995). Recommendations for preventing the spread of vancomycin resistance: Recommendations of the Hospital Infection Control Practices Advisory Committee (HICPAC). *American Journal of Infection Control* **23**:87–94 [*326, 338*]

Anonymous (1996). *Ecologically Based Pest Management*. Washington, DC, USA: National Academy Press [*447*]

Anonymous (1997a). Human monkeypox in Kasai Oriental, Democratic Republic of Congo (former Zaire). *Weekly Epidemiological Record* **72**:365–372 [*345*]

Anonymous (1997b). Human monkeypox in Kasai Oriental, Zaire (1996–1997). *Weekly Epidemiological Record* **72**:101–104 [*345*]

Anonymous (2000a). Inventing Africa. *New Scientist* **2251**:30–33 [*397*]

Anonymous (2000b). Malariasphere. *New Scientist* **2247**:32–35 [*398*]

Antia R & Koella JC (1994). A model of non-specific immunity. *Journal of Theoretical Biology* **168**:141–150 [*300*]

Antia R, Nowak MA & Anderson RM (1996). Antigenic variation and within-host dynamics of parasites. *Proceedings of the National Academy of Sciences of the USA* **93**:985–989 [*177*]

Antonovics J (1976). The nature of limits to natural selection. *Annals of the Missouri Botanical Garden* **63**:224–248 [*106*]

Apanius V, Penn D, Slev PR, Ruff LR & Potts WK (1997). The nature of selection on the major histocompatibility complex. *Critical Reviews in Immunology* **17**:179–224 [*254*]

Armstrong RA & McGehee R (1980). Competitive exclusion. *The American Naturalist* **115**:151–170 [*138*]

Arnot D (1999). Clone multiplicity in *Plasmodium falciparum* infections in individuals exposed to variable levels of disease transmission. *Transactions of the Royal Society of Tropical Medicine and Hygiene* **92**:580–585 [*167*]

Arthur M & Courvalin P (1993). Genetics and mechanisms of glycopeptide resistance in enterococci. *Antimicrobial Agents and Chemotherapy* **37**:1563–1571 [*326*]

Atkinson CT & Van Riper C (1991). Pathogenicity and epizootiology of avian haematozoa: *Plasmodium*, *Leucocytozoon*, and *Haemoproteus*. In *Bird–Parasite Interactions: Ecology, Evolution and Behaviour*, eds. Loye JE & Zuk M. Oxford, UK: Oxford University Press [*165*]

Atsatt PR & O'Dowd DJ (1976). Plant defense guilds. *Science* **193**:24–29 [*315*]

Augner M (1994). Should a plant always signal its defence against herbivores? *Oikos* **70**:322–332 [*315*]

Augner M, Fagerström T & Tuomi J (1991). Competition, defence and games between plants. *Behavioural Ecology and Sociobiology* **29**:231–234 [*315, 317*]

Austin DJ, Kristinsson KG & Anderson RM (1999). The relationship between the volume of antimicrobial consumption in human communities and the frequency of resistance. *Proceedings of the National Academy of Sciences of the USA* **96**:1152–1156 [*383*]

Austrian R (1986). Some aspects of the pneumococcal carrier state. *Journal of Antimicrobial Chemotherapy* **18**(Suppl. A):35–45 [*363*]

Axelrod R & Hamilton WD (1981). The evolution of cooperation. *Science* **211**:1390–1396 [*166*]

Babiker HA & Walliker D (1997). Current views on the population structure of *Plasmodium falciparum*: Implications for control. *Parasitology Today* **13**:262–267 [*167*]

Babiker HA, Ranford-Cartwright L, Currie D, Charlwood JD, Billingsley P, Teuscher T & Walliker D (1994). Random mating in a natural population of the malaria parasite *Plasmodium falciparum*. *Parasitology* **109**:413–421 [*171*]

Badley AD, Dockrell D, Simpson M, Schut R, Lynch DH, Leibson P & Paya CV (1997). Macrophage-dependent apoptosis of CD4(+) T lymphocytes from HIV-infected individuals is mediated by FasL and tumor necrosis factor. *Journal of Experimental Medicine* **185**:55–64 [*190*]

Baer M, Vuento R & Vesikari T (1995). Increase in bacteraemic pneumococcal infections in children. *Lancet* **345**:661 [*365*]

Balardin RS, Jarosz AM & Kelly JD (1997). Virulence and molecular diversity in *Colletotrichum lindemuthianum* from South, Central, and North America. *Phytopathology* **87**:1184–1191 [*445*]

Ballabeni W & Ward PI (1993). Local adaptation of a trematode, *Diplostomum phoxini*, to the European *Minnox phoxinus phoxinus*, its second intermediate host. *Functional Ecology* **7**:84–90 [*422*]

Ballou JD (1993). Assessing the risks of infectious diseases in captive breeding and reintroduction programs. *Journal of Zoo and Wildlife Research* **24**:327–335 [*421*]

Baquero F & Blázquez J (1997). Evolution of antibiotic resistance. *Trends in Ecology and Evolution* **12**:482–487 [*60–61, 65, 163*]

Barbour ML, Mayon-White RT, Coles C, Crook DWM & Moxon ER (1995). The impact of conjugate vaccine on carriage of *Haemophilus influenzae* type b. *Journal of Infectious Diseases* **171**:93–98 [*364–365, 369*]

Barnes DM, Whittier S, Gilligan PH, Soares S, Tomasz A & Henderson FW (1995). Transmission of multidrug-resistant serotype 23F *Streptococcus pneumoniae* in group day care: Evidence suggesting capsular transformation of the resistant strain *in vivo*. *Journal of Infectious Diseases* **171**:890–896 [*371*]

Barrett JA (1980). Pathogen evolution in multilines and variety mixtures. *Zeitschrift für Pflanzenkrankheiten und Pflanzenschutz* **87**:383–396 [*447*]

Barrett JA (1987). The dynamics of genes in populations. In *Populations of Plant Pathogens: Their Dynamics and Genetics*, eds. Wolfe MS & Caten CE, pp. 39–53. Oxford, UK: Blackwell Scientific [*438*]

Baudoin M (1975). Host castration as a parasitic strategy. *Evolution* **29**:335–352 [*452*]

Bazer FW, Geisert RD & Zavy MT (1987). Fertilization, cleavage and implantation. In *Reproduction in Farm Animals*, ed. Hafez ESE, pp. 210–228. Philadelphia, PA, USA: Lea and Febiger [*256*]

Bean WJ, Kawaoka Y, Wood JM, Pearson JE & Webster RG (1985). Characterization of virulent and avirulent

A/Chicken/Pennsylvania/83 influenza A viruses: Potential role of defective interfering RNAs in nature. *Journal of Virology* **54**:151–160   [*268*]

Beattie AJ (1985). *The Evolutionary Ecology of Ant–Plant Mutualisms*. Cambridge, UK: Cambridge University Press   [*298, 318*]

Beck K (1984). Coevolution: Mathematical analysis of host–parasite interactions. *Journal of Mathematical Biology* **19**:63–77   [*211, 231*]

Beck K, Keener J & Ricciardi P (1984). The effects of epidemics on genetic evolution. *Journal of Mathematical Biology* **19**:79–94   [*223–224, 228–229*]

Beck HP, Felger I, Huber W, Steiger S, Smith T, Weiss N, Alonso P & Tanner M (1997). Analysis of multiple *Plasmodium falciparum* infections in Tanzanian children during phase III trial of malaria vaccine SPf66. *Journal of Infectious Diseases* **175**:921–926   [*168–169*]

Becker NG (1989). *Analysis of Infectious Disease Data*. London, UK: Chapman & Hall   [*426, 432*]

Beddington JR & Hammond PS (1977). On the dynamics of host–parasite hyperparasite interactions. *Journal of Animal Ecology* **46**:811–821   [*286, 296*]

Beddington JR, Free CA & Lawton JH (1975). Dynamic complexity in predator–prey models framed in difference equations. *Nature* **225**:58–60   [*199*]

Begon M, Townsend CR & Harper JL (1996). *Ecology: Individuals, Populations and Communities*, Third Edition. Oxford, UK: Blackwell Science   [*197, 385, 449*]

Bentley BL (1977). Extrafloral nectaries and protection by pugnacious bodyguards. *Annual Review of Ecology and Systematics* **8**:407–427   [*298*]

Berenbaum MR & Zangerl AR (1998). Chemical phenotype matching between a plant and its insect herbivore. *Proceedings of the National Academy of Sciences of the USA* **95**:13743–13748   [*396*]

Berendt AR, Ferguson DJ, Gardner J, Turner G, Rowe A, McCormick C, Roberts D, Craig A, Pinches R, Elford BC *et al.* (1994). Molecular mechanisms of sequestration in malaria. *Parasitology* **108b**(Suppl.):S19–28   [*354*]

Bergelson J & Purrington CB (1996). Surveying patterns in the cost of resistance in plants. *The American Naturalist* **148**:536–558   [*245*]

Bergstrom CT, McElhany P & Real LA (1999). Transmission bottlenecks as determinants of virulence in rapidly evolving pathogens. *Proceedings of the National Academy of Sciences of the USA* **96**:5095–5100   [*395*]

Bernasconi ML, Turlings TCJ, Ambrosetti L, Bassetti P & Dorn S (1998). Herbivore-induced emissions of maize volatiles repel the corn leaf aphid, *Rhopalosiphum maidis*. *Entomologia Experimentalis et Applicata* **87**:133–142   [*314*]

Berryman AA, Dennis B, Raffa KF & Stenseth NCh (1985). Evolution of optimal group attack, with particular reference to bark beetles (*Coleoptera: Scolytidae*). *Ecology* **66**:898–903   [*299*]

Bigler F (1994). Quality control in *Trichogramma* production. In *Biological Control with Egg Parasitoids*, eds. Wajnberg E & Hassan SA, pp. 93–111. Wallingford, UK: CAB International   [*455*]

Binder S, Levitt AM, Sacks JJ & Hughes JM (1999). Emerging infectious diseases: Public health issues for the 21st century. *Science* **284**:1311–1313   [*397*]

Bishop JDD (1996). Female control of paternity in the internally fertilizing compound ascidian *Diplosoma listerianum*. I. Autoradiographic investigation of sperm movement in the female reproductive tract. *Proceedings of the Royal Society of London B* **263**:369–376   [*253*]

Bishop JDD, Jones CS & Noble LR (1996). Female control of paternity in the internally fertilizing compound ascidian *Diplosoma listerianum*. II. Investigation of male mating success using RAPD markers. *Proceedings of the Royal Society of London B* **263**:401–407   [*253*]

Bissegger M, Rigling D & Heiniger U (1996). Population structure and disease development of *Cryphonectria parasitica* in European chestnut forests in the presence of natural hypovirulence. *Phytopathology* **87**:50–59   [*292–293*]

Björkman A & Phillips-Howard PA (1990). The epidemiology of drug resistance malaria. *Transactions of the Royal Society of Tropical Medicine and Hygiene* **84**:177–180   [*415*]

Black BC, Brennan LA, Dierks PM & Gard IE (1997). Commercialization of baculoviral insecticides. In *The Baculoviruses*, ed. Miller LK, pp. 341–388. New York, NY, USA: Plenum Press   [*77, 81*]

Black S, Shinefield H, Fireman B, Lewis E, Ray P, Hansen JR, Elvin L, Ensor KM, Hackell J, Siber G, Malinoski F, Madore D, Chang I, Kohberger R, Watson W, Austrian R & Edwards K (2000). Efficacy, safety, and immunogenicity of heptavalent pneumococcal conjugate vaccine in children. Northern California Kaiser Permanente Vaccine Study Center Group [In Process Citation]. *Pediatric Infectious Diseases Journal* **19**:187–195   [*365*]

Blair PJ, Boise LH, Perfetto SP, Levine BL, McCrary G, Wagner KF, St Louis DC, Thompson CB, Siegel JN & June CH (1997). Impaired induction of the apoptosis-protective protein *bcl-x*(l) in activated PBMC from asymptomatic HIV-infected individuals. *Journal of Clinical Immunology* **17**:234–246   [*190*]

Bloom BR & Murray CJL (1992). Tuberculosis: Commentary on a reemergent killer. *Science* **257**:1055–1064   [*326*]

Blower SM, Small PM & Hopewell PC (1996). Control strategies for tuberculosis epidemics: New models for old problems. *Science* **273**:497–500   [*337*]

Bodmer WF (1972). Evolutionary significance of the HL-A system. *Nature* **237**:139–145   [*211*]

Boerlijst MC, Lamers ME & Hogeweg P (1993). Evolutionary consequences of spiral waves in a host parasitoid system. *Proceedings of the Royal Society of London B* **253**:15–18   [*450*]

Bohannan BJM & Lenski RE (1997). Effect of resource enrichment on a chemostat community of bacteria and bacteriophage. *Ecology* **78**:2303–2315   [*206*]

Bolker BM & Grenfell BT (1995). Space, persistence and dynamics of measles epidemics. *Philosophical Transactions of the Royal Society of London B* **348**:308–320   [*352*]

Bolker BM, Pacala SW & Levin SA (2000). Moment methods for ecological processes in continuous space. In *The Geometry of Ecological Interactions: Simplifying Spatial Complexity*, eds. Dieckmann U, Law R & Metz JAJ, pp. 388–411. Cambridge, UK: Cambridge University Press   [*98*]

Bonhoeffer S & Nowak MA (1994a). Intra-host versus inter-host selection: Viral strategies of immune function impairment. *Proceedings of the National Academy of Sciences of the USA* **91**:8062–8066   [*152, 177*]

Bonhoeffer S & Nowak MA (1994b). Mutation and the evolution of virulence. *Proceedings of the Royal Society of London B* **258**:133–140   [*177*]

Bonhoeffer S & Nowak MA (1995). Can live attenuated virus work as post-exposure treatment? *Immunology Today* **16**:131–135   [*188*]

Bonhoeffer S, Lenski R & Ebert D (1996). The curse of the pharaoh: The evolution of virulence in pathogens with long-living propagules. *Proceedings of the Royal Society of London B* **263**:715–721   [*19, 23, 25, 158, 160, 453, 455*]

Bonhoeffer S, Lipsitch M & Levin BR (1997). Evaluating treatment protocols to prevent antibiotic resistance. *Proceedings of the National Academy of Sciences of the USA* **94**:12106–12111   [*60–61, 65, 327*]

Boot WJ, Calis JNM, Beetsma J, Minh Hai D, Kim Lan N, Van Toan T, Quang Trung L & Hung Minh N (1999). Natural selection of *Varroa jacobsoni* explains the differential reproductive strategies in colonies of *Apis cerana* and *Apis mellifera*. In *Ecology and Evolution of the Acari*, eds. Bruin J, Van der Geest LPS & Sabelis MW, pp. 349–357. Dordrecht, Netherlands: Kluwer Academic Publishers   [*448*]

Boots M & Sasaki A (1999). "Small worlds" and the evolution of virulence: Infection occurs locally and at a distance. *Proceedings of the Royal Society of London B* **266**:1933–1938   [*89, 101*]

Booy R (1998). Getting Hib vaccine to those who need it [comment]. *Lancet* **351**:1446–1447   [*364*]

Booy R & Kroll JS (1997). Is *Haemophilus influenzae* finished? *Journal of Antimicrobial Chemotherapy* **40**:149–153   [*364*]

Booy R, Heath P, Willocks L, Mayon-White D, Slack M & Moxon R (1995). Invasive pneumococcal infections in children. *Lancet* **345**:1245–1246   [*365*]

Borghans JAM & de Boer RJ (2001). Diversity in the immune system. In *Design Principles for the Immune System and Other Distributed Autonomous Systems*, eds. Segel LA & Cohen IR, pp. 161–183. Oxford, UK: Oxford University Press   [*221*]

Borghans JAM, Noest AJ & de Boer RJ (1999). How specific should immunological memory be? *Journal of Immunology* **163**:569–575 [*213*]

Borghans JAM, Beltman JB & de Boer RJ. Extensive MHC polymorphism requires frequency-dependent selection by coevolving pathogens. Unpublished [*213, 221*]

Bosch FX, Garten W, Klenk HD & Rott R (1981). Proteolytic cleavage of influenza virus hemagglutinins: Primary structure of the connecting peptide between HA1 and HA2 determines proteolytic cleavability and pathogenicity of avian influenza viruses. *Virology* **113**:725–735 [*268*]

Both GW, Sleigh MJ, Cox NJ & Kendal AP (1983). Antigenic drift in influenza virus H3 hemagglutinin from 1968 to 1980: Multiple evolutionary pathways and sequential amino acid changes at key antigenic sites. *Journal of Virology* **48**:52–60 [*210, 268*]

Boughton DA (1999). Empirical evidence for complex source–sink dynamics with alternative states in a butterfly metapopulation. *Ecology* **80**:2727–2739 [*105*]

Bouma JE & Lenski RE (1988). Evolution of a bacteria/plasmid association. *Nature* **335**:351–352 [*330, 441*]

Bouma A, de Jong MCM & Kimman TG (1997a). The influence of maternally derived antibodies on the transmission of pseudorabies virus and the effectiveness of vaccination. *Vaccine* **15**:287–294 [*426, 434*]

Bouma A, de Jong MCM & Kimman TG (1997b). Comparison of two pseudorabies virus vaccines, that differ in their capacity to reduce virus excretion after a challenge infection, in their capacity of reducing transmission of pseudorabies virus. *Veterinary Microbiology* **54**:113–122 [*426, 434*]

Bradshaw AD (1959). Population differentiation in *Agrostis tenuis* Sibth. II. The incidence and significance of infection by *Epichloe typhina*. *New Phytologist* **58**:310–315 [*452*]

Bradshaw AD (1991). Genostasis and the limits to evolution. *Philosophical Transactions of the Royal Society of London B* **333**:289–305 [*104*]

Brasier CM (2000). The rise of the hybrid fungi. *Nature* **405**:134–135 [*381*]

Brasier CM, Cooke DEL & Duncan JM (1999). Origin of a new *Phytophthora* pathogen through interspecific hybridization. *Pro-ceedings of the National Academy of Sciences of the USA* **96**:5878–5883 [*381, 448*]

Bremermann HJ (1980). Sex and polymorphism as strategies in host–pathogen interactions. *Journal of Theoretical Biology* **87**:671–702 [*223, 244*]

Bremermann HJ & Pickering J (1983). A game-theoretical model of parasite virulence. *Journal of Theoretical Biology* **100**:411–426 [*65, 86, 88, 92, 124, 150, 166–167, 172, 287*]

Bremermann HJ & Thieme HR (1989). A competitive-exclusion principle for pathogen virulence. *Journal of Mathematical Biology* **27**:179–190 [*138, 140, 166, 172, 348, 359*]

Bronstein JL (1994a). Our current understanding of mutualism. *The Quarterly Review of Biology* **69**:31–51 [*318, 396*]

Bronstein JL (1994b). Conditional outcomes in mutualistic interactions. *Trends in Ecology and Evolution* **9**:214–217 [*318, 396*]

Brookfield JFY (1998). Quorum sensing and group selection. *Evolution* **52**:1263–1269 [*394*]

Brooks WM (1993). Host–parasitoid–pathogen interactions. In *Parasites and Pathogens of Insects Volume 2: Pathogens*, eds. Beckage NE, Thompson SN & Federici BA, pp. 231–272. San Diego, CA, USA: Academic Press [*456*]

Brown JL (1997). A theory of mate choice based on heterozygosity. *Behavioral Ecology* **8**:60–65 [*254*]

Brown SP (1999). Cooperation and conflict in host-manipulating parasites. *Proceedings of the Royal Society of London B* **266**:1899–1904 [*394*]

Brown JL & Eklund A (1994). Kin recognition and the major histocompatibility complex: An integrative review. *The American Naturalist* **143**:435–461 [*255*]

Brown JS & Pavlovic NB (1992). Evolution in heterogeneous environments: Effects of migration on habitat specialization. *Evolutionary Ecology* **6**:360–382 [*104*]

Brown JH, Stevens GC & Kaufman DM (1996). The geographic range: Size, shape, boundaries and internal structure. *Annual Review of Ecology and Systematics* **27**:597–623 [*197*]

Brown J, Higo H, Mckalip A & Herman B (1997). Human papillomavirus (HPV) 16 E6 sensitizes cells to atractyloside-induced apoptosis: Role of *p53*, ice-like proteases and the mitochondrial permeability transition. *Journal of Cellular Biochemistry* **66**:245–255 [*190*]

Bruin J, Dicke M & Sabelis MW (1992). Plants are better protected against spider mites after exposure to volatiles from infested conspecifics. *Experientia* **48**:525–529 [*316*]

Bruin J, Sabelis MW & Dicke M (1995). Do plants tap SOS-signals from their infested neighbours. *Trends in Ecology and Evolution* **10**:167–170 [*316*]

Buck KW (1986). *Fungal Virology*. Boca Raton, FL, USA: CRC Press [*286*]

Buckley RC (1986). Ant–plant–*Homoptera* interactions. *Advances in Ecological Research* **16**:53–85 [*298*]

Buckley RC (1987). Interactions involving plants, *Homoptera* and ants. *Annual Review of Ecology and Systematics* **18**:111–135 [*298*]

Buckling AGJ, Taylor LH, Carlton JMR & Read AF (1997). Adaptive changes in *Plasmodium* transmission strategies following chloroquine chemotherapy. *Proceedings of the Royal Society of London B* **264**:553–559 [*406*]

Bull JJ (1994). Virulence. *Evolution* **48**:1423–1437 [*106, 166, 250, 379, 451*]

Bull JJ & Molineux IJ (1992). Molecular genetics of adaptation in an experimental of cooperation. *Evolution* **46**:882–895 [*389*]

Bull JJ, Molineux IJ & Rice WR (1991). Selection of benevolence in a host–parasite system. *Evolution* **45**:875–882 [*391*]

Bull PC, Lowe BS, Kortok M & Marsh K (1998). Antibody recognition of *Plasmodium falciparum* red cell surface antigens in Kenya: Evidence for immune selection of surface phenotype. *Nature Medicine* **4**:358–360 [*354*]

Burdon JJ (1987). *Diseases and Plant Population Biology*. Cambridge, UK: Cambridge University Press [*233–236, 245, 415*]

Burdon JJ (1991). Fungal pathogens as selective forces in plant populations and communities. *Australian Journal of Ecology* **16**:423–432 [*452*]

Burdon JJ (1992). Host population subdivision and the genetic structure of natural pathogen populations. *Advances in Plant Pathology* **8**:81–94 [*437, 439–440*]

Burdon JJ & Jarosz AM (1991). Host–pathogen interactions in natural populations

of *Linum marginale* and *Melampsora lini*: I. Patterns of resistance and racial variation in a large host population. *Evolution* **45**:205–217 [*33, 415, 440*]

Burdon JJ & Jarosz AM (1992). Temporal variation in the racial structure of flax rust (*Melampsora lini*) populations growing on natural stands of wild flax (*Linum marginale*): Local versus metapopulation dynamics. *Plant Pathology* **41**:165–179 [*440–441*]

Burdon JJ & Roelfs AP (1985). The effect of sexual and asexual reproduction on the isozyme structure of populations of *Puccinia graminis*. *Phytopathology* **75**:1068–1073 [*441*]

Burdon JJ & Thrall PH (1999). Spatial and temporal patterns in coevolving plant and pathogen associations. *The American Naturalist* **153**:S15–S33 [*396, 448*]

Burdon JJ, Oates JD & Marshall DR (1983). Interactions between *Avena* and *Puccinia* species. I. The wild hosts: *Avena barbata* Pott ex link, *A. fatua* L., *A. ludoviciana* Durieu. *Journal of Applied Ecology* **20**:571–584 [*206*]

Bush AO & Holmes JC (1986). Intestinal helminths of lesser scaup duck: An interactive community. *Canadian Journal of Zoology* **64**:142–152 [*29*]

Buss LW & Green DR (1985). Histoincompatibility in vertebrates: The relict hypothesis. *Developmental and Comparative Immunology* **9**:191–201 [*220*]

Butler JC (1997). Epidemiology of pneumococcal serotypes and conjugate vaccine formulations. *Microbial Drug Resistance* **3**:125–129 [*374*]

Callan RJ, Bunch TD, Workman GW & Mock RE (1991). Development of pneumonia in desert bighorn sheep after exposure to a flock of exotic wild and domestic sheep. *Journal of the American Veterinary Association* **198**:1052–1056 [*36*]

Calow P, ed. (1999). *The Encyclopedia of Ecology and Environmental Management*. Oxford, UK: Blackwell Science Ltd [*388*]

Campbell CL & Madden LV (1990). *Introduction to Plant Disease Epidemiology*. New York, NY, USA: John Wiley and Sons [*439*]

Cann RL & Douglas JD (1999). Parasites and conservation of Hawaiian birds. In *Genetics and the Extinction of Species*, eds. Landweber L & Dobson AP, pp. 121–136. Princeton,

NJ, USA: Princeton University Press [*30*]

Carbonari M, Pesce AM, Cibati M, Modica A, Dellanna L, Doffizi G, Angelici A, Uccini S, Modesti A & Fiorilli M (1997). Death of bystander cells by a novel pathway involving early mitochondrial damage in human immunodeficiency virus-related lymphadenopathy. *Blood* **90**:209–216 [*190*]

Carlson J, Helmby H, Hill AVS, Brewster D, Greenwood BM & Wahlgren M (1990). Human cerebral malaria: Association with erythrocyte rosetting and lack of anti-rosetting antibodies. *Lancet* **336**:1457–1460 [*171*]

Carman WF, Zanetti AR & Karayiannis P (1990). Vaccine-induced escape mutant of hepatitis B virus. *Lancet* **336**:325–329 [*345*]

Carpenter MA, Appel MJG, Roelke-Parker ME, Munson L, Hofer H, East M & O'Brien SJ (1998). Genetic characterization of canine distemper virus in Serengeti carnivores. *Veterinary Immunology and Immunopathology* **65**:259–266 [*382*]

Carrington M, Nelson GW, Martin MP, Kissner T, Vlahov D, Goedert JJ, Kaslow R, Buchbinder S, Hoots K & O'Brien SJ (1999). HLA and HIV-1: Heterozygote advantage and B*35-Cw*04 disadvantage. *Science* **283**:1748–1752 [*211*]

Cassirer EF, Beecham J, Coggins V, Whittaker D, Fowler P, Johnson R, Martin K, Schommer T, Taylor E, Thomas A, Gilchrist D & Oldenberg L (1997a). *Restoration of Bighorn Sheep to Hells Canyon*. Idaho Department of Fish & Game, Clearwater Region [*36*]

Cassirer EF, Oldenberg L, Coggins L, Fowler P, Rudolph K, Hunter D & Foreyt WJ (1997b). *Overview and Preliminary Analysis of a Bighorn Sheep Die-off, Hells Canyon 1995–1996*. Idaho Department of Fish & Game, Clearwater Region [*36*]

Castro MEB, Souza ML, Araujo S & Bilimoria SL (1997). Replication of *Anticarsia gemmatalis* nuclear polyhedrosis virus in four lepidopteran cell lines. *Journal of Invertebrate Pathology* **69**:40–45 [*190*]

Cavalli-Sforza LL & Bodmer WF (1971). *The Genetics of Human Populations*. San Francisco, CA, USA: WH Freeman and Company [*223*]

CDC (Center for Disease Control) (1989). A strategic plan for the elimination of tuberculosis in the United States. *Morbidity and Mortality Weekly Report* **38**:38–42 [*337*]

CDC (Center for Disease Control) (1995). Progress toward elimination of *Haemophilus influenzae* type b disease among infants and children: United States, 1993–1994. *Morbidity and Mortality Weekly Report* **44**:545–550 [*363*]

CDC (Centers for Disease Control) (1996). Dengue fever at the US–Mexico border, 1995–1996. *Morbidity and Mortality Weekly Report* **45**:841–844 [*408*]

CDC (Center for Disease Control) (1997). Prevention of pneumococcal disease: Recommendations of the Advisory Committee on Immunization Practices (ACIP). *Morbidity and Mortality Weekly Report* **46**(RR-8) [*363*]

Chao L, Hanley KA, Burch CL, Dahlberg C & Turner PE (2000). Kin selection and parasite evolution: Higher and lower virulence with hard and soft selection. *Quarterly Review of Biology* **75**:261–275 [*88, 394, 455*]

Chen B, Choi GH & Nuss DL (1994a). Attenuation of fungal virulence by synthetic infectious hypovirus transcripts. *Science* **264**:1762–1764 [*293*]

Chen RT, Weierbach R & Bisoffi Z (1994b). A "post-honeymoon period" measles outbreak in Muyinga sector, Burundi. *International Journal of Epidemiology* **23**:185–193 [*340*]

Chen B, Chen C-H, Bowman B & Nuss DL (1996). Phenotypic changes associated with wild-type and mutant hypovirus RNA transfection of plant pathogenic fungi phylogenetically related to *Cryphonectria parasitica*. *Phytopathology* **86**:301–310 [*293*]

Chen Q, Barragan A, Fernandez V, Sundstrom A, Schlichtherle M, Sahlen A, Carlson J, Datta S & Wahlgren M (1998). Identification of *Plasmodium falciparum* erythrocyte membrane protein 1 (PfEMP1) as the rosetting ligand of the malaria parasite *P. falciparum*. *Journal of Experimental Medicine* **187**:15–23 [*171*]

Choi GH & Nuss DL (1992). Hypovirulence of chestnut blight fungus conferred by an infectious viral cDNA. *Science* **257**:800–803 [*293*]

Chou J & Roizman B (1992). The *gamma-1-34.5* gene of Herpes simplex virus-1 precludes neuroblastoma cells from triggering total shutoff of protein-synthesis characteristic of programmed cell death in neuronal cells. *Proceedings of the National Academy* of Sciences of the USA 89:3266–3270 [190–191]

Chung P-H, Bedker PJ & Hillman BI (1994). Diversity of Cryphonectria parasitica hypovirulence-associated double-stranded RNAs within a chestnut population in New Jersey. Phytopathology 84:984–990 [293, 295]

Claessen D & de Roos AM (1995). Evolution of virulence in a host–pathogen system with local pathogen transmission. Oikos 74:401–413 [85, 94, 124]

Clarke B (1976). The ecological genetics of host–parasite relationships. Symposia of the British Society for Parasitology 14:87–103 [246]

Clarke DK, Duarte EA, Elena SF, Moya, A, Domingo E & Holland J (1994). The Red Queen reign in the kingdom of RNA viruses. Proceedings of the National Academy of Sciences of the USA 91:4821–4824 [249]

Clarke RT, Thomas JA, Elmes GW & Hochberg ME (1997). The effects of spatial patterns in habitat quality on community dynamics within a site. Proceedings of the Royal Society of London B 264:347–354 [199]

Clarke RT, Thomas JA, Elmes GW, Wardlaw JC, Munguira ML & Hochberg ME (1998). Population modelling of the spatial interactions between Maculinea rebeli, their initial foodplant Gentiana cruciata and Myrmica ant hosts. Journal of Insect Conservation 2:29–38 [197]

Clay K (1990). Insects, endophytic fungi and plants. In Pests, Pathogens and Plant Communities, eds. Burdon JJ & Leather SR, pp. 111–130. Oxford, UK: Blackwell Scientific Publications [452]

Clay K & Kover PX (1996). The red queen hypothesis and plant/pathogen interactions. Annual Review of Phytopathology 34:29–50 [396]

Clem RJ, Fechheimer M & Miller LK (1991). Prevention of apoptosis by a baculovirus gene during infection of insect cells. Science 254:1388–1390 [184, 192]

Clements CJ, Strassburg M, Cutts FT & Torel C (1992). The epidemiology of measles. World Health Statistics Quarterly 45:285–291 [340]

Clouston WM & Kerr JFR (1985). Apoptosis, lymphocytotoxicity, and the containment of viral infections. Medical Hypothe-

ses 18:399–404   [194]

Clutton-Brock TH & Parker GA (1992). Potential reproductive rates and the operation of sexual selection. Quarterly Review of Biology 67:437–456   [250]

Clutton-Brock TH & Vincent A (1991). Sexual selection and the potential reproductive rates of males and females. Nature 351:58–60   [250]

Coetzee M, Horne DWK, Brooke BD & Hunt RH (1999). DDT, dieldrin and pyrethroid insecticide resistance in African malaria vector mosquitoes: An historical review and implications for future malaria control in southern Africa. South African Journal of Science 95:215–218   [398]

Cohen ML (1994). Antimicrobial resistance: Prognosis for public health. Trends in Microbiology 2:422–425   [326]

Contamin H, Fandeur T, Rogier C, Bonnefoy S, Konate L, Trape J-F & Mercereau-Puijalon O (1996). Different genetic characteristics of Plasmodium falciparum isolates collected during successive clinical malaria episodes in Senegalese children. American Journal of Tropical Medicine and Hygiene 54:632–643   [168]

Conway DJ & Roper C (2000). Microevolution and emergence of pathogens. International Journal for Parasitology 30:1423–1430   [398]

Conway DJ, Greenwood BM & McBride JS (1991). The epidemiology of multiple-clone Plasmodium falciparum infections in Gambian patients. Parasitology 103:1–6   [168]

Cornell HV & Lawton JH (1992). Species interactions, local and regional processes, and limits to the richness of ecological communities: A theoretical perspective. Journal of Animal Ecology 61:1–12   [197]

Correa-Victoria FJ, Zeigler RS & Levy M (1994). Virulence characteristics of genetic families of Pyricularia grisea in Colombia. In Rice Blast Disease, eds. Zeigler RS, Leong S & Teng P, pp. 211–229. Wallingford, UK: Commonwealth Agriculture Bureaux International, and Los Banos, Philippines: IRRI   [446]

Cory JS, Hirst ML & Williams T (1994). Field trial of a genetically improved baculovirus insecticide. Nature 370:138–140   [83]

Cory JS, Hails RS & Vasconcelos JC (1997). The ecology of insect viruses. In The Baculoviruses, ed. Miller LK, pp. 651–703. New

York, NY, USA: Plenum Press   [*79*]

Cox MJ, Kum D, Tavul L, Narara A, Raiko A, Baisor M, Alpers M, Medley G & Day KP (1994). Dynamics of malaria parasitaemia associated with febrile illness in children from a rural area of Madang, Papua New Guinea. *Transactions of the Royal Society of Tropical Medicine and Hygiene* **88**:191–197 [*358*]

Crawley MJ (1989). Insect herbivores and plant population dynamics. *Annual Review of Entomology* **34**:531–564   [*452*]

Crawley MJ & Pacala SW (1991). Herbivores, plant parasites, and plant diversity. In *Parasite–Host Associations: Coexistence or Conflict?*, eds. Toft CA, Aeschlimann A & Bolis L, pp. 158–173. Oxford, UK: Oxford University Press   [*449*]

Croft BA (1992). IPM systems that conserve pesticides, pest-resistant plants and biological controls, including genetically altered forms. *Journal of the Entomological Society of South Africa* **55**:107–121   [*457*]

Crofton HD (1971). A quantitative approach to parasitism. *Parasitology* **62**:179–193 [*417*]

Cromroy HL (1979). Eriophyoidea in biological control of weeds. In *Recent Advances in Acarology*, ed. Rodriguez JG, pp. 473–475. New York, NY, USA: Academic Press [*452*]

Cross AP & Singer B (1991). Modelling the development of resistance of *Plasmodium falciparum* to antimalaria drugs. *Transactions of the Royal Society of Tropical Medicine and Hygiene* **85**:349–355   [*415*]

Cunningham AA (1996). Disease risks of wildlife translocations. *Conservation Biology* **10**:349–353   [*413, 416, 421, 423*]

Cutts FT & Markowitz L (1994). Successes and failures in measles control. *Journal of Infectious Diseases* **170**:S32–S41   [*340*]

Dagan R, Melamed R, Muallem M, Piglansky L & Yagupsky P (1996a). Nasopharyngeal colonization in southern Israel with antibiotic-resistant pneumococci during the first two years of life: Relation to serotypes likely to be included in pneumococcal conjugate vaccines. *Journal of Infectious Diseases* **174**:1352–1355   [*364–365, 374*]

Dagan R, Melamed R, Muallem M, Piglansky L, Greenberg D, Abramson O, Mendelman PM, Bohidar N & Yagupsky P (1996b). Reduction of nasopharyngeal carriage of pneumococci during the second year of life by a heptavalent conjugate pneumococcal vaccine. *Journal of Infectious Diseases* **174**:1271–1278   [*374*]

Dagan R, Givon N, Yagupsky P, Porat N, Janco J, Chang I, Kimura A & Hackell J (1998). Effect of a 9-valent pneumococcal vaccine conjugated to CRM_{197} (PncCRM9) on nasopharyngeal (NP) carriage of vaccine type and non-vaccine type *S. pneumoniae* (Pnc) strains among day-care-center (DCC) attendees. In *38th Interscience Conference on Antimicrobial Agents and Chemotherapy Abstracts*, p. G52. Washington, DC, USA: American Society for Microbiology [*364–365*]

Daly TM & Long CA (1993). A recombinant 15-kilodalton carboxyl-terminal fragment of *Plasmodium yoelii yoelii* 17XL merozoite surface protein 1 induces a protective immune response in mice. *Infection and Immunology* **61**:2462–2467   [*358*]

Dangl JL (1994). The enigmatic avirulence genes of phytopathogenic bacteria. In *Current Topics in Microbiology and Immunology*, Vol. 192: *Bacterial Pathogenesis of Plants and Animals*. ed. Dangl JL. Berlin, Germany: Springer-Verlag   [*440*]

Davidson WR & Nettles VF (1992). Relocation of wildlife: Identifying and evaluating disease risks. *Transactions of the North American Wildlife and Natural Resource Conference* **57**:466–473   [*421–422*]

Dawson M, Johnson PT, Feldman L, Glover R, Koehler J, Blake P & Toomey KE (1997). Probable locally acquired mosquito-transmitted *Plasmodium vivax* infection – Georgia, 1996. *Journal of the American Medical Association* **277**:1191–1193 [*409*]

Day JF & Edman JD (1983). Malaria renders mice susceptible to mosquito feeding when gametocytes are most infective. *Journal of Parasitology* **69**:163–170   [*165*]

Day KP & Marsh K (1991). Naturally acquired immunity to *Plasmodium falciparum*. *Immunology Today* **12**:68–71   [*169*]

Day JF, Ebert KM & Edman JD (1983). Feeding patterns of mosquitoes (*Diptera: Culicidae*) simultaneously exposed to malarious and healthy mice, including a method for separating blood meals from conspecific hosts. *Journal of Medical Entomology* **20**:120–127   [*18*]

Day KP, Koella JC, Nee S, Gupta S & Read AF (1992). Population genetics and dynamics of *Plasmodium falciparum*: An ecological view. *Parasitology* **104**:S35–52 [*167*]

Day KPF, Karamalis F, Thompson J, Barnes DA, Peterson C, Brown H, Brown GV & Kemp DJ (1993). Genes necessary for expression of a virulence determinant and for transmission of *Plasmodium falciparum* are located on a 0.3 megabase region of chromosome 9. *Proceedings of the National Academy of Sciences of the USA* **89**:6015–6019 [*389*]

Dearsly AL, Sinden RE & Self IA (1990). Sexual development in malarial parasites, gametocyte production, fertility and infectivity to the mosquito vector. *Parasitology* **100**:359–368 [*389*]

Debets AJM, Yang X & Griffith AJF (1994). Vegetative incompatibility in *Neurospora*: Its effect on horizontal transfer of mitochondrial plasmids and senescence in natural populations. *Current Genetics* **26**:113–119 [*282, 284*]

De Boer RJ (1995). The evolution of polymorphic compatibility molecules. *Molecular Biology and Evolution* **12**:494–502 [*220*]

De Jong MCM (1995). Mathematical modelling in veterinary epidemiology: Why model-building is important. *Preventive Veterinary Medicine* **25**:183–194 [*426*]

De Jong MCM & Kimman TG (1994). Experimental quantification of vaccine-induced reduction in virus transmission. *Vaccine* **8**:761–766 [*426, 434*]

De Jong MCM & Sabelis MW (1988). How bark beetles avoid interference with squatters: An ESS for colonization by *Ips typographus*. *Oikos* **51**:88–96 [*299*]

De Jong MCM & Sabelis MW (1989). How bark beetles avoid interference with squatters: A correction. *Oikos* **54**:128 [*299*]

De Jong MCM, Diekmann O & Heesterbeek JAP (1995). How does transmission of infection depend on population size? In *Epidemic Models: Their Structure and Relation to Data*, ed. Mollison D, pp. 84–94. Cambridge, UK: Cambridge University Press [*426*]

De Jong MCM, Van der Poel WHM, Kramps JA, Brand A & Van Oirschot JT (1996). A quantitative investigation of population persistence and recurrent outbreaks of bovine respiratory syncytial virus (BRSV) on dairy farms. *American Journal of Veterinary Research* **57**:628–633 [*433*]

De Leo GA & Guberti V. The effect of culling in the evolution of virulence evolution: Management implications in the control and eradication of wildlife diseases. Unpublished [*420*]

De Melker HE, Conyn van Spaendonck MA, Runke HC, Van Wjngaarden JK, Mooi F & Schellekens JF (1997). Pertussis in the Netherlands: An outbreak despite high levels of immunization with whole-cell vaccine. *Emerging Infectious Diseases* **3**:175–178 [*345*]

De Roos AM, McCauley E & Wilson WG (1991). Mobility versus density-limited predator–prey dynamics on different spatial scales. *Proceedings of the Royal Society of London B* **246**:117–122 [*89, 199*]

DeSantis R & Pinto MR (1991). Gamete self-discrimination in ascidians: A role for the follicle cells. *Molecular Reproduction and Development* **29**:47–50 [*253*]

De Wit PJGM (1992). Molecular characterization of gene-for-gene systems in plant–fungus interactions and the application of avirulence genes in control of plant pathogens. *Annual Review of Phytopathology* **30**:391–418 [*380*]

Diamond J (1997). *Guns, Germs and Steel: Fates of Human Societies*. New York, NY, USA: WW Norton [*397*]

Dias PC (1996). Sources and sinks in population biology. *Trends in Ecology and Evolution* **11**:326–330 [*104–105*]

Dicke M & Sabelis MW (1988). How plants obtain predatory mites as bodyguards. *Netherlands Journal of Zoology* **38**:148–165 [*298*]

Dicke M & Sabelis MW (1989). Does it pay plants to advertise for bodyguards? Towards a cost–benefit analysis of induced synomone production. In *Variation in Growth Rate and Productivity of Higher Plants*, eds. Lambers H, Cambridge ML, Konings H & Pons TL, pp. 341–358. The Hague, Netherlands: SPB Academic Publishing BV [*298, 318*]

Dicke M & Sabelis MW (1992). Costs and benefits of chemical information conveyance: Proximate and ultimate factors. In *Insect Chemical Ecology: An Evolutionary Approach*, eds. Roitberg B & Isman M, pp. 122–155. Hants, UK: Chapman & Hall [*298*]

Dicke M, Sabelis MW, Takabayashi J, Bruin J & Posthumus MA (1990). Plant strategies of manipulating predator–prey interactions through allelochemicals: Prospects for application in pest control. *Journal of Chemical Ecology* **16**:3091–3118   [*298*]

Dieckmann U (1994). *Coevolutionary Dynamics of Stochastic Replicator Systems*. Jülich, Germany: Central Library of the Research Center Jülich   [*46*]

Dieckmann U & Law R (1996). The dynamical theory of coevolution: A derivation from stochastic ecological processes. *Journal of Mathematical Biology* **34**:579–612   [*41, 46, 99*]

Dieckmann U & Law R (2000). Relaxation projections and the method of moments. In *The Geometry of Ecological Interactions: Simplifying Spatial Complexity*, eds. Dieckmann U, Law R & Metz JAJ, pp. 412–455. Cambridge, UK: Cambridge University Press   [*93, 98*]

Dieckmann U, Marrow P & Law R (1995). Evolutionary cycling in predator–prey interactions: Population dynamics and the Red Queen. *Journal of Theoretical Biology* **176**:91–102   [*234, 396*]

Dieckmann U, Law R & Metz JAJ, eds. (2000). *The Geometry of Ecological Interactions: Simplifying Spatial Complexity*. Cambridge, UK: Cambridge University Press   [*397*]

Diekmann O & Heesterbeek JAP (1999). *Mathematical Epidemiology of Infectious Diseases: Model Building, Analysis and Interpretation*. Chichester, UK: Wiley & Sons Ltd   [*397*]

Diekmann O, Heesterbeek JAP & Metz JAJ (1990). On the definition and the computation of the basic reproductive ratio R_0 in models for infectious diseases in heterogeneous populations. *Journal of Mathematical Biology* **28**:365–382   [*81, 124, 387, 426*]

Diekmann O, Heesterbeek JAP & Metz JAJ (1995). The legacy of Kermack and McKendrick. In *Epidemic Models: Their Structure and Relation to Data*, ed. Mollison D, pp. 95–115. Cambridge, UK: Cambridge University Press   [*426*]

Diekmann O, Gyllenberg M, Metz JAJ & Thieme HR (1998). On the formulation and analysis of general deterministic structured population models: I. Linear theory. *Journal of Mathematical Biology* **36**:349–388 [*387*]

Dietz K (1975). Transmission and control of arboviruses. In *Epidemiology*, eds. Ludwig D & Cooke KL, pp. 104–121. Philadelphia,

PA, USA: Society for Industrial and Applied Mathematics   [*11, 14*]

Dietz K (1976). The incidence of infectious disease under the influence of seasonal fluctuations. In *Mathematical Models in Medicine*, eds. Berger J, Repges, R, Buhler W & Tautu P, *Lecture Notes in Biomathematics*, Vol. 11, pp. 1–15. Berlin, Germany: Springer-Verlag   [*12*]

Diffley P, Scott JO, Mama K & Tsen TN-R (1987). The rate of proliferation among African trypanosomes is a stable trait that is directly related to virulence. *Journal of Tropical Medicine and Hygiene* **36**:533–540 [*389*]

Dixon SM & Baker RL (1988). Effect of size on predation risk, behavioral response to fish and cost of reduced feeding in larval *Ischnura verticalis* (Odonata: Coenagrionidae). *Oecologia* **76**:2000–2005   [*450*]

Dobson AP (1988). Behavioural and life history adaptations of parasites for living in desert environments. *Journal of Arid Environments* **17**:185–192   [*31*]

Dobson AP & Crawley MJ (1994). Pathogens and the structure of plant communities. *Trends in Ecology and Evolution* **9**:393–398 [*30, 32*]

Dobson AP & Hudson PJ (1986). Parasites, disease, and the structure of ecological communities. *Trends in Ecology and Evolution* **1**:11–15   [*30*]

Dobson AP & Keymer AE (1990). Population dynamics and community structure of parasitic helminths. In *Living in a Patchy Environment*, eds. Shorrochs B & Swingland IR, pp. 107–126. Oxford, UK: Oxford University Press   [*29*]

Dobson AP & McCallum H (1997). The role of parasites in bird conservation. In *Host–Parasite Evolution: General Principles and Avian Models*, eds. Clayton DH & Moore J, pp. 155–173. Oxford, UK: Oxford University Press   [*197*]

Dobson AP & Meagher M (1996). The population dynamics of brucellosis in Yellowstone National Park. *Ecology* **77**:1026–1036   [*36*]

Dobson AP & Roberts MG (1994). The population dynamics of parasitic helminth communities. *Parasitology* **109**:S97–S108   [*29*]

Doebeli M (1996). Quantitative genetics and population dynamics. *Evolution* **50**:532–546 [*234*]

Doebeli M (1997). Genetic variation and the

persistence of predator–prey interactions in the Nicholson–Bailey model. *Journal of Theoretical Biology* **188**:109–120 [*234*]

Doern GV, Brueggemann A, Holley HPJ & Rauch AM (1996). Antimicrobial resistance of *Streptococcus pneumoniae* recovered from outpatients in the United States during the winter months of 1994 to 1995: Results of a 30-center national surveillance study. *Antimicrobial Agents and Chemotherapy* **40**:1208–1213 [*326*]

Doherty PC & Zinkernagel RM (1975). Enhanced immunological surveillance in mice heterozygous at the H-2 gene complex. *Nature* **256**:50–52 [*211, 254*]

Drukker B, Scutareanu P & Sabelis MW (1995). Do anthocorid predators respond to synomones from *Psylla*-infested pear trees under field conditions? *Entomologia Experimentalis et Applicata* **77**:193–203 [*298*]

Durrant BS (1995). Reproduction in mammals: Captive perspectives. In *Conservation of Endangered Species in Captivity: An Interdisciplinarian Approach*, eds. Gibbon EF Jr, Durrant BS & Demarest J, pp. 331–376. Albany, NY, USA: State University of New York Press [*260*]

Dwyer G & Elkinton JS (1993). Predicting virus epizootics in gypsy moth populations. *Journal of Animal Ecology* **62**:1–11 [*79*]

Dwyer G, Levin SA & Buttel L (1990). A simulation model of the population dynamics and evolution of myxomatosis. *Ecological Monographs* **60**:423–447 [*56, 453*]

Dwyer G, Buonaccorsi JP & Elkinton JS (1997). Host heterogeneity in susceptibility and disease dynamics: Tests of a mathematical model. *The American Naturalist* **150**:687–707 [*31, 74, 78, 80–82*]

Dwyer G, Dushoff J, Elkinton JS & Levin SA (2000). Host–pathogen dynamics revisited: Discrete host generations and host heterogeneity in susceptibility. *The American Naturalist* **156**:105–120 [*74–78*]

Eberhard WG (1996). *Female Control: Sexual Selection by Cryptic Female Choice*. Princeton, NJ, USA: Princeton University Press [*253*]

Ebert D (1994). Virulence and local adaptation of a horizontally transmitted parasite. *Science* **265**:1084–1086 [*36, 250, 390, 422, 448*]

Ebert D (1998a). Experimental evolution of parasites. *Science* **282**:1432–1435 [*36, 391, 394–395*]

Ebert D (1998b). The evolution and expression of parasite virulence. In *Evolution in Health and Disease*, ed. Stearns SC. Oxford, UK: Oxford University Press [*166*]

Ebert D & Hamilton WD (1996). Sex against virulence: The coevolution of parasitic diseases. *Trends in Ecology and Evolution* **11**:79–82 [*32, 36, 250*]

Ebert D & Herre EA (1996). The evolution of parasitic diseases. *Parasitology Today* **2**:96–101 [*32, 250, 391, 422, 452*]

Ebert D & Mangin KL (1997). The influence of host demography on the evolution of virulence of a microsporidian gut parasite. *Evolution* **51**:1828–1837 [*161, 166, 392*]

Ebert O, Finke S, Salahi A, Herrmann M, Trojaneck B, Lefterova P, Wagner E, Kircheis R, Huhn D, Schriever F & Schmidtwolf IGH (1997). Lymphocyte apoptosis: Induction by gene transfer techniques. *Gene Therapy* **4**:296–302 [*195, 453*]

Edmunds GF & Alstad DN (1978). Coevolution in insect herbivores and conifers. *Science* **199**:941–945 [*422*]

Edmunds WJ, O'Callaghan CJ & Nokes DJ (1997). Who mixes with whom? A method to determine the contact patterns of adults that may lead to the spread of airborne infections. *Proceedings of the Royal Society of London B* **264**:949–957 [*102*]

Egas M, Van Baalen M & Sabelis MW. Avoidance and tolerance of predation. Unpublished [*313–314*]

Egid K & Brown JL (1989). The major histocompatibility complex and female mating preferences in mice. *Animal Behaviour* **38**:548–550 [*253*]

Elhassan IM, Hviid L, Jakobsen PH, Giha H, Satti GMH, Arnot DE, Jensen JB & Theander TG (1995). High proportion of subclinical *Plasmodium falciparum* infections in an area of seasonal and unstable malaria in Sudan. *American Journal of Tropical Medicine and Hygiene* **53**:78–83. [*406*]

Elkinton JS & Liebhold AM (1990). Population dynamics of the gypsy moth in North America. *Annual Review of Entomology* **35**:571–596 [*79*]

Elliot SL, Sabelis MW, Janssen A, Van der Geest LPS, Beerling EAM & Fransen JJ (2000). Can arthropod pathogens be bodyguards of plants? *Ecology Letters* **3**:228–235 [*454*]

Elliot SL, Adler FR & Sabelis MW. How virulent should a parasite be to its vector? Unpublished   [*456*]

Elliston JE (1985). Characteristics of dsRNA-free and dsRNA-containing strains of *Endothia parasitica* in relation to hypovirulence. *Phytopathology* **75**:151–158   [*293*]

Endler J (1977). *Geographic Variation, Speciation, and Clines*. Princeton, NJ, USA: Princeton University Press   [*106*]

Enebak SA, MacDonald WL & Hillman BI (1994a). Effect of dsRNA associated with isolates of *Cryphonectria parasitica* from the central Appalachians and their relatedness to other dsRNAs from North America and Europe. *Phytopathology* **84**:528–534   [*293–295*]

Enebak SA, Hillman BI & MacDonald WL (1994b). A hypovirulent isolate of *Cryphonectria parasitica* with multiple, genetically unique dsRNA segments. *Molecular Plant Microbe Interactions: MPMI* **7**:590–595   [*296*]

Engelbrecht F, Felger I, Genton B, Alpers M & Beck H-P (1995). *Plasmodium falciparum*: Malaria morbidity is associated with specific merozoite surface antigen 2 genotypes. *Experimental Parasitology* **81**:90–96   [*168*]

Epstein PR (2000). Is global warming harmful to health? *Scientific American* **289**:36–43   [*398*]

Eshel I (1977). On the founder effect and the evolution of altruistic traits: An ecogenetical approach. *Theoretical Population Biology* **11**:410–424   [*87–88, 150, 166*]

Eshel I (1983). Evolutionary and continuous stability. *Journal of Theoretical Biology* **103**:99–111   [*42–43*]

Eshel I & Motro U (1981). Kin selection and strong stability of mutual help. *Theoretical Population Biology* **19**:420–433   [*42*]

Eskola J, Kilpi T, Palmu A, Jokinen J, Haapakoski J, Herva E, Takala A, Kayhty H, Karma P, Kohberger R, Siber G & Makela PH (2001). Efficacy of a pneumococcal conjugate vaccine against acute otitis media. *New England Journal of Medicine* **344**:403–409   [*365–366*]

Espiau C, Riviere D, Burdon JJ, Gartner S, Daclinat B, Hasan S & Chaboudez P (1998). Host–pathogen diversity in a wild system: *Chondrilla juncea–Puccinia chondrillina*. *Oecologia* **113**:133–139   [*234*]

Evans HF & Entwistle PF (1987). Viral diseases. In *Epizootiology of Insect Diseases*, eds. Fuxa JR & Tanada Y, pp. 257–322. New York, NY, USA: John Wiley and Sons   [*79*]

Ewald PW (1983). Host–parasite relations, vectors, and the evolution of disease severity. *Annual Review of Ecology and Systematics* **14**:465–485   [*10–11, 15, 17, 150, 391*]

Ewald PW (1987a). Pathogen-induced cycling of outbreaks in insect populations. In *Insect Outbreaks*, eds. Barbosa PA & Schultz JC, pp. 269–286. London, UK: Academic Press   [*18, 157*]

Ewald PW (1987b). Transmission modes and evolution of the parasitism–mutualism continuum. *Annals of the New York Academy of Sciences* **503**:295–306   [*456*]

Ewald PW (1988). Cultural vectors, virulence, and the emergence of evolutionary epidemiology. *Oxford Surveys in Evolutionary Biology* **5**:215–245   [*18, 25, 399*]

Ewald PW (1991a). Waterborne transmission and the evolution of virulence among gastrointestinal bacteria. *Epidemiology and Infection* **106**:83–119   [*17, 24, 391, 399–400*]

Ewald PW (1991b). Transmission modes and the evolution of virulence, with special reference to cholera, influenza and AIDS. *Human Nature* **2**:1–30   [*18*]

Ewald PW (1993). The evolution of virulence. *Scientific American* **268**:56–62   [*60, 87*]

Ewald PW (1994a). *Evolution of Infectious Disease*. Oxford, UK: Oxford University Press   [*10–11, 15, 17, 25, 60, 87, 104, 107, 118, 138, 157, 160, 165, 167, 250, 391, 399–400, 406, 411, 427, 431, 433, 435, 453, 456*]

Ewald PW (1994b). The evolutionary ecology of virulence. *Quarterly Review of Biology* **69**:381–384   [*87*]

Ewald PW (1995). Response to Van Baalen and Sabelis. *Trends in Microbiology* **3**:416–417   [*11*]

Ewald PW (1996). Vaccines as evolutionary tools: The virulence-antigen strategy. In *Concepts in Vaccine Development*, ed. Kaufmann SHE, pp. 1–25. Berlin, Germany: de Gruyter & Co.   [*371, 411*]

Ewald PW (1999). Using evolution as a tool for controlling infectious diseases. In *Evolutionary Medicine*, eds. Trevathan WR, Smith EO & McKenna JJ, pp. 245–269. New York, NY, USA: Oxford University Press   [*411*]

Ewald PW, Sussman JB, Distler MT, Libel C, Dirita VJ, Salles CA, Vicente AC, Heitmann I & Cabello F (1998). Evolutionary control of infectious disease: Prospects for vector-borne and waterborne pathogens. *Memorias do Instituto Oswaldo Cruz* **93**:567–576 [*401–403*]

Falkow S (1975). *Infectious Multiple Drug Resistance*. London, UK: Pion Press [*335*]

Farley MM, Stephens DS & Brachman PS Jr (1992). Invasive *Haemophilus influenzae* disease in adults: A prospective, population-based surveillance. *Annals of Internal Medicine* **116**:806–812 [*365*]

Farmer EE & Ryan CA (1990). Interplant communication: Airborne methyl jasmonate induces synthesis of proteinase inhibitors in plant leaves. *Proceedings of the National Academy of Sciences of the USA* **87**:7713–7716 [*316*]

Faust EC, Beaver PC & Jung RC (1968). *Animal Agents and Vectors of Human Disease*, Third Edition. Philadelphia, PA, USA: Lea & Febiger [*405*]

Feavers IM, Heath AB, Bygraves JA & Maiden MCJ (1992). Role of horizontal genetic exchange in the antigenic variation of the class 1 outer membrane protein of *Neisseria meningitidis*. *Molecular Microbiology* **6**:489–495 [*351*]

Fekety FRJ (1964). The epidemiology and prevention of staphylococcal infection. *Medicine* **45**:593–613 [*337*]

Fellowes MDE, Kraaijeveld AR & Godfray HCJ (1998). Trade-off associated with selection for increased ability to resist parasitoid attack in *Drosophila melanogaster*. *Proceedings of the Royal Society of London B* **265**:1553–1558 [*458*]

Felsenstein J (1981). Evolutionary trees from DNA sequences: A maximum likelihood approach. *Journal of Molecular Evolution* **17**:368–376 [*263, 265–266, 270, 272*]

Fenner R (1983). Biological control as exemplified by smallpox eradication and myxomatosis. *Proceedings of the Royal Society of London B* **218**:259–285 [*74*]

Fenner F (1994). Myxomatosis. In *Parasitic and Infectious Diseases: Epidemiology and Ecology*, eds. Scott ME & Smith G, pp. 337–346. London, UK: Academic Press [*33*]

Fenner F & Fantini B (1999). *Biological Control of Vertebrate Pests: The History of Myxomatosis – An Experiment in Evolution*. Wallingford, UK: CAB International [*56, 383, 385, 389, 396, 448, 452–453, 457*]

Fenner F & Ratcliffe RN (1965). *Myxomato-sis*. Cambridge, UK: Cambridge University Press [*33–34, 56, 222, 228*]

Fenner F & Ross J (1994). Myxomatosis. In *The European Rabbit: History and Biology of a Successful Colonizer*, eds. Thompson HV & King CM, pp. 205–239. Oxford, UK: Oxford University Press [*55–56*]

Ferguson NM, May RM & Anderson RM (1998). Measles: Persistence and synchronicity in disease dynamics. In *Spatial Ecology: The Role of Space in Population Dynamics and Interspecific Interactions*, eds. Tilman D & Kareiva P. Princeton, NJ, USA: Princeton University Press [*352*]

Ferrière R (2000). Adaptive responses to environmental threats: Evolutionary suicide, insurance, and rescue. *Options* Spring 2000, pp. 12–16. Laxenburg, Austria: International Institute for Applied Systems Analysis [*56*]

Fine PEP (1975). Vectors and vertical transmission: An epidemiological perspective. *Annals of the New York Academy of Science* **266**:173–194 [*10*]

Finkel SE & Kolter R (1999). Evolution of microbial diversity during prolonged starvation. *Proceedings of the National Academy of Sciences of the USA* **96**:4023–4027 [*383*]

Finlay BB & Cossart P (1997). Exploitation of mammalian host cell functions by bacterial pathogens. *Science* **276**:718–725 [*194*]

Fischer C, Jock B & Vogel F (1998). Interplay between humans and infective agents: A population genetic study. *Human Genetics* **102**:415–422 [*223*]

Fisher RA (1930). *The Genetical Theory of Natural Selection*. Oxford, UK: Clarendon Press [*251*]

Fitch WM, Bush RM, Bender CA & Cox NJ (1997). Long-term trends in the evolution of H(3) HA1 human influenza type A. *Proceedings of the National Academy of Sciences of the USA* **94**:7712–7718 [*268*]

Flor HH (1956). The complementary genetics system in flax and flax rust. *Advanced Genetics* **8**:29–54 [*236, 437*]

Folstad I & Karter AK (1992). Parasites, bright males, and the immunocompetence handicap. *The American Naturalist* **139**:603–622 [*251*]

Foreyt WJ, Snipes KP & Kasten RW (1994). Fatal pneumonia following innoculation of healthy bighorn sheep with *Pasteurella haemolytica* from healthy domestic sheep. *Journal of Wildlife Diseases* **30**:137–145

[37]
Forrest S (1993). Genetic algorithms: Principles of natural selection applied to computation. *Science* **261**:872–878   [*212*]

Fox LR & Eisenbach R (1992). Contrary voices: Possible exploitation of enemy-free space by herbivorous insects in cultivated vs wild crucifers. *Oecologia* **89**:574–579   [*314*]

Frank SA (1992a). A kin selection model for the evolution of virulence. *Proceedings of the Royal Society of London B* **250**:195–197   [*88, 124, 150, 152, 166–167, 172, 287*]

Frank SA (1992b). Models of plant–pathogen coevolution. *Trends in Genetics* **8**:213–219   [*396*]

Frank SA (1993a). Evolution of host–parasite diversity. *Evolution* **47**:1721–1732   [*205*]

Frank SA (1993b). Coevolutionary genetics of plants and pathogens. *Evolutionary Ecology* **7**:45–75   [*233–234, 244, 246, 396*]

Frank SA (1994a). Coevolutionary genetics of hosts and parasites with quantitative inheritance. *Evolutionary Ecology* **8**:74–94   [*201, 234, 396*]

Frank SA (1994b). Kin selection and virulence in the evolution of protocells and parasites. *Proceedings of the Royal Society of London B* **258**:153–161   [*88, 150–152, 154*]

Frank SA (1996a). Statistical properties of polymorphism in host–parasite genetics. *Evolutionary Ecology* **10**:307–317   [*244, 396*]

Frank SA (1996b). Problems inferring the specificity of plant–pathogen genetics: Reply. *Evolutionary Ecology* **10**:323–325   [*244*]

Frank SA (1996c). Models of parasite virulence. *Quarterly Review of Biology* **71**:37–78   [*10–11, 35, 88, 106, 109, 150–152, 154, 160, 166, 205, 222, 250, 287, 396*]

Frank SA (1997). Spatial processes in host–parasite genetics. In *Metapopulation Biology: Ecology, Genetics and Evolution*. eds. Hanski IA & Gilpin ME, pp. 325–352. San Diego, CA, USA: Academic Press   [*438*]

Frankham R (1997). Do island populations have less genetic variation than mainland populations? *Heredity* **78**:311–327   [*205*]

Franklin-Tong NVE & Franklin CFCH (1993). Gametophytic self-incompatibility: Contrasting mechanisms for *Nicotiana* and *Papaver*. *Trends in Cell Biology* **3**:340–345   [*253*]

Freed L (1999). Extinction and endangerment of Hawaiian honeycreepers: A comparative approach. In *Genetics and the Extinction of Species*, eds. Landweber L & Dobson AP, pp. 137–162. Princeton, NJ, USA: Princeton University Press   [*30*]

Fretwell SD (1977). The regulation of plant communities by the food chains exploiting them. *Perspectives in Biology and Medicine* **20**:169–185   [*286*]

Fritz RS (1995). Direct and indirect effect of plant genetic variation on enemy impact. *Ecological Entomology* **20**:18–26   [*315*]

Fritz RS & Nobel J (1990). Host plant variation in mortality of the leaf-folding sawfly on the arryo willow. *Ecological Entomology* **15**:25–35   [*315*]

Fu YX & Galán JE (1999). A *Salmonella* protein antagonizes Rac-1 and Cdc42 to mediate host-cell recovery after bacterial invasion. *Nature* **401**:293–297   [*380*]

Fulbright DW, Weidlich WH, Haufler KZ, Thomas CS & Paul CP (1983). Chestnut blight and recovering chestnut trees in Michigan. *Canadian Journal of Botany* **61**:3164–3171   [*293, 443*]

Fuxa JR & Tanada Y, eds. (1987). *Epizootiology of Insect Diseases*. New York, NY, USA: John Wiley and Sons   [*77*]

Galyov EE, Hakansson S, Forsberg A & Wolf-Watz H (1993). A secreted protein kinase of *Yersinia* tuberculosis is an indispensable virulence determinant. *Nature* **361**:730–732   [*382*]

Gandon S (1998). The curse of the pharaoh hypothesis. *Proceedings of the Royal Society of London B* **265**:1545–1552   [*88, 152–154, 156, 159, 453, 455*]

Gandon S & Michalakis Y (1999). Evolutionary stable dispersal rate in a metapopulation with extinctions and kin competition. *Journal of Theoretical Biology* **199**:275–290   [*155–156*]

Gandon S, Jansen VAA & Van Baalen M (2001). Host life-history and the evolution of parasite virulence. *Evolution* **55**:1056–1062   [*152, 161, 164, 453*]

Gao F, Bailes E, Robertson DL, Chen YL, Rodenburg CM, Michael SF, Cummins LB, Arthur LO, Peeters M, Shaw GM, Sharp PM & Hahn BH (1999). Origin of HIV-1 in the chimpanzee *Pan troglodytes troglodytes*. *Nature* **397**:436–441   [*382*]

García-Guzmán G, Burdon JJ & Nicholls AO

(1996). Effects of the systemic flower infecting smut *Ustilago bullata* on the growth and competitive ability of the grass *Bromus catharticus*. *Journal of Ecology* **84**:657–665 [*452*]

García-Ramos G & Kirkpatrick, M (1997). Genetic models of adaptation and gene flow in peripheral populations. *Evolution* **51**:21–28 [*199, 205*]

Gardner JP, Pinches RA, Roberts DJ & Newbold CI (1996). Variant antigens and endothelial receptor adhesion in *Plasmodium falciparum*. *Proceedings of the National Academy of Sciences of the USA* **93**:3503–3508 [*354*]

Garnham PCC (1966). *Malaria Parasites and Other Haemosporidia*. Oxford, UK: Blackwell [*406*]

Gatto M & De Leo GA (1998). Models of interspecific competition among macroparasites. *Journal of Mathematical Biology* **27**:467–490 [*29*]

Gee GF (1995). Avian reproductive physiology. In *Conservation of Endangered Species in Captivity: An Interdisciplinarian Approach*, eds. Gibbon EF Jr, Durrant BS & Demarest J, pp. 241–262. Albany, NY, USA: State University of New York Press [*260*]

Gemmill AW, Viney ME & Read AF (1997). Host immune status determines sexuality in a parasitic nematode. *Evolution* **51**:393–401 [*249, 257*]

Gepts P & Bliss FA (1985). F1 hybrid weakness in the common bean: Differential geographic origin suggests two gene pools in cultivated bean germplasm. *Journal of Heredity* **76**:447–450 [*444*]

Geritz SAH, Metz JAJ, Kisdi É & Meszéna G (1997). Dynamics of adaptation and evolutionary branching. *Physical Review Letters* **78**:2024–2027 [*42–43, 99–100*]

Geritz SAH, Kisdi É, Meszéna E & Metz JAJ (1998). Evolutionary singular strategies and the adaptive growth and branching of the evolutionary tree. *Evolutionary Ecology* **12**:35–57 [*144, 387*]

Gibbons A (1993). Where are "new" diseases born? *Science* **261**:6 [*37*]

Gibbons EF Jr, Durrant BS & Demarest J, eds. (1995). *Conservation of Endangered Species in Captivity*. Albany, NY, USA: State University of New York Press [*260–261*]

Gibson FD (1958). Malaria parasite survey

of some area in Southern Cameroon under United Kingdom Administration. *West African Medicine Journal* **VII**:170–178 [*355*]

Gilbert AN, Yamazaki K, Beauchamp GK & Thomas L (1986). Olfactory discrimination of mouse strains (*Mus musculus*) and major histocompatibility types by humans (*Homo sapiens*). *Journal of Comparative Psychology* **100**:262–265 [*254*]

Gill TJ III (1994). Reproductive immunology and immunogenetics. In *The Physiology of Reproduction*, Second Edition, eds. Knobil E & Neill JD, pp. 783–812. New York, NY, USA: Raven Press, Ltd. [*254*]

Gillespie J (1975). Natural selection for resistance to epidemics. *Ecology* **56**:493–495 [*223–224, 227*]

Gilpin ME (1975). *Group Selection in Predator–Prey Communities*. Princeton, NJ, USA: Princeton University Press [*308*]

Goater CP & Bush AO (1988). Intestinal helminth communities in long-billed curlews: The importance of congeneric host specialists. *Holartic Ecology* **11**:140–145 [*29*]

Goater CP, Esch GW & Bush AO (1987). Helminth parasites of sympatric salamanders: Ecological concepts at infracommunity, component and compound community levels. *American Midland Naturalist* **118**:289–300 [*29*]

Godfray HCJ (1994). *Parasitoids: Behavioural and Evolutionary Ecology*. Princeton, NJ, USA: Princeton University Press [*167*]

Goldman N (1993). Statistical tests of models of DNA substitution. *Journal of Molecular Evolution* **36**:182–198 [*263, 265–266*]

Goldman N (1998). Effects of sequence alignment procedures on estimates of phylogeny. *Bioessays* **20**:287–290 [*267*]

Gomulkiewicz R, Holt RD & Barfield M (1999). The effects of density-dependence and immigration on local adaptation and niche evolution in a "black-hole" sink environment. *Theoretical Population Biology* **55**:283–296 [*106*]

Gomulkiewicz R, Thompson JN, Holt RD, Nuismer SL & Hochberg ME (2000). Hot spots, cold spots, and the geographic mosaic theory of coevolution. *The American Naturalist* **156**:156–174 [*104, 108–109, 112, 117*]

Gonzalez JM, Olano V, Vergara J, Arevalo-

Herrera M, Carrasquilla G, Herrera S & Lopez JA (1997). Unstable, low-level transmission of malaria on the Colombian Pacific Coast. *Annals of Tropical Medicine and Parasitology* **91**:349–358   [*406*]

Gorer PA & Mikulska ZB (1959). Some further data on the H-2 system of antigens. *Proceedings of the Royal Society of London B* **151**:57–69   [*256*]

Gosgen R, Dunbar R, Haig D, Heyer E, Mace R, Milinski M, Pison G, Richner H, Strassmann B, Thaler D, Wedekind C & Stearns SC (1998). Evolutionary interpretation of the diversity of reproductive health and disease. In *Evolution in Health and Disease*, ed. Stearns SC. Oxford, UK: Oxford University Press   [*259–260*]

Gougeon ML, Garcia S, Heeney J, Tschopp R, Lecoeur H, Guetard D, Rame V, Dauguet C & Montagnier L (1993). Programmed cell death in AIDS-related HIV and SIV infections. *AIDS Research and Human Retroviruses* **9**:553–563   [*184*]

Gould F (1991). The evolutionary potential of crop pests. *American Scientist* **79**:496–507   [*457*]

Grafen A (1990). Biological signals as handicaps. *Journal of Theoretical Biology* **144**:517–546   [*251*]

Grahn M, Langesfors A & von Schantz T (1998). The importance of mate choice in improving viability in captive populations. In *Behavioral Ecology and Conservation Biology*, ed. Caro TM. Oxford, UK: Oxford University Press   [*252*]

Gravatt F (1914). *The Chestnut Blight in Virginia*. Richmond, VA, USA: Virginia Department of Agriculture   [*292*]

Gravenor MB, McLean AR & Kwiatkowski D (1995). The regulation of malaria parasitaemia: Parameter estimates for a population model. *Parasitology* **110**:115–122   [*172, 176*]

Gray BM & Dillon HC Jr (1988). Epidemiological studies of *Streptococcus pneumoniae* in infants: Antibody to types 3, 6, 14, and 23 in the first two years of life. *Journal of Infectious Diseases* **158**:948–955   [*372*]

Greene GR (1978). Meningitis due to *Haemophilus influenzae* other than type b: Case report and review. *Pediatrics* **62**:1021–1025   [*365*]

Greenwood BM, Bradley AK, Greenwood AM, Byars P, Jammeh K, Marsh K, Tulloch

S, Oldfield FSJ & Hayes RJ (1987). Mortality and morbidity from malaria among children in a rural area of The Gambia, West Africa. *Transactions of the Royal Society of Tropical Medicine and Hygiene* **81**:478–486   [*358*]

Greenwood B, Marsh K & Snow R (1991). Why do some African children develop severe malaria? *Parasitology Today* **7**:277–281   [*165, 172*]

Gregory CD, Dive C, Henderson S, Smith CA, Williams GT, Gordon J & Rickinson AB (1991). Activation of Epstein–Barr-virus latent genes protects human B-cells from death by apoptosis. *Nature* **349**:612–614   [*184, 190–191*]

Grenfell BT (1988). Gastrointestinal nematode parasites and the stability and productivity of intensive ruminant grazing systems. *Philosophical Transactions of the Royal Society of London B* **321**:541–563   [*31*]

Grenfell BT & Gulland FM (1995). Introduction: Ecological impact of parasitism on wildlife host populations. *Parasitology* **111**(Suppl.):S3–14   [*31, 415*]

Grenfell BT, Wilson K, Isham VS, Boyd HE & Dietz K (1995). Modelling the patterns of parasite aggregation in natural population: *Trichostrongylid nematode*–ruminant interactions as a case study. *Parasitology* **111**(Suppl.):S135–151   [*31, 33, 415*]

Griffiths AJF (1992). Fungal senescence. *Annual Review of Genetics* **26**:351–372   [*281*]

Griffiths AJF, Kraus SR, Barton R, Court DA, Myers CJ & Bertrand H (1990). Heterokaryotic transmission of senescence plasmid DNA in *Neurospora*. *Current Genetics* **17**:139–145   [*284*]

Grossman CJ (1985). Interactions between the gonadal steroids and the immune system. *Science* **227**:257–261   [*251*]

Gulland FMD (1995). The impact of infectious diseases on wild animal populations: A review. In *The Ecology of Infectious Diseases in Natural Populations*, eds. Grenfell BT & Dobson AP, pp. 20–51. Cambridge, UK: Cambridge University Press   [*416–417, 421*]

Gulland FMD, Albon SD, Pemberton JM, Moorcroft PR & Clutton-Brock TH (1993). Parasite-associated polymorphism in a cyclic ungulate population. *Proceedings of the Royal Society of London B* **254**:7–13   [*421*]

Gupta S & Day KP (1994a). A theoretical framework for the immunoepidemiology of *Plasmodium falciparum* malaria. *Parasite Immunology* **16**:361–370 [*358*]

Gupta S & Day KP (1994b). A strain theory of malarial transmission. *Parasitology Today* **10**:476–481 [*358*]

Gupta S & Hill A (1995). Dynamic interactions in malaria: Host heterogeneity meets parasite polymorphism. *Proceedings of the Royal Society of London B* **261**:271–277 [*230, 357*]

Gupta S, Hill AVS, Kwiatkowski D, Greenwood AM, Greenwood BM & Day KP (1994a). Parasite virulence and disease patterns in *P. falciparum* malaria. *Proceedings of the National Academy of Sciences of the USA* **91**:3715–3719 [*341, 355, 357*]

Gupta S, Swinton J & Anderson RM (1994b). Theoretical studies of the effects of genetic heterogeneity in the parasite population on the transmission dynamics of malaria. *Proceedings of the Royal Society of London B* **256**:231–238 [*348, 359*]

Gupta S, Trenholme K, Anderson RM & Day KP (1994c). Antigenic diversity and the transmission dynamics of *Plasmodium falciparum*. *Science* **263**:961–963 [*169, 171, 354, 358–359*]

Gupta S, Maiden M, Feavers I, Nee S, May RM & Anderson RM (1996). The maintenance of strain structure in populations of recombining infectious agents. *Nature Medicine* **2**:437–442 [*341, 352, 373, 394*]

Gupta S, Ferguson N & Anderson RM (1997). Vaccination and the population structure of antigenically diverse pathogens that exchange genetic material. *Proceedings of the Royal Society of London B* **264**:1435–1443 [*351, 360, 373*]

Gupta S, Ferguson N & Anderson RM (1998). Chaos, persistence, and evolution of strain structure in antigenically diverse infectious agents. *Science* **240**:912–915 [*352, 358*]

Gustafson L, Qarnström A & Sheldon B (1995). Trade-offs between life-history traits and a secondary sexual character in male collared flycatcher. *Nature* **375**:311–313 [*251*]

Guzmán P, Gilbertson RL, Nodari R, Johnson WC, Temple SR, Mandala D, Mkandawire ABC & Gepts P (1995). Characterization of variability in the fungus *Phaeoisariopsis griseola* suggests coevolution with the common bean (*Phaseolus vulgaris*). *Phytopathology* **85**:600–607 [*444*]

Gwaltney JM Jr, Sande MA, Austrian R & Hendley JO (1975). Spread of *Streptococcus pneumoniae* in families. II. Relation of transfer of *S. pneumoniae* to incidence of colds and serum antibody. *Journal of Infectious Diseases* **132**:62–68 [*372*]

Gyllenberg M & Metz JAJ (2001). On fitness in structured metapopulations. *Journal of Mathematical Biology* **43**:545–560 [*387, 397*]

Hairston NG, Smith FE & Slobodkin LB (1960). Community structure, population control and competition. *The American Naturalist* **94**:421–424 [*286, 297*]

Haldane JBS (1949). Disease and evolution. *La Ricerca Scientifica* **19**(Suppl.):68–76 [*222*]

Halloran ME, Haber M & Longini IM (1992). Interpretation and estimation of vaccine efficacy under heterogeneity. *American Journal of Epidemiology* **136**:328–343 [*426*]

Hamer JE, Talbot NJ & Levy M (1993). Genome dynamics and pathotype evolution in the rice blast fungus. In *Advances in Molecular Genetics of Plant–Microbe Interactions*, eds. Nester EW & Verma DPS, pp. 299–311. Dordrecht, Netherlands: Kluwer Academic Publishers [*445–446*]

Hamilton WD (1964). The genetical evolution of social behaviour. I and II. *Journal of Theoretical Biology* **7**:1–16, 17–52 [*89, 95*]

Hamilton WD (1972). Altruism and related phenomena, mainly in social insects. *Annual Review of Ecology and Systematics* **3**:193–232 [*166*]

Hamilton WD (1980). Sex versus non-sex versus parasite. *Oikos* **35**:282–290 [*223, 233–234, 244, 249*]

Hamilton WD (1993). Haploid dynamics polymorphism in a host with matching parasite: Effect of mutation/subdivision, linkage and patterns of selection. *Journal of Heredity* **84**:328–338 [*244*]

Hamilton WD & Howard JD (1994). Infection, polymorphism, and evolution. *Philosophical Transactions of the Royal Society of London B* **346**:267–385 [*223*]

Hamilton WD & Howard JC (1997). *Infection, Polymorphism and Evolution*. London, UK: Chapman & Hall [*398*]

Hamilton WD & Zuk M (1982). Heritable true fitness and bright birds: A role for parasites? *Science* **218**:384–387 [*251–252*]

Hamilton WD, Axelrod R & Tanese R (1990). Sexual reproduction as an adaptation to resist parasites (a review). *Proceedings of the National Academy of Sciences of the USA* **87**:3566–3573   [*233–234, 244, 249*]

Hammerschmidt R & Dann EK (1997). Induced resistance to disease. In *Environmentally Safe Approaches to Crop Disease Control*, eds. Rechcigl NA & Rechcigl JE, pp. 177–199. Boca Raton, FL, USA: CRC Press   [*437*]

Hanski I (1997). Metapopulation dynamics from concepts and observations to predictive models. In *Metapopulation Biology: Ecology, Genetics and Evolution*, eds. Hanski IA & Gilpin ME, pp. 69–91. San Diego, CA, USA: Academic Press   [*441*]

Haraguchi Y & Sasaki A (1996). Host–parasite arms race in mutation modifications: Indefinite escalation despite a heavy load? *Journal of Theoretical Biology* **183**:121–137   [*234, 396*]

Haraguchi Y & Sasaki A (1997). Evolutionary pattern of intra-host pathogen antigenic drift: Effect of cross-reactivity in immune response. *Philosophical Transactions of the Royal Society of London B* **352**:11–20   [*233, 246*]

Hartl DL & Clark AG (1997). *Principles of Population Genetics*, Third Edition. Sunderland, MA, USA: Sinauer Associates Inc. [*224*]

Hasegawa M, Kishino H & Yano T (1985). Dating the human–ape splitting by a molecular clock of mitochondrial DNA. *Journal of Molecular Evolution* **22**:160–174   [*263, 265, 270*]

Hassell MP (1978). *Dynamics of Arthropod Predator–Prey Systems*. Monographs in Population Biology. Princeton, NJ, USA: Princeton University Press   [*299*]

Hassell MP, Comins HN & May RM (1991). Spatial structure and chaos in insect population dynamics. *Nature* **353**:255–258   [*198–199*]

Hassell MP, Comins HN & May RM (1994). Species coexistence via self-organising spatial dynamics. *Nature* **370**:290–292   [*104, 198–199*]

Hastings A (1980). Disturbance, coexistence, history, and competition for space. *Theoretical Population Biology* **18**:363–373   [*138*]

Hay ME (1986). Associational defenses and the maintenance of species diversity: Turn-ing competitors into accomplices. *The American Naturalist* **128**:617–641   [*315*]

Hayden TP, Bidochka MJ & Khachatourians GC (1992). Entomopathogenicity of several fungi toward the English grain aphid (Homoptera: Aphididae) and enhancement of virulence with host passage of *Paecilomyces farinosus*. *Journal of Economic Entomology* **85**:58–64   [*455*]

Heads PA (1986). The costs of reduced feeding due to predator avoidance: Potential effects on growth and fitness in *Ischnura elegans* larvae (Odonata: Zygoptera). *Ecological Entomology* **11**:369–377   [*450*]

Hedrick PW (1988). HLA-sharing, recurrent spontaneous abortion, and the genetic hypothesis. *Genetics* **119**:199–204   [*259*]

Hedrick PW & Thomson G (1983). Evidence for balancing selection at HLA. *Genetics* **104**:449–456   [*254*]

Heesterbeek JAP & Dietz K (1996). The concept of R_0 in epidemic theory. *Statistica Neelandica* **50**:89–110   [*11*]

Heesterbeek JAP & Roberts MG (1995). Mathematical model for microparasites of wildlife. In *Ecology of Infectious Diseases in Natural Populations*, eds. Grenfell BT & Dobson AP, pp. 90–122. Cambridge, UK: Cambridge University Press   [*31*]

Heide-Jorgensen MP & Harkonen T (1992). Epizootology of the seal disease in the eastern North Sea. *Journal of Applied Ecology* **29**:99–107   [*228*]

Heino M, Metz JAJ & Kaitala V (1997). Evolution of mixed maturation strategies in semelparous life histories: The crucial role of dimensionality of feedback environment. *Philosophical Transactions of the Royal Society of London B* **352**:1647–1655   [*41*]

Heithoff DM, Sinsheimer RL, Low MA & Mahan MJ (1999). An essential role for DNA adenine methylation in bacterial virulence. *Science* **284**:967–970   [*380*]

Hellriegel B (1992). Modelling the immune response to malaria with ecological concepts: Short-term behaviour against long-term equilibrium. *Proceedings of the Royal Society of London B* **250**:249–256   [*176*]

Henter HJ (1995). The potential for coevolution in a host–parasitoid system. 2. Genetic variation within a population of wasps in the ability to parasitize an aphid host. *Evolution* **49**:439–445   [*448, 455, 457*]

Henter HJ & Via S (1995). The potential for

coevolution in a host–parasitoid system. 1. Genetic variation within an aphid population in susceptibility to a parasitic wasp. *Evolution* **49**:427–438 [*396, 448, 457–458*]

Herms DA & Mattson WJ (1992). The dilemma of plants: To grow or to defend. *Quarterly Review of Biology* **67**:283–335 [*318*]

Herre EA (1993). Population structure and the evolution of virulence in nematode parasites of fig wasps. *Science* **259**:1442–1445 [*166, 287, 392, 448*]

Herre EA (1995). Factors affecting the evolution of virulence: Nematode parasites of fig wasps as a case study. *Parasitology* **111**(Suppl.):S179–191 [*166, 172, 392, 448*]

Herre EA, Knowlton N, Mueller UG & Rehner SA (1999). The evolution of mutualisms: Exploring the paths between conflict and cooperation. *Trends in Ecology and Evolution* **14**:49–53 [*396*]

Hetzel C & Anderson RM (1996). The within-host cellular dynamics of bloodstage malaria: Theoretical and experimental studies. *Parasitology* **113**:25–38 [*176*]

Hewings AD (1989). Soybean dwarf virus. In *Compendium of Soybean Diseases*, eds. Sinclair JB & Blackman PA, pp. 54–55. St. Paul, MN, USA: APS Press [*439*]

Higgins DG, Bleasby AJ & Fuchs R (1991). CLUSTAL V: Improved software for multiple sequence alignment. *Computer Applications in the Biosciences* **8**:189–191 [*269, 273*]

Hill AVS, Allsopp CEM, Kwiatkowski D, Anstey NM, Twumasi P, Rowe PA, Bennet S, Brewster D, McMichael AJ & Greenwood BM (1991). Common West African HLA antigens are associated with protection from severe malaria. *Nature* **352**:595–600 [*354, 357*]

Hill AVS, Elvin J, Willis AC, Aidoo M, Allsopp CEM, Gotch FM, Gao XM, Takiguchi M, Greenwood BM, Townsend ARM, McMichael AJ & Whittle HC (1992). Molecular analysis of the association of HLA-B53 and resistance to severe malaria. *Nature* **360**:434–439 [*358*]

Hings IM & Billingham RE (1981). Splenectomy and sensitization of Fischer female rats favors histoincompatibility of R2 backcross progeny. *Transplantation Proceedings* **13**:1253–1255 [*256*]

Hings IM & Billingham RE (1983). Parity-induced changes in the frequency of RT1 heterozygotes in an R2 backcross. *Transplantation Proceedings* **15**:900–902 [*256*]

Hings IM & Billingham RE (1985). Maternal–fetal immune interactions and the maintenance of major histocompatibility complex polymorphism in the rat. *Journal of Reproductive Immunology* **7**:337–350 [*256*]

Hirayama K, Matsushita S, Kikuchi I, Iuchi M, Ohta N & Sasazuki T (1987). HLA-DQ is epistatic to HLA-DR in controlling the immune response to schistomal antigen in humans. *Nature* **327**:426–430 [*255*]

Hjältén J & Price PW (1997). Can plants gain protection from herbivory by association with unpalatable neighbours? A field experiment in a willow–sawfly system. *Oikos* **78**:317–322 [*315*]

Hobbs NT & Miller MW (1991). Interactions between pathogens and hosts: Simulation of pasteurellosis epizootics in bighorn sheep populations. In *Wildlife 2001: Populations*, eds. McCullough DR & Barrett RH, pp. 997–1007. New York, NY, USA: Elsevier Applied Science [*36*]

Hochberg ME (1989). The potential role of pathogens in biological control. *Nature* **337**:262–265 [*453, 455*]

Hochberg ME (1991). Non-linear transmission rates and the dynamics of infectious disease. *Journal of Theoretical Biology* **153**:301–321 [*199*]

Hochberg ME (1996). An integrative paradigm for the dynamics of monophagous parasitoid–host interactions. *Oikos* **77**:556–560 [*197*]

Hochberg ME (1997). Hide or fight? The competitive evolution of camouflage and encapsulation in host–parasitoid interactions. *Oikos* **80**:342–352 [*205*]

Hochberg ME (1998). Establishing genetic correlations involving parasite virulence. *Evolution* **52**:1865–1868 [*205*]

Hochberg ME (2000). What, conserve parasitoids? In *Parasitoid Population Biology*, eds. Hochberg ME & Ives AR. Princeton, NJ, USA: Princeton University Press [*197, 207*]

Hochberg ME & Holt RD (1990). The coexistence of competing parasites. I. The role of cross-species infection. *The American Naturalist* **136**:517–541 [*286, 288*]

Hochberg ME & Holt RD (1995). Refuge evolution and the population dynamics of coupled host–parasitoid associations. *Evolutionary Ecology* **9**:633–661   [*205*]

Hochberg ME & Ives AR (1999). Can natural enemies enforce geographical range limits? *Ecography* **22**:268–276   [*207–208*]

Hochberg ME & Van Baalen M (1998). Antagonistic coevolution over productivity gradients. *The American Naturalist* **152**:620–634   [*106, 113, 116, 198–200, 202, 205–206, 208*]

Hochberg ME, Elmes GW, Thomas JA & Clarke RT (1998). Effects of habitat reduction on the persistence of *Ichneumon eumerus*, the specialist parasitoid of *Maculinea rebeli*. *Journal of Insect Conservation* **2**:59–66   [*207*]

Hofbauer J & Sigmund K (1988). *The Theory of Evolution and Dynamical Systems*. Cambridge, UK: Cambridge University Press   [*238*]

Hofbauer J & Sigmund K (1998). *Evolutionary Games and Population Dynamics*. Cambridge, UK: Cambridge University Press   [*129–130*]

Hofmann SL, Oster CN, Plowe CV, Wollett GR, Beier JC, Chulay JD, Wirtz RA, Hollingdale MR & Mugambi M (1987). Naturally acquired antibodies to sporozoites do not prevent malaria: Vaccine development implications. *Science* **237**:639–642   [*358*]

Hokkanen H & Pimentel D (1984). New approach for selecting biological control agents. *Canadian Entomologist* **116**:1109–1121   [*208*]

Holland JH (1975). *Adaptation in Natural and Artificial Systems*. Ann Arbor, MI, USA: University of Michigan Press   [*212*]

Hollings M (1982). Mycoviruses and plant pathology. *Plant Disease* **66**:1106–1112   [*286*]

Holmes JC (1982). Impact of infectious disease agents on the population growth and geographical distribution of animals. In *Population Biology of Infectious Diseases*, eds. Anderson RM & May RM, pp. 37–51. Berlin, Germany: Springer-Verlag   [*416*]

Holmes EE (1997). Basic epidemiological concepts in a spatial context. In *Spatial Ecology: The Role of Space in Population Dynamics and Interspecific Interactions*, eds. Tilman D & Kareiva P, pp. 111–136. Princeton, NJ, USA: Princeton University Press   [*104*]

Holt RD (1977). Predation, apparent competition and the structure of prey communities. *Theoretical Population Biology* **12**:197–229   [*172, 317*]

Holt RD (1984). Spatial heterogeneity, indirect interactions, and the coexistence of prey species. *The American Naturalist* **124**:377–406   [*199*]

Holt RD (1985). Population dynamics in two-patch environments: Some anomalous consequences of an optimal habitat distribution. *Theoretical Population Biology* **28**:181–208   [*104–105, 199*]

Holt RD (1993). Ecology at the mesoscale: The influence of regional processes on local communities. In *Species Diversity in Ecological Communities*, eds. Ricklefs R & Schluter D, pp. 77–88. Chicago, IL, USA: University of Chicago Press   [*105*]

Holt RD (1996). Demographic constraints in evolution: Towards unifying the evolutionary theories of senescence and niche conservatism. *Evolutionary Ecology* **10**:1–11   [*104, 106, 108, 197, 205*]

Holt RD (1997). Rarity and evolution: Some theoretical considerations. In *The Biology of Rarity*, eds. Kunin WE & Gaston KJ, pp. 210–234. London, UK: Chapman & Hall   [*205*]

Holt RD & Gaines MS (1992). Analysis of adaptation in heterogeneous landscapes: Implications for the evolution of fundamental niches. *Evolutionary Ecology* **6**:433–447   [*106*]

Holt RD & Gomulkiewicz R (1997). How does immigration influence local adaptation? A reexamination of a familiar paradigm. *The American Naturalist* **149**:563–572   [*104, 106, 112, 199, 208*]

Holt RD & Hochberg ME (1997). When is biological control evolutionarily stable (or is it)? *Ecology* **78**:1673–1683   [*209, 457*]

Holt RD & Hochberg ME (1998). The coexistence of competing parasites. II. Hyperparasitism and food chain dynamics. *Journal of Theoretical Biology* **193**:485–495   [*286, 288*]

Holt RD & Kiett TH (2000). Alternative causes for range limits: A metapopulation perspective. *Ecology Letters* **3**:41–47   [*201*]

Holt RD & Lawton JH (1993). Apparent competition and enemy-free space in insect host–parasitoid communities. *The American Naturalist* **142**:623–645   [*205, 207*]

Holt RD & Polis GA (1997). A theoretical framework for intraguild predation. *The American Naturalist* **14**:745–764 [*317*]

Holt RD, Lawton JH, Gaston KJ & Blackburn TM (1997). On the relationship between range-size and local abundance: Back to basics. *Oikos* **78**:183–190 [*197, 201*]

Holt RD, Hochberg ME & Barfield M (1998). Population dynamics and the evolutionary stability of biological control. In *The Theory of Biological Control*, eds. Hawkins BA & Cornell HV. Cambridge, UK: Cambridge University Press [*205–206*]

Hopper KR, Roush RT & Powell W (1993). Management of genetics of biological-control introductions. *Annual Review of Entomology* **38**:27–51 [*455*]

Horimoto T, Rivera E, Pearson JE, Senne D, Krauss S, Kawaoka Y & Webster RG (1995). Origin and molecular changes associated with emergence of a highly pathogenic H5N2 influenza virus in Mexico. *Virology* **213**:223–230 [*268–269, 271*]

Howard RS & Lively CM (1994). Parasitism, mutation accumulation, and the maintenance of sex. *Nature* **367**:554–557 [*248–249*]

Hsieh TC, Aguerorosenfeld ME, Wu JM, Ng CY, Papanikolaou NA, Varde SA, Schwartz I, Pizzolo JG, Melamed M, Horowitz HW, Nadelman RB & Wormser GP (1997). Cellular changes and induction of apoptosis in human promyelocytic HL-60 cells infected with the agent of human granulocytic ehrlichiosis (HGE). *Biochemical and Biophysical Research Communications* **232**:298–303 [*190*]

Hsin H & Kenyon C (1999). Signals from the reproductive system regulate the lifespan of *C. elegans*. *Nature* **399**:362–366 [*452*]

Hudson PJ & Dobson AP (1995). Macroparasites: Observed patterns in naturally fluctuating animal populations. In *Ecology of Infectious Diseases in Natural Populations*, eds. Grenfell BT & Dobson AP, pp. 52–89. Cambridge, UK: Cambridge University Press [*31*]

Hudson PJ, Dobson AP & Newborn D (1992). Regulation and stability of a free-living host–parasite system: *Trichostrongylus tenuis* in red grouse. I. Monitoring and parasite reduction experiments. *Journal of Animal Ecology* **61**:477–486 [*31, 421*]

Huelsenbeck JP & Bull JJ (1996). A likelihood ratio test to detect conflicting phylogenetic signal. *Systematic Biology* **45**:92–98 [*273, 275*]

Huelsenbeck JP & Rannala B (1997). Phylogenetic methods come of age: Testing hypotheses in an evolutionary context. *Science* **276**:227–232 [*265–267*]

Hughes AL & Nei M (1988). Pattern of nucleotide substitution at major histocompatibility complex class I loci reveals overdominant selection. *Nature* **335**:167–170 [*210–211, 221*]

Hughes AL & Nei M (1989). Nucleotide substitution at major histocompatibility complex class II loci: Evidence for overdominant selection. *Proceedings of the National Academy of Sciences of the USA* **86**:958–962 [*210–211*]

Hughes AL & Nei M (1992). Models of host–parasite interaction and MHC polymorphism. *Genetics* **132**:863–864 [*211*]

Hurd H (1998). Parasite manipulation of insect reproduction: Who benefits? *Parasitology* **116**:S13–S21 [*452*]

Hurst LD (1991). The incidence and evolution of cytoplasmic male killers. *Proceedings of the Royal Society of London B* **244**:91–99 [*391*]

Hurst LD & Peck JR (1996). Recent advances in understanding of the evolution and maintenance of sex. *Trends in Ecology and Evolution* **11**:46–52 [*248–249*]

Hutchins M, Foose T & Seal US (1991). The role of veterinarian medicine in endangered species conservation. *Journal of Zoo and Wildlife Medicine* **22**:277–281 [*413, 421*]

Huynen MA & Hogeweg P (1989). Genetic algorithms and information accumulation during the evolution of gene regulation. In *Proceedings of the Third International Conference on Genetic Algorithms*, ed. Schaffer JD. San Mateo, CA, USA: Morgan Kaufmann Publishers [*212*]

Ina Y & Gojobori T (1994). Statistical analysis of nucleotide sequences of the hemagglutinin gene of human influenza A viruses. *Proceedings of the National Academy of Sciences of the USA* **91**:8388–8392 [*269*]

Ito M, Watanabe M, Ihara T, Kamiya H & Sakurai M (1997). Measles virus induces apoptotic cell death in lymphocytes activated with phorbol 12-myristate 13-acetate (PMA) plus calcium ionophore. *Clinical and Experimental Immunology* **108**:266–

271   [*190*]

Iwasa Y (2000). Lattice models and pair approximation in ecology. In *The Geometry of Ecological Interactions: Simplifying Spatial Complexity*, eds. Dieckmann U, Law R & Metz JAJ, pp. 227–251. Cambridge, UK: Cambridge University Press   [*89*]

Iwasa Y, Pomiankowski A & Nee S (1991). The evolution of costly mate preferences. II. The "handicap" principle. *Evolution* **45**:1431–1442   [*251*]

Jackson AC & Rossiter JP (1997). Apoptotic cell death is an important cause of neuronal injury in experimental Venezuelan equine encephalitis virus infection of mice. *Acta Neuropathologica* **93**:349–353   [*190*]

Jacoby GA (1996). Antimicrobial-resistant pathogens in the 1990s. *Annual Review of Medicine* **47**:169–179   [*326*]

Jaenike J (1978). An hypothesis to account for the maintenance of sex within populations. *Evolutionary Theory* **3**:191–194   [*244, 249*]

Jaenike J (1993). Suboptimal virulence of an insect-parasitic nematode. *Evolution* **50**:2241–2247   [*395*]

Jaenike J (1996). Suboptimal virulence of an insect-parasitic nematode. *Evolution* **50**:2241–2247   [*448*]

Jaenike J (1998). On the capacity of macroparasites to control insect populations. *The American Naturalist* **151**:84–96   [*31, 448, 452*]

Jansen VAA (1995). Effects of dispersal in a tri-trophic metapopulation model. *Journal of Mathematical Biology* **34**:195–224   [*317*]

Jansen VAA & Sabelis MW (1992). Prey dispersal and predator persistence. *Experimental and Applied Acarology* **14**:215–231   [*317*]

Jansen VAA & Sabelis MW (1995). Outbreaks of colony-forming pests in tri-trophic systems: Consequences for pest control and the evolution of pesticide resistance. *Oikos* **74**:172–176   [*317*]

Janssen A & Sabelis MW (1992). Phytoseiid life-histories, local predator–prey dynamics and strategies for control of tetranychid mites. *Experimental and Applied Acarology* **14**:233–250   [*302, 305*]

Janssen A, Bruin J, Jacobs G, Schraag R & Sabelis MW (1997a). Predators use volatiles to avoid prey patches with conspecifics. *Journal of Animal Ecology* **66**:223–232   [*300*]

Janssen A, Van Gool E, Lingeman R, Jacas J & Van de Klashorst G (1997b). Metapopulation dynamics of a persisting predator–prey system in the laboratory: Time series analysis. *Experimental and Applied Acarology* **21**:415–430   [*305*]

Janssen A, Pallini A, Venzon M & Sabelis MW (1998). Behaviour and indirect interactions in food webs of plant-inhabiting arthropods. *Experimental & Applied Acarology* **22**:497–521   [*455*]

Janzen DH (1966). Coevolution of mutualism between ants and acacias in Central America. *Evolution* **20**:249–275   [*297*]

Jarosz AM & Burdon JJ (1991). Host–pathogen interactions in natural populations of *Linum marginale* and *Melampsora lini*: II. Local and regional variation in patterns of resistance and racial structure. *Evolution* **45**:1618–1627   [*440*]

Jarosz AM & Burdon JJ (1992). Host–pathogen interactions in a natural population of *Linum marginale* and *Melampsora lini*: III. Influence of pathogen epidemics on host survivorship and flower production. *Oecologia* **89**:53–61   [*440*]

Jarosz AM & Davelos AL (1995). Effects of disease in wild plant populations and the evolution of pathogen aggressiveness. *New Phytologist* **129**:371–387   [*448*]

Jarra W & Brown KN (1989). Protective immunity to malaria: Studies with cloned lines of rodent malaria in CBA/Ca mice. IV. The specificity of mechanisms resulting in crisis and resolution of the primary acute phase parasitaemia of *Plasmodium chabaudi chabaudi* and *P. yoelii yoelii*. *Parasite Immunology* **11**:1–13   [*170*]

Jayaker SD (1970). A mathematical model for interaction of gene frequencies in a parasite and its host. *Theoretical Population Biology* **1**:140–164   [*237, 244*]

Jaynes RA & Elliston JE (1980). Pathogenicity and canker control by mixtures of hypovirulent strains of *Endothia parasitica* in American chestnut. *Phytopathology* **70**:453–456   [*443*]

Jelachich ML & Lipton HL (1996). Theilers murine encephalomyelitis virus kills restrictive but not permissive cells by apoptosis. *Journal of Virology* **70**:6856–6861   [*190*]

Jeltsch F, Müller MS, Grimm V, Wissel C & Brandl R (1997). Pattern formation triggered by rare events: Lessons from the spread of rabies. *Proceedings of the Royal Society of London B* **264**:495–503 [*89*]

Jeon KW (1972). Development of cellular dependence in infective organisms: Microsurgical studies in amoebas. *Science* **176**:1122–1123 [*391*]

Jeon KW (1983). Integration of bacterial endosymbionts in amoebae. *International Review of Cytology* **14**:29–47 [*391*]

Jernigan DB, Cetron MS & Breiman RF (1996). Minimizing the impact of drug-resistant *Streptococcus pneumoniae* (DRSP): A strategy from the DRSP working group. *Journal of the American Medical Association* **275**:206–209 [*326*]

Johanson WG Jr, Blackstock R, Pierce AK & Sanford JP (1970). The role of bacterial antagonism in pneumococcal colonization of the human pharynx. *Journal of Laboratory and Clinical Medicine* **75**:946–952. [*366*]

Johnstone RA (1995). Sexual selection, honest advertisement, and the handicap principle: Reviewing the evidence. *Biological Reviews* **70**:1–65 [*251*]

Jolivet P (1996). *Ants and Plants: An Example of Coevolution*. Leiden, Netherlands: Backhuys Publishers [*298*]

Jones TR (1997). Quantitative aspects of the relationship between the sickle-cell gene and malaria. *Parasitology Today* **13**:107–111 [*223*]

Joosten MHAJ, Cozijnsen TJ & De Wit PJGM (1994). Host resistance to a fungal pathogen lost by a single base-pair change in an avirulence gene. *Nature* **367**:384–386 [*380*]

Jukes TH & Cantor CR (1969). Evolution of protein molecules. In *Mammalian Protein Metabolism*, ed. Munro HN, pp. 21–123. New York, NY, USA: Academic Press [*263–264*]

Kaltz O & Shykoff JA (1998). Local adaptation in host–parasite systems. *Heredity* **81**:361–370 [*106*]

Kam KM, Luey KY, Fung SM, Yiu PP, Harden TJ & Cheung MM (1995). Emergence of multiple-antibiotic-resistant *Streptococcus pneumoniae* in Hong Kong. *Antimicrobial Agents and Chemotherapy* **39**:2667–2670 [*337*]

Karaolis DKR, Somara S, Maneval Jr DR, Johnson JA & Kaper JB (1999). A bacteriophage encoding a pathogenicity island, a type-IV pilus and a phage receptor in cholera bacteria. *Nature* **399**:375–379 [*381*]

Kawaoka Y, Nestorowicz A, Alexander DJ & Webster RG (1987). Molecular analysis of the hemagglutinin genes of H5 influenza viruses: Origin of a virulent turkey strain. *Virology* **158**:218–227 [*268*]

Kawecki TJ (1995). Demography of source–sink populations and the evolution of ecological niches. *Evolutionary Ecology* **9**:38–44 [*104, 106, 108*]

Keddy PA (1981). Experimental demography of the sand-dune annual, *Cakile edentula*, growing along an environmental gradient in Nova Scotia. *Journal of Ecology* **69**:615–630 [*105*]

Keeling MJ (1999). The effects of local spatial structure on epidemiological invasions. *Proceedings of the Royal Society of London B* **266**:859–867 [*85, 89, 97*]

Keeling MJ (2000). Evolutionary dynamics in spatial host–parasite systems. In *The Geometry of Ecological Interactions: Simplifying Spatial Complexity*, eds. Dieckmann U, Law R & Metz JAJ, pp. 271–291. Cambridge, UK: Cambridge University Press [*89*]

Keeling MJ & Gilligan CA (2000). Metapopulation dynamics of bubonic plague. *Nature* **407**:903–906 [*382*]

Keeling MJ, Rand DA & Morris A (1997). Correlation models for childhood epidemics. *Proceedings of the Royal Society of London B* **264**:1149–1156 [*89, 97, 103*]

Keller SM, Wolfe MS, McDermott JM & BA McDonald (1997). High genetic similarity among populations of *Phaeosphaeria nodorum* across wheat cultivars and regions of Switzerland. *Phytopathology* **87**:1134–1139 [*444*]

Kelly T, Dillard JP & Yother J (1994). Effect of genetic switching of capsular type on virulence of *Streptococcus pneumoniae*. *Infection and Immunity* **62**:1813–1819 [*371*]

Kermack WO & McKendrick AG (1927). A contribution to the mathematical theory of epidemics. Part I. *Proceedings of the Royal Society of London A* **115**:700–721 [*11, 75–76, 96, 225, 426*]

Kermack WO & McKendrick AG (1932). A contribution to the mathematical theory of epidemics. II. The problem of endemicity. *Proceedings of the Royal Society of London A* **138**:55–83 [*426*]

Kermack WO & McKendrick AG (1933). A contribution to the mathematical theory of epidemics. III. Further studies of the problem of endemicity. *Proceedings of the Royal Society of London A* **139**:94–122   [*426*]

Kermack WO & McKendrick AG (1937). A contribution to the mathematical theory of epidemics. IV. Analysis of experimental epidemics of the virus disease mouse ectromelia. *Journal of Hygiene* **37**:172–187 [*426*]

Kermack WO & McKendrick AG (1939). A contribution to the mathematical theory of epidemics. V. Analysis of experimental epidemics of mouse typhoid: A bacterial disease conferring incomplete immunity. *Journal of Hygiene* **39**:271–288   [*426*]

Kerr PJ & Best SM (1998). Myxoma virus in rabbits. *Reviews of the Science and Technology Office for International Epizootiology* **17**:256–268   [*452*]

Kholkute SD & Dukelow WR (1995). Comparative reproductive physiology of female and male nonhuman primates. In *Conservation of Endangered Species in Captivity: An Interdisciplinarian Approach*, eds. Gibbon EF Jr, Durrant BS & Demarest J, pp. 407–424. Albany, NY, USA: State University of New York Press   [*260*]

Kimura M (1980). A simple method for estimating evolutionary rates of base substitutions through comparative studies of nucleotide sequences. *Journal of Molecular Evolution* **16**:111–120   [*263*]

Kirby GC & Burdon JJ (1997). Effects of mutation and random drift on Leonard's gene-for-gene coevolution model. *Phytopathology* **87**:488–493   [*438*]

Kirkpatrick M (1982). Sexual selection and the evolution of female choice. *Evolution* **30**:1–12   [*251*]

Kirkpatrick M & Barton NH (1997). Evolution of a species' range. *The American Naturalist* **150**:1–23   [*104*]

Kirkwood TBL & Holliday R (1979). The evolution of ageing and longevity. *Proceedings of the Royal Society of London B* **205**:531–546   [*284*]

Kisdi É & Meszéna G (1993). Density-dependent life-history evolution in fluctuating environments. In *Adaptation in a Stochastic Environment*, eds. Yoshimura J & Clark C, *Lecture Notes in Biomathematics*, Vol. 98, pp. 26–62. Berlin, Germany: Springer-Verlag   [*42*]

Kitchen SF (1949). Symptomatology: General considerations. In *Malariology: A Compre-*hensive Survey of All Aspects of This Group of Diseases from a Global Standpoint*, ed. Boyd MF, pp. 966–994. Philadelphia, PA, USA: Saunders   [*406*]

Klebanoff A & Hastings A (1994). Chaos in three-species food chains. *Journal of Mathematical Biology* **32**:427–451   [*317*]

Klein J (1980). Generation of diversity to MHC loci: Implications for T cell receptor repertoires. In *Immunology 80*, eds. Fougereau M & Dausset J. London, UK: Academic Press [*211, 221*]

Klein J (1986). *Natural History of the Major Histocompatibility Complex*. New York, NY, USA: John Wiley and Sons   [*255*]

Klein J & Klein D (1991). *Molecular Evolution of the MHC Complex*. Heidelberg, Germany: Springer   [*211, 221*]

Kloosterman A, Permentier HK & Ploeger HW (1992). Breeding cattle and sheep for resistance to gastrointestinal nematodes. *Parasitology Today* **8**:30–335   [*415*]

Knolle H (1989). Host density and the evolution of parasite virulence. *Journal of Theoretical Biology* **136**:199–207   [*166, 172*]

Koella JC (1999). An evolutionary view of the interactions between anophiline mosquitoes and malaria parasites. *Microbes and Infection* **1**:303–308   [*456*]

Koella JC & Doebeli M (1999). Population dynamics and the evolution of virulence in epidemiological models with discrete host generations. *Journal of Theoretical Biology* **198**:461–475   [*106*]

Kolberg R (1994). Finding "sustainable" ways to prevent parasitic diseases. *Science* **264**:1859–1861   [*398*]

Kondrashov AS (1993). Classification of hypotheses on the advantage of amphimixis. *Journal of Heredity* **84**:372–387 [*248–249*]

Koonin EV & Dolja VV (1993). Evolution and taxonomy of positive-strand RNA viruses – implications of comparative-analysis of amino-acid sequences. *CRC Critical Reviews in Biochemistry and Molecular Biology* **28**:375–430   [*380*]

Koopman JS & Longini IM Jr (1994). The ecological effects of individual exposure and nonlinear disease dynamics in populations. *American Journal of Public Health* **84**:836–842   [*426*]

Koptur S (1992). Extrafloral nectary-mediated interactions between insects and plants. In *Insect–Plant Interactions IV*, ed. Bernays

EA, pp. 81–129. Boca Raton, FL, USA: CRC Press [*298*]

Koralnik IJ (1994). A wide spectrum of simian T-cell leukemia/lymphotropic virus type I variants in nature: Evidence for interspecies transmission in Equatorial Africa. *Journal of Virology* **68**:2693–2707 [*382*]

Kover PX (2000). Effects of parasitic castration on plant resource allocation. *Oecologia* **123**:48–56 [*452*]

Kraaijeveld AR & Godfray HCJ (1997). Trade-off between parasitoid resistance and larval competitive ability in *Drosophila melanogaster*. *Nature* **389**:278–280 [*245, 396, 458*]

Kraaijeveld AR & Van Alphen JJM (1994). Geographical variation in resistance of the parasitoid *Asobara tabida* against encapsulation by *Drosophila melanogaster* larvae: The mechanism explored. *Physiological Entomology* **19**:9–14 [*206*]

Kraaijeveld AR & Van Alphen JJM (1995). Geographical variation in encapsulation ability of *Drosophila melanogaster* larvae and evidence for parasitoid-specific components. *Evolutionary Ecology* **9**:10–17 [*206*]

Krakauer DC & Payne RJH (1997). The evolution of virus-induced apoptosis. *Proceedings of the Royal Society of London B* **264**:1757–1762 [*184*]

Kretzschmar M (1996). Graphs and line graphs as a model for contact patterns. *Zeitschrift für Angewandte Mathematik und Mechanik* **76**:433–436 [*102*]

Kretzschmar M & Morris M (1996). Measures of concurrency in networks and the spread of infectious disease. *Mathematical Biosciences* **133**:165–195 [*102*]

Kretzschmar HA, Giese A, Brown DR, Herms J, Keller B, Schmidt B & Groschup M (1997). Cell death in prion disease. *Journal of Neural Transmission* **50**(Suppl.):191–210 [*190*]

Kristinsson KG, Hjalmarsdottir MA & Steingrimsson O (1992). Increasing penicillin-resistance in pneumococci in Iceland. *Lancet* **339**:1606–1607 [*383, 398*]

Krotoski W (1985). Discovery of the hypnozoite and a new theory of malarial relapse. *Transactions of the Royal Society of Tropical Medicine Hygiene* **79**:1–11 [*406*]

Kumar S (1995). Bubonic plague in Surat. *Lancet* **345**:714 [*381*]

Kun JFJ, Kremsner PG & Kretschmer H (1997). Malaria acquired 13 times in two years in Germany. *New England Journal of Medicine* **337**:1636 [*409*]

Kun JFJ, Schmidt-Ott RJ, Lehman LG, Lell B, Luckner D, Greve B, Matousek P & Kremsner PG (1998). Merozoite surface antigen 1 and 2 genotypes and rosetting of *Plasmodium falciparum* in severe and mild malaria in Lambaréné, Gabon. *Transactions of the Royal Society of Tropical Medicine and Hygiene* **92**:110–114 [*168, 407*]

Kuznetsov YuA & Rinaldi S (1996). Remarks on food chain dynamics. *Mathematical Biosciences* **134**:1–33 [*317*]

Kyes S, Harding R, Black G, Craig A, Peshu N, Newbold C & Marsh K (1997). Limited spatial clustering of individual *Plasmodium falciparum* alleles in field isolates from coastal Kenya. *American Journal of Tropical Medicine and Hygiene* **57**:205–215 [*168*]

Ladle RJ (1992). Parasites and sex: Catching the Red Queen. *Trends in Ecology and Evolution* **7**:405–408 [*249–250*]

Lam KM (1996). Growth of Newcastle disease virus in chicken macrophages. *Journal of Comparative Pathology* **115**:253–263 [*190*]

Lande R (1981). Models of speciation by sexual selection on polygenic traits. *Proceedings of the National Academy of Sciences of the USA* **78**:3721–3725 [*251*]

Laugé R & de Wit PJGM (1998). Fungal avirulence genes: Structure and possible functions. *Fungal Genetics and Biology* **24**:285–297 [*440*]

Law R & Dieckmann U (1998). Symbiosis through exploitation and the merger of lineages in evolution. *Proceedings of the Royal Society of London B* **265**:1245–1253 [*396*]

Lawlor DA, Ward FE, Ennis PD, Jackson AP & Parham P (1988). HLA-A and B polymorphisms predate the divergence of humans and chimpanzees. *Nature* **335**:268–271 [*211, 221*]

Lawlor BJ, Read AF, Keymer AE, Perveen G & Crompton DWT (1990). Non-random mating in a parasitic worm: Mate choice by males? *Animal Behaviour* **40**:870–876 [*258*]

LeClerc JE, Li B, Payne WL & Cebula TA (1996). High mutation frequencies among *Escherichia coli* and *Salmonella* pathogens. *Science* **274**:1208–1211 [*380*]

Lehmann R (1992). Ectoparasite impacts on

Gerbillus andersoni allenbyi under natural conditions. *Parasitology* **104**:479–488 [*421*]

Leibold MA (1996). A graphical model of keystone predators in food webs: Trophic regulation of abundance, incidence, and diversity patterns in communities. *The American Naturalist* **147**:784–812 [*199*]

Leister D, Kurth J, Laurie DA, Yano M, Sasaki T, Devos K, Graner A & Schulze-Lefert P (1998). Rapid reorganization of resistance gene homologues in cereal genomes. *Proceedings of the National Academy of Sciences of the USA* **95**:370–375 [*234*]

Lenski RE (1988). Evolution of plague virulence. *Nature* **334**:473–474 [*287*]

Lenski RE & May RM (1994). The evolution of virulence in parasites and pathogens: Reconciliation between two competing hypotheses. *Journal of Theoretical Biology* **169**:253–265 [*86, 92, 106, 250, 287–288, 295, 404*]

Leonard KJ (1977). Selection pressures and plant pathogens. *Annals of the New York Academy of Science* **287**:207–222 [*237*]

Leonard KJ (1994). Stability of equilibria in a gene-for-gene coevolution of host–parasites interactions. *Phytopathology* **84**:70–77 [*438*]

Leopardi R, Vansant C & Roizman B (1997). The Herpes simplex virus 1 protein kinase U(S)3 is required for protection from apoptosis induced by the virus. *Proceedings of the National Academy of Sciences of the USA* **94**:7891–7896 [*190*]

Lesna I & Sabelis MW (1999). Diet-dependent female choice for males with "good genes" in a soil predatory mite. *Nature* **401**:581–584 [*456*]

Leung B & Forbes MR (1998). The evolution of virulence: A stochastic simulation model examining parasitism at individual and population levels *Evolutionary Ecology* **12**:165–177 [*166, 172*]

Levin SA (1970). Community equilibria and stability, and an extension of the competitive exclusion principle. *The American Naturalist* **104**:413–423 [*138*]

Levin SA (1983a). Coevolution. In *Population Biology*, eds. Freedman H & Stroeck C, *Lecture Notes in Biomathematics*, Vol. 52, pp. 328–334. Berlin, Germany: Springer-Verlag [*124*]

Levin SA (1983b). Some approaches to modelling of coevolutionary interactions. In *Coevolution*, ed. Nitecki M, pp. 21–65. Chicago, IL, USA: University of Chicago Press [*124*]

Levin BR & Bull JJ (1994). Short-sighted evolution and the virulence of pathogenetic microorganisms. *Trends in Microbiology* **2**:76–81 [*430*]

Levin BR & Bull JJ (1996). Phage therapy revisited: The population biology of a bacterial infection and its treatment with bacteriophage and antibiotics. *The American Naturalist* **147**:881–898 [*286*]

Levin BR & Lenski RE (1983). Coevolution in bacteria and their viruses and plasmids. In *Coevolution*, eds. Futuyuma DJ & Slatkin M, pp. 99–127. Sunderland, MA, USA: Sinauer Associates Inc. [*286*]

Levin SA & Pimentel D (1981). Selection of intermediate rates of increase in parasite–host systems. *The American Naturalist* **117**:308–315 [*10–11, 35, 88, 124, 138, 166, 172, 176, 230, 288*]

Levin SA & Udovic JD (1977). A mathematical model of coevolving populations. *The American Naturalist* **111**:657–675 [*223*]

Levin BR, Allison A, Bremermann J, Cavalli-Sforza LL, Clarke BC, Frentzel-Beyme R, Hamilton WD, Levin SA, May RM & Thieme HR (1982). Evolution of parasites and hosts. In *Population Biology of Infectious Diseases*, eds. Anderson RM & May RM, pp. 213–243. Berlin, Germany: Springer-Verlag [*10–11*]

Levin BR, Lipsitch M, Perrot V, Schrag S, Antia R, Simonsen L, Walker NM & Stewart FM (1997). The population genetics of antibiotic resistance. *Clinical Infectious Diseases* **24**:S9–S16 [*338*]

Levin BR, Lipsitch M & Bonhoeffer S (1999). Population biology, evolution, and infectious disease: Convergence and synthesis. *Science* **283**:806–809 [*393, 450*]

Levine AJ (1992). *Viruses*. New York, NY, USA: WH Freeman and Company [*268*]

Levine RJ, Khan MR, D'Souza S & Nalin DR (1976). Cholera transmission near a cholera hospital. *Lancet* **ii**:84–86 [*18*]

Levine BL, Huang Q, Isaacs JT, Reed JC, Griffin DE & Hardwick JM (1993). Conversion of lytic to persistent alphavirus infection by the *bcl-2* cellular oncogene. *Nature* **361**:739–742 [*189–190, 193*]

Levine BL, Goldman JE, Jiang HH, Griffin

DE & Hardwick JM (1996). *bcl-2* protects mice against fatal alphavirus encephalitis. *Proceedings of the National Academy of Sciences of the USA* **93**:4810–4815 [*184, 189–190*]

Levins R (1969). Some demographic and genetic consequences of environmental heterogeneity for biological control. *Bulletin of the Entomological Society of America* **15**:237–240 [*441*]

Levins R (1970). Extinction. In *Some Mathematical Problems in Biology*, ed. Gerstenhaber M, pp. 75–107. Providence, RI, USA: American Mathematical Society [*441*]

Levy SB (1998). The challenge of antibiotic resistance. *Scientific American* **278**:46–53 [*60–61, 65*]

Levy M, Correa-Victoria FJ, Zeigler RS, Xu S & Hamer JE (1993). Genetic diversity of the rice blast fungus in a disease nursery in Colombia. *Phytopathology* **83**:1427–1433 [*444–445*]

Lewis JW (1981). On the coevolution of pathogen and host: II. Selfing hosts and haploid pathogens. *Journal of Theoretical Biology* **93**:953–985 [*223*]

Lewis WM Jr (1987). The cost of sex. In *The Evolution of Sex and Its Consequences*, ed. Stearns SC, pp. 33–57. Boston, MA, USA: Birkhäuser [*248*]

Li W (1997). *Molecular Evolution*. Sunderland, MA, USA: Sinauer Associates Inc. [*262*]

Li XS, Chao CY, Gao HM, Zhang YQ, Ishida M, Kanegae Y, Endo A, Nerome R, Omoe K & Nerome K (1992). Origin and evolutionary characteristics of antigenic reassortant influenza A (H1N2) viruses isolated from man in China. *Journal of General Virology* **73**:1329–1337 [*268*]

Lima SL & Dill LM (1990). Behavioral decisions made under the risk of predation: A review and prospectus. *Canadian Journal of Zoology* **68**:619–640 [*450*]

Lipsitch M (1997). Vaccination against colonizing bacteria with multiple serotypes. *Proceedings of the National Academy of Science of the USA* **94**:6571–6576 [*360, 366–368, 372–373*]

Lipsitch M (1999). Bacterial vaccines and serotype replacement: Lessons from *Haemophilus influenzae* and prospects for *Streptococcus pneumoniae*. *Emerging Infectious Diseases* **5**:336–345 [*369*]

Lipsitch M & Moxon ER (1997). Virulence and transmissibility of pathogens: What is the relationship? *Trends in Microbiology* **5**:31–37 [*35, 371, 395*]

Lipsitch M & Nowak MA (1995). The evolution of virulence in sexually transmitted HIV/AIDS. *Journal of Theoretical Biology* **174**:427–440 [*160*]

Lipsitch M, Herre EA & Nowak MA (1995a). Host population structure and the evolution of virulence: A "law of diminishing returns." *Evolution* **49**:743–748 [*106, 124*]

Lipsitch M, Nowak MA, Ebert D & May RM (1995b). The population dynamics of vertically and horizontally transmitted parasites. *Proceedings of the Royal Society of London B* **260**:321–327 [*395–396*]

Lipsitch M, Siller S & Nowak MA (1996). The evolution of virulence in pathogens with vertical and horizontal transmission. *Evolution* **50**:1729–1741 [*395–396*]

Lipsitch M, Dykes, JK, Johnson SE, Ades EW, King J, Briles DE & Carlone GM (2000). Competition among *Streptococcus pneumoniae* for intranasal colonization in a mouse model. *Vaccine* **18**:2895–2901 [*366, 372*]

Little TJ & Ebert D (2000). The cause of parasite infections in natural populations of *Daphnia (Crustacea: Cladocera)*. *Proceedings of the Royal Society of London B* **267**:2037–2042 [*396*]

Liu Y-C & Milgroom MG (1996). Correlation between hypovirus transmission and the number of vegetative incompatibility (*vic*) genes different among isolates from a natural population of *Cryphonectria parasitica*. *Phytopathology* **86**:79–86 [*292–293*]

Liu Y-C, Bortesi P, Double ML, MacDonald WL & Milgroom MG (1996). Diversity and multilocus genetic structure in populations of *Cryphonectria parasitica*. *Phytopathology* **86**:1344–1351 [*443*]

Lively CM (1989). Adaptation by a parasitic trematode to local populations of its snail host. *Evolution* **43**:1663–1671 [*422*]

Lively CM (1992). Parthenogenesis in a freshwater snail: Reproductive assurance versus parasitic release. *Evolution* **46**:907–913 [*33*]

Lloyd DG (1980). Benefits and handicaps of sexual reproduction. *Evolutionary Biology* **13**:69–107 [*248*]

Lo RYC & Macdonald LE (1991). *Pasteurella haemolytica* is highly sensitive to UV irradiation. *Mutation Research* **263**:159–164 [*37*]

Longini IM (1983). Models of epidemics and endemicity in genetically variable host populations. *Journal of Mathematical Biology* **17**:289–304 [*223, 227–228*]

Longini IM, Datta S & Halloran ME (1996). Measuring vaccine efficacy for both susceptibility to infection and reduction in infectiousness for prophylactic HIV-1 vaccines. *Journal of AIDS* 1–17 [*426*]

Lowe SW & Ruley HE (1993). Stabilization of the *p53* tumor suppressor is induced by adenovirus-5 E1A and accompanies apoptosis. *Genes and Development* **7**:535–545 [*184*]

Luxemburger C, Ricci F, Nosten F, Raimond D, Bathet S & White N (1997). The epidemiology of severe malaria in an area of low transmission in Thailand. *Transactions of the Royal Society of Tropical Medicine and Hygiene* **91**:256–262 [*406*]

MacArthur RH & Wilson EO (1967). *The Theory of Island Biogeography*. Princeton, NJ, USA: Princeton University Press [*198*]

Macdonald G (1952). The analysis of equilibrium in malaria. *Tropical Diseases Bulletin* **49**:813–828 [*11, 339*]

Macdonald G (1957). *The Epidemiology and Control of Malaria*. London, UK: Oxford University Press [*456*]

Mackinnon MJ & Read AF (1999a). Genetic relationships between parasite virulence and transmission in the rodent malaria *Plasmodium chabaudi*. *Evolution* **53**:689–703 [*170–171, 389*]

Mackinnon MJ & Read AF (1999b). Selection for high and low virulence in the malaria parasite *Plasmodium chabaudi*. *Proceedings of the Royal Society of London B* **266**:741–748 [*392*]

Maclean DJ, Braithwaite KS, Irwin JAG, Manners JM & Groth JV (1995). Random amplified polymorphic DNA reveals relationships among diverse genotypes in Australian and American collections of *Uromyces appendiculatus*. *Phytopathology* **85**:757–765 [*444*]

Maekawa T, Kimura S, Hirakawa K, Murakami A, Zon G & Abe T (1995). Sequence specificity on the growth suppression and induction of apoptosis of chronic myeloid-leukemia cells by *bcr-abl* antisense oligodeoxynucleoside phosphorothioates. *International Journal of Cancer*

62:63–69 [*195*]

Magowan C, Wollish W, Anderson L & Leech J (1988). Cytoadherence by *Plasmodium falciparum* infected erythrocytes is correlated with the expression of a family of variable proteins on infected erythrocytes. *Journal of Experimental Medicine* **168**:1307–1320 [*354*]

Marina CF, Arredondo-Jiménez JI, Castillo A & Williams T (1999). Sublethal effects of iridovirus disease in a mosquito. *Oecologia* **119**:383–388 [*454–455*]

Marquis RJ & Alexander HM (1992). Evolution of resistance and virulence in plant herbivore and plant–pathogen interactions. *Trends in Ecology and Evolution* **7**:126–129 [*415*]

Marsh K & Snow RW (1997). Host–parasite interaction and morbidity in malaria endemic areas. *Philosophical Transactions of the Royal Society of London* **352**:1385–1394 [*165, 171*]

Marsh K, Hayes RH, Otoo L, Carson DC & Greenwood BM (1989). Antibodies to blood stage antigens of *Plasmodium falciparum* in rural Gambians and their relation to protection against infection. *Transactions of the Royal Society of Tropical Medicine and Hygiene* **83**:293–303 [*358*]

Martignoni ME & Iwai PJ (1986). *A Catalog of Viral Diseases of Insects, Mites and Ticks*, Fourth Edition, revised. USDA Forest Service General Technical Report PNW-195. Corvallis, OR, USA: USDA Forest Service [*77*]

Martinez JP, Groth JP & Young ND (1994). Segregation and phylogenetic data suggest speciation may be occurring within *Uromyces appendiculatus*. *Phytopathology* **84**:1096–1097 (Abstract) [*444*]

Martinez JP, Groth JP & Young ND (1996). Non-Mendelian and skewed segregation of DNA markers in wide crosses of the bean rust, *Uromyces appendiculatus*. *Current Genetics* **29**:159–167 [*444*]

Matsuda H & Abrams PA (1994). Timid consumers: Self-extinction due to adaptive change in foraging and antipredator effort. *Theoretical Population Biology* **45**:76–91 [*56, 234*]

Matsuda H, Ogita N, Sasaki A & Satō K (1992). Statistical mechanics of population: The lattice Lotka–Volterra model. *Progress in Theoretical Physics* **88**:1035–1049 [*89,*

93, 96, 98]

May RM (1977). Togetherness among schistosomes: Its effect on the dynamics of infection. *Mathematical Biosciences* **35**:301–343 [*29*]

May RM (1983). Parasitic infections as regulators of animal populations. *American Scientist* **71**:36–45 [*39, 49*]

May RM (1988). Conservation and disease. *Conservation Biology* **2**:28–30 [*413*]

May RM & Anderson RM (1978). Regulation and stability of host–parasite population interactions. II. Destabilizing processes. *Journal of Animal Ecology* **47**:248–267 [*28, 31, 75*]

May RM & Anderson RM (1979). Population biology of infectious diseases: Part II. *Nature* **280**:455–461 [*426*]

May RM & Anderson R (1983a). Epidemiology and genetics in the coevolution of parasites and hosts. *Proceedings of the Royal Society of London B* **219**:281–313 [*11, 166, 199, 223, 227, 348, 359*]

May RM & Anderson RM (1983b). Parasite–host coevolution. In *Coevolution*, eds. Futuyuma DJ & Slatkin M, pp. 186–206. Sunderland, MA, USA: Sinauer Associates Inc. [*453*]

May RM & Dobson AP (1986). Population dynamics and the rate of evolution of pesticide resistance. In *Pesticide Resistance Management*. Washington, DC, USA: NAS-NRC Publications [*414*]

May RM & Hassell MP (1981). The dynamics of multiparasitoid–host interactions. *The American Naturalist* **117**:234–261 [*286*]

May RM & Leonard W (1975). Nonlinear aspects of competition between three species. *SIAM Journal of Applied Mathematics* **29**:243–252 [*130*]

May RM & Nowak MA (1994). Superinfection, metapopulation dynamics, and the evolution of diversity. *Journal of Theoretical Biology* **170**:95–114 [*124, 136–139, 144, 152, 161*]

May RM & Nowak MA (1995). Coinfection and the evolution of parasite virulence. *Proceedings of the Royal Society of London B* **261**:209–215 [*83, 124, 133–136, 138, 140, 150, 152, 166, 172*]

Mayer WE, Jonker M, Klein D, Ivanyi P, Van Seventer, G. & Klein J (1988). Nucleotide sequences of chimpanzee MHC class I alleles: Evidence for trans-species mode of evolution. *The EMBO Journal* **7**:2765–2774 [*211, 221*]

Maynard Smith J (1964). Group selection and kin selection. *Nature* **201**:1145–1147 [*95*]

Maynard Smith J (1976). A short term advantage of sex and recombination through sib competition. *Journal of Theoretical Biology* **63**:245–258 [*249, 258*]

Maynard Smith J (1978). *The Evolution of Sex*. Cambridge, UK: Cambridge University Press [*248*]

Maynard Smith J (1982). *Evolution and the Theory of Games*. Cambridge, UK: Cambridge University Press [*15, 42, 63, 309*]

Maynard Smith J (1989). *Evolutionary Genetics*. Oxford, UK: Oxford University Press [*246, 255*]

Maynard Smith J & Price GR (1973). The logic of animal conflict. *Nature* **246**:15–18 [*63*]

Maynard Smith J, Smith NH, O'Rourke M & Spratt BG (1993). How clonal are bacteria? *Proceedings of the National Academy of Sciences of the USA* **90**:4384–4388 [*347, 442*]

Mbelle N, Huebner RE, Wasas A, Kimura A, Chang I & Klugman KP (1999). Immunogenicity and impact on nasopharyngeal carriage of a nonvalent pneumococcal conjugate vaccine. *Journal of Infectious Diseases* **180**:1171–1176 [*364–365, 372, 374*]

McCallum HI (1990). Covariance in parasite burdens: The effect of predisposition to infection. *Parasitology* **100**:153–159 [*415*]

McCallum HI & Dobson AP (1995). Detecting disease and parasite threats to endangered species and ecosystems. *Trends in Ecology and Evolution* **10**:190–194 [*30, 417–418, 420, 424*]

McDonald BA, McDermott JM, Allard RW & Webster RK (1989). Coevolution of host and pathogen populations in the *Hordeum vulgare–Rhynchosporium secalis* pathosystem. *Proceedings of the National Academy of Sciences of the USA* **86**:3924–3927 [*443*]

McDonald BA, Miles J, Nelson LR & Pettway RE (1994). Genetic variability in nuclear DNA in field populations of *Stagonospora nodorum*. *Phytopathology* **84**:250–255 [*443*]

McGowan JEJ (1986). Minimizing antimicrobial resistance in hospital bacteria: Can switching or cycling drugs help? *Infection Control* **7**:573–576 [*326, 338*]

McLaughlin JF & Roughgarden J (1992). Predation across spatial scales in heterogeneous environments. *Theoretical Population Biology* **41**:277–299 [*199*]

McLean AR (1995a). After the honeymoon in measles control. *Lancet* **345**:272 [*340*]

McLean AR (1995b). Vaccination, evolution, and changes in the efficacy of vaccines: A theoretical framework. *Proceedings of the Royal Society of London B* **261**:389–393 [*341, 344, 360, 373*]

McLean AR & Anderson RM (1988). Measles in developing countries. Part II. The predicted impact of mass vaccination. *Epidemiology and Infection* **100**:419–442 [*340*]

McNeill WH (1976). *Plagues and Peoples*. New York, NY, USA: Anchor Press [*69, 103, 381, 397*]

McNicholl JM, Downer MV, Udhayakumar V, Alper CA & Swerdlow DL (2000). Host–pathogen interactions in emerging and re-emerging infectious diseases: Genomic perspective of tuberculosis, malaria, human immunodeficiency virus infection, hepatitis B and cholera. *Annual Review of Public Health* **21**:15–46 [*382*]

Medema GJ, Teunis PF, Havelaar AH & Haas CN (1996). Assessment of the dose–response relationship of *Campylobacter jejuni*. *International Journal of Food Microbiology* **30**:101–111 [*427*]

Medley GF (1994). Chemotherapy. In *Infectious Diseases*, eds. Scott ME & Smith G, pp. 141–157. New York, NY, USA; Academic Press [*65*]

Melchinger W, Strauss S, Zucker B & Bauer G (1996). Antiapoptotic activity of bovine papilloma-virus: Implications for the control of oncogenesis. *International Journal of Oncology* **9**:927–933 [*190*]

Mendis KN & Carter R (1995). Clinical disease and pathogenesis in malaria. *Parasitology Today* **11**(Suppl.):PTI 2–15 [*169*]

Mercereau-Puijalon O (1996). Revisiting host/parasite interactions: Molecular analysis of parasites collected during longitudinal and cross-sectional surveys in humans. *Parasite Immunology* **18**:173–180 [*168–169*]

Messenger SL, Molineux IJ & Bull JJ (1999). Virulence evolution in a virus obeys a trade-off. *Proceedings of the Royal Society of London B* **266**:397–404 [*392, 451*]

Meszéna G, Kisdi É, Dieckmann U, Geritz SAH & Metz JAJ (2000). *Evolutionary Optimisation Models and Matrix Games in the* Unified Perspective of Adaptive Dynamics. IIASA Interim Report IR-00-039. Laxenburg, Austria: International Institute for Applied Systems Analysis [*44*]

Metz JAJ & Diekmann O, eds. (1986). *The Dynamics of Physiologically Structured Populations*, Lecture Notes in Biomathematics, Vol. 68, pp. xii, 511. Berlin, Germany: Springer-Verlag [*301, 397*]

Metz JAJ & Gyllenberg M (2001). How should we define fitness in structured metapopulation models? Including an application to the calculation of ES dispersal strategies. *Proceedings of the Royal Society of London B* **268**:499–508 [*387, 397*]

Metz JAJ, Nisbet RM & Geritz SAH (1992). How should we define "fitness" for general ecological scenarios. *Trends in Ecology and Evolution* **7**:198–202 [*42–44, 46, 100, 387*]

Metz JAJ, Geritz SAH, Meszéna G, Jacobs FJA & Van Heerwaarden JS (1996a). Adaptive dynamics, a geometrical study of the consequences of nearly faithful reproduction. In *Stochastic and Spatial Structures of Dynamical Systems*, eds. Van Strien SJ & Verduyn Lunel SM, pp. 183–231. Amsterdam, Netherlands: North-Holland. [*42–43, 46, 99–100*]

Metz JAJ, Mylius SD & Diekmann O (1996b). *When Does Evolution Optimize? On the Relation between Types of Density Dependence and Evolutionarily Stable Life History Parameters*. IIASA Working Paper WP-96-004. Laxenburg, Austria: International Institute for Applied Systems Analysis [*40–41, 52, 54, 67*]

MHC Sequencing Consortium (1999). Complete sequence and gene map of a human major histocompatibility complex. *Nature* **401**:921–923 [*211*]

Michalakis Y, Olivieri I, Renaud F & Raymond M (1992). Pleiotropic action of parasites: How to be good for the host. *Trends in Ecology and Evolution* **7**:59–62 [*396*]

Michod RE & Levin BR, eds. (1987). *The Evolution of Sex: An Examination of Current Ideas*. Sunderland, MA, USA: Sinauer Associates Inc. [*248–249*]

Milgroom MG (1994). Population Biology of the chestnut blight fungus, *Cryphonectria parasitica*. *Canadian Journal of Botany* **73**:S311–S319 [*292–294*]

Milgroom MG (1995). Population biology of the chestnut blight fungus, *Cryphonectria parasitica. Canadian Journal of Botany* **73**(Suppl. 1):S311–319 [*442*]

Milgroom MG & Lipari SE (1995). Population differentiation in the chestnut blight fungus, *Cryphonectria parasitica*, in eastern North America. *Phytopathology* **85**:155–160 [*443*]

Milgroom MG, Lipari SE & Powell WA (1992). DNA fingerprinting and analysis of population structure in the chestnut blight fungus, *Cryphonectria parasitica*. *Genetics* **131**:297–306 [*443*]

Minchella DJ (1985). Host life-history variation in response to parasitism. *Parasitology* **90**:205–216 [*452*]

Minchella DJ & Scott ME (1991). Parasitism: A cryptic determinant of animal community structure. *Trends in Ecology and Evolution* **6**:250–254 [*30*]

Miralles R, Moya A & Elena SF (1997). Is group selection a factor modulating the virulence of RNA viruses? *Genetical Research* **69**:165–172 [*450*]

Mirold S, Rabsch W, Rohde, M, Stender S, Tscgäpe H, Rüssmann H, Igwe E & Hardt W-D (1999). Isolation of a temperate bacteriophage encoding the type III effector protein SopE from an epidemic *Salmonella typhimurium* strain. *Proceedings of the National Academy of Sciences of the USA* **96**:9845–9850 [*381*]

Mitchell MB & Mitchell HK (1952). A case of "maternal" inheritance in *Neurospora crassa. Proceedings of the National Academy of Sciences of the USA* **38**:442–449 [*281*]

Moerman A, Straver PJ, de Jong MCM, Quak J, Baanvinger T & Van Oirschot JT (1993). A long-term epidemiological study of bovine virus diarrhea infections in a large herd of diary cattle. *Veterinary Research* **132**:622–626 [*428*]

Moerman A, Straver PJ, de Jong MCM, Quak J, Baanvinger T & Van Oirschot JT (1994). Clinical consequences of a bovine virus diarrhea virus infection in a longitudinally studied dairy herd. *Veterinary Quarterly* **16**:115–119 [*428, 433*]

Molineaux L & Gramiccia G (1980). *The Garki Project*. Geneva, Switzerland: The World Health Organization [*456*]

Mollema C (1988). Cellular immune response of *Drosophila melanogaster* against *Asobara tabida*. PhD thesis, University of Lei-

den. [*206*]

Mollison D, ed. (1995). *Epidemic Models: Their Structure and Relation to Data*. New York, NY, USA: Cambridge University Press [*75*]

Momoi Y, Mizuno T, Nishimura Y, Endo Y, Ohno K, Watari T, Goitsuka R, Tsujimoto H & Hasegawa A (1996). Detection of apoptosis induced in peripheral-blood lymphocytes from cats infected with feline immunodeficiency virus. *Archives of Virology* **141**:1651–1659 [*190*]

Moran JS & Levine WC (1995). Drugs of choice for the treatment of uncomplicated gonococcal infections. *Clinical Infectious Diseases* **20**:S47–S65 [*337*]

Moran NA & Wernegreen JJ (2000). Lifestyle evolution in symbiotic bacteria: Insights from genomics. *Trends in Ecology and Evolution* **15**:321–326 [*396*]

Moritz C, McCallum H, Donnellan S & Roberts JD (1991). Parasite loads in parthenogenetic and sexual lizards (*Heteronotia binoei*): Support for the Red Queen hypothesis. *Proceedings of the Royal Society of London B* **244**:145–149 [*33*]

Morris A (1997). *Representing Spatial Interactions in Simple Ecological Models*. PhD thesis. Coventry, UK: University of Warwick [*97*]

Morris M & Kretzschmar M (1997). Concurrent partnerships and the spread of HIV. *AIDS* **11**:641–648 [*102*]

Morse SS (1993). *Emerging Viruses*. New York, NY, USA: Oxford University Press [*103*]

Moscardi F (1999). Assessment of the application of baculoviruses for control of Lepidoptera. *Annual Review of Entomology* **44**:257–289 [*457*]

Mosquera J & Adler FR (1998). Evolution of virulence: A unified framework for coinfection and superinfection. *Journal of Theoretical Biology* **195**:293–313 [*88, 106, 125, 136, 138–139, 141–142, 146–147*]

Moxon ER (1986). The carrier state: *Haemophilus influenzae. Journal of Antimicrobial Chemotherapy* **18**(Suppl. A):17–24 [*363*]

Moxon ER & Vaughn KA (1981). The type b capsular polysaccharide as a virulence determinant of *Haemophilus influenzae*: Studies using clinical isolates and laboratory transformants. *Journal of Infectious Diseases*

14:517–524   [*371*]

Moxon ER, Rainey PB, Nowak MA & Lenski RE (1994). Adaptive evolution of highly mutable loci in pathogenic bacteria. *Current Biology* 4:24–33   [*380, 395*]

Muller HJ (1932). Some genetic aspects of sex. *The American Naturalist* 66:118–138   [*248*]

Mullis K (1986). Specific enzymatic amplification of DNA in vitro: The polymerase chain reaction. *Cold Spring Harbor Symposia on Quantitative Biology* 51:263–273   [*263*]

Munger JC & Karasov WH (1991). Sublethal parasites in white-footed mice: Impact on survival and reproduction. *Canadian Journal of Zoology* 69:398–404   [*421*]

Murray BE (1994). Can antibiotic resistance be controlled? *New England Journal of Medicine* 330:1229–1230   [*326*]

Murray KD & Elkinton JS (1989). Environmental contamination of egg masses as a major component of trans-generational transmission of gypsy moth nuclear polyhedrosis virus (LdMNPV). *Journal of Invertebrate Pathology* 53:324–334   [*79*]

Musher DM, Groover JE, Reichler MR, Riedo FX, Schwartz B, Watson DA, Baughn RE & Breiman RF (1997). Emergence of antibody to capsular polysaccharides of *Streptococcus pneumoniae* during outbreaks of pneumonia: Association with nasopharyngeal colonization. *Clinical Infectious Diseases* 24:441–446   [*372*]

Myers G, Macinnes K & Korber B (1992). The emergence of simian/human immunodeficiency viruses. *AIDS Research and Human Retroviruses* 8:373–386   [*382*]

Mylius SD & Diekmann O (1995). On evolutionarily stable life histories, optimization and the need to be specific about density dependence. *Oikos* 74:218–224   [*40, 52, 67, 69, 86*]

Mylius SD & Metz JAJ. When does evolution optimize? On the relationships between evolutionary stability, optimization and density dependence. In *Elements of Adaptive Dynamics*, eds. Dieckmann U & Metz JAJ. Cambridge, UK: Cambridge University Press. In press   [*67*]

Nagelkerke CJ & Sabelis MW (1998). Precise control of sex allocation in pseudo-arrhenotokous phytoseiid mites. *Journal of Evolutionary Biology* 11:649–684   [*300*]

Nagylaki T (1979). The island model with stochastic migration. *Genetics* 91:163–176   [*106*]

Nåsell I (1985). *Hybrid Models of Tropical Infections*. Berlin, Germany: Springer-Verlag   [*29*]

Nee S & May RM (1992). Dynamics of metapopulations: Habitat destruction and competitive coexistence. *Journal of Animal Ecology* 61:37–40   [*131*]

Newbold CI, Pinches R, Roberts DJ & Marsh K (1992). *Plasmodium falciparum*: The human agglutinating antibody response to the infected red cell surface is predominantly variant specific. *Experimental Parasitology* 75:281–292   [*354*]

Nguyen DM, Wiehle SA, Koch PE, Branch C, Yen N, Roth JA & Cristiano RJ (1997). Delivery of the *p53* tumor suppressor gene into lung cancer cells by an adenovirus/DNA complex. *Cancer Gene Therapy* 4:191–198   [*194*]

Nichol ST, Spiropoulou CF, Morzunov S, Rollin PE, Kziazek TG, Feldmann H, Sanchez A, Childs J, Zaki S & Peters CJ (1993). Genetic identification of hantavirus associated with an outbreak of acute respiratory illness. *Science* 262:914–917   [*262*]

Nicot C, Astiergin T & Guillemain B (1997). Activation of *bcl-2* expression in human endothelial cells chronically expressing the human T-cell lymphotropic virus type I. *Virology* 236:47–53   [*190*]

Nitta DM, Jackson MA, Burry VF & Olson LC (1995). Invasive *Haemophilus influenzae* type f disease. *Pediatric Infectious Disease Journal* 14:157–160   [*365*]

Norton HTJ (1928). Natural selection and Meldenian variation. *Proceedings of the London Mathematical Society (Ser. 2)* 28:1–45   [*229*]

Nowak MA & May RM (1991). Mathematical biology of HIV infection: Antigenic variation and diversity threshold. *Mathematical Biosciences* 106:1–21   [*246*]

Nowak MA & May RM (1992). Coexistence and competition in HIV infections. *Journal of Theoretical Biology* 155:329–342   [*88*]

Nowak M & May RM (1994). Superinfection and the evolution of parasite virulence. *Proceedings of the Royal Society of London B* 255:81–89 [*87–88, 124, 127, 129–132, 138, 150, 152, 166, 172, 176, 287, 298, 312, 453–454*]

Nowak M & Sigmund K (1989). Oscillations in the evolution of reciprocity. *Journal of*

Theoretical Biology **137**:21–26 [*43*]

Nowak MA, May RM & Anderson RM (1990). The evolutionary dynamics of HIV-1 quasi-species and the development of immunodeficiency disease. *AIDS* **4**:1095–1103 [*88*]

Nowak MA, Anderson RM, McLean AR, Wolfs T, Goudsmit J & May RM (1991). Antigenic diversity thresholds and the development of AIDS. *Science* **254**:963–969 [*210, 300, 380, 451*]

Ntoumi F, Contamin H, Rogier C, Bonnefoy S, Trape J-F & Mercereau-Puijalon O (1995). Age-dependent carriage of multiple *Plasmodium falciparum* merozoite surface antigen-2 alleles in asymptomatic malaria infections. *American Journal of Tropical Medicine and Hygiene* **52**:81–88 [*168*]

Nuss DL (1992). Biological control of chestnut blight: An example of virus-mediated attenuation of fungal pathogenesis. *Microbiological Reviews* **56**:561–576 [*286, 296*]

Oates JD, Burdon JJ & Brouwer JB (1983). Interactions between *Avena* and *Puccinia* species. II. The pathogens: *Puccinia coronata* CDA and *P. graminis* Pers. F. sp. *avenae* Eriks. and Henn. *Journal of Applied Ecology* **20**:585–596 [*206*]

Obaro SK, Adegbola RA, Banya WAS & Greenwood BM (1996). Carriage of pneumococci after pneumococcal vaccination. *Lancet* **348**:271–272 [*364–365, 372*]

Ober C, Elias S, O'Brien E, Kostyu DD, Hauck WW & Bombard A (1988). HLA sharing and fertility in Hutterite couples: Evidence for prenatal selection against compatible fetuses. *American Journal of Reproductive Immunology and Microbiology* **18**:111–115. [*254*]

Ober C, Weitkamp LR, Cox N, Dytch H, Kostyu D & Elias S (1997). HLA and mate choice in humans. *American Journal of Human Genetics* **61**:497–504 [*254*]

Oberhaus SM, Smith RL, Clayton GH, Dermody TS & Tyler KL (1997). Reovirus infection and tissue injury in the mouse central nervous system are associated with apoptosis. *Journal of Virology* **71**:2100–2106 [*190*]

Obrebski S (1975). Parasite reproductive strategy and evolution of castration of hosts by parasites. *Science* **188**:1314–1316 [*452*]

O'Brien KL, Bronsdon MA, Carlone GM, Facklam RR, Schwartz B, Reid RR & Santosham M (2001). Effect of a 7-valent pneumococcal conjugate vaccine on nasopharyngeal (NP) carriage among Navajo and White Mountain Apache (N/WMA) infants. Abstract 1463. Baltimore, MD, USA: Society for Pediatric Research [*364, 370*]

Oduor GI, Sabelis MW, Lingeman R, de Moraes GJ & Yaninek JS (1997). Modelling fungal (*Neozygites* cf. *floridana*) epizootics in local populations of cassava green mites (*Mononychellus tanajoa*). *Experimental & Applied Acarology* **21**:485–506 [*450*]

Ohsaki N & Sato Y (1994). Food plant choice of *Pieris* butterflies as a trade-off between parasitoid avoidance and quality of plants. *Ecology* **75**:59–68 [*314*]

Ojeda F, Skardova I, Guarda MI, Ulloa J & Folch H (1997). Proliferation and apoptosis in infection with infectious bursal disease virus: A flow cytometric study. *Avian Diseases* **41**:312–316 [*190*]

Oksanen L, Fretwell SD, Arruda J & Niemela P (1981). Exploitation ecosystems in gradients of primary productivity. *The American Naturalist* **118**:240–261 [*199, 286*]

Oldroyd BP (1999). Coevolution while you wait: *Varroa jacobsoni*, a new parasite of western honeybees. *Trends in Ecology and Evolution* **14**:312–315 [*448*]

Olivieri I, Michalakis Y & Gouyon P-H (1995). Metapopulation genetics and the evolution of dispersal. *The American Naturalist* **146**:202–228 [*154*]

Olsson M, Shine R, Madsen T, Gullberg A & Tegelström H (1996). Sperm selection by females. *Nature* **383**:585 [*253*]

O'Neill SL, Hoffmann AA & Werren JH (1997). *Influential Passengers: Inherited Microorganisms and Arthropod Reproduction.* Oxford, UK: Oxford University Press [*391*]

Oppliger A, Christe P & Richner H (1996). Clutch size and malaria resistance. *Nature* **381**:565 [*251*]

O'Reilly DR (1997). Auxiliary genes of baculoviruses. In *The Baculoviruses*, ed. Miller LK, pp. 267–300. New York, NY, USA: Plenum Press [*79, 81*]

Osterhaus ADME, Rimmelzwaan GF, Martina BEE, Bestebroer TM & Fouchier RAM (2000). Influenza B virus in seals. *Science* **288**:1051–1053 [*382*]

Oxford English Dictionary (Compact Edition) (1971). Oxford, UK: Oxford University Press [*105*]

Pacala SW & Dobson AP (1988). The relation between the number of parasites/host and host age: Population dynamic causes and maximum likelihood estimation. *Parasitology* **96**:197–210 [*417*]

Pagie L & Hogeweg P (1997). Evolutionary consequences of coevolving targets. *Evolutionary Computation* **5**:401–418 [*212*]

Pallini A, Janssen A & Sabelis MW (1997). Odour-mediated responses of phytophagous mites to conspecific and heterospecific competitors. *Oecologia* **110**:179–185 [*315*]

Pallini A, Janssen A & Sabelis MW (1998). Predators induce interspecific herbivore competition for food in refuge space. *Ecology Letters* **1**:171–177 [*450*]

Pallini A, Janssen A & Sabelis MW (1999). Spider mites avoid plants with predators. *Experimental and Applied Ecology* **23**:803–815 [*315*]

Palm J (1969). Association of maternal genotype and excess heterozygosity for Ag-B histocompatibility antigens among male rats. *Transplantation Proceedings* **1**:82–84 [*256*]

Palm J (1970). Maternal–fetal interaction and histocompatibility antigen polymorphisms. *Transplantation Proceedings* **2**:162–173 [*256*]

Pappenheimer AM (1982). Diphtheria: Studies on the biology of an infectious disease. *Harvey Lectures* **76**:45–73 [*411*]

Parer I (1995). Relationship between survival rate and survival time of rabbits, *Oryctolagus cuniculus* (L.), challenged with myxoma virus. *Australian Journal of Zoology* **43**:303–311 [*452*]

Parer I, Sobey WR, Conolly D & Morton R (1994). Virulence of strains of myxoma virus and the resistance of wild rabbits, *Oryctolagus cuniculus* (L.), from different locations in Australasia. *Australian Journal of Zoology* **42**:347–362 [*452*]

Parham P & Ohta T (1996). Population biology of antigen presentation by MHC class I molecules. *Science* **272**:67–74 [*210, 221*]

Parham P, Benjamin RJ, Chen BP, Clayberger C, Ennis PD, Krensky AM, Lawlor DA, Littman DR, Norment AM, Orr HT, Salter RD & Zemmour J (1989a). Diversity of class I HLA molecules: Functional and evolutionary interactions with T cells. *Cold Spring Harbor Symposia on Quantitative Biology* **54**:529–543 [*210*]

Parham P, Lawlor DA, Lomen CE & Ennis PD (1989b). Diversity and diversification of HLA-A,B,C alleles. *Journal of Immunology* **142**:3937–3950 [*210–211, 221*]

Park EJ, Yin C-M & Burand JP (1996). Baculovirus replication alters hormone-regulated host development. *Journal of General Virology* **77**:3–15 [*81*]

Parker MA (1994). Pathogens and sex in plants. *Evolutionary Ecology* **8**:560–584 [*234, 236, 244*]

Parker MA (1996). The nature of plant–parasite specificity: Comment. *Evolutionary Ecology* **10**:319–322 [*244*]

Parker GG, Hill SM & Kuehnel LA (1993). Decline of understory American chestnut (*Castanea dentata*) in a southern Appalachian forest. *Canadian Journal of Forest Research* **23**:259–265 [*291*]

Parvinen K, Dieckmann U, Gyllenberg M & Metz JAJ (2000). *Evolution of Dispersal in Metapopulations with Local Density Dependence and Demographic Stochasticity*. IIASA Interim Report IR-00-035. Laxenburg, Austria: International Institute for Applied Systems Analysis [*56*]

Paton JC (1998). Novel pneumococcal surface proteins: Role in virulence and vaccine potential. *Trends in Microbiology* **6**:85–87 [*374*]

Paul WE (1999). *Fundamental Immunology*. New York, NY, USA: Raven Press [*211, 221*]

Paul REL & Day KP (1998). Mating patterns of *Plasmodium falciparum*. *Parasitology Today* **14**:197–202 [*167*]

Paul CP & Fulbright DW (1988). Double-stranded RNA molecules from Michigan hypovirulent isolates of *Endothia parasitica* vary in size and sequence homology. *Phytopathology* **78**:751–755 [*293*]

Paul REL, Packer MJ, Walmsley M, Lagog M, Ranford-Cartwright LC, Paru R & Day KP (1995). Mating patterns in malaria parasite populations of Papua New Guinea. *Science* **269**:1709–1711 [*171*]

Pell JK, Pluke R, Clark SJ, Kenward MG & Alderson PG (1997). Interactions between two aphid natural enemies, the entomopathogenic fungus *Erynia neoaphidis* Remaudière & Hennebert (Zygomycetes: Entomophthorales) and the predatory beetle *Coccinella septempunctata* L. (Coleoptera: Coccinellidae). *Journal of Invertebrate*

Pathology **69**:261–268 [*456*]

Pels B & Sabelis MW (1999). Local dynamics, overexploitation and predator dispersal in an acarine predator–prey system. *Oikos* **86**:573–583 [*305–306, 311, 320, 450, 456*]

Pels B, de Roos AM & Sabelis MW. Evolutionary dynamics of prey exploitation in a metapopulation of predators. *The American Naturalist.* In press [*310–311*]

Penn DJ & Potts WK (1999). The evolution of mating preferences and major histocompatibility complex genes. *The American Naturalist* **153**:145–164 [*398*]

Peters W (1982). Anti-malarial drug resistance. *British Medical Bulletin* **38**:187–192 [*415*]

Peters W (1985). The problem of drug resistance in malaria. *Parasitology* **90**:705–716 [*398*]

Peters W (1987). *Chemotherapy and Drug Resistance in Malaria*, Second Edition. London, UK: Academic Press [*415*]

Petras F, Chidambaram M, Illyes EF, Froshauer S, Weinstock G & Reese CP (1995). Antigenic and virulence properties of *Pasteurella haemolytica* leukotoxin mutants. *Infection and Immunity* **63**:1033–1039 [*37*]

Petrie M & Williams A (1993). Peahens lay more eggs for peacocks with larger trains. *Proceedings of the Royal Society of London B* **251**:127–131 [*261*]

Pfister CA & Hay ME (1988). Associational plant refuges: Convergent patterns in marine and terrestrial communities result from different mechanisms. *Oecologia* **77**:118–129 [*315*]

Pickering J (1980). *Sex Ratio, Social Behaviour and Ecology in Polistes* (Hymenoptera, Vespidae), *Pachysomoides* (Hymenoptera, Ichneumonidae) and *Plasmodium* (Protozoa, Haemosporidia). PhD dissertation, Harvard University, USA [*167*]

Pickering J, Read AF, Guerrero S & West SA (2000). Sex ratio and virulence in two species of lizard malaria parasites. *Evolutionary Ecology Research* **2**:171–184 [*170*]

Pier GB, Grout M & Colledge WH (1998). *Salmonella typhi* uses CFTR to enter intestinal epithelial cells. *Nature* **393**:79–82 [*223*]

Plowright W (1985). La peste bovine aujourd'hui dans le monde. Contrôle et possibilité d'eradication par la vaccination. *Annales de Medecine Veterinaire* **129**:9–32

[*31*]

Polis GA & Holt RD (1992). Intraguild predation: The dynamics of complex trophic interactions. *Trends in Ecology and Evolution* **7**:151–154 [*317*]

Polis GA, Myers CA & Holt RD (1989). The ecology and evolution of intraguild predation: Potential competitors that eat each other. *Annual Review of Ecology and Systematics* **20**:297–330 [*317*]

Pomiankowski A, Iwasa Y & Nee S (1991). The evolution of costly mate preferences. I. Fisher and biased mutation. *Evolution* **45**:1422–1430 [*251*]

Potts WK & Wakeland EK (1993). Evolution of MHC genetic diversity: A tale of incest, pestilence, and sexual preference. *Trends in Genetics* **9**:408–412 [*254*]

Potts WK, Manning CJ & Wakeland EK (1991). Mating patterns in seminatural populations of mice influenced by MHC genotype. *Nature* **352**:619–621 [*253, 256*]

Potts WK, Manning CJ & Wakeland EK (1994). The role of infectious disease, inbreeding and mating preferences in maintaining MHC genetic diversity: An experimental test. *Philosophical Transactions of the Royal Society of London B* **346**:369–378 [*255*]

Poulin R (1998). *Evolutionary Ecology of Parasites: From Individuals to Communities.* London, UK: Chapman & Hall [*379*]

Poulin R & Combes C (1999). The concept of virulence: Interpretations and implications. *Parasitology Today* **15**:474–475 [*388*]

Power AG (1992). Patterns of virulence and benevolence in insect-borne pathogens of plants. *Critical Reviews in Plant Sciences* **11**:351–372 [*453*]

Powers ME (1992). Top-down and bottom-up forces in food webs: Do plants have primacy? *Ecology* **73**:733–744 [*286*]

Prescott SC & Horwood MP (1935). *Sedgwick's Principles of Sanitary Science and Public Health.* New York, NY, USA: Macmillan [*18*]

Price PW (1980). *Evolutionary Biology of Parasites.* Princeton, NJ, USA: Princeton University Press [*29*]

Price PW, Bouton CE, Gross P, McPheron BA, Thompson JN & Weiss AE (1980). Interactions among three trophic levels: Influence of plants on interactions between insect herbivores and natural enemies. *Annual Review of Ecology and Systematics* **11**:41–65 [*297, 299, 318*]

Price PW, Westoby M & Rice B (1988). Parasite-mediated competition: Some predictions and tests. *The American Naturalist* **131**:544–555 [*30*]

Prins HHT & van der Jeugd HP (1993). Herbivore population crashes and woodland structure in East Africa. *Journal of Ecology* **8**:305–314 [*32*]

Prusiner SB (1982). Novel proteinaceous infection particles cause scrapie. *Science* **216**:136–144 [*432*]

Prusiner SB (1995). The prion diseases. *Scientific American* **272**:48–51 [*432*]

Pugliese A (2000). Evolutionary dynamics of virulence. In *Advances in Adaptive Dynamics*, eds. Dieckmann U & Metz JAJ. Cambridge, UK: Cambridge University Press [*139*]

Pugliese A & Rosà R (1995). Epidemic 2-dimensional model with logistic growth for the host–macroparasite system. *Journal of Biological Systems* **3**:833–849 [*28*]

Pulliam HR (1988). Sources, sinks and population regulation. *The American Naturalist* **132**:652–661 [*105*]

Pulliam HR (1996). Sources and sinks: Empirical evidence and population consequences. In *Population Dynamics in Ecological Space and Time*, eds. Rhodes E, Chesser RK & Smith MH, pp. 45–70. Chicago, IL, USA: University of Chicago Press [*104*]

Quinn TC, Wawer MJ, Sewankambo N, Serwadda D, Li CJ, Wabwire-Mangen F, Meehan MO, Lutalo T & Gray RH (2000). Viral load and heterosexual transmission of human immunodeficiency virus type 1. *New England Journal of Medicine* **342**:921–929 [*390*]

Radrizzani M, Accornero P, Delia D, Kurrle R & Colombo MP (1997). Apoptosis induced by HIV-gp120 in a TH1 clone involves the generation of reactive oxygen intermediates downstream cd95 triggering. *FEBS Letters* **411**:87–92 [*195*]

Rambaut A (1996). *The Use of Temporally Sampled DNA Sequences in Phylogenetic Analysis*. PhD Thesis. Oxford, UK: Oxford University [*266, 270–272*]

Rambaut A & Grassly NC (1996). *SPATULA Version 1.0*. Oxford, UK: Oxford University Press [*271, 275*]

Rand DA, Wilson HB & McGlade JM (1994).

Dynamics and evolution: Evolutionarily stable attractors, invasion exponents and phenotype dynamics. *Philosophical Transactions of the Royal Society of London B* **343**:261–283 [*100*]

Rand DA, Keeling M & Wilson HB (1995). Invasion, stability and evolution to criticality in spatially extended, artificial host–pathogen ecologies. *Proceedings of the Royal Society of London B* **259**:55–63 [*85, 89, 199*]

Rannala B & Yang Z (1996). Probability distribution of evolutionary trees: A new method of phylogenetic inference. *Journal of Molecular Evolution* **43**:304–311 [*263*]

Ray CA, Black RA, Kronheim SR, Greenstreet TA, Sleath PR, Salvesen GS & Pickup DJ (1992). Viral inhibition of inflammation: Cows encode an inhibitor of the interleukin-1β converting enzyme. *Cell* **69**:597–604 [*184*]

Read AF (1994). The evolution of virulence. *Trends in Microbiology* **73**:73–76 [*165–166, 250*]

Read AP (1995). Genetics and evolution of infectious diseases in natural populations: Group report. In *Ecology of Infectious Diseases in Natural Populations*, eds. Grenfell BT & Dobson AP, pp. 450–477. Cambridge, UK: Cambridge University Press [*33, 415–416*]

Read AF & Anwar MA. Clonal population dynamics in mixed-clone infections of the rodent malaria *Plasmodium chabaudi*. Unpublished [*170, 173–175*]

Read AF & Viney ME (1996). Helminth immunogenetics: Why bother? *Parasitology Today* **12**:337–343 [*249, 257*]

Reichler MR, Allphin AA, Breiman RF, Schreiber JR, Arnold JE, McDougal LK, Facklam RR, Boxerbaum B, May D, Walton RO & Jacobs MR (1992). The spread of multiply resistant *Streptococcus pneumoniae* at a day care center in Ohio. *Journal of Infectious Diseases* **166**:1346–1353 [*366*]

Reid AH, Fanning TG, Hultin JV & Taubenberger JK (1999). Origin and evolution of the 1918 "Spanish" influenza virus hemagglutinin gene. *Proceedings of the National Academy of Sciences of the USA* **96**:1651–1656 [*382*]

Rhodes CJ & Anderson RM (1996). Persistence and dynamics in lattice models of epidemic spread. *Journal of Theoretical*

Biology **180**:125–133 [*89*]

Rice WR (1983). Parent–offspring pathogen transmission: A selective agent promoting sexual reproduction. *The American Naturalist* **121**:187–203 [*249*]

Richner H, Christe P & Oppliger A (1995). Paternal investment affects prevalence of malaria. *Proceedings of the National Academy of Sciences of the USA* **92**:1192–1194 [*251*]

Ridgway SH (1995). The tides of change: Conservation of marine mammals. In *Conservation of Endangered Species in Captivity: An Interdisciplinarian Approach*, eds. Gibbon EF Jr, Durrant BS & Demarest J, pp. 407–424. Albany, NY, USA: State University of New York Press [*260*]

Roane MK, Griffin GJ & Elkins JR (1986). *Chestnut Blight, Other Endothia Diseases, and the Genus* Endothia. St. Paul, MN, USA: American Phytopathological Society Press [*291, 293–294*]

Robert F, Ntoumi F, Angel G, Candito D, Rogier C, Fandeur T, Sarthou J-L & Mercereau-Puijalon O (1996a). Extensive genetic diversity of *Plasmodium falciparum* isolates collected from patients with severe malaria in Dakar, Senegal. *Transactions of the Royal Society of Tropical Medicine and Hygiene* **90**:704–711 [*168–169*]

Robert V, Read AF, Essong J, Tchuinkam T, Mulder B, Verhave J-P & Carnevale P (1996b). Effect of gametocyte sex ratio on infectivity of *Plasmodium falciparum* to *Anopheles gambiae*. *Transaction of the Royal Society of Tropical Medicine and Hygiene* **90**:621–624 [*171*]

Roberts DR, Manguin S & Mouchet J (2000). DDT house spraying and re-emerging malaria. *Lancet* **356**:330–332 [*398*]

Robertson DL, Sharp PM, McCutchan FE & Hahn BH (1995). Recombination in HIV-1. *Nature* **374**:124–126 [*351*]

Roelfs AP (1982). Effects of Barberry eradication on stem rust in the United States. *Plant Disease* **66**:177–181 [*441*]

Roelfs AP (1986). Development and impact of regional cereal rust epidemics. In *Plant Disease Epidemiology*, eds. Leonard KJ & Fry WE, pp. 129–150. New York, NY, USA: Macmillan [*441*]

Roelke-Parker ME, Munson L, Packer C, Kock R, Cleaveland S, Carpenter M, O'Brien SJ, Pospischil A, Hofmann-Lehmann R, Lutz H, Mwamengele GLM, Mgasa MN, Machange GA, Summers BA & Appel MJG (1996). A canine distemper virus epidemic in Serengeti lions (*Panthera leo*). *Nature* **379**:441–444 [*382*]

Rohm C, Horimoto T, Kawaoka Y, Suss J & Webster RG (1995). Do hemagglutinin genes of highly pathogenic avian influenza viruses constitute unique phylogenetic lineages? *Virology* **209**:664–670 [*268*]

Room PM (1990). Ecology of a simple plant–herbivore system: Biological control of *Salvinia*. *Trends in Ecology and Evolution* **5**:74–79 [*422*]

Roper C, Richardson W, Elhassan IM, Giha H, Hviid L, Satti GMH, Theander TG & Arnot DE (1998). Seasonal changes in the *Plasmodium falciparum* population in individuals and their relationship to clinical malaria: A longitudinal study in a Sudanese village. *Parasitology* **116**:501–510 [*168–169, 172*]

Rosati F & DeSantis R (1978). Studies on fertilization in the ascidians I. Self-sterility and specific recognition between gametes of *Ciona intestinalis*. *Experimental Cell Research* **112**:111–119 [*253*]

Rosenthal SS (1996). *Aceria, Epitrimerus* and *Aculus* species and biological control of weeds. In *Eriophyoid Mites: Their Biology, Natural Enemies and Control*, eds. Lindquist EE, Sabelis MW & Bruin J, pp. 729–739. Amsterdam, Netherlands: Elsevier [*452*]

Rosenzweig ML, Brown JS & Vincent TL (1987). Red Queens and ESS: The coevolution of evolutionary rates. *Evolutionary Ecology* **1**:59–94 [*234*]

Rosqvist R, Skurnik M & Wolf-Watz H (1988). Increased virulence of *Yersinia pseudotuberculosis* by 2 independent mutations. *Nature* **334**:522–525 [*382*]

Ross R (1911). *The Prevention of Malaria*. London, UK: Murray [*11*]

Rota JS, Hummel KB & Rota PA (1992). Genetic variability of the glycoprotein genes of current wild-type measles isolates. *Virology* **188**:135–142 [*344*]

Rothman LD & Myers JH (1996). Debilitating effects of viral diseases on host lepidoptera. *Journal of Invertebrate Pathology* **67**:1–10 [*79*]

Roush RT & Tabashnik BE, eds. (1990). *Pesticide Resistance in Arthropods*. New York, NY, USA: Chapman & Hall [*457*]

Rowe JA, Moulds JM, Newbold CI & Miller LH (1997). *P. falciparum* rosetting mediated by a parasite-variant erythrocyte membrane protein and complement-receptor 1. *Nature* **388**:292–295 [*171*]

Roy HE, Pell JK, Clark SJ & Alderson PG (1998). Implications of predator foraging on aphid pathogen dynamics. *Journal of Invertebrate Pathology* **71**:236–247 [*456*]

Rülicke T, Chapuisat M, Homberger FR, Macas E & Wedekind C (1998). MHC-genotype of progeny influenced by parental infection. *Proceedings of the Royal Society of London B* **265**:711–716 [*256, 260*]

Saadati HA, Gibbs HA, Parton R & Coote JG (1997). Characterization of the leukotoxin produced by different strains of *Pasteurella haemolytica*. *Journal of Medical Microbiology* **46**:276–284 [*37*]

Sabelis MW (1990). Life history evolution in spider mites. In *The Acari: Reproduction, Development and Life-History Strategies*, eds. Schuster R & Murphy PW, pp. 23–50. New York, NY, USA: Chapman and Hall [*301*]

Sabelis MW (1992). Arthropod predators. In *Natural Enemies: The Population Biology of Predators, Parasites and Diseases*, ed. Crawley MJ, pp. 225–264. Oxford, UK: Blackwell [*299, 302, 305*]

Sabelis MW & Bruin J (1996). Evolutionary ecology: Life history patterns, food plant choice and dispersal. In *Eriophyoid Mites: Their Biology, Natural Enemies and Control*, eds. Lindquist EE, Sabelis MW & Bruin J, pp. 329–366. Amsterdam, Netherlands: Elsevier [*452*]

Sabelis MW & de Jong MCM (1988). Should all plants recruit bodyguards? Conditions for a polymorphic ESS of synomone production in plants. *Oikos* **53**:247–252 [*315–317*]

Sabelis MW & Janssen A (1994). Evolution of life-history patterns in the Phytoseiidae. In *Mites: Ecological and Evolutionary Analyses of Life-History Patterns*, ed. Houck MA, pp. 70–98. New York, NY, USA: Chapman & Hall [*301, 305*]

Sabelis MW & Van der Meer J (1986). Local dynamics of the interaction between predatory mites and two-spotted spider mites. In *Dynamics of Physiologically Structured Populations*, eds. Metz JAJ & Diekmann O, *Lecture Notes in Biomathematics*, Vol. 68, pp. 322–344. Berlin, Germany: Springer-Verlag [*305–306, 311, 456*]

Sabelis MW & Van Rijn PCJ (1997). Predation by insects and mites. In *Thrips as Crop Pests*, ed. Lewis T, pp. 259–354. New York, NY, USA: Oxford University Press/CAB International [*305–306*]

Sabelis MW, Diekmann O & Jansen VAA (1991). Metapopulation persistence despite local extinction: Predator–prey patch models of the Lotka–Volterra type. *Biological Journal of the Linnean Society* **42**:267–283 [*317*]

Sabelis MW, Janssen A, Pallini A, Venzon M, Bruin J, Drukker B & Scutareanu P (1999a). Behavioral responses of predatory and herbivorous arthropods to induced plant volatiles: From evolutionary ecology to agricultural applications. In *Induced Plant Defenses against Pathogens and Herbivores*, eds. Agrawal A, Tuzun S & Bent E, pp. 269–298. St. Paul, MI, USA: APS Press, The American Phytopathological Society [*298, 318, 454*]

Sabelis, MW, Janssen A, Bruin J, Bakker FM, Drukker B, Scutareanu P & Van Rijn PCJ (1999b). Interactions between arthropod predators and plants: A conspiracy against herbivorous arthropods? In *Ecology and Evolution of the Acari*, eds. Bruin J, Van der Geest LPS & Sabelis MW, pp. 207–230, Series Entomologica, Vol. 55. Dordrecht, Netherlands: Kluwer Academic Publishers [*298, 318*]

Sabelis MW, Van Baalen M, Bruin J, Egas M, Jansen VAA, Janssen A & Pels B (1999c). The evolution of overexploitation and mutualism in plant–herbivore–predator interactions and its impact on population dynamics. In *Theoretical Approaches to Biological Control*, eds. Hawkins BA & Cornell HV, pp. 259–282. Cambridge, UK: Cambridge University Press [*298, 450, 454*]

Sabelis MW, Van Baalen M, Bakker FM, Bruin J, Drukker B, Egas M, Janssen A, Lesna I, Pels B, Van Rijn PCJ & Scutareanu P (1999d). Evolution of direct and indirect plant defence against herbivorous arthropods. In *Herbivores: Between Plants and Predators*, eds. Olff H, Brown VK & Drent RH, pp. 109–166. Oxford, UK: Blackwell Science [*298, 450, 454*]

Salles CA & Momen H (1991). Identification of *Vibrio cholerae* by enzyme electrophoresis. *Transactions of the Royal Society of*

Tropical Medicine and Hygiene **85**:544–547 [*402*]

Saloniemi I (1993). A coevolutionary predator–prey model with quantitative characters. *The American Naturalist* **141**:880–896 [*234*]

Sanders CC, Sanders WE Jr & Harrowe DJ (1976). Bacterial interference: Effects of oral antibiotics on the normal throat flora and its ability to interfere with group A streptococci. *Infection and Immunity* **13**:808–812 [*366*]

Sasaki A (1994). Evolution of antigen drift and switching: Continuously evading pathogens. *Journal of Theoretical Biology* **168**:291–308 [*246*]

Sasaki A (2000). Host–parasite coevolution in a multilocus gene-for-gene system. *Proceedings of the Royal Society of London B* **267**:2183–2188 [*396*]

Sasaki A & Godfray HCJ (1999). A model for the coevolution of resistance and virulence in coupled host–parasitoid interactions. *Proceedings of the Royal Society of London B* **266**:455–463 [*234, 396, 458*]

Sasaki A & Haraguchi Y (2000). Antigenic drift of viruses within a host: A finite site model with demographic stochasticity. *Journal of Molecular Evolution* **51**:245–255 [*246*]

Sasaki A & Iwasa Y (1991). Optimal growth schedule of pathogens within a host: Switching between lytic and latent cycles. *Theoretical Population Biology* **39**:201–239 [*166, 300*]

Satō K & Iwasa Y (2000). Pair approximation methods for lattice-based ecological models. In *The Geometry of Ecological Interactions: Simplifying Spatial Complexity*, eds. Dieckmann U, Law R & Metz JAJ, pp. 341–358. Cambridge, UK: Cambridge University Press [*89*]

Satō K, Matsuda H & Sasaki A (1994). Pathogen invasion and host extinction in lattice structured populations. *Journal of Mathematical Biology* **32**:251–268 [*89, 99*]

Satta Y, O'hUigin C, Takahata N & Klein J (1993). The synonymous substitution rate of the major histocompatibility complex loci in primates. *Proceedings of the National Academy of Sciences of the USA* **90**:7480–7484 [*210*]

Saunders IW (1981). Epidemics in competition. *Journal of Mathematical Biology*
11:311–318 [*230*]

Sayyed AH, Ferre J & Wright DJ (2000). Mode of inheritance and stability of resistance to *Bacillus thuringiensis VAR kurstaki* in a diamondback moth (*Plutella xylostella*) population from Malaysia. *Pest Management Science* **56**:743–748 [*457*]

Schall JJ (1989). The sex ratio of *Plasmodium* gametocytes. *Parasitology* **98**:343–350 [*170*]

Schall JJ (1996). Malarial parasites of lizards: Diversity and ecology *Advances in Parasitology* **37**:255–333 [*165*]

Schärer L & Wedekind C (1999). Lifetime reproductive output in a hermaphrodite cestode when reproducing alone or in pairs: A time cost of mating. *Evolutionary Ecology* **13**:381–394 [*257*]

Schmid-Hempel P (1998). *Parasites in Social Insects*. Princeton, NJ, USA: Princeton University Press [*448, 450*]

Schmid-Hempel P & Schmid-Hempel R (1993). Transmission of a pathogen in *Bombus terrestris* with a note on division of labour in social insects. *Behavioural Ecology and Sociobiology* **33**:319–327 [*396*]

Schmitz OJ & Nudds TD (1993). Parasite mediated competition in deer and moose: How strong is the effect of meningeal worm on moose? *Ecological Applications* **4**:91–103 [*30*]

Schrag SJ & Perrot V (1996). Reducing antibiotic resistance. *Nature* **381**:120–121 [*330*]

Schrag SJ & Wiener P (1995). Emerging infectious disease: What are the relative roles of ecology and evolution? *Trends in Ecology and Evolution* **10**:319–324 [*37*]

Scofield VL, Schlumpberger JM, West LA & Weissman IL (1982). Protochordate allorecognition is controlled by a MHC-like gene system. *Nature* **295**:499–502 [*253*]

Scott GR (1964). Rinderpest. *Advances in Veterinary Science* **9**:113–224 [*31*]

Scott ME (1987). Regulation of mouse colony abundance by *Heligmosomoides polygyrus*. *Parasitology* **95**:111–124 [*421*]

Scott ME (1988). The impact of infection and disease on animal populations: Implications for conservation biology. *Conservation Biology* **2**:40–56 [*416*]

Scott GR (1990). The role of parasites in regulating host abundance. *Parasitology Today* **5**:176–183 [*31*]

Scott ME & Smith G (1994). *Parasitic*

and *Infectious Diseases: Epidemiology and Ecology*. London, UK: Academic Press [29]

Scutareanu P, Drukker B, Bruin J, Posthumus MA & Sabelis MW (1997). Isolation and identification of volatile synomones involved in the interaction between Psylla-infested pear trees and two anthocorid predators. *Journal of Chemical Ecology* 23:2241–2260   [298]

Seger J (1992). Evolution of exploiter–victim relationships. In *Natural Enemies: The Population Biology of Predators, Parasites and Diseases*, ed. Crawley MJ, pp. 3–25. Oxford, UK: Blackwell Science   [199]

Seger J & Brockmann HJ (1987). What is bet-hedging? *Oxford Surveys in Evolutionary Biology* 4:182–211   [244]

Seger J & Hamilton WD (1988). Parasite and sex. In *The Evolution of Sex*, ed. Michod RE, Levin BR, pp. 176–193. Sunderland, MA, USA: Sinauer Associates Inc.   [237–238, 244, 246]

Sela-Donnenfeld D, Korner M, Pick M, Eldor A & Panet A (1996). Programmed endothelial-cell death induced by an avian hemangioma retrovirus is density-dependent. *Virology* 223:233–237   [190, 192]

Shaner G, Stromberg EL, Lacy G, Barker KR & Pirone TP (1992). Nomenclature and concepts of pathogenicity and virulence. *Annual Review of Phytopathology* 30:47–66   [388]

Sheldon RW & Kerr SR (1972). The population density of monsters in Loch Ness. *Limnology and Oceanography* 17:796–798   [386]

Sheldon BC & Verhulst S (1996). Ecological immunology: Costly parasite defences and trade-offs in evolutionary ecology. *Trends in Ecology and Evolution* 11:317–321   [251]

Shonle I & Bergelson J (1995). Interplant communication revisited. *Ecology* 76:2660–2663   [316]

Shulaev V, Silverman P & Raskin I (1997). Airborne signalling by methyl salicylate in plant pathogen resistance. *Nature* 385:718–721   [316]

Shutler D, Bennett GF & Mullie A (1995). Sex proportions of haemoproteus blood parasites and local mate competition. *Proceedings of the National Academy of Sciences of the USA* 92:6748–6752   [171]

Shykoff JA & Schmid-Hempel P (1991). Para-

sites and the advantage of genetic variability within social insect colonies. *Proceedings of the Royal Society of London B* 243:55–58   [249, 396]

Silflow R & Foreyt WJ (1994). Susceptibility of phagocytes from elk, deer, bighorn sheep, and domestic sheep to *Pasteurella haemolytica* cytotoxins. *Journal of Wildlife Diseases* 30:529–535   [36]

Silflow R, Foreyt WJ & Leid R (1993). *Pasteurella haemolytica* cytotoxin-dependent killing of neutrophils from bighorn and domestic sheep. *Journal of Wildlife Diseases* 29:30–35   [36]

Simmons LW, Stockley P, Jackson RL & Parker GA (1996). Sperm competition or sperm selection: No evidence for female influence over paternity in yellow dung flies *Scatophaga stercoraria*. *Behavioural Ecology and Sociobiology* 38:199–206   [253]

Simms EL (1992). Costs of plant resistance to herbivory. In *Plant Resistance to Herbivores and Pathogens: Ecology, Evolution and Genetics*, eds. Fritz RS & Simms EL, pp. 392–425. Chicago, IL, USA: University of Chicago Press   [318]

Simms EL & Fritz RS (1990). The ecology and evolution of host-plant resistance to insects. *Trends in Ecology and Evolution* 5:356–360   [415]

Simms EL & Rausher MD (1987). Costs and benefits of plant resistance to herbivory. *The American Naturalist* 130:570–581   [318]

Simms EL & Rausher MD (1989). Natural selection by insects and costs of plant resistance to herbivory. *Evolution* 43:573–585   [318]

Simon AE & Bujarski JJ (1994). RNA-RNA recombination and evolution in virus-infected plants. *Annual Review of Phytopathology* 32:337–362   [442]

Sinden RE (1991). Asexual blood stages of malaria modulate gametocyte infectivity to the mosquito vector: Possible implications for control strategies. *Parasitology* 103:191–196   [406]

Singh SP, Gepts P & Debouck DG (1991). Races of common bean (*Phaseolus vulgaris*, *Fabaceae*). *Economic Botany* 45:379–396   [444]

Singh N, Agrawal S & Rastogi AK (1997). Infectious diseases and immunity: Special reference to major histocompatibility complex. *Emerging Infectious Diseases* 3:41–49

[223]
Slade RW & McCallum HI (1992). Overdominant versus frequency-dependent selection at MHC loci. *Genetics* **132**:861–864 [*211*]

Slavicek JM, Popham HJR & Riegel CI (1999). Deletion of the *Lymantria dispar* multicapsid nucleopolyhedrovirus ecdysteroid UDP-glucosyltransferase gene enhances viral killing speed in the last instar of the gypsy moth. *Biological Control* **16**:91–103 [*81, 83*]

Smart CD & Fulbright DW (1995). Characterization of a strain of *Cryphonectria parasitica* doubly infected with hypovirulence-associated dsRNA viruses. *Phytopathology* **85**:491–494 [*296*]

Smith VH & Holt RD (1996). Resource competition and within-host disease dynamics. *Trends in Ecology and Evolution* **11**:386–389 [*176*]

Smith T, Lehmann D, Montgomery J, Gratten M, Riley ID & Alpers MP (1993). Acquisition and invasiveness of different serotypes of *Streptococcus pneumoniae* in young children. *Epidemiology and Infection* **111**:27–39 [*364, 371–372*]

Smith DJ, Forrest S, Ackley DH & Perelson AS (1999). Variable efficacy of repeated annual influenza vaccination. *Proceedings of the National Academy of Sciences of the USA* **96**:14001–14006 [*210*]

Smits MA, Bossers A & Schreuder BEC (1997). Prion protein and scrapie susceptibility. *Veterinary Quarterly* **19**:101–105 [*432*]

Snell GD (1968). The H-2 locus of the mouse: Observations and speculations concerning its comparative genetics and its polymorphism. *Folia Biologica (Prague)* **14**:335–358 [*211*]

Snow RW & Marsh K (1995). Will reducing *Plasmodium falciparum* transmission alter malaria mortality among African children? *Parasitology Today* **11**:188–190 [*407*]

Snow RW, Azevedo B, Lowe BS, Kabiru EW, Nevill CG, Mwankusye S, Kassiga G, Marsh K & Teuscher T (1994). Severe childhood malaria in 2 areas of markedly different falciparum transmission in East Africa. *Acta Tropica* **57**:289–300 [*353*]

Snow RW, Omumbo JA, Lowe B, Molyneux CS, Obiero JO, Palmer A, Weber MW, Pinder M, Nahlen B, Obonyo C, Newbold C, Gupta S & Marsh K (1997). Relation between severe malaria morbidity in children and level of *Plasmodium falciparum* transmission in Africa. *Lancet* **349**:1650–1654 [*406*]

Snow RW, Nahlen B, Palmer A, Donnelly CA, Gypta S & Marsh K (1998). Risk of severe malaria among African infants: Direct evidence of clinical protection during early infancy. *Journal of Infectious Diseases* **177**:819–822 [*358, 407*]

Sokurenko EV, Hasty DL & Dykhuizen DE (1999). Pathoadaptive mutations: Gene loss and variation in bacterial pathogens. *Trends in Microbiology* **7**:191–195 [*453*]

Song WY, Pi LY, Wang GL, Gardner J, Holsten T & Ronald PC (1997). Evolution of the rice Xa21 disease resistance gene family. *Plant Cell* **9**:1279–1287 [*234*]

Sorci G, Moller AP & Boulinier T (1997). Genetics of host–parasite interactions. *Trends in Ecology and Evolution* **12**:196–200 [*249–250, 396*]

Sørensen TIA, Nielsen GG, Andersen PK & Teasdale TW (1988). Genetic and environmental influences on premature death in adult adoptees. *New England Journal of Medicine* **318**:727–732 [*222*]

Stamp NE & Bowers MD (1993). Presence of predatory wasps and stinkbugs alters foraging behavior of cryptic and non-cryptic caterpillars on plantain (*Plantago lanceolata*). *Oecologia* **95**:376–384 [*450*]

Stearns SC, ed. (1987). *The Evolution of Sex and Its Consequences.* Boston, MA, USA: Birkhäuser [*248–249*]

Stearns SC (1999). *Evolution in Health and Disease.* Oxford, UK: Oxford University Press [*379*]

Stinger SM, Hunter N & Woolhouse MEJ (1998). A mathematical model of the dynamics of scrapie in a sheep flock. *Mathematical Biosciences* **153**:79–98 [*230*]

Stockley P (1997). No evidence of sperm selection by female common shrews. *Proceedings of the Royal Society of London B* **264**:1497–1500 [*253*]

Strong DR, Lawton JH & Southwood R (1984). *Insects on Plants.* Cambridge, MA, USA: Harvard University Press [*297*]

Subbarao K, Klimov A, Katz J, Regnery H, Lim W, Hall H, Perdue M, Swayne D, Bender C, Huang J, Hemphill M, Rowe T, Shaw M, Xu X, Fukuda K & Cox N (1998). Characterization of an influenza A (H5N1) virus isolated from a child with a fatal respiratory illness. *Science* **279**:393–396   [*272, 276*]

Superti F, Ammendolia MG, Tinari A, Bucci B, Giammarioli AM, Rainaldi G, Rivabene R & Donelli G (1996). Induction of apoptosis in HT-29 cells infected with SA-11 rotavirus. *Journal of Medical Virology* **50**:325–334 [*190*]

Swartz MN (1994). Hospital-acquired infections: Diseases with increasingly limited therapies. *Proceedings of the National Academy of Sciences of the USA* **91**:2420–2427   [*326*]

Sweeney S, Silflow R & Foreyt WJ (1994). Comparative leukotoxicities of *Pasteurella haemolytica* isolates from domestic sheep and free-ranging bighorn sheep (*Ovis canadensis*). *Journal of Wildlife Diseases* **30**:523–528   [*36*]

Swofford DL (1998). *PAUP* 4d63.* Sunderland, MA, USA: Sinauer Associates Inc. [*269, 273*]

Swofford DL, Olsen GJ, Waddell PJ & Hillis DM (1996). Phylogenetic inference. In *Molecular Systematics*, eds. Hillis DM, Moritz C & Mable BK. Sunderland, MA, USA: Sinauer Associates Inc.   [*263*]

Tabashnik BE (1994). Evolution of resistance to *Bacillus thuringiensis*. *Annual Review of Entomology* **39**:47–79   [*457*]

Taddei F, Matic I, Godelle B & Radman F (1997a). To be a mutator, or how pathogenic and commensal bacteria can evolve rapidly. *Trends in Microbiology* **5**:427–429   [*37*]

Taddei F, Radman M, Maynard Smith J, Toupance B, Gouyon PH & Godelle B (1997b). Role of mutator alleles in adaptive evolution. *Nature* **387**:700–702   [*380, 395*]

Takabayashi J & Dicke M (1996). Plant-carnivore mutualism through herbivore-induced carnivore attractants. *Trends in Plant Science* **1**:109–113   [*298*]

Takahata N & Nei M (1990). Frequency-dependent selection and polymorphism of major histocompatibility complex loci. *Genetics* **124**:967–978   [*211–212, 221, 246*]

Takala AK, Eskola J, Leinonen M, Kayhty H, Nissinen A, Pekkanen E & Makela PH (1991). Reduction of oropharyngeal carriage of *Haemophilus influenzae* type b (Hib) in children immunized with an Hib conjugate vaccine. *Journal of Infectious Diseases* **164**:982–986   [*365*]

Takala AK, Vuopio-Varkila J, Tarkka E,

Leinonen M & Musser JM (1996). Subtyping of common pediatric pneumococcal serotypes from invasive disease and pharyngeal carriage in Finland. *Journal of Infectious Diseases* **173**:128–135   [*371*]

Takasu F (1998). Why do all host species not show defense against avian brood parasites: Evolutionary lag or equilibrium? *The American Naturalist* **151**:193–205   [*205*]

Takken W & Knols BGJ (1999). Odor-mediated behavior of Afrotropical malaria mosquitoes. *Annual Review of Entomology* **44**:131–157   [*398*]

Tamin A, Rota PA, Wang ZD, Heath JL, Anderson LJ & Bellini WJ (1994). Antigenic analysis of current wild-type and vaccine strains of measles virus. *Journal of Infectious Diseases* **170**:795–801   [*344*]

Tartaglia J, Paul CP, Fulbright DW & Nuss DL (1986). Structural properties of double-stranded RNAs associated with biological control of chestnut blight fungus. *Proceedings of the National Academy of Sciences of the USA* **83**:9109–9113   [*293*]

Taubenberger JK, Reid AH, Krafft AE, Bijwaard KE & Fanning TG (1997). Initial genetic characterization of the 1918 "Spanish" influenza virus. *Science* **275**:1793–1796 [*272–273, 276*]

Tauxe RV, Mintz ED & Quick RE (1995). Epidemic cholera in the new world: Translating field epidemiology in new prevention strategies. *Emerging Infectious Diseases* **1**:141–146   [*401, 403*]

Taylor PD (1988). An inclusive fitness model for dispersal of offspring. *Journal of Theoretical Biology* **130**:363–378   [*152*]

Taylor PD (1989). Evolutionary stability in one-parameter models under weak selection. *Theoretical Population Biology* **36**:125–143 [*42*]

Taylor LH (1997). *Epidemiological and Evolutionary Consequences of Mixed-Genotype Infections of Malaria Parasites.* PhD dissertation, University of Edinburgh, UK   [*171*]

Taylor PD & Frank SA (1996). How to make a kin selection model? *Journal of Theoretical Biology* **180**:27–37   [*151–152, 154*]

Taylor LH & Read AF (1998). Determinants of transmission success of individual clones from mixed-clone infections of the rodent malaria parasite, *Plasmodium chabaudi*. *International Journal for Parasitology* **28**:719–725   [*173–174, 394*]

Taylor LH, Walliker D & Read AF (1997a). Mixed-genotype infections of malaria parasites: Within-host dynamics and transmission success of competing clones. *Proceedings of the Royal Society of London B* **264**:927–935   [*170, 173–174, 394*]

Taylor LH, Walliker D & Read AF (1997b). Mixed-genotype infections of the rodent malaria *Plasmodium chabaudi* are more infectious to mosquitoes than single-genotype infections. *Parasitology* **115**:121–132   [*170, 173–175, 249*]

Taylor DR, Jarosz AM, Lenski RE & Fulbright DW (1998a). The acquisition of hypovirulence in host–pathogen systems with three trophic levels. *The American Naturalist* **151**:343–355   [*286–288, 290, 295–296*]

Taylor LH, Mackinnon MJ & Read AF (1998b). Virulence of mixed-clone and single-clone infections of the rodent malaria *Plasmodium chabaudi*. *Evolution* **52**:583–591   [*170, 172–173, 394*]

Tchuem Tchuenté LA, Southgate VR, Imbert-Establet D & Jourdane J (1995). Change of mate and mating competition between males of *Schistosoma intercalatum* and *S. mansoni*. *Parasitology* **110**:45–52   [*258*]

Tchuem Tchuenté L-A, Southgate VR, Combes C & Jourdane J (1996). Mating behaviour in schistosomes: Are paired worms always faithful? *Parasitology Today* **12**:231–236   [*258*]

Te Beest DO, Yang XB & Cisar CR (1992). The status of biological control of weeds with fungal pathogens. *Annual Review of Phytopathology* **30**:637–657   [*451–452*]

Thomas MB (1999). Ecological approaches and the development of "truly integrated" pest management. *Proceedings of the National Academy of Sciences of the USA* **96**:5944–5951   [*450–451*]

Thomas CD, Singer MS & Boughter DA (1996). Catastrophic extinction of population sources in a butterfly metapopulation. *The American Naturalist* **148**:957–975   [*105*]

Thompson JN (1994). *The Coevolutionary Process.* Chicago, IL, USA: Chicago University Press   [*198, 422*]

Thompson CB (1995). Apoptosis in the pathogenesis and treatment of disease. *Science* **267**:1456–1462   [*183*]

Thompson SN (1999). Nutrition and culture of entomophagous insects. *Annual Review of Entomology* **44**:561–592   [*455*]

Thompson JN & Burdon JJ (1992). Gene-for-gene coevolution between plants and parasites. *Nature* **360**:121–125   [*396*]

Thompson RCA & Lymbery AJ (1996). Genetic variability in parasite and host–parasite interactions. *Parasitology* **112**:S7–S22   [*31*]

Thorne ET & Williams ES (1988). Disease and endangered species: The black-footed ferret as a recent example. *Conservation Biology* **2**:66–74   [*30*]

Thrall PH & Antonovics J (1995). Theoretical and empirical studies of metapopulations: Population and genetic dynamics of the Silene–Ustilago system. *Canadian Journal of Botany* **73**(Suppl. 1):S1249–1258   [*440–441, 446*]

Thrall PH & Burdon JJ (1997). Host–pathogen dynamics in a metapopulation context: The ecological and evolutionary consequences of being spatial. *Journal of Ecology* **85**:743–753   [*440*]

Thrall PH & Burdon JJ (1999). The spatial scale of pathogen dispersal: Consequences for disease dynamics and persistence. *Evolutionary Ecology Research* **1**:681–701   [*440–441*]

Tikhomirov E (2001). *Human Plague in 1998 and 1999.* Geneva, Switzerland: WHO-CSR Report   [*381*]

Tilman D (1982). *Resource Competition and Community Structure.* Princeton, NJ, USA: Princeton University Press   [*176*]

Tilman D (1994). Competition and biodiversity in spatially structured habitats. *Ecology* **75**:2–16   [*131, 138–139, 144*]

Tilman D & Kareiva P, eds. (1997). *Spatial Ecology: The Role of Space in Population Dynamics and Interspecific Interactions.* Princeton, NJ, USA: Princeton University Press   [*199*]

Tilman D, May RM, Lehman CL & Nowak MA (1994). Habitat destruction and the extinction debt. *Nature* **371**:65–66   [*131*]

Toft CA & Karter AJ (1990). Parasite–host coevolution. *Trends in Ecology and Evolution* **5**:326–329   [*396*]

Tollefson AE, Ryerse JS, Scaria A, Hermiston TW & Wold WSM (1996). The E3-11.6-kDa adenovirus death protein (ADP) is required for efficient cell death: Characterization of cells infected with ADP mutants. *Virology* **220**:152–162   [*184, 190, 192*]

Tollrian R & Harvell CD (1999). *The Ecology*

and Evolution of Inducible Defenses. Princeton, NJ, USA: Princeton University Press [*458*]

Topley WWC (1919). The spread of bacterial infection. *Lancet* (July 5)1–5 [*371*]

Tos AG, Cignetti A, Rovera G & Foa R (1996). Retroviral vector-mediated transfer of the tumor-necrosis-factor-alpha gene into human cancer cells restores an apoptotic cell death program and induces a bystander-killing effect. *Blood* **87**:2486–2495 [*194*]

Trager W & Jensen JB (1976). Human malaria parasites in continuous culture. *Science* **193**:673–675 [*354*]

Trivers RL (1972). Parental investment and sexual selection. In *Sexual Selection and the Descent of Man*, ed. Campbell B, pp. 136–179. Chicago, IL, USA: Aldine [*250*]

Trivers RL (1985). *Social Evolution*. Menlo Park, CA, USA: Benjamin/Cummings [*166*]

Tuomi J & Augner M (1993). Synergistic selection of unpalatability in plants. *Evolution* **47**:668–672 [*315*]

Tuomi J, Augner M & Nilsson J (1994). A dilemma of plant defences: Is it really worth killing the herbivore? *Journal of Theoretical Biology* **170**:427–430 [*316–317, 319*]

Turlings TCJ, Loughrin JH, McCall PJ, Rse USR, Lewis WJ & Tumlinson JH (1995). How caterpillar-damaged plants protect themselves by attracting parasitic wasps. *Proceedings of the National Academy of Sciences of the USA* **92**:4169–4174 [*298*]

Turner PE, Cooper VS & Lenski RE (1998). Trade-off between horizontal and vertical modes of transmission in bacterial plasmids. *Evolution* **52**:315–329 [*392*]

Urwin G, Krohn JA, Deaver-Robinson K, Wenger JD, Farley MM & Group HIS (1996). Invasive disease due to *Haemophilus influenzae* serotype f: Clinical and epidemiological characteristics in the *H. influenzae* serotype b vaccine era. *Clinical Infectious Diseases* **22**:1069–1076 [*365*]

Van Baalen M (1998). Coevolution of recovery ability and virulence. *Proceedings of the Royal Society of London B* **265**:317–325 [*61–64, 68*]

Van Baalen M (2000). Pair approximations for different geometries. In *The Geometry of Ecological Interactions: Simplifying Spatial Complexity*, eds. Dieckmann U, Law R & Metz JAJ, pp. 359–387. Cambridge, UK: Cambridge University Press [*89, 93, 97–98*]

Van Baalen M & Rand DA (1998). The unit of selection in viscous populations and the evolution of altruism. *Journal of Theoretical Biology* **193**:631–648 [*89, 93–94, 97, 160*]

Van Baalen M & Sabelis MW (1993). Coevolution of patch selection strategies of predator and prey and the consequences for ecological stability. *The American Naturalist* **142**:646–670 [*203*]

Van Baalen M & Sabelis MW (1995a). The dynamics of multiple infection and the evolution of virulence. *The American Naturalist* **146**:881–910 [*11, 15, 35, 66, 86–88, 92, 106, 124, 137, 140, 150, 152, 162, 164, 166–167, 172, 176, 287, 298, 312, 420, 448, 453–454*]

Van Baalen M & Sabelis MW (1995b). The scope for virulence management: A comment on Ewald's view on the evolution of virulence. *Trends in Microbiology* **3**:414–416 [*19, 35, 60–61, 87–88, 164, 166, 178, 448, 453–454*]

Van Baalen M & Sabelis MW (1995c). The milker–killer dilemma in spatially structured predator–prey interactions. *Oikos* **74**:391–413 [*308–311, 320, 450–451*]

Van de Klashorst G, Readshaw JL, Sabelis MW & Lingeman R (1992). A demonstration of asynchronous local cycles in an acarine predator–prey system. *Experimental and Applied Acarology* **14**:179–185 [*305*]

Van Lenteren JC & Nicoli G (1999). Quality control of mass produced beneficial insects. In *Biological Control of Arthropod Pests in Protected Cultures*, eds. Heinz KM, Van Driesche RG & Parrella MP. Batavia, IL, USA: Ball Publishing [*455*]

Van Riper C III, Van Riper SG, Goff ML & Laird M (1986). Epizootiology and ecological significance of malaria in Hawaiian land birds. *Ecological Monographs* **56**:327–344 [*30*]

Van Tienderen PH & de Jong G (1986). Sex-ratio under the haystack model – polymorphism may occur. *Journal of Theoretical Biology* **122**:69–81 [*42*]

Varley GC, Gradwell GR & Hassell MP (1973). *Insect Population Ecology: An Analytical Approach*. Berkeley, CA, USA: University of California Press [*77*]

Vasconcelos SD, Williams T & Hails RS (1996). Prey selection and baculovirus dissemination by carabid predators of *Lepidoptera*. *Ecological Entomology* **21**:98–104 [*456*]

Vaughn DW, Green S, Kalayanarooj S, Innis BL, Nimmannitya S, Suntayakorn S, Rothman AL, Ennis FA & Nisalak A (1997). Dengue in the early febrile phase: Viremia and antibody responses. *Journal of Infectious Diseases* **176**:322–330 [*405*]

Vaux DL, Haecker G & Strasser A (1994). An evolutionary perspective on apoptosis. *Cell* **76**:777–779 [*183*]

Venezia RA & Robertson RG (1975). Bactericidal substance produced by *Haemophilus influenzae* type b. *Canadian Journal of Microbiology* **21**:1587–1594 [*366*]

Verrell PA & McCabe NR (1990). Major histocompatibility antigens and spontaneous abortion: An evolutionary perspective. *Medical Hypotheses* **32**:235–238 [*259*]

Via S, Gomulkiewicz R, de Jong G, Scheiner SM, Schlichting CD & Van Tienderen PH (1995). Adaptive phenotypic plasticity: Consensus and controversy. *Trends in Ecology and Evolution* **10**:212–217 [*167*]

Viggers KL, Lindenmayer DB & Spratt DM (1993). The importance of diseases in reintroduction programs. *Wildlife Research* **20**:678–698 [*421, 423*]

Vincent TL & Brown JS (1989). The evolutionary response of systems to a changing environment. *Applied Mathematics and Computation* **32**:185–206 [*203*]

Vogel TU, Evans DT, Urvater JA, O'Connor DH, Hughes AL & Watkins DI (1999). Major histocompatibility complex class I genes in primates: Coevolution with pathogens. *Immunological Reviews* **167**:327–337 [*210, 221*]

Voyles BA (1993). *The Biology of Viruses.* New York, NY, USA: Mosby [*267*]

Waage JK (1990). Ecological theory and the selection of biological control agents. In *Critical Issues in Biological Control*, eds. Mackauer M, Ehler LE & Roland, pp. 135–157. Andover, UK: Intercept [*208*]

Waage JK & Nondo J (1982). Host behaviour and mosquito feeding success: An experimental study. *Transactions of the Royal Society of Tropical Medicine and Hygiene* **76**:119–122 [*18*]

Wachsman JT (1996). The beneficial effects of dietary restriction – Reduced oxidative damage and enhanced apoptosis. *Mutation Research: Fundamental and Molecular Mechanisms of Mutagenesis* **350**:25–34 [*195*]

Wakelin D & Blackwell J (1988). *Genetics of Resistance to Bacterial and Parasitic Infection.* London, UK: Taylor and Francis [*415*]

Wallace WA & Carman WF (1997). Surface gene variation of HBV: Scientific and medical relevance. *Viral Hepatitis Reviews* **3**:5–16 [*345*]

Walliker D (1989). Genetic recombination in malaria parasites. *Experimental Parasitology* **69**:303–309 [*351*]

Wallinga J, Edmunds WJ & Kretzschmar M (1999). Perspective: Human contact patterns and the spread of airborne infectious diseases. *Trends in Microbiology* **7**:372–377 [*85*]

Walter DE (1996). Living on leaves: Mites, tomenta, and leaf domatia. *Annual Review of Entomology* **41**:101–114 [*298*]

Walther BA & Ewald PW. Pathogen survival in the external environment and the evolution of virulence. Unpublished [*18*]

Warrell DA, Molyneux ME & Beales PF (1990). Severe and complicated malaria. *Transactions of the Royal Society of Tropical Medicine and Hygiene* **84**(Suppl. 2):1–65 [*168*]

Watkinson AR & Sutherland WJ (1995). Sources, sinks, and pseudosinks. *Journal of Animal Ecology* **64**:126–130 [*105*]

Watson RB (1949). Location and mosquito-proofing of dwellings. In *Malariology: A Comprehensive Survey of All Aspects of This Group of Diseases from a Global Standpoint*, ed. Boyd MF, pp. 1184–1202. Philadelphia, PA, USA: Saunders [*407–408*]

Watts DJ & Strogatz SH (1998). Collective dynamics of "small-world" networks. *Nature* **393**:440–442 [*85*]

Webster JP & Woolhouse MEJ (1999). Costs of resistance: Relationship between reduced fertility and increased resistance in a snail-schistosome host–parasite system. *Proceedings of the Royal Society of London B* **266**:391–396 [*396*]

Weckstein LN, Patrizio P, Balmaceda JP, Asch RH & Branch DW (1991). Human leukocyte antigen compatibility and failure to achieve a viable pregnancy with assisted reproductive technology. *Acta Europaea Fertilitatis* **22**:103–107 [*254*]

Wedekind C (1994a). Handicaps not obligatory

in sexual selection for resistance genes. *Journal of Theoretical Biology* **170**:57–62 [*251, 254*]

Wedekind C (1994b). Mate choice and maternal selection for specific parasite resistances before, during and after fertilization. *Philosophical Transactions of the Royal Society of London B* **346**:303–311 [*251–252, 254, 259*]

Wedekind C (1998). Pathogen-driven sexual selection and the evolution of health. In *Evolution in Health and Disease*, ed. Stearns SC, pp. 102–107. Oxford, UK: Oxford University Press [*249, 257*]

Wedekind C & Folstad I (1994). Adaptive or non-adaptive immunosuppression by sex hormones? *The American Naturalist* **143**:936–938 [*251*]

Wedekind C & Füri E (1997). Body odour preferences in men and women: Do they aim for specific MHC combinations or simply heterozygosity? *Proceedings of the Royal Society of London B* **264**:1471–1479 [*254–255*]

Wedekind C & Rüetschi A (2000). Parasite heterogeneity affects infection success and the occurrence of within-host competition: An experimental study with a cestode. *Evolutionary Ecology Research* **2**:1031–1043 [*249, 257–258*]

Wedekind C, Seebeck T, Bettens F & Paepke AJ (1995). MHC-dependent mate preferences in humans. *Proceedings of the Royal Society of London B* **260**:245–249 [*254–255*]

Wedekind C, Chapuisat M, Macas E & Rülicke T (1996). Non-random fertilization in mice correlates with the MHC and something else. *Heredity* **77**:400–409 [*255, 260*]

Wedekind C, Strahm D & Schärer L (1998). Evidence for strategic egg production in a hermaphroditic cestode. *Parasitology* **117**:373–382 [*257–258*]

Weissman JB, Murton KI, Lewis JN, Friedemann CHT & Gangarosa EJ (1974). Impact in the US of the Shiga dysentery pandemic of Central America and Mexico: A review of surveillance data through 1972. *Journal of Infectious Diseases* **129**:218–223 [*409*]

Wenger JD, Pierce R, Deaver K, Franklin R, Bosley G, Pigott N & Broome CV (1992). Invasive *Haemophilus influenzae* disease: A population-based evaluation of the role of capsular polysaccharide serotype. *Journal of Infectious Diseases* **165**(Suppl. 1):S34–35

[*365, 371*]

Whitlock M & McCauley DE (1990). Some population genetic consequences of colony formation and extinction: Genetic correlations within founding groups. *Evolution* **44**:1717–1724 [*153*]

Williams GC (1957). Pleiotropy, natural selection, and the evolution of senescence. *Evolution* **22**:406–421 [*284*]

Williams GC (1966). *Adaptation and Natural Selection*. Princeton, NJ, USA: Princeton University Press [*39*]

Williams GC (1975). *Sex and Evolution*. Princeton, NJ, USA: Princeton University Press [*248–249, 258*]

Williams P, Camara M, Hardman A, Swift S, Milton D, Hope VJ, Winzer K, Middleton B, Pritchard DI & Bycroft BW (2000). Quorum sensing and the population-dependent control of virulence. *Philosophical Transactions of the Royal Society of London B* **355**:667–680 [*394, 454–455*]

Williamson MH (1981). *Island Populations*. Oxford, UK: Oxford University Press [*104*]

Willimans CK & Moore RJ (1991). Inheritance of acquired immunity to myxomatosis. *Australian Journal of Zoology* **39**:307–311 [*34*]

Wills C (1991). Maintenance of multiallelic polymorphism at the MHC region. *Immunological Reviews* **124**:165–220 [*211–212, 221*]

Wills C & Green DR (1995). A genetic herd-immunity model for the maintenance of MHC polymorphism. *Immunological Reviews* **143**:263–292 [*212, 221*]

Wilson ME, Levins R, Spielman & A, eds. (1994). *Diseases in Evolution: Global Changes and Emergence of Infectious Diseases*, New York, NY, USA: Academy of Science [*37–38*]

Wolgemuth DJ (1983). Synthetic activities of the mammalian early embryo: Molecular and genetic alterations following fertilization. In *Mechanism and Control of Animal Fertilization*, ed. Hartmann JF, pp. 415–452. New York, NY, USA: Academic Press [*256*]

Woodford MH & Rossiter PB (1993). Disease risks associated with wildlife translocation projects. *Revue Scientifique et Technique de l'Office International des Epizzoties* **12**:115–135 [*421*]

Woods SA & Elkinton JS (1987). Bimodal patterns of mortality from nuclear polyhedrosis virus in gypsy moth (*Lymantria dispar*) populations. *Journal of Invertebrate Pathology* **50**:151–157 [*80*]

Woods SA, Elkinton JS, Murray KD, Liebhold AM, Gould JR & Podgwaite JD (1991). Transmission dynamics of a nuclear polyhedrosis virus and predicting mortality in gypsy moth (Lepidoptera: Lymantriidae) populations. *Journal of Economic Entomology* **84**:423–430 [*80*]

Wrensch DL & Ebbert MA, eds. (1993). *Evolution and Diversity of Sex Ratio in Insects and Mites*. London, UK: Chapman & Hall [*167*]

Wyler DJ (1993). Malaria: Overview and update. *Clinical Infectious Diseases* **16**:449–458 [*409*]

Wyllie A (1987). Apoptosis: Cell death in tissue regulation. *Journal of Pathology* **153**:313–316 [*183*]

Yamaguchi Y & Gojobori T (1997). Evolutionary mechanisms and population dynamics of the third variable envelope region of HIV within single hosts. *Proceedings of the National Academy of Sciences of the USA* **94**:1264–1269 [*246*]

Yamamura N (1993). Vertical transmission and evolution of mutualism from parasitism. *Theoretical Population Biology* **44**:95–109 [*395–396*]

Yamazaki K, Boyse EA, Miké V, Thaler HT, Mathieson BJ, Abbott J, Boyse J, Zayas ZA & Thomas L (1976). Control of mating preference in mice by genes in the major histocompatibility complex. *Journal of Experimental Medicine* **144**:1324–1335 [*253, 260*]

Yamazaki K, Yamaguchi M, Baranoski L, Bard J, Boyse EA & Thomas L (1979). Recognition among mice. Evidence from the use of a Y-maze differentially scented by congenic mice of different major histocompatibility types. *Journal of Experimental Medicine* **150**:755–760 [*253*]

Yamazaki K, Beauchamp GK, Ecorov IK, Bard J, Thomas L & Boyse EA (1983a). Sensory distinction between H-2b and H-2bml mutant mice. *Proceedings of the National Academy of Sciences of the USA* **80**:5685–5688 [*253–254*]

Yamazaki K, Beauchamp GK, Wysocki CJ, Bard J, Thomas L & Boyse EA (1983b). Recognition of H-2 types in relation to the blocking of pregnancy in mice. *Science* **221**:186–188 [*253–254*]

Yamazaki K, Beauchamp GK, Kupniewski D, Bard J, Thomas L & Boyse EA (1988). Familial imprinting determines H-2 selective mating preferences. *Science* **240**:1331–1332 [*253*]

Yamazaki K, Beauchamp GK, Shen F-W, Bard J & Boyse EA (1994). Discrimination of odor types determined by the major histocompatibility complex among outbred mice. *Proceedings of the National Academy of Sciences of the USA* **91**:3735–3738 [*253*]

Yang Z (1994). Maximum likelihood phylogenetic estimation from DNA sequences with variable rates over sites: Approximate methods. *Journal of Molecular Evolution* **39**:306–314 [*263, 270*]

Yang Z (1997). PAML: A program package for phylogenetic analysis by maximum likelihood. *Computer Applications in the Biosciences* **13**:555–556 [*270*]

Yaninek JS, Sabelis MW, de Moraes GJ, Hanna R & Guttierez AP. Host plant enhances introduced predator of exotic cassava green mite pests in Africa. Unpublished [*451*]

Yasuda J, Shortridge KF, Shimizu Y & Kida H (1991). Molecular evidence for a role of domestic ducks in the introduction of avian H3 influenza viruses to pigs in southern China, where the A/Hong Kong/68 (H3N2) strain emerged. *Journal of General Virology* **72**:2007–2010 [*268*]

Yoeli M, Hargreaves B, Carter R & Walliker D (1975). Sudden increase in virulence in a strain of *Plasmodium berghei yoelii*. *Annals of Tropical Medicine and Parasitology* **69**:173–178 [*171*]

Young E (1969). The significance of infectious diseases in African game populations. *Zoologica Africana* **4**:275–281 [*416*]

Young JPW (1981). Sib competition can favour sex in two ways. *Journal of Theoretical Biology* **88**:755–756 [*249, 258*]

Zahavi A (1975). Mate selection: A selection for a handicap. *Journal of Theoretical Biology* **53**:205–214 [*251*]

Zakeri Z, Bursch W, Tenniswood M & Lockshin RA (1995). Cell death: Programmed, apoptosis, necrosis, or other. *Cell Death and Differentiation* **2**:87–96 [*183*]

Zeh JA & Zeh DW (1997). The evolution of polyandry II: Postcopulatory defence against genetic incompatibility. *Proceedings of the Royal Society of London B* **264**:69–75

[*253*]
Zeigler RS, Tohme J, Nelson R, Levy M & Correa-Victoria FJ (1994). Lineage exclusion: A proposal for linking blast population analysis to resistance breeding. In *Rice Blast Disease*, eds. Zeigler RS, Leong S & Teng P, pp. 267–292. Wallingford, UK: Commonwealth Agriculture Bureaux International, and Los Banos, Philippines: IRRI [*445*]
Zwaan B, Bijlsma R & Hoekstra RF (1995). Direct selection on life span in *Drosophila melanogaster*. *Evolution* **49**:649–659 [*284*]
Zwetyenga J, Rogier C, Tall A, Fontenille D, Snounou G, Trape J-F & Mercereau-Puijalon O (1998). No influence of age on infection complexity and allelic distribution in *Plasmodium falciparum* infections in Ndiop, a Senegalese village with seasonal, mesoendemic malaria. *American Journal of Tropical Medicine and Hygiene* **59**:726–735 [*168–169, 172*]

Index

abortion, spontaneous, *252, 254*
acquired immunodeficiency disease *see* AIDS
Acyrthosiphon pisum, 457
adaptive dynamics, *3, 9, 39–59*
  canonical equation, *46*
  models, *46–47*
  pathogen evolution, *45–54*
  pathogen–host coevolution, *54–57*
  relating theory to experiment, *379–398*
  theory, *42–43, 44–45*
adenovirus (Ad), *190, 192*
  death protein (ADP), *190, 192*
  E1A protein, *184*
  E1B gene, *184*
Aedes aegypti, 37
age
  distribution
    malaria, *353, 354, 355, 358, 361*
    pertussis, *345, 346*
    virulence evolution and, *285*
  at first infection, *33*
  in malaria, *358–359*
aggregated (clustered) distribution, *33*
  contact networks and, *93–94, 95–97, 101*
  plant–herbivore–predator interactions and,
    299–300
aggressive virulence, *2, 234, 389, 438*
  evolution, *439–440*
agricultural crop pathogens *see*
  plant–pathogen systems
AIDS, *1, 41, 211*
  see also human immunodeficiency virus
alpha virus chimera, *184, 189, 193*
anastomosis, hyphal, *281, 284, 292, 380–381*
ancestor, most recent common (MRCA),
  269–270
anemia, malarial, *170, 353, 354*
anthrax, *32*
animal husbandry, *377, 425–335*
antibiotic resistance, *61, 324, 326–338, 410*
  conjugate vaccine design and, *374*
  de novo, 332, 333, 334, 336
  fitness costs, *328, 330, 331–332, 333,*
    334–335, 337
  large-scale therapy and, *163*
  models, *327–335*
  multiple (MDR), *233–234, 247, 383*
  plasmids, *441–442*

prevention, *60, 233–234, 247, 326*
  primary, *332*
  time to evolve, *414*
  transmission, *333–334, 335, 336*
antibiotic therapy
  combination (COT), *334, 335, 336–337*
  comparing policies using two agents,
    336–337
  cycling (CAT), *334, 335, 336–337, 338*
  ethical dilemmas, *61, 69, 338, 411*
  evaluation of strategies, *327*
  intermittent, *330*
  large-scale, *162, 163*
  models, *327–335*
    extended, *331–335*
    simple, *327–331*
    steady state, *328*
  multiple, *333–335*
  optimal strategies, *328, 335–338, 462*
  pathogen evolutionary responses, *65–68,*
    390
  simultaneous single (SSAT), *334, 335,*
    336–337
  with two agents, *331–333*
antibodies
  anti-cell-death, *195*
  in malaria, *354, 357–358*
  Streptococcus pneumoniae, 372
antigenic determinants, polymorphic,
  356–357
antigenic diversity, *325, 347–361, 362*
  despite cross-immunity, *356–357, 358*
  Haemophilus influenzae, 363–364
  malaria case study, *353–358*
  multilocus model with genetic exchange,
    349–352
  in single-locus systems, *347–349*
  Streptococcus pneumoniae, 363–364
  vaccination impact, *358–360, 373*
  see also genetic diversity
antigenic drift, *268*
antigenic types, *347*
  in malaria, *354–355*
antimalarial drugs, *405–406, 415*
antioxidants, *195*
antisense oligonucleotides, *195*
aphids, corn leaf, *314–315*
Aphidus ervi, 457

515

Lyme disease, *38*
lymphocytes, *184, 213*
   cytotoxic (CTL), *353, 358*
   see also T lymphocytes
lysis, virus-induced, *184, 193*
   experimental studies, *189, 191–192*
   modeling, *185–186, 187*

Macrocyclops albidus, *257–258*
macroparasites, *8–9, 27–29, 383*
   detecting impact in wildlife, *416, 418*
   basic reproduction ratio, *27, 28–29*
   regulation of host population, *31*
Magnaporthe grisea, *444, 445–446*
major histocompatibility complex (MHC)
   evasion of presentation on, *217–218*
   in mate choice, *253–255, 256–257*
   odor preferences and, *253–254*
   peptide presentation, *212–213*
   polymorphism, *181, 210–221, 223, 398*
      coevolutionary simulation model,
         212–221
      mechanisms of evolution, *211–212,*
         220–221
      population diversity, *210–211*
      rare alleles, *211–212, 220*
malaria, *1, 60*
   antigenic diversity, *353–358, 361*
   avian, *30, 171*
   cerebral (CM), *353, 354, 355*
   drug resistance, *415*
   epidemiological modeling, *11*
   evolutionary explanations, *165–178*
      conditional virulence strategies, *167–171*
      genetically fixed virulence strategies,
         171–172
      management implications, *177–178*
      within-host competition and
         between-host fitness, *172–175*
   immunity, *169, 170, 176–177, 353–358,*
      407
   mouse (rodent), *170–171, 173–175, 389,*
      392
   sickle-cell disease and, *223, 225–226, 355,*
      398
   therapy of neurosyphilis, *172*
   vaccination impact, *358–360*
   virulence management, *398, 405–409*
   see also Plasmodium falciparum
Marburg virus, *37*
mass action assumption, *4, 12, 51, 184–185*
mass rearing
   biocontrol agents, *449*

field virulence and, *455–456*
matching genotype, *234, 236, 246*
matching virulence, *2, 234, 389, 437–438*
   evolution, *438–440, 442*
   strategies to overcome, *444–445*
mate choice, *251, 252, 260–261*
   conditional preferences, *255–257*
   genetically variable preferences, *253–255*
   in parasites, *258, 261*
   preference for health and vigor, *252–253*
   virulence management implications,
      259–260
mates, competition for, *251*
maximum likelihood (ML), *263–265*
   estimates (MLEs), *265*
mean-field dynamics, *91–992*
measles, *410*
   basic reproduction ratio, *340*
   vaccine resistance, *344–345*
   virus (MV), *190*
mechanistic resource–consumer theory, *176*
medical treatment
   costs, *63*
   ethical dilemmas, *60–69, 162–163*
   in multiple infections, *162–163*
   optimal strategies, *62–65*
   parasite evolutionary responses, *65–68*
   see also antibiotic therapy; drug treatment;
      vaccination
meiosis, second division, *255, 256*
Melampsora lini, *440–441*
meningitis, *363, 395, 430*
merozoites, malaria, *353–354*
metapopulation
   biocontrol, *458*
   fitness, *387*
   models, *131*
      tritrophic interactions, *307–318, 319*
   parasites, *151*
   structure, *72, 307, 384, 440–441*
   virulence, *388*
   see also population(s)
Metarhizium flavoviride, *450*
Mexican chicken flu, *268–272, 275–276*
MHC *see* major histocompatibility complex
microparasites, *8–9, 27–29, 382–383*
   detecting impact in wildlife, *416, 418*
   host population regulation, *30–31*
   sexual selection, *257*
   within-host interactions, *122–123*
   see also bacteria; fungi; viruses; *specific*
      organisms
Microphallus, *422*

International Institute for Applied Systems Analysis

IIASA is an interdisciplinary, nongovernmental research institution founded in 1972 by leading scientific organizations in 12 countries. Situated near Vienna, in the center of Europe, IIASA has been for more than two decades producing valuable scientific research on economic, technological, and environmental issues.

IIASA was one of the first international institutes to systematically study global issues of environment, technology, and development. IIASA's Governing Council states that the Institute's goal is: *to conduct international and interdisciplinary scientific studies to provide timely and relevant information and options, addressing critical issues of global environmental, economic, and social change, for the benefit of the public, the scientific community, and national and international institutions*. Research is organized around three central themes:

- Energy and Technology;
- Environment and Natural Resources;
- Population and Society.

The Institute now has national member organizations in the following countries:

Austria
The Austrian Academy of Sciences

Bulgaria*
Ministry of Environment and Waters

Czech Republic
The Academy of Sciences of the
Czech Republic

Finland
The Finnish Committee for IIASA

Germany**
The Association for the
Advancement of IIASA

Hungary
The Hungarian Committee for
Applied Systems Analysis

Japan
The Japan Committee for IIASA

Republic of Kazakhstan*
The Ministry of Science –
The Academy of Sciences

Netherlands
The Netherlands Organization for
Scientific Research (NWO)

Norway
The Research Council of Norway

Poland
The Polish Academy of Sciences

Russian Federation
The Russian Academy of Sciences

Slovak Republic*
The Executive Slovak National
Committee for IIASA

Sweden
The Swedish Research Council for
Environment, Agricultural Sciences
and Spatial Planning (FORMAS)

Ukraine*
The Ukrainian Academy of Sciences

United States of America
The American Academy of
Arts and Sciences

*Associate member
**Affiliate